I0479054

Animal Genetics and Genomics

Animal Genetics and Genomics

Edited by George Gardner

SYRAWOOD
PUBLISHING HOUSE

New York

Published by Syrawood Publishing House,
750 Third Avenue, 9th Floor,
New York, NY 10017, USA
www.syrawoodpublishinghouse.com

Animal Genetics and Genomics
Edited by George Gardner

International Standard Book Number: 978-1-68286-754-9 (Hardback)

Cataloging-in-Publication Data

Animal genetics and genomics / edited by George Gardner.
    p. cm.
Includes bibliographical references and index.
ISBN 978-1-68286-754-9
1. Animal genetics. 2. Genomics. 3. Animal genome mapping. I. Gardner, George.
QH432 .A55 2019
591.35--dc23

TABLE OF CONTENTS

PREFACE

This book has been a concerted effort by a group of academicians, researchers and scientists, who have contributed their research works for the realization of the book. This book has materialized in the wake of emerging advancements and innovations in this field. Therefore, the need of the hour was to compile all the required researches and disseminate the knowledge to a broad spectrum of people comprising of students, researchers and specialists of the field.

Animals are multicellular eukaryotic organisms. They are part of the biological kingdom Animalia. The study of genes and the processes of reproduction are vital to the understanding of animal diversity and characteristics. In animals, the DNA is arranged in multiple linear chromosomes, which can be very long. Most animals are diploid, which means they have two homologous copies of each chromosome, one from the mother and one from the father. This also means that they possess two copies of every gene. The two alleles specific to a particular gene are present on identical loci of the two chromosomes. In many animals including humans, the genes that are responsible for the inheritance of the male and female characteristics are the Y and X chromosomes respectively. This book elucidates the concepts and innovative models around prospective developments with respect to the fields of animal genetics and genomics. The topics included in this book are of the utmost significance and bound to provide incredible insights to readers. The extensive content of this book provides the readers with a thorough understanding of the subject.

At the end of the preface, I would like to thank the authors for their brilliant chapters and the publisher for guiding us all-through the making of the book till its final stage. Also, I would like to thank my family for providing the support and encouragement throughout my academic career and research projects.

Editor

Impact of ploidy level on the distribution of *Pokey* element insertions in the *Daphnia pulex* complex

Roland Vergilino[1,2,3*], Shannon HC Eagle[2], Teresa J Crease[2] and France Dufresne[1,3]

Abstract

Background: Transposable elements (TEs) play a major role in genome evolution. Their capacity to move and/or multiply in the genome of their host may have profound impacts on phenotypes and dramatic consequences on genome structure. The population dynamics and distribution of TEs are influenced by their mode of transposition, the availability of niches in host genomes, and host population dynamics. Theories predict an increase in the number of TE insertions following hybridization or polyploidization. Evolution of TEs in hybrids and polyploids has mostly been studied in plants; few studies have examined the impacts of hybridization and/or polyploidization on TEs in animals. Hybrids and polyploids have arisen multiple times in the *Daphnia pulex* complex and are thought to reproduce by obligate parthenogenesis. Our study examines the effects of ploidy level on polymorphism and number of *Pokey* element insertions in diploid and polyploid hybrid isolates from the *Daphnia pulex* complex.

Results: The polymorphism of *Pokey* insertion sites did not depend solely on either the ploidy level or the genetic background of their host; therefore, it may be the result of interactions between these parameters and other parameters such as *Pokey* activity, selection and/or drift. No significant effect of ploidy level was found on the number of *Pokey* insertions using TE display and qPCR. However, the load of *Pokey* insertion sites and the number of unique insertion sites were slightly (but not significantly) higher in polyploids than in diploids.

Conclusions: These results suggest a lack of increase in the number of *Pokey* insertions following polyploidization but higher availability of *Pokey* insertion sites in polyploids than in diploids. Compared to previous TE display and qPCR results, the load of *Pokey* insertions in hybrid diploids was higher than in non-hybrid sexual and asexual diploids, which suggests an increase in the density of *Pokey* insertions following hybridization.

Keywords: *Daphnia pulex*, Hybrids, Insertion site polymorphism, Load, *Pokey*, Polyploids, Transposable element

Background

Transposable elements (TEs) are genetic components that are able to move and multiply within and between genomes. They are found in the genomes of almost all living organisms [1], although there are exceptions in endosymbiont organisms [2]. There is large variation in the proportion of TEs across genomes [3]. TE populations are impacted by host population dynamics, such as effective population size, mode of reproduction, hybridization, and polyploidization [4-6]. A decrease in effective population size of the host [7] or an increase in its level of selfing [8] are expected to lead to an increase in the density of TE insertions. The mode of reproduction of the host also has a substantial impact on the dynamics and density of TE insertions in the genome [9-12]. For example, TEs may spread via recombination and out-crossing in sexual populations [11], whereas the spread of TEs among lineages is prevented in asexual populations except by horizontal transmission [13,14]. Empirical studies have tested and are generally in accordance with the theoretical prediction that the genomes of sexual organisms will contain a higher number of TE insertions compared to asexual ones [10,12-15]. Hybridization and polyploidization, which play a significant role in the diversification of plants and animals [16-18], might also have an impact on the load and insertion site polymorphism of TEs. Activation of TEs has been observed in hybrid genomes [19-22] (however, there are contradictory results in hybrid sunflowers [23]), and

* Correspondence: rvergilino@gmail.com
[1]Département de Biologie, Chimie et Géographie, Université du Québec à Rimouski, Rimouski, Québec G5L 3A1, Canada
[2]Department of Integrative Biology, University of Guelph, Guelph, Ontario N1G 2W1, Canada
Full list of author information is available at the end of the article

polyploidization may lead to an increase in the density of TE insertions [24,25], although there are contradictory results in allopolyploid plants [26]. Bursts of TE activity are thought to have a substantial impact on genome rearrangement [27] and may lead to phenotypic diversification in hybrids and polyploids [24,28]. Many studies have explored the effects of hybridization and polyploidization on TE dynamics in plants [22,23,26,28-34], but few studies have focused on these effects in animals, with the exception of studies in carp [35], Drosophila [36] and wallaby [20], all of which have been reviewed [37]. Studying the dynamics of TEs in hybrids and polyploids may provide insight on the evolution of their genomes and their propensity to adapt to various environments.

The *Daphnia pulex* (*D. pulex*) species complex has been intensively studied due to its dominance in freshwater habitats in North America and its variation in reproductive mode and ploidy level. *Daphnia* usually reproduce by cyclic parthenogenesis, which is clonal reproduction interrupted by bouts of sexual reproduction. However, some lineages reproduce by obligate parthenogenesis (i.e., without any sexual reproduction) [38-42]. The *D. pulex* complex includes numerous lineages that have been distinguished on the basis of morphological, ecological, and genetic data [43-47]. Analyses of mitochondrial DNA variation have revealed the presence of three major groups in this complex. The pulicaria group consists of five different lineages; North American *D. pulicaria* (with three sublineages, Eastern *D. pulicaria*, Western *D. pulicaria* and Polar *D. pulicaria*), *D. pulex*, *D. melanica*, *D. middendorffiana sensu stricto*, and *D. arenata*, an endemic species inhabiting Oregon ponds [43,48]. The tenebrosa group includes two lineages, European *D. pulicaria* and *D. tenebrosa* [43]. The third group includes European *D. pulex*. Mitochondrial lineages in the pulicaria group may have diverged during the Pleistocene (between 1.2 and 2.2 million years ago) [43,49,50] while the pulicaria and tenebrosa groups seem to have diverged during the Pliocene (around 3 million years ago) [43]. Relationships between lineages based on nuclear genes are less clear and may be confounded by incomplete lineage sorting and a highly reticulate history [45,51,52]. In North America, two lineages, *D. pulex* and *D. pulicaria* (considered to be ecological species), are dominant in freshwater habitats. They are morphologically similar but ecologically distinct [44], although they hybridize in nature [42,53-55]. *D. pulex* and F1 hybrids are usually found in fishless shallow ponds whereas *D. pulicaria* inhabits lakes. Variation in the *Lactate dehydrogenase* gene (*Ldh*) is diagnostic [54,55]; *D. pulex* is fixed for the S allele whereas the F allele is fixed in *D. pulicaria* [55]. Diploid hybrids of these two lineages possess an SF genotype at the *Ldh* locus, always have *D. pulex* mitochondrial genomes, and have been found to reproduce by obligate

parthenogenesis in nature [42,56], although laboratory-produced hybrids may be able to reproduce by cyclical parthenogenesis [56,57]. It has been suggested that hybridization may play a role in the spread of meiosis suppressing genetic elements in the obligate parthenogenetic populations of *D. pulex* with SS *Ldh* genotypes via introgression [58,59].

Polyploidy has evolved repeatedly in the *D. pulex* complex [49,60-62] and shows a geographical pattern [49,62-66]. Polyploid populations are obligate parthenogens and are found at high latitudes and altitudes, and diploid populations (hybrid or not) are prevalent in temperate regions [47,60,62]. A polyphyletic assemblage of polyploids collectively known as *D. middendorffiana* (and which we term *D. middendorffiana sensu lato* in this study) is thought to have arisen from hybridization between *D. pulex* males and *D. pulicaria* females, or females of another species which no longer exists as a cyclic parthenogen [49,60,61]. Other polyploids are thought to have arisen from crosses between *D. pulex* females and *D. pulicaria* males and are encountered in the Northeast of Quebec and in Ontario (Canada) [45,67]. Moreover, *D. tenebrosa*, a circumarctic species [62], includes both diploids and polyploids [67], but the hybrid nature of the polyploids in this species is still unclear [45]. A study using microsatellite data, flow cytometry, and mitochondrial sequences has shown that most polyploids of the *D. pulex* complex are triploids, although some tetraploids have also been observed [67].

The *D. pulex* genome of one cyclically parthenogenetic isolate from Oregon has been sequenced [68], and numerous class II TEs have been identified in it [9]. Previous studies have reported that the class II TE load is lower in the genomes of obligate compared to cyclical parthenogenetic *D. pulex* lineages [10,15], as theoretically predicted if sexual reproduction helps TEs to spread [11]. *Pokey*, a class II TE from the *piggyBac* superfamily, has been extensively studied in diploid populations of *Daphnia*. It inserts in the tandemly repeated rRNA genes [69] and in other parts of the genome [15,70]. Based on patterns of polymorphism in *Pokey* insertion sites observed among natural populations, previous studies have suggested that *Pokey* may be active in cyclically parthenogenetic populations of *D. pulex* but not in obligate parthenogens [15,70]. The diversity, and potentially the activity, of *Pokey* in rRNA genes is greatly influenced by recombination events, especially in hybrids [71]. The *D. pulex* complex and *Pokey* represent an interesting model to study the effect of hybridization and polyploidization on the evolution and dynamics of a class II TEs *in natura*.

The aim of this study is to compare the polymorphism of *Pokey* insertion sites between diploid and polyploid hybrid genomes in obligately parthenogenetic isolates of the *D. pulex* complex. If *Pokey* was not active during and following hybridization events, the similarity between *Pokey* insertion profiles should be congruent with host evolutionary relationships. To test this prediction, the polymorphism of *Pokey* insertion site profiles was compared with the ploidy level and the genetic similarity of the hosts determined using microsatellite multilocus genotypes. Moreover, we test the prediction that the number of *Pokey* per haploid genome (hereafter called density) is similar in polyploid and diploid hybrids using two complementary techniques, TE display and quantitative PCR (qPCR). TE display allows us to compare the diversity of *Pokey* insertion sites in polyploid and diploid isolates. This technique also provides an estimate of the number of *Pokey* insertion sites (*Pokey* load) but it cannot distinguish between homozygosity and heterozygosity at a particular site. Conversely, qPCR allows us to estimate the total number of *Pokey* insertions per haploid genome (*Pokey* density) regardless of location, including those that occur in rDNA, which appear as a single peak in a TE display analysis. A higher density of *Pokey* insertions per haploid genome in polyploids than in diploids may be evidence of an increase in *Pokey* activity after polyploidization.

Methods
Daphnia samples
In the laboratory, we established parthenogenetic lines of *Daphnia* (hereafter called isolates) from 27 individual obligately parthenogenetic females (14 diploid hybrids and 13 polyploid hybrids) sampled from ponds in North America between 2004 and 2008 (Additional file 1). *Daphnia* were sampled from ponds accessed via public roadsides or on private land with the permission of the land owner. No specific permissions are required to sample *Daphnia* as they are not endangered or protected species. The lines were cultured using standard techniques [72]. The isolates represent six mitochondrial lineages (*D. pulex*, Polar *D. pulicaria*, Western *D. pulicaria*, Eastern *D. pulicaria*, *D. middendorffiana sensu stricto*, and *D. tenebrosa*). Due to the geographical polyploidy pattern, all the polyploids come from two subarctic regions (Churchill, MB, Canada and Kuujjuarapik, QC, Canada), although the diploids come from both temperate and subarctic regions (Additional file 1). For each isolate, genomic DNA from 10 to 30 individuals, weighing approximately 100 mg (wet weight), was extracted using the DNeasy Tissue kit (QIAGEN Inc., Mississauga, ON, Canada) according to the supplier's protocol. Origin of the putative parental species of each isolate (Additional file 1) was determined by combining information on morphology, haplotype of the mitochondrial ND5 gene, and genotype at the nuclear *Ldh* gene [45]. Ploidy levels were previously assessed using nine microsatellite loci and flow cytometry [45].

TE display
We used a PCR-based approach called TE display [73], which generates dominant AFLP-like markers, to test the effect of ploidy level on insertion site polymorphism, and on the load of *Pokey* insertions in the genomes of 14 diploid and 13 polyploid isolates. We followed a modified version of the TE display protocol of Valizadeh and Crease [15] that involves digestion of genomic DNA using the restriction enzyme BfaI followed by ligation of BfaI linkers and two rounds of PCR amplification using a *Pokey*-specific forward primer (Additional file 2). The ligated DNA was used as a template for a primary (pre-selective) PCR with the primer Pok6456F, located near the 3′ end of *Pokey*, and the primer BfaI-R that anneals to the BfaI linker sequence followed by a secondary (selective) PCR using fluorescent labeled primer Pok6464F and the primer BfaI-R (Table 1). Our TE display protocol, unlike that of Valizadeh and Crease [15], used an annealing temperature of 50°C instead of 55°C for both the primary and secondary PCR. This allows amplification of *Pokey* insertions in *Daphnia* species with genomes that are divergent from *D. pulex*. Only fragments ≥160 bp were included in our analyses and primary PCR were repeated three times in each individual followed by a secondary PCR on the product of each primary reaction to ensure that *Pokey* profiles were reproducible and to remove possible artefacts from our analysis (Additional file 2).

Comparison of genetic distance based on *Pokey* profiles, microsatellites and the *ND5* gene
We used the results of TE display to generate a binary matrix of presence (1) or absence (0) of peaks, which represents the *Pokey* insertion profile (Additional file 3). We then generated a matrix of Jaccard distance estimates from the *Pokey* profiles. The Jaccard distance was chosen because it does not use shared absence of an allele as a shared characteristic [74]. A distance matrix was also calculated for each locus of the microsatellite dataset previously obtained for our isolates [45] using a modified version of the Bruvo distance [75], implemented in the PolySat package [76] using the R software [77]. The Bruvo distance allowed us to estimate relationships of mixed-ploidy level genotypes using co-dominant markers. The Bruvo distance takes into account stepwise mutation models between alleles. In the non-modified version (equation 2 in [75]), the algorithm adds "virtual allele" with an "infinite" value to lowest ploidy-level genotypes to compare them to the highest ploidy-level genotypes. This may lead to group artificially genotypes with the same ploidy level [78]. Thus, we used a modified version of the Bruvo distance (Bruvo2.distance implemented in PolySat set with the parameters add = TRUE and loss = TRUE) that allows genome "addition" and "loss", simulating gene addition by polyploidization but

Table 1 TE display and qPCR primers and linkers used in this study

Purpose	Annealing temperature	Primer name	Sequence (Dye)	Percent amplification efficiency	Amplicon size
Linkers for TED	/	BfaI Linker F	5'-TACTCAGGACTCAT	/	/
		BfaI Linker R	5'-GACGATGAGTCCTGAG		
Primary PCR for TED	50°C/55°C	Pok6456F	5'-GACAACGGTGGCCGAAACGCGG	/	/
		BfaIR	5'-GACGATGAGTCCTGAGTAG		
Secondary PCR for TED	50°C/55°C	Pok6464F	5'-TGGCCAAAACACGGTTTGGCCG (HEX)	/	/
		BfaIR	5'-GACGATGAGTCCTGAGTAG		
18S genes for qPCR	60°C	18S1864F	5'-CCGCGTGACAGTGAGCAATA	0.9556	50
		18S1913R	5'-CCCAGGACATCTAAGGGCATC		
28S genes for qPCR	60°C	28S3054F	5'-GGTAGCCAAATGCCTCGTCA	0.9246	150
		28S3204R	5'-GAGTCAAGCTCAACAGGGTCTTCTTTCCC		
Total *Pokey* for qPCR	60°C	Pok6456F	5'-GACAACGGTGGCCGAAACGCGG	0.9136	122
		Pok6578R	5'-GATGGTCGGATTCGATTGAATGCTCG		
Pokey in rDNA for qPCR	60°C	Pok6456F	5'-GACAACGGTGGCCGAAACGCGG	0.8957	192
		28S3104R	5'-GTTAATCCATTCGTGCGCG		
Tif for qPCR	60°C	TIF392F	5'-GACATCATCCTGGTTGGCCT	0.9493	50
		TIF442R	5'-AACGTCAGCCTTGGCATCTT		
Gtp for qPCR	60°C	GTP385R	5'-TATTCAGCATGGAGAGACGGC	0.9369	50
		GTP435R	5'-GATGTCGACTGACGCTGGAA		

also possible gene loss via diploidization. This modified version of the Bruvo distance does not lead to artificially grouping genotypes with the same ploidy level altogether. In addition, we generated a matrix of sequence divergence between *ND5* sequences from previous studies by Vergilino et al. [45,67] (Table 1 for Genbank accession number) from our isolates using the maximum composite likelihood model implemented in MEGA5.1 [79].

To determine if the *Pokey* insertion sites profiles differed depending on the genetic background and ploidy level between isolates, a principal coordinate analysis (PCoA) [80] and a K-means cluster analysis were conducted to represent affinities between the different *Pokey* insertion profiles or multilocus microsatellite genotypes using R software version 2.15.2 [77,81]. Each PCoA was constructed using the pco module of the labdsv library in the R software on the Jaccard distance matrix for the *Pokey* profiles and the modified Bruvo distance matrix for the microsatellites after transforming these distance matrices in Euclidean distances [82]. The K-means analyses were conducted on binary matrices representing either the *Pokey* profiles or microsatellite genotypes (transformed into a binary matrix), and the number of clusters for each analysis was set using an iterative method, CascadeKM with the calinski criterion [83], implemented in the vegan package available with R software. We also performed a Mantel test according to Legendre and Legendre (section 10.5 in [74]) using the Pearson method with 10,000 replicates (package vegan

in R software) to compare the Jaccard distance matrix based on *Pokey* insertion profiles with both the Bruvo distance matrix based on microsatellite data and the distance matrix based on *ND5* mitochondrial haplotypes.

To test the hypothesis that the load of *Pokey* insertions increases with ploidy level, we compared the number of *Pokey* insertion sites estimated by TE display to the ploidy level after taking into account the heterozygosity of the isolates. Heterozygosity was weighted by ploidy level and was calculated from the variability of nine microsatellite loci [45]. Theoretically, polyploids may arise from independent hybridization events and those with different parental genomes may possess a higher diversity of TE insertion sites than polyploids with similar parental genomes due to increased probability of homozygosity of some TE insertions in the latter case. If we do not account for the different genomes that form polyploids, then we may overestimate the effect of ploidy level since the number of TE insertions could be more strongly correlated with heterozygosity level than ploidy level. Thus, we introduce a ploidy-weighted heterozygosity index (H_{pl}) for a comparison between diploids and triploids, which takes into account the ploidy level of each genotype such that:

$$H_{pl} = \frac{n_3 + 0.5 * n_2}{n_L}$$

where n_L is the total number of microsatellite loci analyzed (9), n_3 is the number of loci with 3 different alleles, and n_2 is the number of loci with only 2 different alleles. Genotypes that are homozygous for all microsatellite loci have an H_{pl} of 0, diploid isolates that are heterozygous for every locus and triploid isolates with two different alleles at every locus have an H_{pl} of 0.5, and triploid isolates that have three different alleles at every locus have an H_{pl} of 1. Therefore, triploids with low H_{pl} values (under 0.5) can be compared to diploid hybrids. To disentangle the effect of adding different genomes from the effect of increased ploidy level, we performed an ANCOVA (Analysis of Covariance) using R software [77] with the number of Pokey insertions as the dependent variable and the ploidy level and H_{pl} as the independent variables. The number of singletons (i.e., Pokey insertion sites encountered in only one isolate) between diploid and polyploid isolates was compared using a Fisher exact test performed on a 2×2 contingency table similar to the approach of Wright et al. [73].

Direct comparison of our results to those obtained on cyclic and obligate non-hybrid diploid populations previously studied by Valizadeh and Crease [15] was not possible as these authors used a higher annealing temperature (55°C instead of 50°C). Therefore, we performed additional TE display assays on six diploid hybrid and six polyploid hybrid isolates using the 55°C annealing temperature of Valizadeh and Crease [15].

qPCR assays

Because TE display generates dominant markers, it provides more information about the polymorphism of Pokey insertion sites than their density within the genome. This difference may be significant especially if a significant proportion of Pokey insertions are homozygous, which may be possible in polyploids [24,25]. Therefore, to help resolve this problem, the number of Pokey insertions per haploid genome was estimated using qPCR. We performed qPCR assays on Pokey inserted in 28S rRNA genes (rPokey) and in the entire genome (tPokey) of 9 diploid and 10 polyploid isolates as described by Eagle and Crease [84]. We also estimated the number of 18S and 28S rRNA genes as the number of rPokey may be correlated to the number of rRNA genes [84,85]. Briefly, we used the ΔC_T qPCR method as described in Eagle and Crease [84] (Additional file 2) to estimate the density of multicopy genes (18S, 28S, tPokey, rPokey) relative to two single-copy genes (Table 1); Tif, a transcription initiation factor and Gtp, a member of the RAB subfamily of small GTPases. Assuming that diploids have two copies and triploids have three copies of these two genes, these estimates correspond to the haploid number of multicopy genes in each genome. Reaction conditions were run in triplicate as described in Eagle and Crease [84] (Additional file 2). The mean haploid copy number, rounded to the nearest 0.5 for diploids and 0.34

for triploids, and standard deviations were calculated for each multicopy gene in each isolate. The number of Pokey insertions outside 28S rRNA genes per haploid genome (gPokey) was calculated as [tPokey number − rPokey number].

We used modules available in the R software package to perform correlation and regression analyses between the haploid number of 18S rRNA genes, 28S rRNA genes, rPokey, and gPokey in diploids and polyploids. Levene's tests (equality of variances) and Student's t-tests (equality of means) were used to test for possible significant differences in 18S, 28S, rPokey, and gPokey haploid numbers between diploids and polyploids. The sequential Bonferroni technique proposed by Rice [86] was used to adjust the significance level (0.05) for the multiple Student's t-tests comparing 18S and 28S number within isolates.

Under the assumption that the same gPokey elements are amplified using qPCR and TE display techniques, we can estimate the average heterozygosity for these elements in diploids by using:

$$H_{Pokey} = \frac{2\left(n_{TED-1} - n_{gPokey}\right)}{n_{TED-1}}$$

where n_{TED-1} represents the number of different Pokey insertion sites estimated by TE display minus the peak representing rPokey, and n_{gPokey} is the haploid number of gPokey estimated by qPCR. However, due to partial heterozygosity in triploids, we were not able to calculate their exact heterozygosity level. The ratio $(k^* n_{gPokey})/n_{TED-1}$, where k is the ploidy level of the isolate, allows us to evaluate if TE display and/or qPCR techniques underestimate or overestimate Pokey insertions. If every gPokey insertion is in a heterozygous state, this ratio will be 1. If the ratio is below 1, qPCR underestimates or TE display overestimates the number of Pokey insertions. If all insertions are in a homozygous state, the ratio equals 2 for diploids and 3 for triploids. If the ratio is greater than 2 or 3 in diploids or triploids, respectively, qPCR overestimates or TE display underestimates the number of Pokey insertions.

Results

Polymorphism of Pokey insertion site profiles

Using the TE display technique, the average number of Pokey insertion sites in 14 diploid isolates was 16.64 (±4.94), with values from 6 to 26, whereas the average number of Pokey insertion sites in 13 polyploid isolates was 19.00 (±4.36), with values from 12 to 27. The two means are not significantly different (Student t-test, $t = -1.3105$, $df = 25$, $P = 0.202$; Table 2; Additional file 1). Overall, 88 different Pokey insertion sites were detected (Additional file 3). Such polymorphism allowed us to

Table 2 Summary of TE display and qPCR analyses of *Pokey* number in diploid and polyploid isolates in the *Daphnia pulex* complex

Isolates	TE display		qPCR					
	Anneal at 50°C	Anneal at 55°C	18S genes	28S genes	Total Pokey	rDNA Pokey	Genome Pokey (Total-rDNA)	Tif:Gtp
Diploids with known hybrid status	19.09 ± 3.99 [11]	13.83 ± 4.22 [6]	293.25 ± 111.19 [8]	486.25 ± 201.36 [8]	17.81 ± 4.35 [8]	5.19 ± 5.03 [8]	12.63 ± 5.04 [8]	0.91 ± 0.06 [8]
Polyploids with known hybrid status	21.50 ± 3.54 [10]	16.33 ± 1.75 [6]	214.57 ± 62.34 [7]	346.67 ± 113.86 [7]	15.57 ± 3.43 [7]	3.10 ± 1.07 [7]	12.48 ± 3.20 [7]	0.92 ± 0.07 [7]
Total diploids	16.64 ± 4.94 [14]	12.71 ± 4.86 [7]	292.78 ± 104.02 [9]	488.28 ± 188.45 [9]	16.67 ± 5.32 [9]	5.05 ± 4.72 [9]	11.61 ± 5.61 [9]	0.90 ± 0.06 [9]
Total polyploids	19.00 ± 4.36 [13]	14.44 ± 3.17 [9]	248.03 ± 75.21 [10]	398.10 ± 127.42 [10]	14.33 ± 3.45 [10]	3.77 ± 1.40 [10]	10.57 ± 4.03 [10]	0.92 ± 0.08 [10]

Results from qPCR are average ± standard deviation of *Pokey* inserts per haploid genome whereas results from TE display are for whole genomes. Numbers in brackets are the number of isolates tested. *Tif:Gtp* ratio is the number of *Tif* relative to the number of *Gtp* single copy reference genes.

analyze the similarity between isolates based on these *Pokey* profiles and on microsatellite genotypes using PCoA. In addition, the *Pokey* profiles and the genetic similarity of the hosts based on microsatellite loci and mitochondrial haplotypes were compared using Mantel tests.

The first two axes of the PCoA accounted for 23.3%, with 12.4% for axis 1 and 10.9% for axis 2, of the total variability in *Pokey* profiles (Figure 1A; Additional file 4A). For the host genetic backgrounds, the first two axes of the PCoA accounted for 24.0%, with 16.0% for axis 1 and 8.0% for axis 2, of the total microsatellite variability (Figure 1B and Additional file 4B). *Pokey* profiles and their host microsatellite genotypes ordinate differently in each PCoA according to the two first axes (Figure 1). Although according to the PCoA on the host genetic background, axis 2 differentiates the hybrids with *D. pulex* mitochondria from the *D. middendorffiana sensu lato* (Figure 1B); it is axis 1 that differentiates these isolates according the PCoA from *Pokey* profiles. The second PCoA axis of the *Pokey* profiles does not show clear differentiation pattern according to the mitochondrial haplotype or the ploidy level (Figure 1A). The K-means analyses of both *Pokey* profiles and microsatellite genotypes separate the isolates into two clusters (K = 2, Calinski criterion), but each cluster represents a different set of isolates in the two datasets (Figure 1). One cluster based on *Pokey* profiles contains only polyploid isolates from Kuujjuarappik (Quebec) and Churchill (Manitoba) possessing *D. pulex* (PX3-QC-1 and PX3-QC-2), Eastern *D. pulicaria* (PC3-QC-1, PC3-QC-2

and PC3-QC-3) or Western *D. pulicaria* (PC3-MB-6) mitochondrial haplotypes. The second cluster based on *Pokey* profiles contains diploid hybrids with *D. pulex* mitochondrial haplotypes, *D. tenebrosa* isolates, and polyploids with Polar *D. pulicaria* (PC3-MB-4 and PC3-MB-5) or *D. middendorffiana sensu stricto* (MI3-MB-2) mitochondrial haplotypes. In contrast, one cluster based on microsatellites contains all the *D. tenebrosa* isolates (both diploids and polyploids) while the other cluster contains all the other isolates. Despite these differences, the distance matrices of *Pokey* insertion site profiles and both microsatellite genotype and mitochondrial haplotype datasets are partially correlated according to Mantel tests (Mantel test; $r = 0.3957$, $P = 0.0001$ and $r = 0.3047$, $P = 0.003$, respectively). The third axis of the PCoA constructed from *Pokey* profiles (9.3%, Additional file 4) may explain why this distance matrix is partially correlated with the distance matrix based on microsatellite genotypes. *Pokey* profiles from *D. tenebrosa* isolates ordinate together according to axis 3 in the PCoA of *Pokey* (Additional file 4A) as they ordinate altogether according to axis 1 in the PCoA of microsatellites (Additional file 4B).

As the purpose of our study was to test the effect of ploidy level and not the effect of hybridization, six isolates with *D. tenebrosa* mitochondrial haplotypes and unknown hybrid origin (TE2-MB-1, TE2-MB-2, TE2-MB-3, TE3-MB-1, TE3-MB-2 and TE3-MB-3; Additional file 1) were excluded from the ANCOVA analysis. Excluding these isolates, 88 different *Pokey* insertion sites

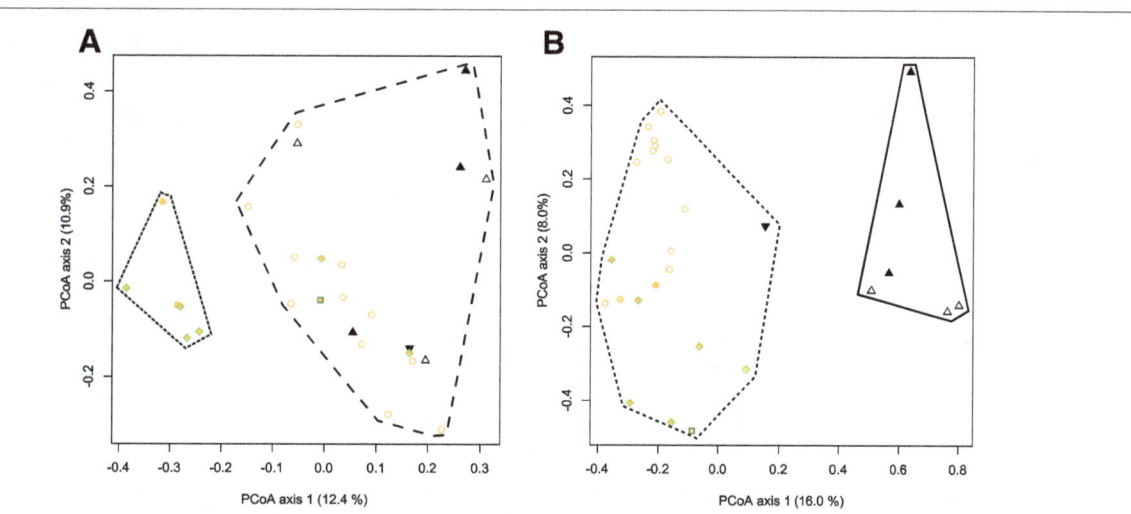

Figure 1 Principal Coordinate Analyses of Jaccard distance matrix of *Pokey* profiles and Bruvo distance matrix of microsatellite diversity in diploid and polyploid isolates of the *Daphnia pulex* complex. (A) *Pokey* profiles generated using TE display; **(B)** Microsatellite genotypes determined by Vergilino et al. [45]. The first two axes are represented in each graph. Empty symbols are diploids and solid symbols are polyploids. Empty orange circles: diploid hybrids with *D. pulex* mitochondrial haplotypes; solid orange circles: triploid hybrids with *D. pulex* mitochondrial haplotypes; solid square: *D. middendorffiana sensu stricto*; empty black triangles: diploid *D. tenebrosa*; solid black triangles: triploid *D. tenebrosa*; solid red triangle: introgressed *D. tenebrosa* with a *D. pulex* nuclear genome; solid green diamond filled with orange: triploid hybrids with *D. pulicaria* mitochondrial haplotypes.

were detected using TE display (Additional file 3). The mode of reproduction is not a confounding effect in this analysis as both diploid and polyploid hybrids are obligate parthenogens. The mean number of *Pokey* insertion sites in diploid hybrid isolates is 18.09 (±3.99), with values from 13 to 26, whereas the mean number of *Pokey* insertion sites in polyploid isolates is 20.60 (±3.47), with values from 16 to 27. These means are not significantly different (Student *t*-test, $t = -1.531$, $df = 19$, $P = 0.1423$; Table 2; Additional file 1). Correlations between average number of *Pokey* insertion sites and the average heterozygosity weighted by host ploidy level (H_{pl}) were positive but not significant (Table 3). H_{pl}, ploidy level and H_{pl}:*ploidy* interaction had no significant effect on the number of *Pokey* insertion sites (ANCOVA, $F = 2.132$, $df = 1$, $P = 0.162$; $F = 1.162$, $df = 1$, $P = 0.296$ and $F = 0.205$, $df = 1$, $P = 0.657$, respectively; Figure 2; Additional file 5). The 11 diploid hybrid isolates displayed 65 of the 88 *Pokey* insertion sites, of which 8 (12.3%) are singletons. The 10 polyploid hybrid isolates displayed 68 of the 88 *Pokey* insertion sites, of which 14 (20.6%) are singletons. Twenty-one *Pokey* insertions sites were only sampled in polyploid hybrids while 18 were only sampled in diploid hybrids, and this difference is not statistically significant (Fisher exact test, $P = 0.8559$). The difference in the number of singletons

between diploid and polyploid hybrids is not statistically significant (Fisher exact test, $P = 0.3579$). The number of *Pokey* insertion sites observed with TE display using an annealing temperature of 55°C, as in Valizadeh and Crease [15], was lower than that observed using an annealing temperature of 50°C (12.71 ±4.86 *vs.* 16.75 ±6.07, paired Student's *t*-test, $t = 3.9506$, $df = 6$, $P = 0.008$ for diploids and 14.44 ±3.17 *vs.* 18.78 ±4.86; paired Student's *t*-test, $t = 5.3072$, $df = 8$, $P = 0.0007$ for polyploids; Additional file 1). The number of *Pokey* insertion sites was about two times lower with an annealing temperature of 55°C in some isolates. For example, 13 sites were detected using an annealing temperature of 50°C but only 6 were detected using 55°C in the diploid hybrid PX2-MB-1 (Additional file 1). Although we are aware that artefacts can be produced during the TE display process, these differences do not seem to be due to a higher frequency of artefacts at 50°C than in 55°C as most artifacts were encountered using both annealing temperatures and were excluded from analysis as indicated in the Methods section.

qPCR analysis of rRNA gene and *Pokey* copy number

Using the qPCR technique, we estimated the haploid number of 18S genes, 28S genes (Additional file 6), and

Table 3 Correlations between *Pokey* and rRNA gene number in diploid and polyploid *Daphnia* from North America

Cytotypes	X-axis	Y-axis	Slope	y-intercept	R^2	P-value	Figure
All diploids	18S	28S	1.8038	−39.8411	0.9901	1.76 e-08*	AF6[8]
All triploids	18S	28S	1.6627	−14.3154	0.9586	5.07 e-07*	AF6
Diploid hybrids	Hpl[1]	*Pokey* (TED)[2]	16.500	11.674	0.0634	0.4549	2
	18S	r*Pokey*[3]	−0.0146	9.4540	0.1035	0.4371	AF7
	g*Pokey*[4]	r*Pokey*	−06251	13.0792	0.3927	0.0964	AF8
	Pokey-1 (TED)[5]	Total g*Pokey*[6]	1.5329	−1.7681	0.4950	0.0515	3
Diploid hybrids - PX2-MB-1	18S	r*Pokey*	0.0147	−1.07571	0.3979	0.1287	AF7
	g*Pokey*	r*Pokey*	−0.3350	8.0216	0.5605	0.0528	AF8
Diploid hybrids – (PX2-QC-9, PX2-MI-7)	*Pokey*-1 (TED)	Total g*Pokey*	1.5741	−2.7883	0.7383	0.0283*	-
	Pokey-1 (TED)	H^7_{Pokey}	−0.0088	0.7623	0.0376	0.7130	AF9
	Hpl	H_{Pokey}	−3.2220	1.8820	0.6592	0.0497*	AF10
Triploid hybrids	Hpl	*Pokey* (TED)	6.6490	17.534	0.0756	0.4418	2
	18S	r*Pokey*	0.0085	1.2673	0.2492	0.2541	AF7
	g*Pokey*	r*Pokey*	0.0201	2.8478	0.0036	0.8978	AF8
	Pokey-1 (TED)	Total g*Pokey*	−1.6300	1.0290	0.3342	0.1740	3
Triploid tenebrosa	*Pokey*-1 (TED)	Total g*Pokey*	−0.1923	20.7692	0.4808	0.5122	3

1. Denotes average heterozygosity estimated using nine microsatellite loci weighted by the ploidy level.
2. Denotes *Pokey* insertion site number estimated using TE display.
3. Denotes haploid *Pokey* number in 28S genes amplified using qPCR.
4. Denotes haploid *Pokey* number outside 28S genes amplified using qPCR.
5. Denotes *Pokey* insertion site number estimated using TE display minus the peak from elements in 28S genes.
6. Denotes total *Pokey* number outside 28S genes amplified using qPCR.
7. Denotes average heterozygosity of *Pokey* insertions outside 28S genes amplified using qPCR.
8. AF refers to Additional Files.
*P values are significant results.

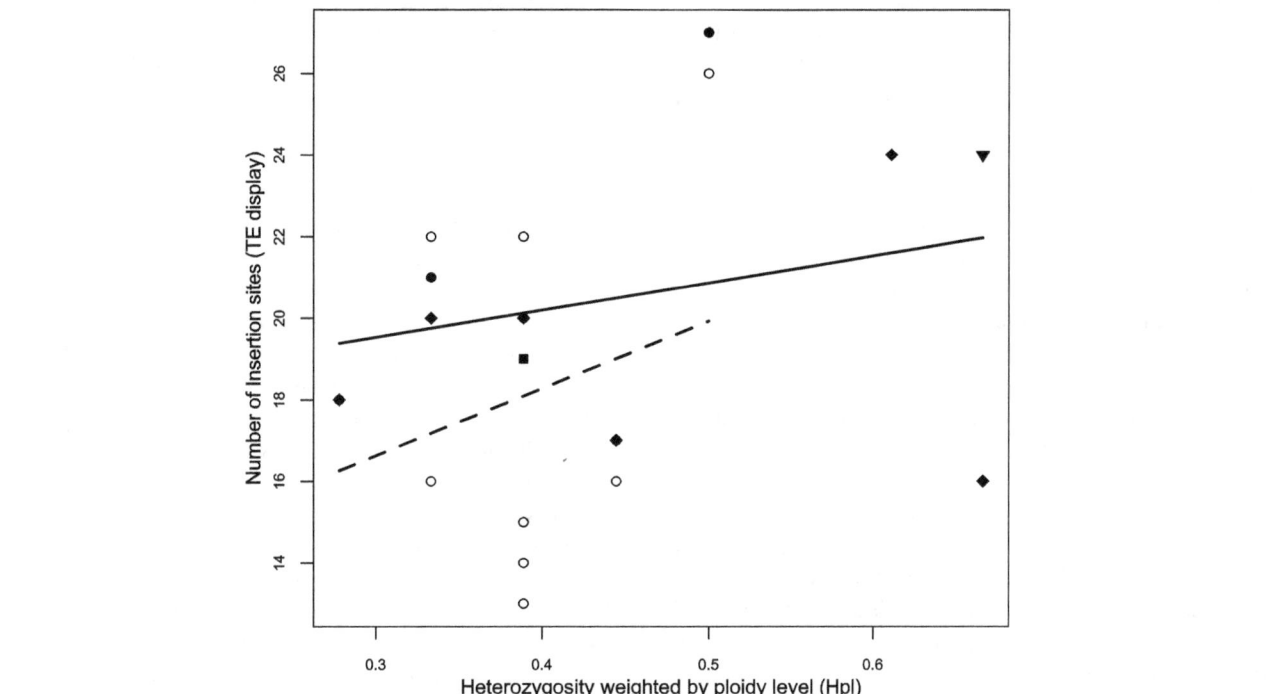

Figure 2 Relationship between *Pokey* insertion site number outside rDNA estimated using TE display and ploidy-weighted heterozygosity (H_{pl}) in diploid and polyploid hybrids of the *Daphnia pulex* complex. Empty circles: diploid hybrids with *D. pulex* mitochondrial haplotypes; solid circles: triploid hybrids with *D. pulex* mitochondrial haplotypes; solid square: *D. middendorffiana sensu stricto*; solid triangle: introgressed *D. tenebrosa* with a *D. pulex* nuclear genome; solid diamond: triploid hybrids with *D. pulicaria* mitochondrial haplotypes. Dashed and solid lines are linear regressions estimated from the data.

Pokey inserted in 28S genes (r*Pokey*) and in the whole genome (t*Pokey*) in 19 isolates including 9 diploids and 10 polyploids (Table 2, Additional file 1). Using these estimates, we calculated the number of *Pokey* insertions outside 28S genes (g*Pokey* = t*Pokey* - r*Pokey*). Under the assumption that *Tif* or *Gtp* reference genes were neither duplicated or lost in any of the isolates, we expect the *Tif:Gtp* ratio to be close to 1, and this was the case with ratios ranging from 0.79 to 1.03 and a mean of 0.90 for diploids and 0.92 for polyploids (Table 2). It is unlikely that correlated losses or duplications of both genes would occur in multiple isolates and so we have assumed that diploids have two copies and triploids have three copies of each reference gene.

Both 18S and 28S genes showed a tendency towards a higher copy number per haploid genome in diploids than in polyploids (292.78 ±111.19 *vs.* 248.04 ±62.34 for 18S genes and 488.28 ±201.36 *vs.* 398.10 ±113.86 for 28S genes; Table 2), but differences between diploids and polyploids were not significant (Student's *t*-test; t =1.0828, df = 17, P = 0.2940 for 18S and Student's *t*-test; t =1.2337, df = 17, P = 0.2341 for 28S). The estimates of 18S and 28S number within each isolate were significantly correlated (Additional file 6) but the slopes of the lines generated by plotting them relative to one another were above the

expected value of 1.0 with values of 1.80 for diploid hybrids and 1.66 for polyploids (Table 3). It is possible that we overestimated the number of 28S genes (Additional file 6) and so the number of 18S genes was used as a proxy of rDNA copy number in all subsequent analyses.

Excluding isolates with unknown hybrid nature, the average haploid number of t*Pokey* insertions was 17.81 ±4.35 for diploid hybrids and 15.58 ±3.43 for polyploid hybrids (Table 2) and the difference was not statistically significant (Student's *t*-test; t = 1.1092, df = 13, P = 0.2947).

The number of r*Pokey* was higher in diploids than in polyploids (mean 5.19 *vs.* 3.10, respectively), but this difference was not significant (Student's *t*-test; t = 1.0735, df = 13, P = 0.3026). Variation in the number of r*Pokey* insertions was higher in diploid than in polyploid hybrids (SD 5.03 *vs.* 1.07) but the difference was not statistically significant (Levene's test, W = 2.0149, P = 0.1793). No correlation was found between the number of r*Pokey* and the number of 18S genes for either ploidy level (Table 3; Additional file 7) even if the outlier PX2-MB-1, which possesses a high number of r*Pokey* (16.5) and a low number of 18S genes compared to other diploids (Table 3), was omitted from the analysis.

The mean number of g*Pokey* was 12.63 in diploid hybrids and 12.48 in polyploid hybrids (Table 2) and the difference was not significant (Student's t-test; $t = 0.0653$, $df = 13$, $P = 0.9489$). Variation in the number of g*Pokey* insertions was higher in diploid than in polyploid hybrids (SD 5.04 *vs.* 3.20) but the difference was not statistically significant (Levene's test, $W = 0.8830$, $P = 0.3645$).

No significant correlation was found between the number of g*Pokey* and the number of r*Pokey* in hybrids of either ploidy level (Table 3; Additional file 8). The negative (but not significant) relationship between r*Pokey* and g*Pokey* in diploids was partly due to the high number of r*Pokey* in the isolate PX2-MB-1. However, the relationship was still negative and was nearly significant when this isolate was discarded from the analysis (Table 3; Additional file 8).

Comparison of TE Display and qPCR

The assumption that the g*Pokey* elements amplified using qPCR and TE display are identical seems to be reasonable (Figure 3). Of all 19 isolates, only one (PX2-MI-7) had a ratio $(k*n_{g\mathrm{Pokey}})/n_{\mathrm{TED}\text{-}1}$ below 1 which may indicate an overestimation of g*Pokey* number using TE display compared to qPCR. One diploid isolate (PX2-QC-2) had a ratio above 2 and a triploid isolate (MI3-MB-2) had a ratio above 3 (Figure 3). This suggests that either the number of g*Pokey* was overestimated by qPCR, underestimated by TE display, or both. The relationship between g*Pokey* number based on qPCR and TE display is positive and significant in diploid hybrids (Table 3; Figure 3). This relationship was negative but not significant in triploid *D. tenebrosa* and in triploid hybrids (Table 3; Figure 3).

After excluding isolates outside the lower and upper limits of possible values of total g*Pokey* insertions estimated with qPCR and TE display (values of the ratio $(k*n_{g\mathrm{Pokey}})/n_{\mathrm{TED}\text{-}1}$ between 1 and 2 for diploids and between 1 and 3 for triploids), the average heterozygosity across *Pokey* insertions loci among diploid hybrids is 59.85%. The relationship between the heterozygosity of *Pokey*-inserted loci and the number of g*Pokey* estimated using TE display is slightly negative but not significant (Table 3; Additional file 9). The slope of the relationship between the ploidy-weighted heterozygosity using nine microsatellite loci and the average heterozygosity of *Pokey*-inserted loci is negative and is significant for diploid hybrids (Table 3, Additional file 10).

Discussion

The polymorphism of *Pokey* insertion sites in *Daphnia* isolates

The polymorphism of TE insertion sites may depend on multiple factors such as selective pressure, drift, recombination rate, ploidy level, genomic background (*i.e.*, the parental origins of the hosts), geographic location, and the characteristics of the element(s) hosted in the genome [8,25,31,73,87-94]. If the diversity of *Pokey* insertion sites is due to the admixture of haploid genomes from different species with different architecture (that is nucleotide variation, number of repetitive genetic structures, etc.), the similarity of *Pokey* profiles is expected to mirror the genetic relationship of their hosts. According to the PCoA (Figure 1), the pairwise distance between *Pokey* profiles of the *Daphnia* isolates is not congruent with their pairwise genetic distance based on nine microsatellite loci if only the two first axes are taken into account. Similarities between the patterns produced by TE display and microsatellite analyses can only be revealed if the third axis of the PCoA is taken into account (Additional file 4). According to the K-means analysis, clusters based on similarity of *Pokey* profiles are not congruent with clusters based on microsatellite genotypes. Conversely, Mantel tests indicated that similarity between *Pokey* profiles is partially correlated with distance matrices constructed from microsatellite diversity and with mitochondrial haplotype diversity ($r = 0.3957$ and $r = 0.3047$, respectively).

The polymorphism of *Pokey* insertion sites in the isolates studied here imperfectly follows their evolutionary relationship with one another. This is concordant with previous results in which sequences from r*Pokey* elements amplified from some of the isolates included in this study show a different reticulation history than the one described by microsatellite data [71]. For example, r*Pokey* sequences from triploids PC3-QC-1 and PX3-QC-1, whose *Pokey* profiles cluster together using the K-mean analysis (Figure 1A), have *Pokey* sequences that are similar (Figure two in [71]). Similarly, r*Pokey* sequences from triploid isolates from Churchill (MI3-MB-2 and PC3-MB-5), whose *Pokey* profiles cluster with *D. tenebrosa* and diploid hybrid isolates in the K-means analysis (Figure 1A), are recombinant and show signatures of hybridization between *D. tenebrosa* and *D. pulex* or *D. pulicaria*. However, there was no indication of hybridization based on the analysis of microsatellite data, which clustered all *D. tenebrosa* isolates with one another (Figure 1B). Weider et al. [66] hypothesized introgression between *D. tenebrosa* and *D. pulex* or *D. pulicaria* based on mitochondrial DNA and allozyme analyses. The polymorphism of *Pokey* profiles may then mirror hybridization or introgression events between these species that microsatellites do not display due to homoplasy or null alleles in the microsatellite dataset. In our study, all *D. tenebrosa* isolates may be of hybrid origins but can still ordinate separately in the PCoA and cluster together in a separate group using the K-means analysis due to the sharing of a specific allele belonging to the *D. tenebrosa* species. Alternatively, *Pokey* insertion profiles may not correspond to the genetic relationships of

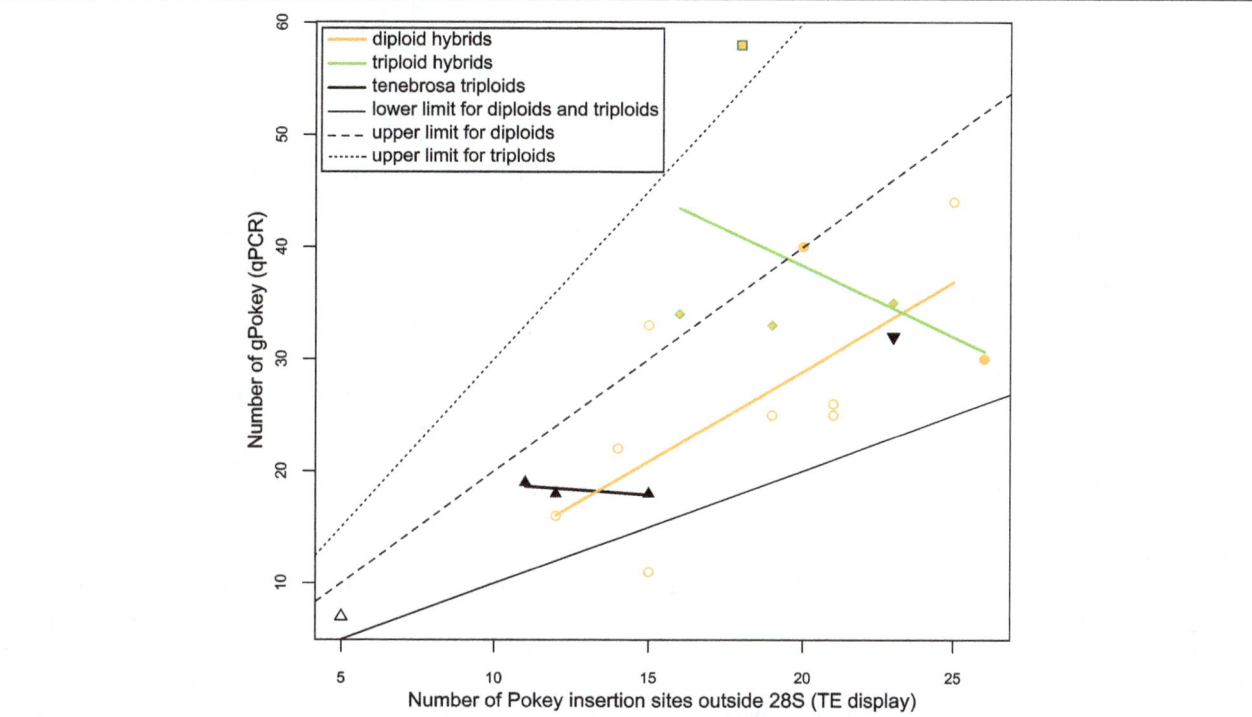

Figure 3 Relationship between the total number of *Pokey* insertions outside rDNA estimated using qPCR and TE display. Empty symbols are diploids and solid symbols are polyploids. Empty orange circles: diploid hybrids with *D. pulex* mitochondrial haplotypes; solid orange circles: triploid hybrids with *D. pulex* mitochondrial haplotypes; solid square: *D. middendorffiana*; empty black triangles: diploid *D. tenebrosa*; solid black triangles: triploid *D. tenebrosa*; solid red triangle: introgressed *D. tenebrosa* with a *D. pulex* nuclear genome; solid green diamond filled with orange: triploid hybrids with *D. pulicaria* mitochondrial haplotypes. Thin lines represent lower (slope = 1 and y-intercept = 0) and upper limits for diploids (slope = 2 and y-intercept = 0) and triploids (slope = 3 and y-intercept = 0) of *Pokey* insertion number (see Methods). Thick lines are linear regressions estimated from the data.

their host due to genomic rearrangements and random loss of copies in the course of evolution.

Patterns of *Pokey* insertion site polymorphism cannot be explained solely by ploidy level as K-means analyses show that individuals with different ploidy levels group in the same cluster whereas triploid individuals may belong to different clusters (Figure 1A). Valizadeh and Crease [15] did not find a relationship between the similarity of *Pokey* profiles and the mode of reproduction using a Neighbor-Joining tree of *Pokey* profiles from cyclic and obligate isolates of *D. pulex*. They concluded that the absence of a relationship was due to the multiple origins of obligate parthenogenetic lineages from multiple cyclical parthenogenetic populations. Similarly, the absence of a relationship between similarity of *Pokey* profiles and ploidy level is likely due to the multiple and independent origins of polyploid isolates.

Is *Pokey* load higher in *Daphnia* polyploids than diploids?
Our study examined the load of *Pokey* insertions in relation to ploidy level in natural populations of *Daphnia*. Both our diploid and polyploid isolates are hybrids (with the exception of some *D. tenebrosa* isolates that were excluded from the analyses of load) and are obligate parthenogens. These characteristics allow us to test the effect of ploidy level on the load of a class II transposable element without the confounding effects of hybridization *per se* and of different modes of reproduction. No significant differences in the density of *Pokey* insertions using either qPCR (per haploid genome) or TE display were found between diploid and polyploid hybrids, suggesting that an increase in ploidy level does not lead to an increase of *Pokey* insertions in the long term. The isolates studied were sampled from natural populations and the age of these clones is unknown. Previous studies have suggested that obligately parthenogenetic populations of *D. pulex* originated some 150,000 years ago [93] and that some polyploids from the *D. pulex* complex were produced during the Pleistocene [49]. It is possible that an increase in *Pokey* insertions occurs shortly after polyploidization as predicted by several hypotheses [25,28,94], but that genomic reorganization results in the loss of *Pokey* insertions with time. Loss of TE insertions following polyploidization (in both the short and long term) seems to be the rule rather the exception in most allopolyploid plants [26], regardless of whether the TEs are active. Loss of TE

insertions is thought to be due to genome rearrangements via unequal and ectopic recombination events between TEs at non-homologous loci. Therefore, the non-significant difference between diploid and polyploid hybrids may be due to loss of *Pokey* following polyploidization. Conversely, the absence of statistical significance may be due to the substantial variability in the number of *Pokey* in both groups, which may be due to high variability of *Pokey* load in the parents. For example, Eagle and Crease [84] surveyed 69 non-hybrid isolates of *D. pulex* and *D. pulicaria* from 22 sampling sites and found that g*Pokey* number can vary from 4 to 24. Thus, g*Pokey* number in hybrids between these species will also vary according to the g*Pokey* load in their ascendants. Alternatively, the presence or absence of active *Pokey* elements in parental species may influence the subsequent proliferation of *Pokey* in the hybrid offspring and increase the variability of *Pokey* insertion sites in hybrids. However, if *Pokey* is not active during apomixis, it cannot proliferate in obligately parthenogenetic hybrid lineages – except perhaps through ameiotic recombination events.

It has been suggested that *Pokey* is not active in nonhybrid obligately parthenogenetic isolates but may be active in cyclically parthenogenetic isolates of *D. pulex* [15,70] and *D. pulicaria* [84]. Even so, it is possible that *Pokey* may be active in hybrids at least in the first generations after their formation due to the presence of active *Pokey* in their ascendants. Therefore, increases in the density of *Pokey* insertions may depend on the activity of *Pokey* and the effectiveness of regulation of *Pokey* in hybrid genomes [21,95-97]. Testing the activity of *Pokey* in diploid hybrids and performing additional studies on a larger number of hybrid and non-hybrid isolates will enhance our understanding of the dynamics and increase, if any, of *Pokey* elements in *D. pulex* × *D. pulicaria* hybrids.

There is no difference between the load of *Pokey* insertion sites based on TE display in the genomes of polyploid hybrids (20.60 ±3.47 at 50°C and 16.33 ±1.75 at 55°C) compared to diploid hybrids (18.09 ±3.99 at 50°C and 13.83 ±4.22 at 55°C). Similarly, there is no difference in the density of g*Pokey* per haploid genome between polyploid (12.48 ±3.20) and diploid (12.63 ±5.04) hybrids based on qPCR (Table 2). Conversely, the number of singletons (TE display at 50°C) is slightly higher in polyploids (20.6%) than in diploids (12.3%). The relationship between *Pokey* number and heterozygosity also differs between the two groups. g*Pokey* number (qPCR) increases with an increase in *Pokey* insertion sites (TE display) in diploids (Figure 3) but decreases in triploids (although not significantly). Moreover, in diploids, *Pokey* heterozygosity tends to decrease, though not significantly, as the number of different *Pokey* insertion sites increases (Table 3, Additional file 9). In contrast, polyploid heterozygosity increases (Table 3) and the number of

total g*Pokey* insertions (qPCR) decreases when *Pokey* insertion sites (TE display) increases (Table 3, Figure 3). Taken together, these results may reflect relaxed selection on insertions at some sites in polyploid compared to diploid hybrids, as suggested by the genomic niche redundancy hypothesis [24,25]. For example, if two functioning copies of a gene are necessary and sufficient for survival of the host, a third copy could become a potential genomic niche for TEs in triploid individuals. If so, then no TEs should be inserted in this gene in diploids but one gene copy could carry an insert in triploids without causing a decrease in host fitness. Alternatively, the difference in the number of singleton sites between the two ploidy levels may simply be a consequence of insertion site polymorphism contributed by the additional chromosome sets carried by polyploids.

Pokey in rDNA

The mean haploid number of 18S is higher (although not significant) in diploid hybrid isolates than in polyploid hybrids, but when the haploid rDNA copy number is multiplied by the ploidy level, the average is equal between diploid (586.5) and polyploid (586.5) hybrids with *D. pulex*, *D. pulicaria* or *D. middendorffiana* mtDNA haplotypes. Previous studies have shown that polyploid plant species, such as natural and artificial allotetraploid populations of *Tragopogon* [98] and allotetraploid and allohexaploid grass species [99], may experience rDNA rearrangement, including loss of rRNA gene copies. As most organisms have many more rRNA genes than they require for survival [100], it is possible that a loss of copies in polyploids is not harmful. Indeed, it is possible that selection could actually favor the loss of copies if polyploidization initially results in high rDNA copy numbers that are somewhat deleterious. It is noteworthy that the average haploid 18S number (293.3 with values from 131.5 to 451) in the diploid hybrid isolates is more than 30% higher than the average haploid 18S number in the non-hybrid diploid isolates of *D. pulex* (221.0 with values from 94 to 489.5) and *D. pulicaria* (217.3 with values from 97 to 444) studied by Eagle and Crease [84] using the same qPCR protocol. This difference is not statistically significant (one-way ANOVA; $F = 2.418$, $df = 2$, $P = 0.0961$). However, the number of *Daphnia* diploid hybrids tested here is low (n = 8) comparing to *D. pulicaria* (n = 37) and *D. pulex* (n = 43) isolates tested in the study by Eagle and Crease [84]. There is a high level of variation within each group and it will be interesting to confirm this pattern after analysis in a larger sample of diploid hybrids and laboratory-produced hybrids.

The number of 28S genes with r*Pokey* insertions accounts for an average of 1.93% of rDNA (with only one isolate, PX2-MB-1, above 5%; Table 2; Additional file 1), which is consistent with the results of Eagle and Crease

[84] who found r*Pokey* insertions in approximately 1% of rDNA units in non-hybrid *D. pulex* and *D. pulicaria* isolates. Moreover, and still in accordance with Eagle and Crease [84], we did not find a correlation between the number of r*Pokey* and rDNA in *Daphnia* hybrid isolates, including the polyploids in which rDNA copy number per haploid genome is lower. This is consistent with the hypothesis that *Pokey* is not highly active in the rDNA of these species, and its number does not increase with the number of rDNA units. Even so, selection is not so efficient that it eliminates deleterious elements present at low copy number in a highly repetitive gene family [84,101,102].

The effect of hybridization on *Pokey* load

Valizadeh and Crease [15] found a significantly lower (one-way ANOVA; F =67.65, df =3, P <0.001) average number of *Pokey* insertion sites in obligately and in cyclically parthenogenetic diploid isolates of *D. pulex* (3.27 ±2.07, n = 22 and 5.18 ±2.24, n = 22 respectively) compared to our survey of 12 obligately parthenogenetic hybrid isolates using the same annealing temperature (55° C) in TE display (13.83 ±4.22 for six diploid hybrid isolates and 16.33 ±1.75 for six polyploid hybrid isolates; Table 2). In addition, the qPCR estimate of g*Pokey* density is higher (although not statistically significant; one-way ANOVA; $F = 2.549$, $df = 2$, $P = 0.085$) in our reduced data set of diploid hybrid isolates (12.63 ±5.04 for eight diploid hybrids) than in the diploid non-hybrid isolates analyzed by Eagle and Crease (9.6 for *D. pulex*, n = 43, and 9.5 for *D. pulicaria*, n = 37) [84]. Therefore, there seems to be an increase in the density of *Pokey* insertions in the genomes of hybrid *Daphnia*. This increase could occur either during the early generations after hybridization by bursts of *Pokey* activity or, if *Pokey* elements are still active in hybrids, over a long period through the slow accumulation of *Pokey* insertions within the genome. Bursts of TE activity in hybrids have been highlighted in numerous homoploid hybrid plants [22,29], fruit flies [19,21,36] and wallabies [20]. In *Drosophila melanogaster* and *D. virilis*, hybrid dysgenic crosses may lead to bursts in activity of various TEs [21,97] due to release from cytoplasmic repression [37,95,96,103]. Interestingly, there is a slight trend of decreasing *Pokey* site heterozygosity (Table 3; Additional file 10) and a trend of increasing *Pokey* insertions sites (Figure 2) as host average heterozygosity increases. These trends suggest there may be increased activity of *Pokey* in hybrids that have the most evolutionarily divergent parents. Alternatively, increased genome and cell size favored by natural selection in new and/or stressful habitats may lead to a slow increase in the number of TEs in the genome [104]. Genetic drift may also lead to a slow increase in the number of TEs as suspected in sunflowers

[104]. Three hybrid species of sunflower inhabiting harsh environments show genome size expansion due to proliferation of numerous class I TEs (retrotransposons) [34]. However, the proliferation of TEs is rare in contemporary natural sunflower hybrid populations and in artificial hybrid crosses [23,33], which suggest an increase of TEs after hybrid establishment via population processes such as genetic drift or natural selection. Selection in marginal habitats or drift following hybridization could also lead to an increase of TE density in hybrid *Daphnia* genomes.

Conclusions

Using TE display and qPCR, we were able to describe insertion site polymorphism and the load of *Pokey* elements in diploid and polyploid hybrid isolates of the *D. pulex* species complex. The polymorphism of *Pokey* insertion sites was not congruent with the evolutionary history and genetic relationships of their hosts. Diploid and polyploid hybrids did not differ significantly in the number of *Pokey* insertions, using either qPCR or TE display, as has been shown in studies comparing diploid and polyploid plants. The number of singletons estimated with TE display is slightly higher in polyploid than in diploid hybrids. Together, these results may reflect a higher number of sites available for *Pokey* insertions in polyploid than in diploid hybrids, or an increase in polymorphism due to the combination of genomes with *Pokey* at different insertion sites. Compared to previous studies on *Pokey* in the *D. pulex* complex, we found the density of *Pokey* insertions per haploid genome to be higher in obligately parthenogenetic hybrids (both diploids and polyploids) than in non-hybrid diploids (either cyclical or obligate parthenogens) leading to the conclusion that hybridization may lead to an overall increase in *Pokey* insertions. The estimation of polymorphism and TE load in laboratory-produced hybrids and the analysis of additional samples of hybrids will provide more insight into the population dynamics of TEs in diploid and polyploid hybrids of *Daphnia*.

Additional files

Additional file 1: Characteristics of the *Daphnia* isolates used in this study. Labels of the isolates are composites of their characteristics. The first two letters represent the mitochondrial haplotype followed by the ploidy level (2x or 3x), a 2-letter country or state/province code and the isolate number. Mitochondrial haplotypes are as follows: EPC = Eastern *D. pulicaria*, WPC = Western *D. pulicaria*, PPC = Polar *D. pulicaria*, PanPX = Panarctic *D. pulex*, MIDD = *D. middendorffiana sensu stricto*, TENE = *D. tenebrosa*. Ldh is the Lactate dehydrogenase genotype and indicates the hybrid nature of each isolate. H$_{pl}$ is the ploidy-weighted heterozygosity. rRNA gene and *Pokey* number were determined using TE display and qPCR; 50°C and 55°C are the annealing temperatures used to generate the PCR amplicons in TE display. Total *Pokey* = all *Pokey* elements in the genome. rDNA *Pokey* = *Pokey* elements in 28S rRNA genes. Genomic *Pokey* = total -rDNA elements.

TG ratio is the number of *Tif* relative to the number of *Gtp* single copy reference genes.

Additional file 2: Supplementary material and methods describing TE display and qPCR protocols.

Additional file 3: TE display profiles of diploid and polyploid isolates in the *D. pulex* species complex. Sum of the amplification signals (peaks) are presented for each isolate.

Additional file 4: Three-dimensional representation of Principal Coordinate Analyses of Jaccard distance matrix of *Pokey* profiles and Bruvo distance matrix of microsatellite diversity in diploid and polyploid isolates of the *Daphnia pulex* complex. (A) *Pokey* profiles generated using TE display; (B) Microsatellite genotypes determined by Vergilino et al. [45]. The three first axes are represented. Empty symbols are diploids and solid symbols are polyploids. Empty orange circles: diploid hybrids with pulex mitochondrial haplotype, solid orange circles: triploid hybrids with *D. pulex* mitochondrial haplotype, solid square: *D. middendorffiana sensu stricto*, empty black triangles: diploid *D. tenebrosa*, solid black triangles: triploid *D. tenebrosa*, solid red triangle: introgressed *D. tenebrosa* with *D. pulex* nuclear genome, solid green diamond filled with orange: triploid hybrids with *D. pulicaria* mitochondrial haplotype; (C) and (D) are screeplots and represent the eigenvalues of the axes of Principal Coordinate Analysis (A) and (B), respectively.

Additional file 5: Results of the covariance analysis (ANCOVA) of TE display results from diploid and polyploid isolates in the *Daphnia pulex* complex. Ploidy level and ploidy-weighted heterozygosity (H_{pl}) were used as the independent variables.

Additional file 6: Relationship between 18S and 28S gene number in diploid and polyploid isolates of the *Daphnia pulex* complex. (A) Histograms of haploid number of 18S and 28S genes in each isolate. (B) Correlation between 18S and 28S gene number. Symbols represent mitochondrial haplotypes: circles for isolates with *D. pulex* mitochondria, diamonds for isolates with *D. pulicaria* mitochondria, squares for isolates with *D. middendorffiana* mitochondria, triangles for isolates with *D. tenebrosa* mitochondria and inverted triangles for introgressed *D. tenebrosa*. Empty symbols represent putative diploids and solid ones indicate polyploids. Dashed and solid lines are linear regressions estimated from the data in diploids and polyploids, respectively. The dotted line was generated by plotting 18S gene number on both axes.

Additional file 7: Correlation between haploid number of *Pokey* in rDNA (r*Pokey*) and haploid 18S gene number in diploid and triploid isolates of the *D. pulex* species complex. Empty circles represent diploid isolates and solid circles represent triploid isolates. Red empty circle represents the diploid isolate PX2-MB-1. Dashed and solid lines are linear regressions estimated from the data. The dashed lines with long strokes represent the linear regression following diploid hybrids without the isolate PX2-MB-1.

Additional file 8: Correlation between the haploid number of *Pokey* in rDNA (r*Pokey*) and the haploid number of *Pokey* in other genomic locations (g*Pokey*) in diploid and triploid isolates of the *D. pulex* species complex. Empty circles represent diploid isolates and solid circles represent triploid isolates. Red empty circle represents the diploid isolate PX2-MB-1. Dashed and solid lines are linear regressions estimated from the data. The dashed lines with long strokes represent the linear regression following diploid hybrids without the isolate PX2-MB-1.

Additional file 9: Correlation between average *Pokey* insertion site heterozygosity (H_{gPokey}) and the number of *Pokey* insertions outside rDNA based on TE display analysis of diploid isolates. The dashed line represents the linear regression estimated from the data.

Additional file 10: Correlation between average *Pokey* insertion site heterozygosity (H_{gPokey}) and heterozygosity (H_{pl}) of their diploid hosts based on microsatellite loci. The dashed line represents the linear regression estimated from the data.

Abbreviations
D. pulex: *Daphnia pulex*; g*Pokey*: Genomic *Pokey* elements inserted outside rDNA; *Ldh*: Lactate dehydrogenase gene; PCoA: Principal coordinate analysis; qPCR: Quantitative PCR; r*Pokey*: *Pokey* elements inserted in rDNA; rDNA: ribosomal DNA; TE: Transposable element; TED: Transposable element display.

Competing interests
The authors declare that they have no competing interests.

Authors' contributions
RV designed the project and wrote the manuscript. RV and SHCE planned the analyses. RV and SHCE conducted the analyses. All authors analyzed the data and contributed to the writing and editing of the manuscript. All authors approved the final manuscript.

Acknowledgments
RV acknowledges a scholarship from Centre d'études Nordiques and a travel fellowship from UQAR. SHCE was supported by an Ontario Graduate Studies Science and Technology Scholarship. We thank the Genomics Facility at the University of Guelph for assistance with the qPCR. Comments from two anonymous reviewers greatly improved the earlier version of this manuscript.

Funding
This work was supported by Discovery Grants from the Natural Sciences and Engineering Research Council (NSERC) of Canada to FD and to TJC.

Author details
[1]Département de Biologie, Chimie et Géographie, Université du Québec à Rimouski, Rimouski, Québec G5L 3A1, Canada. [2]Department of Integrative Biology, University of Guelph, Guelph, Ontario N1G 2W1, Canada. [3]Centre d'Études Nordiques, Université Laval, Québec G1V 0A6, Canada.

References
1. Feschotte C, Pritham EJ: **DNA transposons and the evolution of eukaryotic genomes.** *Ann Rev Gen* 2007, **41**:331–368.
2. Shigenobu S, Watanabe H, Hattori M, Sakaki Y, Ishikawa H: **Genome sequence of the endocellular bacterial symbiont of aphids Buchnera sp. APS.** *Nature* 2000, **407**:81–86.
3. Pidpala O, Yatsishina A, Lukash L: **Human mobile genetic elements: structure, distribution and functional role.** *Cytol Genet* 2008, **42**:420–430.
4. Charlesworth B, Charlesworth D: **The population dynamics of transposable elements.** *Gen Res* 1983, **42**:1–27.
5. Charlesworth D, Charlesworth B: **Transposable elements in inbreeding and outbreeding populations.** *Genetics* 1995, **140**:415–417.
6. Charlesworth D, Wright SI: **Breeding systems and genome evolution.** *Curr Opin Genetics Dev* 2001, **11**:685–690.
7. Brookfield J, Badge R: **Population genetics models of transposable elements.** *Genetica* 1997, **100**:281–294.
8. Wright S, Schoen D: **Transposon dynamics and the breeding system.** *Genetica* 1999, **107**:139–148.
9. Schaack S, Choi E, Lynch M, Pritham E: **DNA transposons and the role of recombination in mutation accumulation in Daphnia pulex.** *Genome Biol* 2010, **11**:R46.
10. Schaack S, Pritham EJ, Wolf A, Lynch M: **DNA transposon dynamics in populations of Daphnia pulex with and without sex.** *Proc R S B* 2010, **277**:2381–2387.
11. Wright S, Finnegan D: **Genome evolution: sex and the transposable element.** *Current Biol* 2001, **11**:R296–R299.
12. Zeyl C, Bell G, Green DM: **Sex and the spread of retrotransposon Ty3 in experimental populations of Saccharomyces cerevisiae.** *Genetics* 1996, **143**:1567–1577.
13. Arkhipova I, Meselson M: **Transposable elements in sexual and ancient asexual taxa.** *Proc Natl Acad Sci USA* 2000, **97**:14473–14477.
14. Gladyshev EA, Meselson M, Arkhipova IR: **Massive horizontal gene transfer in bdelloid rotifers.** *Science* 2008, **320**:1210–1213.
15. Valizadeh P, Crease T: **the association between breeding system and transposable element dynamics in Daphnia pulex.** *J Mol Evol* 2008, **66**:643–654.
16. Mable BK: **'Why polyploidy is rarer in animals than in plants': myths and mechanisms.** *Biol J Linnean Soc* 2004, **82**:453–466.

17. Otto SP: The evolutionary consequences of polyploidy. Cell 2007, 131:452–462.

18. Seehausen O: Hybridization and adaptive radiation. Trends Ecol Evol 2004, 19:198–207.

19. Fontdevila A: Hybrid genome evolution by transposition. Cytogenet Genome Res 2005, 110:49–55.

20. O'Neill RJW, O'Neill MJ, Graves JAM: Undermethylation associated with retroelement activation and chromosome remodelling in an interspecific mammalian hybrid. Nature 1998, 393:68–72.

21. Petrov DA, Schutzman JL, Hartl DL, Lozovskaya ER: Diverse transposable elements are mobilized in hybrid dysgenesis in Drosophila virilis. Proc Natl Acad Sci USA 1995, 92:8050–8054.

22. Shan X, Liu Z, Dong Z, Wang Y, Chen Y, Lin X, Long L, Han F, Dong Y, Liu B: Mobilization of the active MITE transposons mPing and Pong in rice by introgression from wild rice (Zizania latifolia Griseb.). Mol Biol Evol 2005, 22:976–990.

23. Ungerer MC, Kawakami T: Transcriptional dynamics of LTR retrotransposons in early generation and ancient sunflower hybrids. Genome Biol Evol 2013, 5:329–337.

24. Comai L: Genetic and epigenetic interactions in allopolyploid plants. Plant Mol Biol 2000, 43:387–399.

25. Matzke MA, Matzke AJM: Polyploidy and transposons. Trends Ecol Evol 1998, 13:241–251.

26. Parisod C, Alix K, Just J, Petit M, Sarilar V, Mhiri C, Ainouche M, Chalhoub B, Grandbastien MA: Impact of transposable elements on the organization and function of allopolyploid genomes. New Phytol 2010, 186:37–45.

27. McClintock B: The significance of responses of the genome to challenge. Science 1984, 226:792–801.

28. Madlung A, Masuelli RW, Watson B, Reynolds SH, Davison J, Comai L: Remodeling of DNA methylation and phenotypic and transcriptional changes in synthetic arabidopsis allotetraploids. Plant Physiol 2002, 129:733–746.

29. Ainouche M, Fortune P, Salmon A, Parisod C, Grandbastien MA, Fukunaga K, Ricou M, Misset MT: Hybridization, polyploidy and invasion: lessons from Spartina (Poaceae). Biol Invasions 2009, 11:1159–1173.

30. de Araujo PG, Rossi M, de Jesus EM, Saccaro NL, Kajihara D, Massa R, de Felix JM, Drummond RD, Falco MC, Chabregas SM, Ulian EC, Menossi M, Van Sluys MA: Transcriptionally active transposable elements in recent hybrid sugarcane. Plant J 2005, 44:707–717.

31. Hanson RE, Islam-Faridi MN, Crane CF, Zwick MS, Czeschin DG, Wendel JF, McKnight TD, Price HJ, Stelly DM: Ty1-copia-retrotransposon behavior in a polyploid cotton. Chromosome Res 2000, 8:73–76.

32. Kashkush K, Feldman M, Levy AA: Gene loss, silencing and activation in a newly synthesized wheat allotetraploid. Genetics 2002, 160:1651–1659.

33. Kawakami T, Dhakal P, Katterhenry AN, Heatherington CA, Ungerer MC: Transposable element proliferation and genome expansion are rare in contemporary sunflower hybrid populations despite widespread transcriptional activity of LTR retrotransposons. Genome Biol Evol 2011, 3:156.

34. Ungerer MC, Strakosh SC, Zhen Y: Genome expansion in three hybrid sunflower species is associated with retrotransposon proliferation. Current Biol 2006, 16:R872.

35. Dong L, Cuiping Y, Shaojun L, Liangguo L, Wei D, Song C, Jinpeng Y, Yun L: Characterization of a novel Tc1-like transposon from bream (Cyprinidae, Megalobrama) and its genetic variation in the polyploidy progeny of bream-red crucian carp crosses. J Mol Evol 2009, 69:395–403.

36. Labrador M, Fontdevila A: High transposition rates of Osvaldo, a new Drosophila buzzatii retrotransposon. Mol Gen Genet 1994, 245:661–674.

37. Arkhipova IR, Rodriguez F: Genetic and epigenetic changes involving (Retro) transposons in animal hybrids and polyploids. Cytogenet Genome Res 2013, 140:295–311.

38. Hebert PDN: Obligate asexuality in Daphnia. Am Nat 1981, 117:784–789.

39. Hebert PDN, Crease T: Clonal diversity in populations of Daphnia pulex reproducing by obligate parthenogenesis. Heredity 1983, 51:353–369.

40. Innes DJ, Schwartz SS, Hebert PDN: Genotypic diversity and variation in mode of reproduction among populations in the Daphnia pulex group. Heredity 1986, 57:345–355.

41. Innes DJ, Fox CJ, Winsor GL: Avoiding the cost of males in obligately asexual Daphnia pulex (Leydig). Proc R S Series B 2000, 267:991–997.

42. Hebert P, Finston T: Macrogeographic patterns of breeding system diversity in the Daphnia pulex group from the United States and Mexico. Heredity 2001, 87:153–161.

43. Colbourne JK, Crease TJ, Weider LJ, Hebert PDN, Dufresne F, Hobaek A: Phylogenetics and evolution of a circumarctic species complex (Cladocera: Daphnia pulex). Biol J Linnean Soc 1998, 65:347–365.

44. Hebert PDN: The Daphnia of North America: An Illustrated Fauna. Guelph, Ontario: University of Guelph; 1995.

45. Vergilino R, Markova S, Ventura M, Manca M, Dufresne F: Reticulate evolution of the Daphnia pulex complex as revealed by nuclear markers. Mol Ecol 2011, 20:1191–1207.

46. Adamowicz SJ, Gregory TR, Marinone MC, Hebert PDN: New insights into the distribution of polyploid Daphnia: the Holarctic revisited and Argentina explored. Mol Ecol 2002, 11:1209–1217.

47. Aguilera X, Mergeay J, Wollebrants A, Declerck S, de Meester L: Asexuality and polyploidy in Daphnia from the tropical Andes. Limnol Oceanogr 2007, 52:2079–2088.

48. Colbourne JK, Hebert PDN: The systematics of North American Daphnia (Crustacea: Anomopoda): a molecular phylogenetic approach. Philos Trans R Soc London Series B 1996, 351:349–360.

49. Dufresne F, Hebert PDN: Pleistocene glaciations and polyphyletic origins of polyploidy in an arctic cladoceran. Proc R S Series B 1997, 264:201–206.

50. Weider LJ, Hobaek A: Glacial refugia, haplotype distributions, and clonal richness of the Daphnia pulex complex in arctic Canada. Mol Ecol 2003, 12:463–473.

51. Crease TJ, Floyd R, Cristescu MA, Innes D: Evolutionary factors affecting Lactate dehydrogenase A and B variation in the Daphnia pulex species complex. BMC Evol Biol 2011, 11:212.

52. Cristescu ME, Constantin A, Bock DG, Cáceres CE, Crease TJ: Speciation with gene flow and the genetics of habitat transitions. Mol Ecol 2012, 21:1411–1422.

53. Crease TJ, Stanton DJ, Hebert PDN: Polyphyletic origins of asexuality in Daphnia pulex. II. Mitochondrial-DNA variation. Evolution 1989, 43:1016–1026.

54. Hebert PDN, Schwartz SS, Ward RD, Finston TL: Macrogeographic patterns of breeding system diversity in the Daphnia pulex group. I. Breeding systems of Canadian populations. Heredity 1993, 70:148–161.

55. Hebert PDN, Beaton MJ, Schwartz SS, Stanton DJ: Polyphyletic origins of asexuality in Daphnia pulex. I. Breeding-system variation and levels of clonal diversity. Evolution 1989, 43:1004–1015.

56. Innes DJ, Hebert PDN: The origin and genetic basis of obligate parthenogenesis in Daphnia pulex. Evolution 1988, 42:1024–1035.

57. Heier CR, Dudycha JL: Ecological speciation in a cyclic parthenogen: sexual capability of experimental hybrids between Daphnia pulex and Daphnia pulicaria. Limnol Oceanogr 2009, 54:492–502.

58. Tucker AE, Ackerman MS, Eads BD, Xu S, Lynch M: Population-genomic insights into the evolutionary origin and fate of obligately asexual Daphnia pulex. Proc Natl Acad Sci USA 2013, 110:15740–15745.

59. Xu S, Innes DJ, Lynch M, Cristescu ME: The role of hybridization in the origin and spread of asexuality in Daphnia. Mol Ecol 2013, 22:4549–4561.

60. Beaton MJ, Hebert PDN: Geographical parthenogenesis and polyploidy in Daphnia pulex. Am Nat 1988, 132:837–845.

61. Dufresne F, Hebert PDN: Hybridization and origins of polyploidy. Proc R S Series B 1994, 258:141–146.

62. Dufresne F, Hebert PDN: Polyploidy and clonal diversity in an arctic cladoceran. Heredity 1995, 75:45–53.

63. Hobaek A, Weider LJ, Wolf HG: Ecological genetics of Norwegian Daphnia. III. Clonal richness in an Arctic apomictic complex. Heredity 1993, 71:323–330.

64. Ward RD, Bickerton MA, Finston T, Hebert PDN: Geographical cline in breeding systems and ploidy levels in European populations of Daphnia pulex. Heredity 1994, 73:532–543.

65. Weider LJ, Hobaek A, Colbourne JK, Crease TJ, Dufresne F, Hebert PDN: Holarctic phylogeography of an asexual species complex I. Mitochondrial DNA variation in arctic Daphnia. Evolution 1999, 53:777–792.

66. Weider LJ, Hobaek A, Hebert PDN, Crease TJ: Holarctic phylogeography of an asexual species complex – II. Allozymic variation and clonal structure in Arctic Daphnia. Mol Ecol 1999, 8:1–13.

67. Vergilino R, Belzile C, Dufresne F: Genome size evolution and polyploidy in the Daphnia pulex complex (Cladocera: Daphniidae). Biol J Linnean Soc 2009, 97:68–79.

68. Colbourne JK, Pfrender ME, Gilbert D, Thomas WK, Tucker A, Oakley TH, Tokishita S, Aerts A, Arnold GJ, Basu MK, Bauer DJ, Cáceres CE, Carmel L, Casola C, Choi JH, Detter JC, Dong Q, Dusheyko S, Eads BD, Fröhlich T, Geiler-Samerotte KA, Gerlach D, Hatcher P, Jogdeo S, Krijgsveld J,

Kriventseva EV, Kültz D, Laforsch C, Lindquist E, Lopez J, et al: The ecoresponsive genome of Daphnia pulex. Science 2011, 331:555–561.

69. Penton EH, Sullender BW, Crease TJ: Pokey, a new DNA transposon in Daphnia (Cladocera: Crustacea). J Mol Evol 2002, 55:664–673.

70. Sullender BW, Crease TJ: The behavior of a Daphnia pulex transposable element in cyclically and obligately parthenogenetic populations. J Mol Evol 2001, 53:63–69.

71. Vergilino R, Elliott TA, Desjardins-Proulx P, Crease TJ, Dufresne F: Evolution of a transposon in Daphnia hybrid genomes. Mob DNA 2013, 4:7.

72. Hebert PDN, Crease TJ: Clonal coexistence in Daphnia pulex (Leydig): another planktonic paradox. Science 1980, 207:1363–1365.

73. Wright SI, Le QH, Schoen DJ, Bureau TE: Population dynamics of an Ac-like transposable element in self- and cross-pollinating Arabidopsis. Genetics 2001, 158:1279–1288.

74. Legendre P, Legendre L: Numerical Ecology. 2nd English edn. Amsterdam: Elsevier Science; 1998.

75. Bruvo R, Michiels NK, D'Souza TG, Schulenburg H: A simple method for the calculation of microsatellite genotype distances irrespective of ploidy level. Mol Ecol 2004, 13:2101–2106.

76. Clark LV, Jasieniuk M: POLYSAT: an R package for polyploid microsatellite analysis. Mol Ecol Resouces 2011, 11:562–566.

77. R Development Core Team: R: A Language and Environment for Statistical Computing, Vienna: R: Foundation for Statistical Computing. 2004. http://www.R-project.org.

78. Dufresne F, Stift M, Vergilino R, Mable BK: Recent progress and challenges in population genetics of polyploid organisms: an overview of current state-of-the-art molecular and statistical tools. Mol Ecol 2014, 23:40-69.

79. Tamura K, Peterson D, Peterson N, Stecher G, Nei M, Kumar S: MEGA5: Molecular Evolutionary Genetics Analysis using maximum likelihood, evolutionary distance, and maximum parsimony methods. Mol Biol Evol 2011, 28:2731–2739.

80. Gower JC: Some distance properties of latent root and vector methods used in multivariate analysis. Biometrika 1966, 53:325–338.

81. Ihaka R, Gentleman R: R: a language for data analysis and graphics. J Comp Graph Stat 1996, 5:299–314.

82. Jaccard P: Nouvelles recherches sur la distribution florale. Bull de la société Vaudoise des Sci Nat 1908, 44:223–270.

83. Calinski T, Harabasz J: A dendrite method for cluster analysis. Commun Stat Theor M 1974, 3:1–27.

84. Eagle SHC, Crease TJ: Copy number variation of ribosomal DNA and Pokey transposons in natural populations of Daphnia. Mob DNA 2012, 3:4.

85. Eagle SHC: Copy Number Variation of Ribosomal RNA Genes and the Pokey DNA Transposon in the Daphnia pulex Species Complex. Guelph, Ontario: University of Guelph; 2013.

86. Rice WR: Analyzing tables of statistical tests. Evolution 1989, 43:223–225.

87. Bartolomé C, Maside X, Charlesworth B: On the abundance and distribution of transposable elements in the genome of Drosophila melanogaster. Mol Biol Evol 2002, 19:926–937.

88. Charlesworth B, Langley CH, Sniegowski PD: Transposable element distributions in Drosophila. Genetics 1997, 147:1993–1995.

89. Dolgin ES, Charlesworth B: The effects of recombination rate on the distribution and abundance of transposable elements. Genetics 2008, 178:2169–2177.

90. Duret L, Marais G, Biemont C: Transposons but not retrotransposons are located preferentially in regions of high recombination rate in Caenorhabditis elegans. Genetics 2000, 156:1661–1669.

91. Rizzon C, Marais G, Gouy M, Biemont C: Recombination rate and the distribution of transposable elements in the Drosophila melanogaster genome. Genome Res 2002, 12:400–407.

92. Wright SI, Agrawal N, Bureau TE: Effects of recombination rate and gene density on transposable element distributions in Arabidopsis thaliana. Genome Res 2003, 13:1897–1903.

93. Paland S, Lynch M: Transitions to asexuality result in excess amino acid substitutions. Science 2006, 311:990–992.

94. Wendel JF: Genome evolution in polyploids. Plant Mol Biol 2000, 42:225–249.

95. Blumenstiel JP, Hartl DL: Evidence for maternally transmitted small interfering RNA in the repression of transposition in Drosophila virilis. Proc Natl Acad Sci USA 2005, 102:15965–15970.

96. Brookfield J: Models of repression of transposition in P-M hybrid dysgenesis by P cytotype and by zygotically encoded repressor proteins. Genetics 1991, 128:471–486.

97. Lewis AP, Brookfield JFY: Movement of Drosophila melanogaster transposable elements other than P elements in a P-M hybrid dysgenic cross. Mol Gen Genet 1987, 208:506–510.

98. Malinska H, Tate J, Matyasek R, Leitch A, Soltis D, Soltis P, Kovarik A: Similar patterns of rDNA evolution in synthetic and recently formed natural populations of Tragopogon (Asteraceae) allotetraploids. BMC Evol Biol 2010, 10:291.

99. Kotseruba V, Pistrick K, Blattner FR, Kumke K, Weiss O, Rutten T, Fuchs J, Endo T, Nasuda S, Ghukasyan A, Houben A: The evolution of the hexaploid grass Zingeria kochii (Mez) Tzvel. (2n = 12) was accompanied by complex hybridization and uniparental loss of ribosomal DNA. Mol Phylo Evol 2010, 56:146–155.

100. Eickbush TH, Eickbush DG: Finely orchestrated movements: evolution of the ribosomal RNA genes. Genetics 2007, 175:477–485.

101. Glass SK, Moszczynska A, Crease TJ: The effect of transposon Pokey insertions on sequence variation in the 28S rRNA gene of Daphnia pulex. Genome 2008, 51:988–1000.

102. Zhang X, Eickbush MT, Eickbush TH: Role of recombination in the long-term retention of transposable elements in rRNA gene loci. Genetics 2008, 180:1617–1626.

103. Castro JP, Carareto CMA: Drosophila melanogaster transposable elements: mechanisms of transposition and regulation. Genetica 2004, 121:107–118.

104. Baack EJ, Whitney KD, Rieseberg LH: Hybridization and genome size evolution: timing and magnitude of nuclear DNA content increases in Helianthus homoploid hybrid species. New Phytol 2005, 167:623–630.

2

Repeated horizontal transfers of four DNA transposons in invertebrates and bats

Zhou Tang[1†], Hua-Hao Zhang[2†], Ke Huang[3], Xiao-Gu Zhang[2], Min-Jin Han[1] and Ze Zhang[1*]

Abstract

Background: Horizontal transfer (HT) of transposable elements (TEs) into a new genome is considered as an important force to drive genome variation and biological innovation. However, most of the HT of DNA transposons previously described occurred between closely related species or insects.

Results: In this study, we carried out a detailed analysis of four DNA transposons, which were found in the first sequenced twisted-wing parasite, *Mengenilla moldrzyki*. Through the homology-based strategy, these transposons were also identified in other insects, freshwater planarian, hydrozoans, and bats. The phylogenetic distribution of these transposons was discontinuous, and they showed extremely high sequence identities (>87%) over their entire length in spite of their hosts diverging more than 300 million years ago (Mya). Additionally, phylogenies and comparisons of transposons versus orthologous gene identities demonstrated that these transposons have transferred into their hosts by independent HTs.

Conclusions: Here, we provided the first documented example of HT of *CACTA* transposons, which have been so far extensively studied in plants. Our results demonstrated that bats had continuously acquired new DNA elements via HT. This implies that predation on a large quantity of insects might increase bat exposure to HT. In addition, parasite-host interaction might facilitate exchanging of their genetic materials.

Keywords: Horizontal transfer, *CACTA* transposons, Mammals, Recent activity

Background

DNA-mediated or class 2 transposons were one class of transposable elements (TEs). Most DNA transposons transpose via a 'cut and paste' mechanism implemented by transposases. They were generally characterized by terminal inverted repeats (TIRs) and target site duplication (TSD) [1]. Based on their transposases, DNA transposons could be classified into 19 superfamilies, including *Tc1/mariner, hAT, PiggyBac, CACTA, MuDR, Merlin, Transib, P, PIF/Harbinger, Mirage, Zator, Ginger, Kolobok, Chapaev, Novosib, Rehavkus, PHIS, Sola,* and *Academ* [2,3].

Although the possibility of stochastic loss suggests that TEs should be a seemingly inevitable vertical extinction in their original host genomes, TEs are widespread in organisms [1,4-6]. Horizontal transfer (HT) is a process of genetic material exchanging among non-mating species or isolated species. HT of a transposon into a new genome allows the element to evade inevitable extinction, suggesting that HT plays important roles in the persistence of TEs [4]. In addition, HT of TEs into a new genome is also regarded as important forces to drive genome variation and biological innovation.

Generally, there are three criteria used to infer HT events: (1) high sequence similarity of TEs from divergent taxa, (2) incongruence between TE and host phylogeny, and (3) a patchy TE distribution within a group of taxa [7,8]. The first documented example of HT of TEs was the *P* element of *Drosophila* [9]. More than 330 cases (188 cases for DNA transposons and 142 cases for RNA transposons) of eukaryote-to-eukaryote HT events of TEs were described so far [10]. However, no documented example of HT has been described for the *CACTA* superfamily of DNA transposons, which so far has extensively been studied in plants [6,11]. In addition, most of HT of DNA transposons (122 out of 188) previously described occurred between closely related species or insects [10].

* Correspondence: zezhang@cqu.edu.cn
†Equal contributors
[1]School of Life Sciences, Chongqing University, Chongqing 400044, China
Full list of author information is available at the end of the article

In this study, we described four DNA transposons which were present in diverse invertebrate and vertebrate animals. The combination of high identity levels between TEs despite deep divergence times of their host taxa, patchy TE taxonomic distribution, and lower genetic distances for TEs than for host genes clearly demonstrated that they had horizontally transferred into their hosts.

Results

Distribution patterns of four DNA transposons

The twisted-wing parasite, *Mengenilla moldrzyki*, is the first sequenced species of Strepsiptera [12]. Nineteen seventy potential TEs of the twisted-wing parasite were downloaded from Dryad Digital Repository (http://datadryad.org/resource/doi:10.5061/dryad.ts058.2). The screening of the distribution of these transposons revealed that four of these 1970 TEs yielded highly significant (>87%) hits in many diverse species, not only in insects but also in freshwater planarian, hydrozoans, or bats (Table 1, Figures 1

and 2). These four DNA transposons formed the start pointing of this study. They were grouped into *hAT*, *CACTA*, and *piggyBac* superfamilies based on their similarities to known members of these superfamilies (Table 1). Full-length or partial ancestral sequences in each species were reconstructed and compared to each other (Additional file 1: Table S1).

The first DNA transposon, called *Buster1*, was found in *M. moldrzyki, Schmidtea mediterranea, Rhodnius prolixus,* and *Heliconius melpomene* (Table 1 and Figure 2). Except for *H. melpomene*, this element in other three species generated both autonomous and non-autonomous elements (including miniature inverted-repeat transposable elements (MITEs)). Multiple alignments of MITEs and its autonomous ancestors indicated that they originated by internal deletions from master elements. These results supported the hypothesis that MITEs borrowed the machinery of autonomous DNA transposons to transpose [1]. Insertion bias and phylogenetic analysis demonstrated that they belonged

Table 1 Characteristics of four DNA transposons in this study

Superfamily	Family	Organism	TEs	Length (bp)	Copy no.	TIRs (5'-3')	Average divergence ± SE	References
hAT	Buster1	Schmidtea mediterranea	Buster1_SM	2,463	17	CAGGGCTTCTTAAAC	8.22 ± 5.66	hAT-11_SM[13]
			Buster1_NA1_SM	373	270	CAGGGCTTCTTAAAC	1.45 ± 0.65	This study
			Buster1_NA2_SM	449	31	CAGGGCTTCTTAAAC	1.45 ± 0.75	This study
		Mengenilla moldrzyki	Buster1_MM	2,196	3	ND	2.16 ± 3.75	This study
			Buster1_NA1_MM	366	44	CAGGCCTTCTTAAACT	9.94 ± 2.83	This study
		Rhodnius prolixus	Buster1_RP	2,281	3	ND	0.86 ± 1.49	This study
			Buster1_NA1_RP	580	17	CAGGGCTTCTTAAACT	1.82 ± 0.85	This study
		Heliconius melpomene	Buster1_NA1_HM	500	42	CAGGGTTTCTTAAACT	2.32 ± 1.38	nhat-10_Hmel[13]
	Buster2	M. moldrzyki	Buster2_NA1_MM	365	252	CAACGGTGGCCA	13.5 ± 2.16	This study
		R. prolixus	Buster2 _NA1_RP	908	16	CAGGGGGGGGCCAACCT	4.74 ± 1.51	This study
		Nycticeius humeralis	Buster2 _NA1_NH	337	>10	CAGGGGTGGCCAACCT	4.81 ± 1.48	nhAT-5a_Nhu [unpublished]
			Buster2_NA2_NH	246	>104	CAGGGGTGGCCAACCT	4.63 ± 1.87	nhAT-2a_Nhu [unpublished]
CACTA	Spongebob	M. moldrzyki	Spongebob_NA1_MM	496	40	ND	9.88 ± 3.95	This study
		R. prolixus	Spongebob_NA1_RP	433	2	ND	2.74 ± 3.87	This study
		Bombyx mori	Spongebob_NA1_BM	468	17	ND	10.6 ± 3.69	This study
		Hydra magnipapillata	Spongebob_ HMa	5,836	441	CCCAGCCAACATT GAC (17)	5.94 ± 3.48	EnSpm-4N1_HM[2]
piggyBac	Kenshin	M. moldrzyki	Kenshin_MM	2,244	15	CACTAGA	13.2 ± 5.49	This study
		Megachile rotundata	Kenshin_MR	2,520	9	CACTAGA	9.91 ± 3.84	This study
			Kenshin_NA1_MR	235	46	CACTAGA	2.82 ± 2.24	This study
		Myotis davidii	Kenshin_MD	2,108	1	CACTAG	ND	This study
			Kenshin_NA1_MD	1,267	23	CACTAGA	2.16 ± 0.60	This study
			Kenshin_NA2_MD	627	3	CACTAGA	2.46 ± 0.20	This study

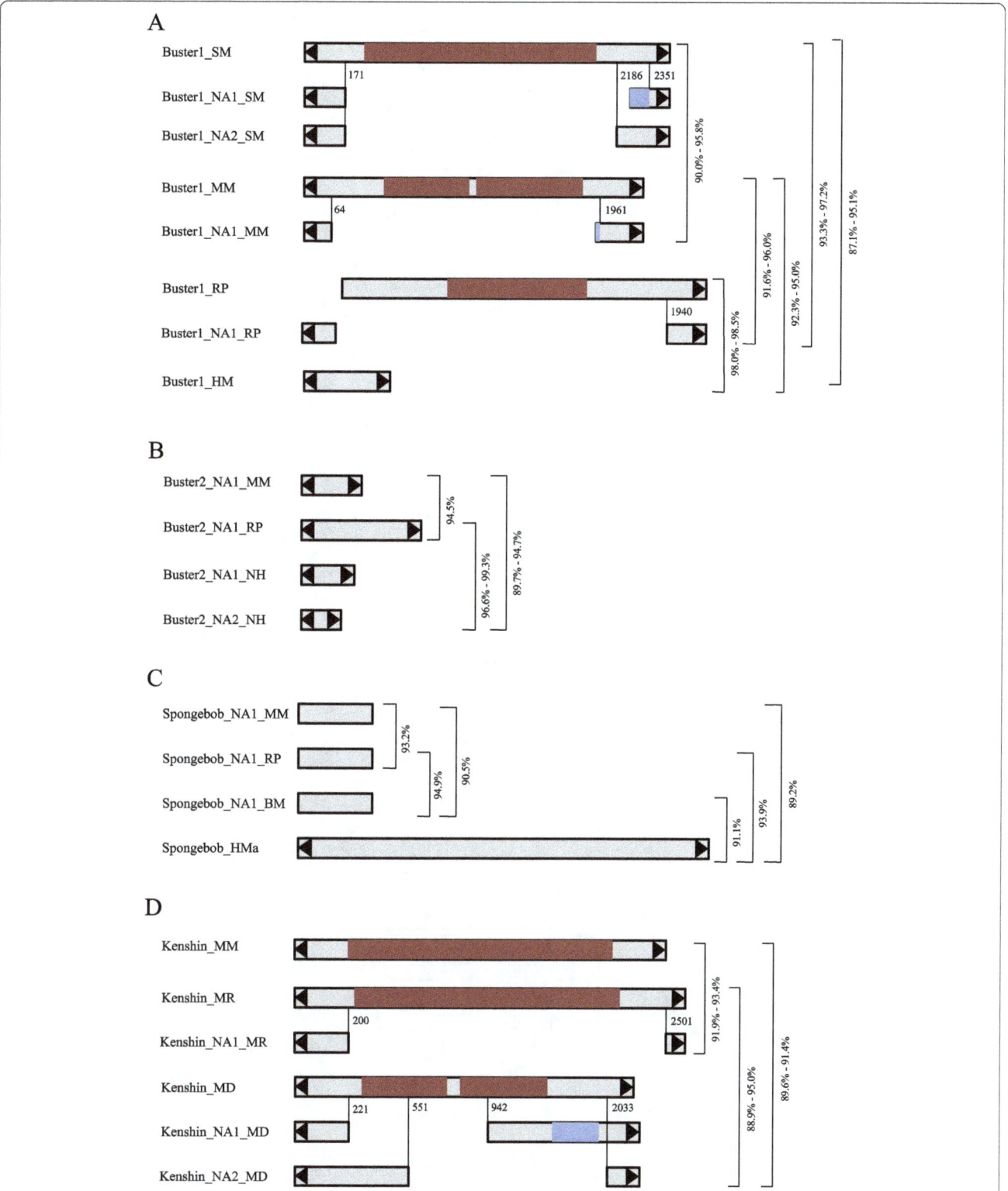

Figure 1 Diagram showing the detail information about transposons of *Buster1* (A), *Buster2* (B), *Spongebob* (C), and *Kenshin* (D). Black triangles represent the TIRs. Gray rectangles represent non-coding regions, and purple rectangles indicate transposase regions. Percentages of identity were calculated using Bioedit. Blue regions represent the variable area of transposons.

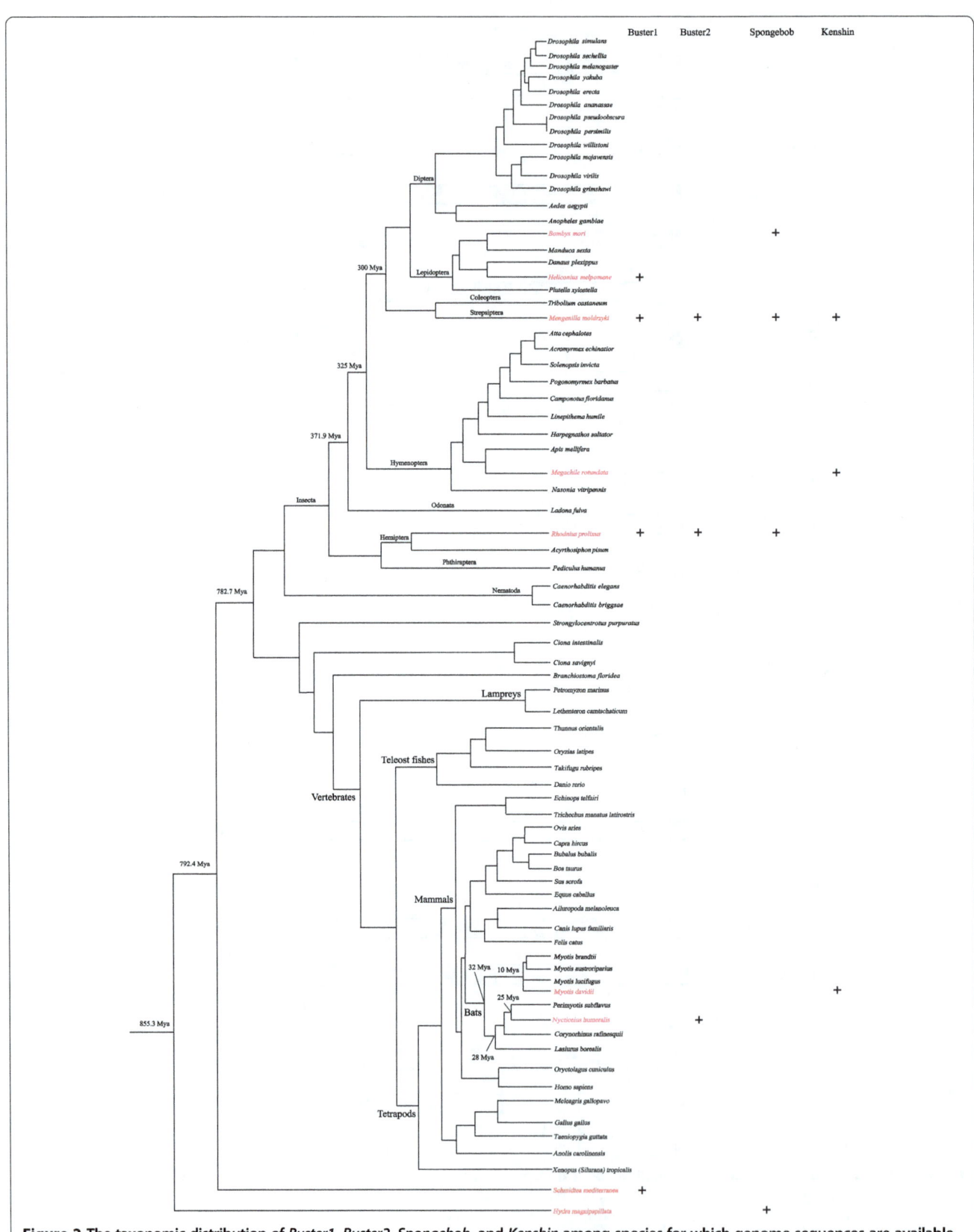

Figure 2 The taxonomic distribution of *Buster1, Buster2, Spongebob,* **and** *Kenshin* **among species for which genome sequences are available.**
Presence of these transposon families in each lineage are denoted by plus sign. Species divergence is taken from previous literatures [14-17].

to one member of the *Buster* family of the *hAT* superfamily (Figures 3A and 4A). Structure analysis indicated that the subterminal regions of the elements contain TGGGTCGCG tandem repeats. Generally, short repeats in subterminal regions have been used to distinguish different *hAT* transposons [18]. Thus, *Buster1* might represent a novel member of the *Buster* family. Moreover, the repetitive motif identified in *Buster1* might have important structural or functional roles during their transposition [19]. These elements identified in these hosts which diverged more than 300 million years ago (Mya) [14] revealed high nucleotide sequence identity (>87%) over almost the full length (Figure 1A), suggesting that these elements were derived from the same active

ancestral element. *Buster1* was found in low copy number (<50) in most species, except for the freshwater planarian *Schmidtea mediterranea* (*Buster1_NA1_SM*) where this element was found more than 250 copies (Table 1). The average sequence divergence between *Buster1_NA1_SM* copies and its consensus sequence was only 1.45%, indicating that this element might have experienced a burst transposition very recently in the freshwater planarian.

The second DNA transposon, called *Buster2*, was found not only in invertebrates (*M. moldrzyki* and *R. prolixus*) but also in one vertebrate (the evening bat *Nycticeius humeralis*) (Table 1 and Figure 2). They have proliferated via amplification of non-autonomous elements in these species. Furthermore, one non-autonomous element was

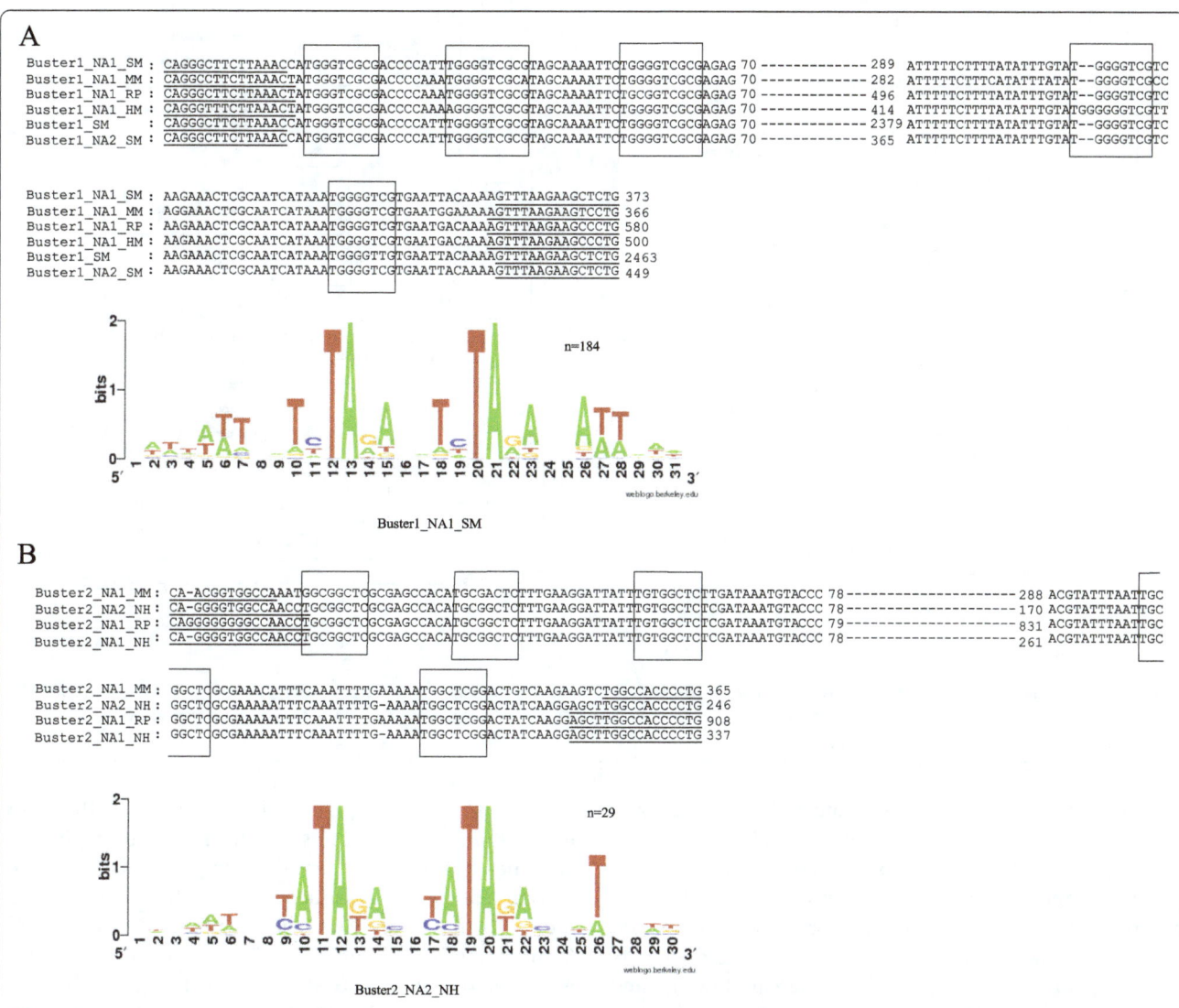

Figure 3 Structure characteristics of *Buster1* (A) and *Buster2* (B) and sequence logo of the regions flanking *Buster1_NA1_SM* and *Buster2_NA2_NH* insertions. The 15 nt upstream and downstream of all full-length copies of these families in *Schmidtea mediterranea* and *Nycticeius humeralis* are presented in each logo. The vertical axis is a measure of sequence information, which has a maximum value of 2 and is proportional to the level of sequence conservation at each position. The rectangles indicate their direct repeats. Their TIRs were shown using underlines, and numbers indicated their alignment positions.

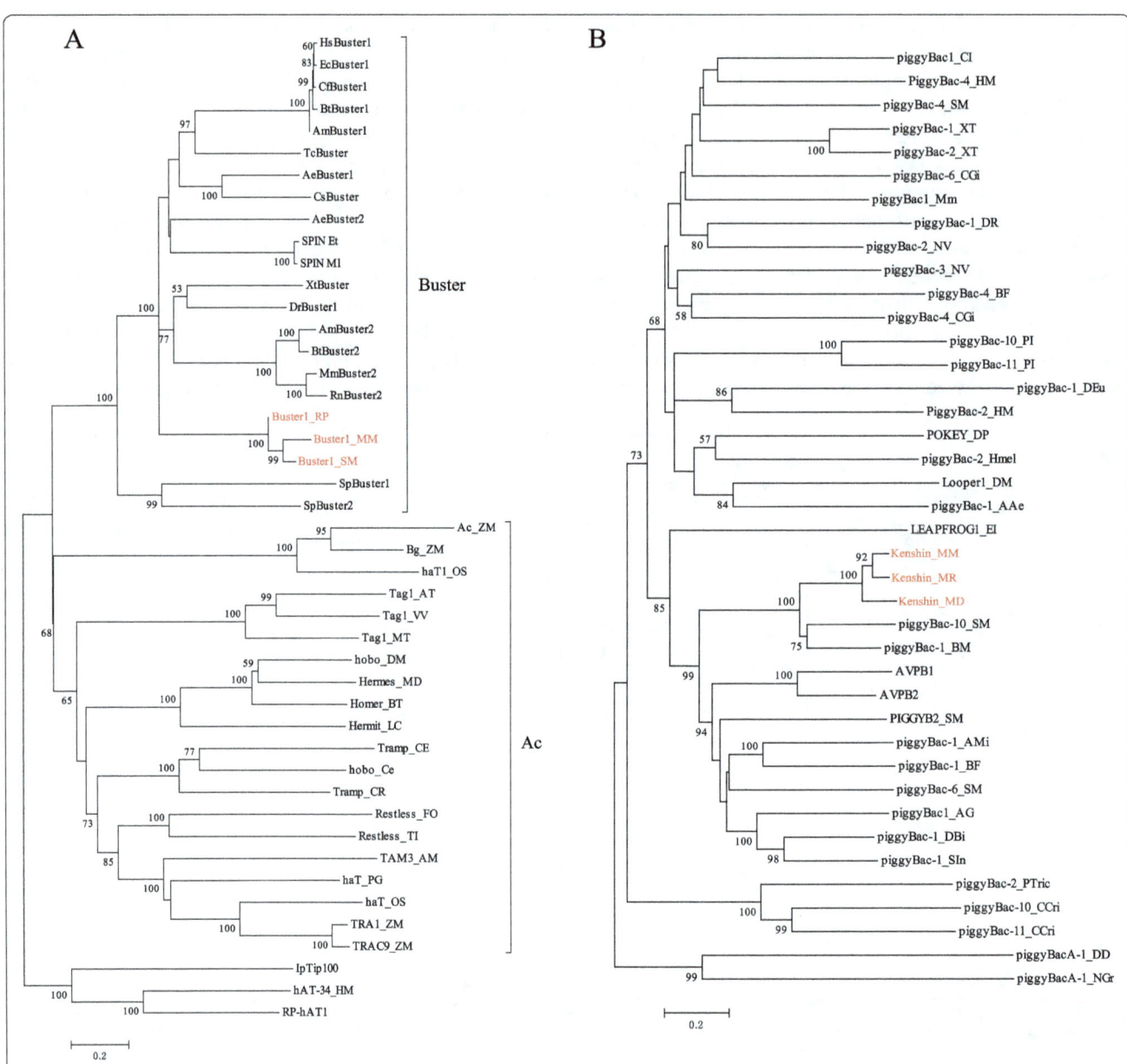

Figure 4 Phylogenetic trees of *Buster1* (A) and *Kenshin* (B). *Buster1* and *Kenshin* identified in this study were shown in red. Representatives of transposons of the *hAT* superfamily were obtained from previous studies [20,21]. Representatives of *piggyBac* transposons were downloaded from Repbase [2]. Bootstrap value <50% was not shown.

identified in the twisted-wing parasite and triatomine bug, and two non-autonomous elements were identified in the evening bat (Table 1). The successful amplification of these non-autonomous elements was surprising because autonomous partners responsible for their transposition were not found in these species. It is possible that its transposition was catalyzed by different but related autonomous elements of their ancient masters, which was known as cross-mobilization [22]. An alternative explanation is that autonomous elements could have remained polymorphic for a long time in the host population without augmenting its copy number and have been lost

through allele sorting. It could also be that the autonomous *Buster2* elements might reside in their host genomes but were not found in this study as a result of incomplete genome sequences. Insertion bias analysis indicated that *Buster2* was also a member of the *Buster* family (Figure 3B). Similar to *Buster1*, *Buster2* elements were flanked by TGCGGCTC tandem repeats. Because *Buster2* elements identified in these species were non-autonomous, we further investigated the similarities of their terminal regions with reported *Buster* elements. Multiple alignments showed that the terminal region of *Buster2* shared high sequence similarities with those of the *Buster* elements (Figure 5), which further demonstrated that it was a *Buster* transposon. Interestingly, similarities of all *Buster* elements were not restricted to their terminal inverted repeats (TIRs) but also extended to about 106 bp of their terminal regions. This also showed that their 5′ terminal regions were more conserved than 3′ terminal regions. These results suggested that these conservative sites in their terminal regions might play important roles during the process of their transposition. This is also consistent with the fact that the *Buster* family might experience a recent burst of amplification based on the phylogeny of their transposases [20].

The third DNA transposon, called *Spongebob*, was found in insects (*M. moldrzyki*, *R. prolixus*, and *Bombyx mori*) and hydrozoans (*Hydra magnipapillata*). Only partial consensus sequences of *Spongebob* could be reconstructed for insects (Table 1 and Figure 1C). In the hydrozoans, *Spongebob* was present in multiple full-length copies (>50), which allowed the reconstruction of a consensus sequence of 5,836 in length. However, it is difficult for us to find the exact transposase encoding by *Spongebob_ HMa* due to stop codons or frameshifts. The first three bases in the TIRs of *Spongebob_ HMa* were CCC, and their copies were flanked by 2 bp target site duplication (TSD) (Additional file 2: Figure S1), suggesting that it was a member of TRC elements of the *CACTA* (also called *En/Spm*) superfamily of DNA transposons [23]. A pairwise comparison of *Spongebob* consensus sequences from the above four species revealed that the elements were more than 89.7% identical over about 430 bp (Figure 1C), suggesting that they should belong to the same family.

The last DNA transposon named as *Kenshin* was shared by the twisted-wing parasite, alfalfa leafcutting bee *Megachile rotundata* and bat *Myotis davidii* (Table 1 and Figure 2). We found one copy of *Kenshin* in the twisted-wing parasite and alfalfa leafcutting bee, which had an intact open reading frame (ORF) encoded a 584- and 581-amino acid (aa) long transposase, respectively. This suggests that the element had an ability of transposition in both species. This element might be also responsible for the amplification of non-autonomous elements in these species. We identified one non-autonomous element in the alfalfa leafcutting bee and

two non-autonomous elements in the bat (Table 1). These non-autonomous elements had experienced successful amplification and largely outnumbered their autonomous masters (Table 1). One explanation might be that non-autonomous elements could avoid defense system of their hosts as a result of short sequence length [1]. *Kenshin* elements identified in these species were very similar to each other and diverged by 5.0%–11.1% (Figure 1D). Phylogenetic analysis based on transposases of autonomous elements demonstrated that it was a member of the *piggyBac* superfamily (Figure 4B).

Evidence for repeated horizontal transfers

Four DNA transposons described here showed extremely high identities (>87%) over the full length at the nucleotide level despite their hosts diverged more than 300 Mya (Figures 1 and 2, respectively). This provided us with convincing evidence to support that these transposons had repeatedly invaded into these species by HTs. However, we should note that these results might result from other evolutionary processes, such as purifying selection acting on transposons or variable rates of the evolution of transposons [24,25]. Therefore, making HT conclusion of these transposons should be cautious.

To obtain more evidence for HTs of these transposons, we investigated the phylogenetic distribution of these transposons. The results indicated that they were discontinuous distribution in species (Table 1 and Figure 2). For example, both *Buster2* and *Kenshin* were only present in two invertebrates and one vertebrate, and they were not identified in all other vertebrate and invertebrate species for which a complete or nearly complete genome is available in the National Center for Biotechnology Information (NCBI) database (>102) [26]. Similar patterns were also observed for *Buster1* and *Spongebob* (Figure 2).

Additionally, in many cases, the sequence identities of these four DNA transposons were extremely high compared with the divergence time of their hosts. For example, there was more than 87% between *Buster2* in the insects and the freshwater planarian, which diverged more than 792 Mya [14] (Figures 1 and 2). Similarly, *Buster2* and *Kenshin* identified in the insects and mammals, which shared the last common ancestor about 782 Mya [14], showed more than 89% identities. Besides, we also

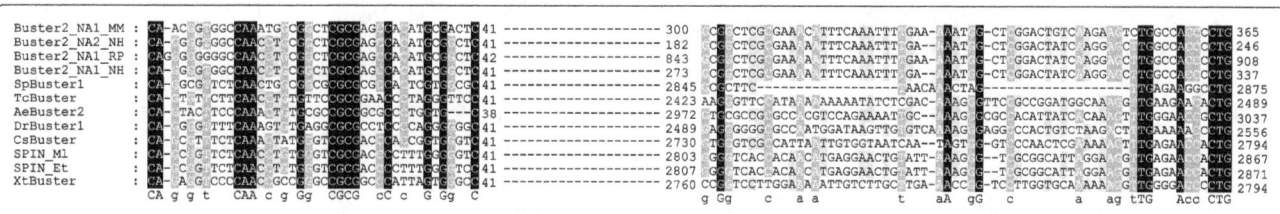

Figure 5 Multiple alignments of *Buster2* identified in this study and previously reported [20], showing portions of the highly conserved 5′ and 3′ termini. Numbers indicated their alignment positions.

found that *Spongebob* in the insects and hydrozoans shared high sequence identities (>89%) at the nucleotide level.

We also observed that phylogenies for *Buster1* and *Buster2* showed a striking lack of structure. For example, phylogenetic analysis based on transposases of *Buster1* showed that this element identified in the twisted-wing parasite was much closer to the freshwater planarian than to another insect, the triatomine bug (Figure 4A). Besides, an unrooted tree based on copies of *Buster2* suggested that *Buster2* elements in the triatomine bug and evening bat were much closer with each other compared with that in the twisted-wing parasite (Additional file 3: Figure S2). All these results were not consistent with vertical inheritance of these transposons.

Finally, our results showed that the nucleotide sequence divergence among four DNA transposons (about 1.5%–13%) was much lower than that observed for three conserved host nuclear genes (about 22%–30%), *heat shock cognate 70*, *Tubulin beta-3*, and *elongation factor 1 alpha*, which were described in our previous study [27]. Therefore, HTs of these transposons might be the only logical explanation for high sequence identities among these transposons in distantly related species.

Discussion

Here, we performed a detailed analysis of characteristics and evolutionary history of four DNA transposons in diverse species. The combination of high identity levels between TEs despite deep divergence times of their host taxa, patchy TE taxonomic distribution, and lower genetic distances for TEs than for host genes clearly demonstrated that these elements had transferred into these species by independent HTs. We also noted that the phenomenon of HT of *Buster1* had previously been reported [13]. However, the detail information about this transposon remains unknown. In this study, both non-autonomous elements and its autonomous partners were found in the twisted-wing parasite, triatomine bug, and freshwater planarian (Table 1), which would provides us with a better understanding for the evolutionary history of *Buster1*. In addition, structural and phylogenetic analyses showed that *Buster1* was a novel member of the *Buster* family.

Although the distribution of these four DNA transposons in species was patchy, their transfer did not randomly happen since the same species have been independently invaded by different, unrelated TEs but others appear to be immune to HT (Figure 2). For example, three transposons are present in the twisted-wing parasite and triatomine bug, but they are not found in other insects (>30) for which genomic sequences are available. This pattern implies that some taxa might be prone to exchanging of genetic materials or are more hospitable to TEs than

others. It is reasonable that species which are vulnerable to HT have a weakened response to TE invasion, which would lead them to lose control of the amplification of the new invader. However, species with a strong resistance would not allow the TEs to amplify in the genome. Similar phenomena have been observed in vertebrates [28].

DNA transposons exist in a wide variety of organisms. However, it was believed that DNA transposons existing in mammals were fossils, and they did not have any ability for mobility in the last 40 Mya [29-31]. This situation has changed when recent DNA transposon activity was discovered in the bats [32-34]. Here, the low average divergence (2.16%–4.81%) between copies of *Buster2* and *Kenshin* and their consensus sequences in the bats strongly suggested that they had been inserted recently (Table 1). Besides, *Buster2* and *Kenshin* were apparently absent from all other mammals (>80) including other seven closely related bats (Figure 2), for which genome sequences are available. Interestingly, *Kenshin* is only present in the genome of the bat *M. davidii* but is not in the other three Myotis genomes sequenced, suggesting that it might be mobilized within the last 10 Mya [15]. These results also implied that bats had continuously acquired new DNA elements via HT. Interestingly, bats belonging to Vespertilionidae family that were the only mammals reported to have recent DNA transposon activity [32-34]. Meanwhile, many of DNA transposons were also horizontally transferred into their hosts [28,35]. However, we should note that HT provides a delivery system for the re-colonization of TEs of genomes and we cannot exclude that DNA transposons might be active in many mammals for which genomes are not sequenced.

Four DNA transposons were found in a wide range of organisms including insects, freshwater planarian, hydrozoans, and bats, suggesting that multiple mechanisms might be involved in their HTs. One interesting finding is the identification of these elements to be present and transferred between insects and bats. The evening bats feed heavily on beetles (Coleoptera), but they also eat moths (Lepidoptera), small flies (Diptera), and other insects [36]. This suggested that predation on a large quantity of insects might increase bat exposure to HT. Another interesting finding is the identification of near identical DNA transposons in insects and the twisted-wing parasite. *M. moldrzyki* is a species of Strepsiptera (Mengenillidae), which infects at least 35 families of insects belonging to seven orders [37]. During the process of parasitism, these parasites obtained nutrients from their hosts [38]. Therefore, parasite-host interaction might facilitate exchanging of their genetic materials.

Conclusions

In this study, we provided the first documented example of HT of *CACTA* transposons. Our results demonstrated that bats had continuously acquired new DNA elements via HT. This implies that predation on a large quantity of insects might increase bat exposure to HT. In addition, parasite-host interaction might facilitate exchanging of their genetic materials.

Methods

Data resources

The silkworm (*Bombyx mori*) assembled genomic sequences were downloaded from Silkworm Genome Database [39] (http://www.silkdb.org/silkdb/). The triatomine bug, *Rhodnius prolixus*, genomic supercontig sequences were downloaded from VectorBase [40] (https://www.vectorbase.org/). Survey sequences from the genomes of five bats (*Myotis austroriparius, Lasiurus borealis, Corynorhinus rafinesquii, Perimyotis subflavus,* and *Nycticeius humeralis*) were downloaded from Dryad Digital Repository [41] (http://datadryad.org/). The postman butterfly (*Heliconius melpomene*) genomic sequences were downloaded from Butterfly Genome Database [42] (http://www.butterflygenome.org/). All of the rest of the genome sequences used in this study were downloaded from the National Center for Biotechnology Information.

Identification of four DNA transposons in *Mengenilla moldrzyki* and other surveyed genomes

Four DNA transposons were identified from the genome of the twisted-wing parasite, *M. moldrzyki*, and they were designated as *Buster1, Buster2, Spongebob,* and *Kenshin,* respectively. Their consensus sequences were reconstructed using the software DAMBE [43]. Then, their consensus sequences were used as queries to search against Repbase [2] (http://www.girinst.org/) to classify them into known superfamilies. To identify related elements in other species, Blastn [44] searches were performed using nucleotide sequences of the above four DNA transposons query against all GenBank databases and Repbase. Significant hits (>85%) were collected and aligned. Their consensus sequences were also reconstructed and compared among species.

Next, we used these respective consensus sequences to mask each genome to estimate copy number. If one autonomous element and its derivatives coexisted in many studied species genomes (Table 1), their copy numbers were calculated using the following criteria. Fragments that were longer than 600 bp were calculated as copies of autonomous elements as miniature inverted-repeat transposable elements are generally shorter than 600 bp [1]. For MITEs or other non-autonomous elements, all fragments with more than 80% identity and coverage to their consensus sequences were calculated as their copies. Meanwhile, fragments were considered to be a single insertion when they were separated by less than 50 bp. If only one autonomous or non-autonomous element was present in one species, all blast hits with more than 100 bp and 80% identity were used to calculate copy number [26].

Sequence analysis

ORF of transposons used in this study was predicted using getorf in EMBOSS-6.3.1 package [45]. These elements were aligned using MUSCLE [46]. Shading and minor manual refinements of multiple alignments were deduced using Genedoc [47] and Illustrator CS5. Then, we used the software Bioedit [48] to calculate each pairwise identity of their consensus sequences after all ambiguous and gapped sites were removed. Sequence logos of *Buster1_NA1_SM* and *Buster2_NA2_NH* were created by WebLogo [49] using 30 bp (15 upstream and 15 downstream) flanking their insertion sites.

To determine the relationship of *Buster1* and *Kenshin* with known DNA transposons, transposase sequences of the *hAT* and *piggybac* superfamilies were downloaded from GeneBank and Repbase. Phylogenies were performed with the neighbor-joining method (NJ) using MEGA 4 [50] (pairwise deletion, Poisson correction model, 1,000 bootstrap replicates) based on their transposase sequences. Besides, we also investigated the relationship of *Buster2* from different species. MEGA 4 [50] (pairwise deletion, maximum composite likelihood, 1,000 bootstrap replicates) was used to build phylogenetic trees based on nucleotide sequences of their full-length or nearly full-length copies.

Additional files

Additional file 1: Table S1. Consensus sequences were used in this study. Sequences multiple alignments were performed using MUSCLE with default parameters. The consensus sequence was generated using DAMBE.

Additional file 2: Figure S1. Neighbor-joining phylogenetic tree of genomic copies of *Buster2* from three species, *Mengenilla moldrzyki, Rhodnius prolixus,* and *Nycticeius humeralis.* The tree is based on a 230-bp long alignment of 5' and 3' termini of genomic copies of *Buster2.* Only bootstrap values >60 are shown. Accession numbers for each element were delineated on each branch.

Additional file 3: Figure S2. Insertion bias of *Spongebob_HMa* (A) and paralogous 'empty' site of transposons identified in this study (B). Their TSD was shown using rectangles.

Competing interests
The authors declare that they have no competing financial interests.

Authors' contributions
ZT designed the study, carried out the data analyses and revised the manuscript. HHZ did the data analyses, drafted, and revised the manuscript. KH did the data analyses. XGZ and MJH revised the manuscript. ZZ designed the study, supervised the study, and revised the manuscript. All authors read and approved the final manuscript.

Acknowledgements
This work is supported by the Hi-Tech R&D Program (863) of China (2013AA102507-2 to ZZ), the National Natural Science Foundation of China (No. 31260632 to XGZ, No. 31401106 to MJH, and No. 31471197 to ZZ), and the National Natural Science Foundation in Jiangxi Province of China (20122BAB204018 to XGZ).

Author details
[1]School of Life Sciences, Chongqing University, Chongqing 400044, China.
[2]College of Pharmacy and Life Science, Jiujiang University, Jiujiang 332000, China. [3]College of Forestry and Life Science, Chongqing University of Sciences and Arts, Yongchuan, Chongqing 40216, China.

References
1. Feschotte C, Pritham EJ. DNA transposons and the evolution of eukaryotic genomes. Annu Rev Genet. 2007;41:331–68.
2. Jurka J. Repbase update: a database and an electronic journal of repetitive elements. Trends Genet. 2000;16:418–20.
3. Han MJ, Xu HE, Zhang HH, Feschotte C, Zhang Z. Spy: a new group of eukaryotic DNA transposons without target site duplications. Genome Biol Evol. 2014;6:1748–57.
4. Schaack S, Gilbert C, Feschotte C. Promiscuous DNA: horizontal transfer oftransposable elements and why it matters for eukaryotic evolution. Trends Ecol Evol. 2010;25:537–46.
5. Hartl DL, Lohe AR, Lozovskaya ER. Modern thoughts on an ancyent marinere: function, evolution, regulation. Annu Rev Genet. 1997;31:337–58.
6. Roberston HM. Evolution of DNA transposons in eukaryotes. In: Craig NL, Craigie M, Lambowitz A, editors. Mobile DNA II. Herndon, VA: ASM; 2002. p. 1093–110.
7. Kidwell MG. Horizontal transfer of P elements and other short inverted repeat transposons. Genetica. 1992;86:275–86.
8. Silva JC, Loreto EL, Clark JB. Factors that affect the horizontal transfer of transposable elements. Curr Issues Mol Biol. 2004;6:57–71.
9. Daniels SB, Peterson KR, Strausbaugh LD, Kidwell MG, Chovnick A. Evidence for horizontal transmission of the P transposable element between Drosophila species. Genetics. 1990;124:339–55.
10. Wallau GL, Ortiz MF, Loreto EL. Horizontal transposon transfer in eukarya: detection, bias, and perspectives. Genome Biol Evol. 2012;4:689–99.
11. Capy P, Bazin C, Higuet D, Langin T. Dynamics and evolution of transposable elements (Molecular Biology Intelligence Unit). New York: Springer; 1998.
12. Niehuis O, Hartig G, Grath S, Pohl H, Lehmann J, Tafer H, et al. Genomic and morphological evidence converge to resolve the enigma of Strepsiptera. Curr Biol. 2012;22:1309–13.
13. Lavoie CA, Platt 2nd RN, Novick PA, Counterman BA, Ray DA. Transposable element evolution in Heliconius suggests genome diversity within Lepidoptera. Mob DNA. 2013;4:21.
14. Hedges SB, Dudley J, Kumar S. TimeTree: a public knowledge-base of divergence times among organisms. Bioinformatics. 2006;22:2971–2.
15. Stadelmann B, Lin LK, Kunz TH, Ruedi M. Molecular phylogeny of New World Myotis (Chiroptera, Vespertilionidae) inferred from mitochondrial and nuclear DNA genes. Mol Phylogenet Evol. 2007;43:32–48.
16. Zhang HH, Shen YH, Xu HE, Liang HY, Han MJ, Zhang Z. A novel hAT element in Bombyx mori and Rhodnius prolixus: its relationship with miniature inverted repeat transposable elements (MITEs) and horizontal transfer. Insect Mol Biol. 2013;22:584–96.
17. Lack JB, Van Den Bussche RA. Identifying the confounding factors in resolving phylogenetic relationships in Vespertilionidae. J Mammal. 2010;91:1435–48.
18. Moreno-Vazquez S, Ning J, Meyers BC. hATpin, afamily of MITE-like hAT mobile elements conserved in diverse plant species that forms highly stable secondary structures. Plant Mol Biol. 2005;58:869–86.
19. Tu Z. Molecular and evolutionary analysis of two divergent subfamilies of a novel miniature inverted repeat transposable element in the yellow fever mosquito, Aedes aegypti. Mol Biol Evol. 2000;17:1313–25.
20. Arensburger P, Hice RH, Zhou L, Smith RC, Tom AC, Wright JA, et al. Phylogenetic and functional characterization of the hAT transposon superfamily. Genetics. 2011;188:45–57.
21. Datzmann T, von Helversen O, Mayer F. Evolution of nectarivory in phyllostomid bats (Phyllostomidae Gray, 1825, Chiroptera: Mammalia). BMC Evol Biol. 2010;10:165.
22. Yang G, Nagel DH, Feschotte C, Hancock CN, Wessler SR. Tuned for transposition: molecular determinants underlying the hyperactivity of a Stowaway MITE. Science. 2009;325:1391–4.
23. DeMarco R, Venancio TM, Verjovski-Almeida S. SmTRC1, a novel Schistosoma mansoni DNA transposon, discloses new families of animal and fungi transposons belonging to the CACTA superfamily. BMC Evol Biol. 2006;6:89.
24. Volff JN. Turning junk into gold: domestication of transposable elements and the creation of new genes in eukaryotes. Bioessays. 2006;28:913–22.
25. Loreto EL, Carareto CM, Capy P. Revisiting horizontal transfer of transposable elements in Drosophila. Heredity. 2008;100:545–54.
26. Gilbert C, Schaack S, Pace II JK, Brindley PJ, Feschotte C. A role for host parasite interactions in the horizontal transfer of transposons across phyla. Nature. 2010;464:1347–50.
27. Zhang HH, Xu HE, Shen YH, Han MJ, Zhang Z. The origin and evolution of six miniature inverted-repeat transposable elements in Bombyx mori and Rhodnius prolixus. Genome Biol Evol. 2013;5:2020–31.
28. Novick P, Smith J, Ray D, Boissinot S. Independent and parallel lateral transfer of DNA transposons in tetrapod genomes. Gene. 2010;449:85–94.
29. Lander ES, Linton LM, Birren B, Nusbaum C, Zody MC, Baldwin J, et al. Initial sequencing and analysis of the human genome. Nature. 2001;409:860–921.
30. Waterston RH, Lindblad-Toh K, Birney E, Rogers J, Abril JF, Agarwal P, et al. Initial sequencing and comparative analysis of the mouse genome. Nature. 2002;420:520–62.
31. Pace JK, Feschotte C, Feschotte C. The evolutionary history of human DNA transposons: Evidence for intense activity in the primate lineage. Genome Res. 2007;17:422–32.
32. Ray DA, Pagan HJ, Thompson ML, Stevens RD. Bats with hATs: evidence for recent DNA transposon activity in genus Myotis. Mol Biol Evol. 2007;24:632–9.
33. Ray DA, Feschotte C, Pagan HJ, Smith JD, Pritham EJ, Arensburger P, et al. Multiple waves of recent DNA transposon activity in the bat, Myotis lucifugus. Genome Res. 2008;18:717–28.
34. Pagán HJ, Macas J, Novák P, McCulloch ES, Stevens RD, Ray DA. Survey sequencing reveals elevated DNA transposon activity, novel elements, and variation in repetitive landscapes among vesper bats. Genome Biol Evol. 2012;4:575–85.
35. Thomas J, Schaack S, Pritham EJ. Pervasive horizontal transfer of rolling-circle transposons among animals. Genome Biol Evol. 2010;2:656–64.
36. Simmons NB. Order chiroptera. In: Wilson DE, Reeder DM, editors. Mammal species of the world. 3rd ed. Johns Hopkins University; 2005. p.312-529.
37. Kathirithamby J. Review of the order Strepsiptera. Syst Ent. 1989;14:41–92.
38. Kathirithamby J. Host-parasitoid associations in Strepsiptera. Annu Rev Ent. 2009;54:227–49.
39. Duan J, Li R, Cheng D, Fan W, Zha X, Cheng T, et al. SilkDB v2.0: a platform for silkworm (Bombyx mori) genome biology. Nucleic Acids Res. 2010;38 (Database issue):D453–6.
40. Lawson D, Arensburger P, Atkinson P, Besansky NJ, Bruggner RV, Butler R, et al. VectorBase: a data resource for invertebrate vector genomics. Nucleic Acids Res. 2009;37(Database issue):D583–7.
41. Pagán HJ, Macas J, Novák P, McCulloch ES, Stevens RD, Ray DA: Data from: Survey sequencing reveals elevated DNA transposon activity, novel elements, and variation in repetitive landscapes among vesper bats. 2012 Dryad Digital Repository. http://dx.doi.org/10.5061/dryad.83164r7v.
42. Dasmahapatra KK, Walters JR, Briscoe AD, Davey JW, Whibley A, Nadeau NJ, et al. Butterfly genome reveals promiscuous exchange of mimicry adaptations among species. Nature. 2012;487:94–8.
43. Xia X, Xie Z. DAMBE: software package for data analysis in molecular biology and evolution. J Hered. 2001;92:371–3.
44. Altschul SF, Gish W, Miller W, Myers EW, Lipman DJ. Basic local alignment search tool. J Mol Biol. 1990;215:403–10.
45. Rice P, Longden I, Bleasby A. EMBOSS: the European molecular biology open software suite. Trends Genet. 2000;16:276–7.
46. Edgar RC. MUSCLE: multiple sequence alignment with high accuracy and high throughput. Nucleic Acids Res. 2004;32:1792–7.
47. Nicholas KB, Nicholas HB, Deerfield DW. GeneDoc: analysis and visualization of genetic variation. EMBNEW News. 1997;4:14.
48. Hall TA. BioEdit: a user-friendly biological sequence alignment editor and analysis program for Windows 95/98/NT. Nucleic Acids Symp Ser. 1999;41:95–8.
49. Crooks GE, Hon G, Chandonia JM, Brenner SE. WebLogo: a sequence logo generator. Genome Res. 2004;14:1188–90.
50. Tamura K, Dudley J, Nei M, Kumar S. MEGA4: molecular evolutionary genetics analysis (MEGA) software version 4.0. Mol Biol Evol. 2007;24:1596–9.

3

Differential SINE evolution in vesper and non-vesper bats

David A Ray[1*], Heidi JT Pagan[2], Roy N Platt II[1], Ashley R Kroll[3], Sarah Schaack[3] and Richard D Stevens[4]

Abstract

Background: Short interspersed elements (SINEs) have a powerful influence on genome evolution and can be useful markers for phylogenetic inference and population genetic analyses. In this study, we examined survey sequence and whole genome data to determine the evolutionary dynamics of Ves SINEs in the genomes of 11 bats, nine from Vespertilionidae.

Results: We identified 41 subfamilies of Ves and linked several to specific lineages. We also revealed substantial differences among lineages including the observation that Ves accumulation and Ves subfamily diversity is significantly higher in vesper as opposed to non-vesper bats. This is especially interesting when one considers the increased transposable element diversity of vesper bats in general.

Conclusions: Our data suggest that survey sequencing and genome mining are valuable tools to investigate SINE evolution among related lineages and can provide substantial information about the ability of SINEs to proliferate in diverse genomes. This method would also be a useful first step in determining which subfamilies would be the best to target when developing SINEs as markers for phylogenetic and population genetic analyses.

Background

Now that it is known that transposable elements (TEs) comprise a significant proportion of most multicellular eukaryotic genomes, there is great interest in understanding their patterns of proliferation and the factors determining their relative success across various lineages. Two classes of TEs are delineated according to mobilization mechanism. Class I elements, the retrotransposons, move through an RNA intermediate, allowing the original copy to stay in place, resulting in replicative gains in copy number. Most Class II elements, the DNA transposons, mobilize in DNA form, with one subclass (hAT, piggyBac, and so on) relying on excision and re-integration (cut-and-paste) and a second subclass (Helitrons and Mavericks) utilizing a DNA-based replication mechanism. In mammals, retrotransposons are by far the most active and largest class of repetitive sequences. This is exemplified by the high prevalence of long and short interspersed elements (LINEs and SINEs) in the human genome, where the primate SINE, *Alu*, has reached over one million copies and continues to multiply [1]. For the last 40 million years, TE activity in mammals has been limited almost exclusively to Class I elements [2-7]. However, exceptions have been identified in several mammals, where multiple horizontal transfers of Class II elements have occurred, and/or activity levels of DNA transposons are high [8-14].

SINEs have been shown to influence genomes in multiple ways including the introduction of CpG islands, regulatory motifs, and as the substrate for homologous and non-homologous recombination events (reviewed in [15]). In addition to their impacts on genome structure and function, they have also proven to be exceptionally useful genetic markers, particularly in the elucidation of phylogenies [16-21]. Once inserted, SINEs are rarely excised [22] and, after fixation in the population, will be vertically inherited, becoming shared derived characters. Further, the absence of a SINE insertion at any particular locus can be safely assumed to represent the ancestral condition. However, because SINE subfamilies will emerge, multiply, and eventually die out over a finite period, it is critical to identify the subfamilies that were active during the period of interest for the phylogeny being inferred. In

---

* Correspondence: david.4.ray@gmail.com
[1]Department of Biological Sciences, Texas Tech University, Lubbock, TX 79409, USA
Full list of author information is available at the end of the article

that way, the researcher is more likely to identify insertions that will be informative in such analyses.

The confident identification of phylogenetically informative patterns requires large numbers of SINE insertions, preferably from multiple representatives of the clade of interest. The most efficient way to identify such patterns would be to query representative genome drafts. While genomes are being assembled at an increasing rate, this is not feasible for all groups. Instead, one can often hope for at best a single genome sequence from the clade of interest, and for most clades even that is not available. An alternative strategy would be to take advantage of high-throughput sequencing technologies and survey sequencing a group of related genomes. Such survey sequencing would provide large amounts of potentially informative data on the identity of TE subfamilies in a range of genomes for a relatively low cost [23-26].

Ves is a tRNA-derived SINE family found in yangochiropteran bats, a clade that includes all microbats with the exception of the yinpterochiropteran microbats of families Megadermatidae, Rhinolophidae, and Rhinopomatidae [27,28]. For this study, we examined Ves accumulations in the draft genomes of the vesper bats *Eptesicus fuscus*, *Myotis brandti*, *M. davidii*, *M. lucifugus*, and the non-vesper bat *Pteronotus parnellii*. We also examined survey data collected from the genomes of six other bats, the verspertilionids, *Corynorhinus rafinesquii*, *Lasiurus borealis*,

M. austroriparius, *Nycticeius humeralis*, *Perimyotis subflavus*, and the phyllostomid *Artibeus literatus*. Our results demonstrate differential activity among these lineages and allow us to identify the subfamilies that are most likely to be informative at various branches within the yangochiropteran phylogeny. We also developed a method for determining lineage specificity of SINE subfamilies and, using this method, were able to establish subfamily identities within each taxon and identified several instances of lineage-specific Ves activity.

Results

To identify patterns of Ves activity in the sampled bats, we re-analyzed the survey-sequence data of Pagan *et al.* [25] and performed a *de novo* analysis of SINEs in the draft genomes of *Eptesicus fuscus*, *Myotis brandtii*, *M. davidii*, *M. lucifugus*, and *Pteronotus parnellii* (AAPE00000000, ANKR00000000, ALWT00000000, ALEH00000000, and AWGZ00000000, respectively). The survey data consisted of approximately 1.3 million 454 reads from six bats. These reads averaged approximately 300 nt in length and represented between 0.76% and 4.75% of the sampled genomes. Complete details are available in Table 1 and Pagan *et al.* [25]. Ves family SINEs were identified in all of the taxa examined. The two non-vesper bats, *A. lituratus* (Phyllostomidae) and *P. parnellii* (Mormoopidae), exhibited lower overall Ves content, suggesting lower levels

Table 1 Taxa examined in this study, data used and basic statistics describing Ves content and Ves insertions used for our analysis of subfamilies

Taxon	Abbreviation	Family, subfamily	Data source	Bases queried	Total Ves bases identified	% Ves-derived bases	Total Ves fragments identified	Full-length Ves insertions analyzed in COSEG
Artibeus lituratus	Alit	Phyllostomidae, Stenodermatinae	SS	101,137,176	2,510,325	2.48%	18,038	5,964
Corynorhinus rafinesquii	Craf	Vespertilionidae, Vespertilioninae	SS	108,013,590	4,191,322	3.88%	33,888	8,068
Eptesicus fuscus	Efus	Vespertilionidae, Vespertilioninae	WGS	388,423,918	17,126,373	4.41%	97,440	15,000
Lasiurus borealis	Lbor	Vespertilionidae, Vespertilioninae	SS	71,255,672	3,431,237	4.82%	25,494	7,624
Myotis austroriparius	Maus	Vespertilionidae, Myotinae	SS	21,992,128	1,078,155	4.90%	8,720	2,143
M. brandtii	Mbra	Vespertilionidae, Myotinae	WGS	533,920,351	28,106,987	5.26%	160,836	15,000
M. davidii	Mdav	Vespertilionidae, Myotinae	WGS	420,014,318	23,723,693	5.65%	137,294	15,000
M. lucifugus	Mluc	Vespertilionidae, Myotinae	WGS	497,001,341	23,480,719	4.72%	129,921	15,000
Nycticeius humeralis	Nhum	Vespertilionidae, Vespertilioninae	SS	34,287,171	2,250,938	6.56%	18,329	4,218
Perimyotis subflavus	Psub	Vespertilionidae, Vespertilioninae	SS	29,578,028	1,290,520	4.36%	10,669	2,419
Pteronotus parnellii	Ppar	Mormoopidae	WGS	265,642,944	5,805,924	2.19%	35,450	15,000

SS = survey sequence data, WGS = draft genome data.

of accumulation in those lineages. Conversely, *N. humeralis* exhibited higher than average Ves accumulation than its fellow vespertilionids. This is likely due to the presence of two novel subfamilies that are not found in other vesper bats (see below). This pattern suggests that the subfamilies evolved after the split between the *N. humeralis* ancestor and the remainder of Vespertilionidae and contributed significantly to genome content in the lineage leading to *Nycticeius*.

Ves elements spanning at least 90% of their respective consensus sequences from each survey data set ranged from 2,143 in *M. austroriparius* to over 8,000 in *C. rafinesquii*. Including extracted Ves elements from genome drafts provided a total of 105,436 insertions to be analyzed. Our iterative approach to defining subfamilies (see 'Methods') resulted in a final Ves library consisting of 41 subfamily consensus sequences.

To determine potential lineage specificity, we developed a novel measure we refer to as the 'Ves score' for each putative subfamily. This score is a statistic consisting of the proportion nucleotides from each subfamily in each taxon compared to the median genome proportion (see 'Methods' for details). When scores for any individual taxon fell outside a range encompassing two standard deviations of the mean score (approximating α = approximately 0.05 on a normal distribution), the subfamily was considered specific to a given lineage. Ves score plots

indicated several levels of lineage specificity and four examples are provided in Figure 1 and Additional file 1. Scores for Ves26 in all taxa fell within two standard deviations of the mean Ves score, indicating that this subfamily is present at approximately equal numbers in all taxa. By contrast, the score for Ves32 in *A. lituratus* fell below the two standard deviation threshold but within the two standard deviation range in all other bats, showing that this subfamily is present in all taxa except our representative phyllostomid. The score for Ves15 is higher than the cutoff in *L. borealis* but not in any other sampled bat, indicating that the subfamily is present in significantly higher numbers in this taxon. In addition to the single *L. borealis*-specific subfamily, Ves15, two *N. humeralis*-specific subfamilies were identified, Ves14 and Ves18 (Additional file 1). In the final example, Ves1, the subfamily is present in all vesper bats but absent in the two representative non-vespers. Not surprisingly given the taxon sampling, the majority of the subfamilies described are specific to vesper bats.

Visual examination of log plots suggested that several subfamilies might border on lineage specificity but not reach the 2.0 sigma cutoff. We relaxed our two standard deviation range requirements by increments down to sigma =1.5 (Additional file 1) and found that several subfamilies could be labeled as borderline lineage specific. For example, Ves23 and Ves31 could be considered as vesper

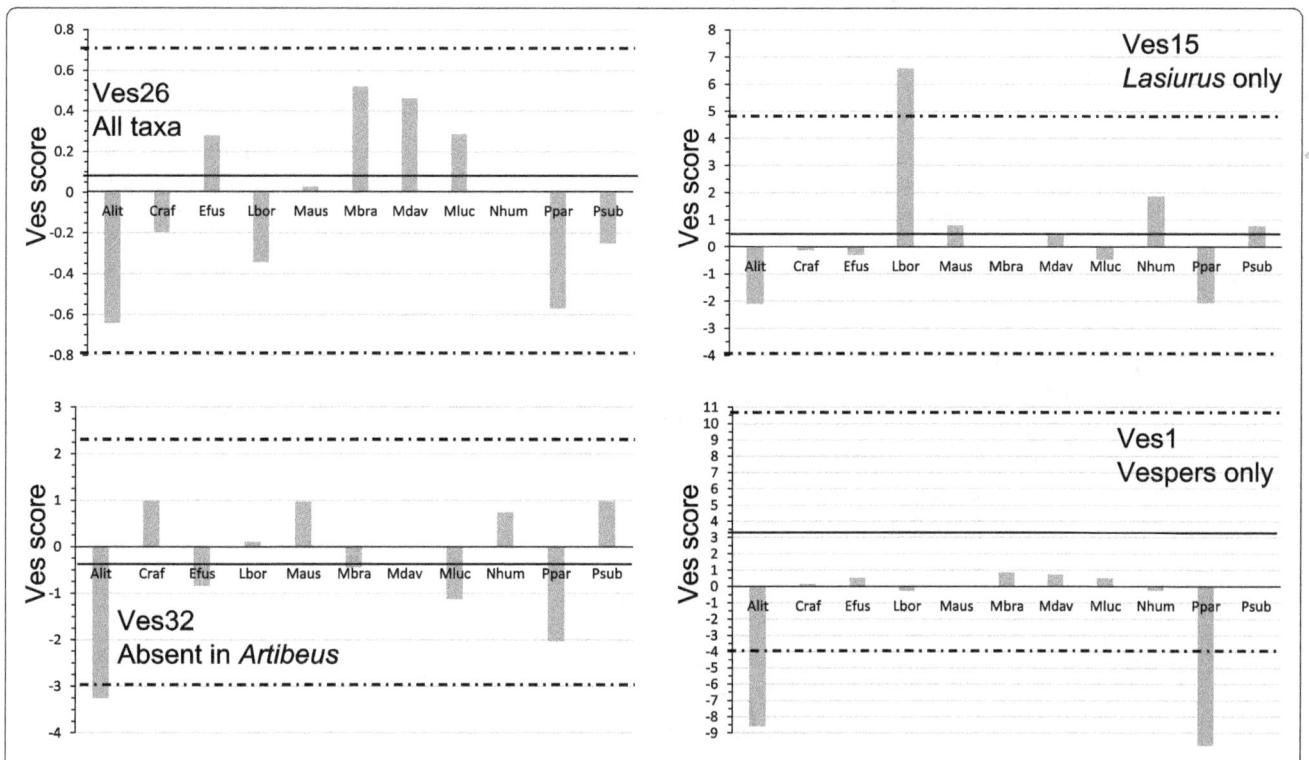

Figure 1 Representative Ves score plots used to identify lineage specific subfamilies. *Solid black horizontal lines* indicate the mean Ves score and *dashed black horizontal lines* indicate the two standard deviation upper and lower bounds. Taxon abbreviations are as in Table 1.

bat specific if one lowers the threshold slightly to 1.9. This suggests that the method can be used as a first approximation to determine likely trends in the data regarding lineage specificity but that individual cases may require special attention.

By examining pairwise divergences among copies of each subfamily and assuming similar neutral mutation rates in bat lineages, it is possible to provide relative estimates of the accumulation periods for various groups of TEs in each lineage. Such estimates assume that element accumulation initially resulted in the formation of multiple identical copies of each retrotransposed element. As time passes, the initially identical elements diverge at a rate determined by the neutral mutation rate. Thus, within a given subfamily, higher average pairwise divergence values among its members indicate more time that has elapsed. So it follows that a subfamily with a higher average pairwise divergence was active in the more distant past than a family with lower average pairwise divergence.

Figure 2 shows the average divergences among subfamilies within and among all taxa. Within the most widely distributed subfamilies, there appear to be three general age categories. For subfamilies that are more restricted, two patterns emerged. First, the three subfamilies that are specific to *L. borealis* and *N. humeralis* (Ves14, 15, and 18) are young compared to most other subfamilies. This is

expected given the relatively recent divergence of these genera from other vesper bats [29]. The converse is true for the subfamilies that are present in non-vesper bats, which diverged from Vespertilionidae approximately 45 mya and from each other approximately 34 mya [29]. These subfamilies show evidence of little recent accumulation, suggesting low rates of novel subfamily evolution and diversity. Indeed, if we consider only *M. lucifugus* and *P. parnellii*, two taxa for which we have full genome drafts, we observe that 28 subfamilies have evolved in the former lineage compared to only seven in the latter. Furthermore, if we consider average sequence divergence as a proxy for subfamily age, the subfamilies that are specific to the *P. parnellii* lineage evolved early and experienced little subsequent diversification, whereas novel subfamilies have been continuously evolving throughout the entire history of vesper bats (Figure 2).

We were also interested in whether or not Ves has been accumulating at similar rates in vesper and non-vesper bats. To examine if the proportion of Ves-derived bases were significantly different between vespertilionid and noctilionoid (*A. lituratus* and *P. parnellii*) groups, we conducted an independent sample t-test based on arcsine transformed proportions. The vespertilionid mean percent Ves-derived bases (Table 1) was larger than the noctilionoid mean, and this difference was significant ($t = 5.33$, df $= 9$, $P < 0.001$). Thus, vesper bats have experienced

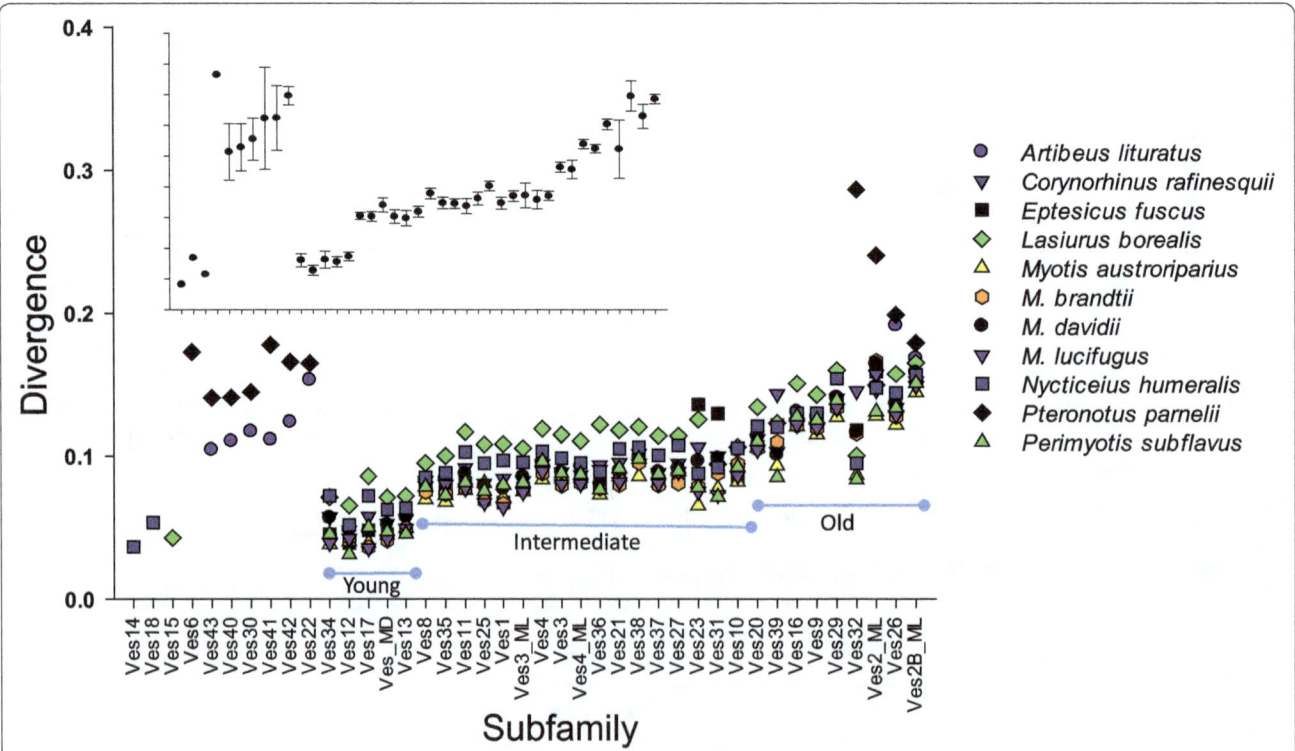

Figure 2 Average divergence values within taxa and among all taxa (inset) for each subfamily. The three major age categories identified among the most widely distributed subfamilies are indicated. *x*-axis values for the inset are the same as for the main figure.

significantly higher rates of Ves accumulation compared to the two non-vesper bats in the study. The temporal accumulation plots in Figure 3 and Additional file 2 illustrate substantial differences in Ves accumulation within and among lineages. Within Vespertilionidae, there is less variability in accumulation patterns, with the exception of *N. humeralis*, whose genome has accumulated substantial mass from the two lineage-specific subfamilies we detected.

Transposition in transposition (TinT) analysis uses nested insertion analysis to independently estimate relative periods of activity among TE families [30]. Because of the limitations of survey sequence data, including the relatively small

average read length, TinT analysis was limited to the genome drafts. Analyses of these genomes using a custom Ves subfamily library confirm the patterns suggested by investigations of lineage specificity and genetic divergence with subfamilies categorized as 'young,' 'intermediate,' or 'old' generally falling at the top, middle, or bottom of the TinT plots, respectively (Figure 4, Additional file 3). Interestingly, Ves18, which was identified as being specific to the *N. humeralis* lineage appears in the *M. brandtii* and *E. fuscus* TinT plots. This can be explained by assuming that the Ves18 lineage began accumulating at low rates in a common ancestor of the broader clade but did not achieve substantial retrotranspositional success until after

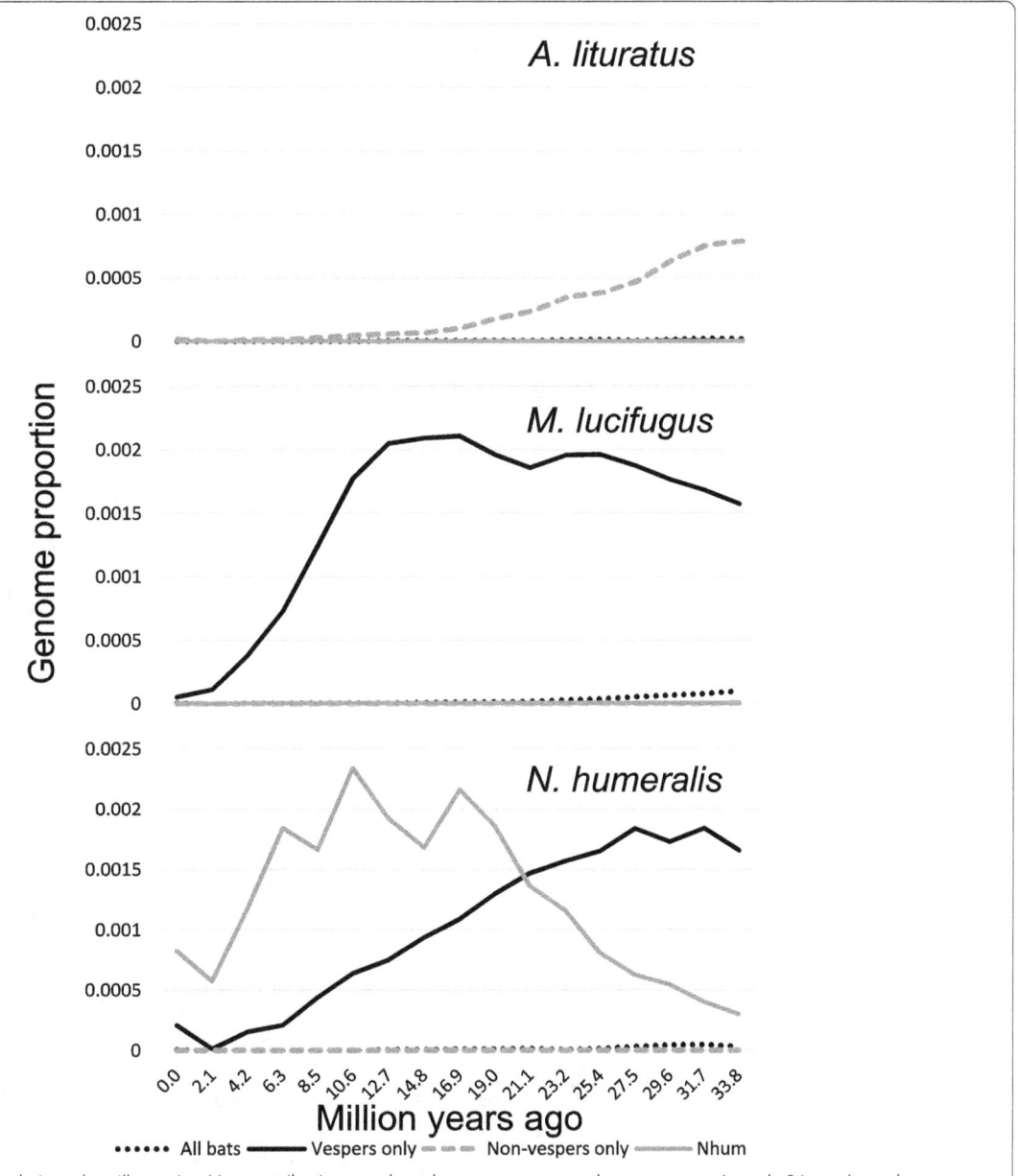

Figure 3 Temporal accumulation plots illustrating Ves contributions to three bat genomes over the past, approximately 34 my. In each case, genome proportions occupied by each category of the Ves subfamily are plotted against periods of accumulation.

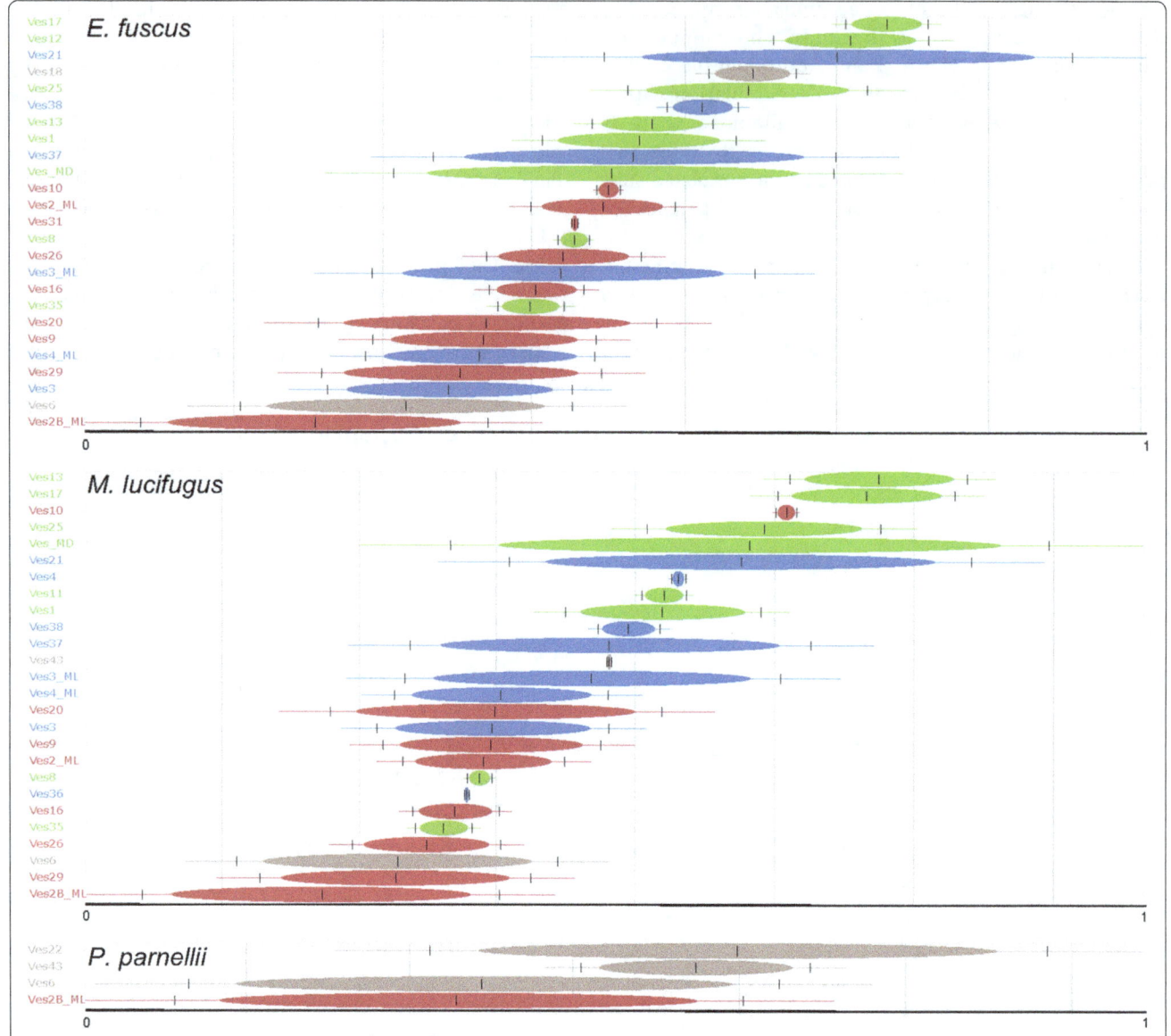

Figure 4 TinT results for three taxa. *Green, blue,* and *red bars* represent subfamilies identified in Figure 4 as young, intermediate, and old, respectively. *Brown bars* indicate subfamilies that are less widely distributed. Plots for *M. brandtii* and *M, davidii* are available in Additional file 3. Numbers on the *x*-axis are relative time periods within each taxon with zero at the origin of detectable activity for these SINE subfamilies and one representing the boundary of current activity.

the *Nycticeius* lineage diverged from the remainder of Vespertilioninae. Indeed, our analysis of lineage-specificity suggests that Ves18 is present at low levels in all of the other bats we investigated, with genome proportions ranging from 0.001% to 0.069%, compared to 0.424% in the *N. humeralis* genome (Additional file 1).

Figure 5 illustrates the relationships among subfamilies recovered by Bayesian analysis of the core consensus sequences of each subfamily. Because subfamilies 26 and 2B_ML are present in all of the bats analyzed, they likely represent basal Ves lineages, and Ves2B_ML was judged to be the older of the two by TinT and our genetic

distance estimates (Figures 2 and 4). We therefore rooted the tree on that subfamily. The resulting topology suggests three well supported Ves lineages. Clade A consists of six subfamilies that are restricted to non-vesper bats. Clade B consists, with two exceptions, of subfamilies that are specific to vesper bats. The two exceptions (Ves2_ML and Ves32) are found in *P. parnellii* but do not show significant accumulations in *A. lituratus* (Additional file 1). This pattern is unexpected given the fact that these taxa belong to the sister families Moomopidae and Phyllostomidae. In both cases, however, these subfamilies exhibit substantially, but not significantly, lower rates of

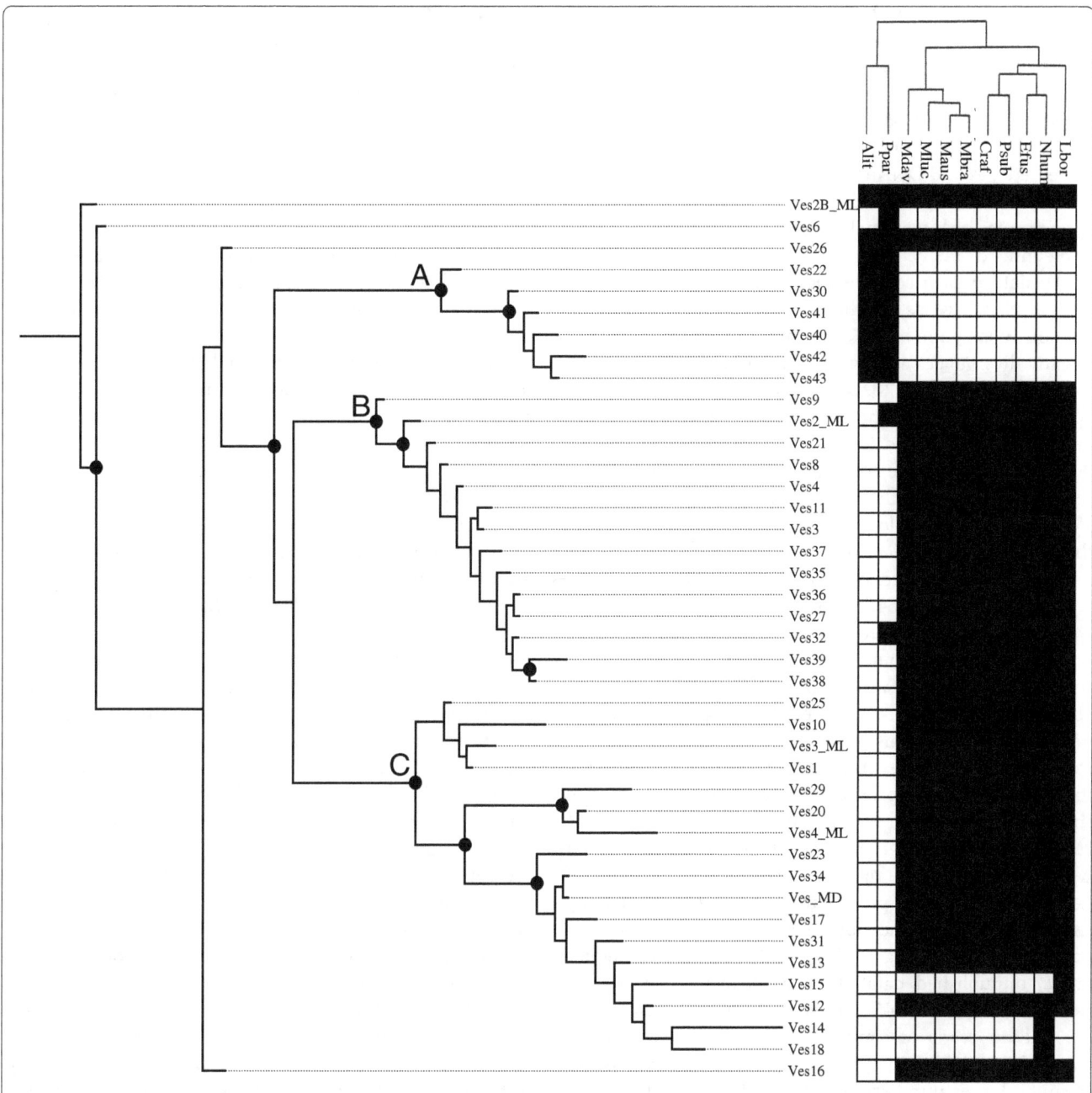

Figure 5 Bayesian tree of Ves subfamily relationships. *Gray circles* represent nodes with posterior probabilities of 0.95 or greater. The chart to the right indicates the species in which each subfamily can be identified as per the lineage specificity plots in Additional file 1. The cladogram at the top illustrates relationships among taxa as per Lack and Van Den Bussche (2010). Taxon abbreviations are as in Table 1.

accumulation in the non-vesper data (see Ves32 in Figure 1). This suggests that the two subfamilies were experiencing reduced retrotransposition in both lineages, consistent with the observation of reduced accumulation described above but that these two subfamilies may have managed to replicate a bit more successfully in the lineage leading to *Pteronotus* after its divergence from the ancestor of *Artibeus*. Clade C is similar in that it also comprises vesper-specific subfamilies, but it also harbors the three

subfamilies that are specific to the *Nycticeius* and *Lasiurus* lineages with the two *Nycticeius* subfamilies being sister to one another. The picture to be gleaned therefore is that our subfamilies generally reflect the phylogeny of the bats whose genomes they inhabit.

Discussion

Two of our observations, increased Ves diversity and increased Ves accumulation in Vespertilionidae, are interesting in the context of overall mammalian and chiropteran TE diversity. As has been repeatedly observed [25,31,11,12], vesper bats are home to an astonishing diversity of DNA transposons not seen in any other mammal to date. Furthermore, these DNA transposons appear to have led to functional evolutionary innovations [14,13]. Increased Ves and DNA transposon accumulation appears to be a characteristic of this family, which is the second most species-rich mammalian clade, and may have played a role in its diversity.

The data presented here suggest that several subfamilies would serve as excellent markers for investigating relationships within bats. For example, for those interested in early divergences among all yangochiropterans, probing for members of Ves2B_ML or Ves26 would be most appropriate. For researchers interested in investigations of relationships within the lineage leading to genus *Nycticeius*, it would be preferable to focus on Ves14 and/or Ves18. Broader interest in the evolution of Vespertilionidae/Vespertilioninae would be served by focusing on any of the intermediate subfamilies.

It should be pointed out that survey sequencing was accomplished using 454/Roche chemistry. When the analysis was first conceived, this chemistry was the only one available that would provide reads long enough to sequence full-length Ves insertions. However, Illumina chemistry has recently achieved read lengths of 300 nt on some of its systems. Using a paired-end sequencing strategy and creating libraries consisting of fragments under 600 nt would produce overlapping reads of more than sufficient length to accomplish the same task but resulting in much larger data sets than the one described here. For example, we recently surveyed several mammal genomes as part of a project to investigate LINE activity using this strategy and obtained just under one million paired-end reads using 1/10th of a MiSeq lane (Mangum *et al.*, unpublished data). In our original study, we required multiple full 454 runs to obtain the just under 1.3 million reads of similar length [25]. This suggests that substantially larger data sets, potentially consisting of much larger numbers of taxa, could be easily and inexpensively obtained.

Indeed, we recently used the information provided by this study to inform a novel experimental protocol to analyze the phylogeny of selected *Myotis* bats (Platt *et al.*, under revision). In that study, a combined, computational and laboratory-based approach based on ME-Scan [32] was used to identify potentially polymorphic SINE insertions in seven species. Probes were designed that match subfamilies in Clade C. That work was successful in inferring previously established relationships among the bats investigated, further suggesting that this method will be useful in informing projects designed to use SINEs as phylogenetic markers.

Conclusions

While genome sequencing costs continue to decline, the huge biological diversity observed still prevents us from achieving the ideal - a complete genome from all taxa. We find that this survey method can provide substantial information about SINE families/subfamilies in a range of taxa at minimal cost and suggest that it may serve as a valuable initial step in guiding SINE-based analyses of a variety of taxa, especially those that are not represented by a draft genome. Furthermore, the Ves score method we developed to identify lineage-specific subfamilies should be easily implemented in studies of SINE families in a wide variety of taxa.

Finally, there is no reason to limit the methods described here to SINEs alone, and substantial information about the overall TE content in a genome can also be gleaned.

Methods

We used RepeatMasker [33] to query all survey sequence data and approximately one quarter of the whole genome drafts. We used a custom Ves library consisting of the VES, Ves2_ML, Ves2B_ML, Ves3_ML, and Ves4_ML subfamilies from RepBase [34]. All Ves insertions spanning at least 90% of the identified consensus were extracted, limiting ourselves to 15,000 hits from the genome drafts. The extracted sequences were combined into a single set of Ves insertions and analyzed using COSEG [35,36] after aligning them to the VES4_ML consensus sequence. A custom Perl script provided by R. Hubley was used to refine the consensus sequence for each Ves subfamily and is available upon request.

Upon identification of Ves subfamily structure using COSEG, a custom RepeatMasker library was constructed and applied to a pseudogenome consisting of all survey sequences and the original subset of WGS data. To verify the presence of each subfamily in the data, 25 random hits identified as belonging to each subfamily were extracted and aligned with their respective consensus. Alignments were examined by eye and, when necessary, new 50% majority-rule consensus sequences were generated. These new consensus sequences were compared among themselves and with the original RepBase Ves elements. Several predicted subfamilies were collapsed into identical subfamilies already defined in RepBase or into other COSEG-derived subfamilies after generating refined consensus sequences. Analysis of two subfamilies, 7 and 33, revealed that these are instances where a Ves element inserted into an active Helitron element, which then deposited copies throughout the genome as it multiplied. Because these two predicted subfamilies were likely disseminated throughout the genome by mechanisms other than retrotransposition, they were not included in subsequent analyses of SINE dynamics. For any COSEG-predicted subfamilies matching those already described in RepBase, the RepBase subfamily designations were used. All newly described subfamily consensus sequences are available in Additional file 4 and have been deposited in RepBase.

The 3' ends of Ves elements consist of an A-rich region preceded by multiple low complexity, pyrimidine-rich regions. These regions are highly variable and, thus, problematic for estimating divergence in downstream analyses. Thus, we created a second library consisting of Ves 'core' sequences (defined as the 5' ends up to but not including the first major poly-pyrimidine tract). These core sequences averaged 159 bp in length compared to the average 212 bp for the full-length library. Relationships among Ves subfamilies were inferred by generating a Bayesian tree of the core consensus sequences in MrBayes v3.2.1 [37]. We used the GTR model of nucleotide substitution and performed one million iterations with a burnin of 1000.

To determine potential lineage specificity of Ves subfamilies, we first determined the genome proportions occupied by each subfamily using RepeatMasker. The total number of bases assigned to each Ves subfamily in each data set was then divided by the total number of bases analyzed from each taxon. For each subfamily, the median genome proportion among the eight taxa was calculated. We next calculated the Ves score, $\log_2$ (proportion/median), for each subfamily within each taxon and compared these by calculating the mean Ves score for each subfamily among the taxa. Ves scores within each taxon were plotted and compared to the mean scores for each subfamily. When scores for any individual taxon fell outside a range encompassing two standard deviations of the mean score (approximating α = approximately 0.05 on a normal distribution), the subfamily was considered a candidate for lineage specificity, with the home lineage(s) being determined based on its presence or absence in the species under consideration.

To calculate approximate periods of accumulation we used a modified version of the calcDivergenceFromAlign.pl script that is included in the RepeatMasker package to calculate Kimura two-parameter distances between each insertion and its respective consensus [10]. The -noCpG option was invoked. We applied the mutation rate estimated by Ray et al. [11], 2.366×10^{-9} substitutions per site/my to calculate average divergences among subfamily insertions and within taxa and to plot relative accumulation periods.

Temporal analyses were supplemented by implementing TinT (Transposition in transposition) analyses using the online server at http://www.bioinformatics.uni-muenster.de/tools/tint/ [30,38]. For this analysis, we queried the full genome drafts of three Myotis species, E. fuscus, and P. parnellii using our custom Ves library and generated bar graphs to illustrate rates of Ves elements inserting into other Ves elements, a proxy for relative activity periods.

Abbreviations
TE: transposable element; LINEs: tong interspersed elements; SINEs: short interspersed elements; TinT: transposition in transposition.

Competing interests
The authors declare that they have no competing interests.

Authors' contributions
DAR and HJTP conducted analyses and participated in writing the manuscript. RNP provided Perl scripts, participated in data analysis, and contributed to writing. SS and ARK participated in writing the manuscript and contributed to data analysis. RDS contributed sequence data, performed statistical analyses and participated in writing the manuscript. All authors read and approved the final manuscript.

Acknowledgements
We thank Federico Hoffmann, Richard Strauss, and Juergen Schmitz for contributing a thoughtful discussion which helped guide the manuscript. Robert Hubley provided help with running COSEG and RepeatMasker. This work was supported by the National Science Foundation [DEB-1355176 (DAR), DEB-1020865 (DAR), DEB-1020890 (RDS), DEB-1411403 (RDS) and MCB-1150213 (SS)], the Helen Stafford Post-Bac Fellowship (ARK), and grants from the M.J. Murdock Charitable Trust (SS). Additional support was provided by the College of Arts and Sciences at Texas Tech University. This is manuscript number T-9-1265, College of Agricultural Sciences and Natural Resources, Texas Tech University.

Author details
[1]Department of Biological Sciences, Texas Tech University, Lubbock, TX 79409, USA. [2]Harbor Branch Oceanographic Institute, Florida Atlantic University, Fort Pierce, FL, USA. [3]Department of Biology, Reed College, Portland, OR 97202, USA. [4]Department of Natural Resources Management and the Museum, Texas Tech University, Lubbock, TX 79409, USA.

References
1. Deininger PL, Batzer MA. Mammalian retroelements. Genome Res. 2002;12(10):1455–65.
2. Waterston RH, Lindblad-Toh K, Birney E, Rogers J, Abril JF, Agarwal P, et al. Initial sequencing and comparative analysis of the mouse genome. Nature. 2002;420(6915):520–62.
3. Gibbs RA, Weinstock GM, Metzker ML, Muzny DM, Sodergren EJ, Scherer S, et al. Genome sequence of the Brown Norway rat yields insights into mammalian evolution. Nature. 2004;428(6982):493–521.
4. Lander ES, Linton LM, Birren B, Nusbaum C, Zody MC, Baldwin J, et al. Initial sequencing and analysis of the human genome. Nature. 2001;409(6822):860–921.
5. Lindblad-Toh K, Wade CM, Mikkelsen TS, Karlsson EK, Jaffe DB, Kamal M, et al. Genome sequence, comparative analysis and haplotype structure of the domestic dog. Nature. 2005;438(7069):803–19.
6. Mikkelsen TS, Wakefield MJ, Aken B, Amemiya CT, Chang JL, Duke S, et al. Genome of the marsupial Monodelphis domestica reveals innovation in non-coding sequences. Nature. 2007;447(7141):167–77.
7. Pace 2nd JK, Feschotte C. The evolutionary history of human DNA transposons: evidence for intense activity in the primate lineage. Genome Res. 2007;17(4):422–32.
8. Ray DA, Pagan HJT, Thompson ML, Stevens RD. Bats with hATs: evidence for recent DNA transposon activity in genus Myotis. Mol Biol Evol. 2007;24:632–9.
9. Pace JK, Gilbert C, Clark MS, Feschotte C. Repeated horizontal transfer of a DNA transposon in mammals and other tetrapods. Evolution. 2008;105(44):17023–8.
10. Pagan HJT, Smith JD, Hubley RM, Ray DA. PiggyBac-ing on a primate genome: novel elements, recent activity and horizontal transfer. Genome Biol Evol. 2010;2:293–303. doi:10.1093/gbe/evq021.
11. Ray DA, Feschotte C, Pagan HJ, Smith JD, Pritham EJ, Arensburger P, et al. Multiple waves of recent DNA transposon activity in the bat, Myotis

lucifugus. Genome Res. 2008;18(5):717–28. doi:10.1101/gr.071886.107.

12. Thomas J, Sorourian M, Ray D, Baker RJ, Pritham EJ. The limited distribution of Helitrons to vesper bats supports horizontal transfer. Gene. 2011;474(1–2):52–8. doi:10.1016/j.gene.2010.12.007.

13. Thomas J, Phillips CD, Baker RJ, Pritham EJ. Rolling-circle transposons catalyze genomic innovation in a mammalian lineage. Genome Biol Evol. 2014;6(10):2595–610. doi:10.1093/Gbe/Evu204.

14. Platt RN, Vandewege MW, Kern C, Schmidt CJ, Hoffmann FG, Ray DA. Large numbers of novel miRNAs originate from DNA transposons and are coincident with a large species radiation in bats. Mol Biol Evol. 2014;31(6):1536–45. doi:10.1093/molbev/msu112.

15. Schmitz J. SINEs as driving forces in genome evolution. Genome Dynam. 2012;7:92–107.

16. Konkel MK, Walker JA, Batzer MA. LINEs and SINEs of primate evolution. Evol Anthropol. 2010;19(6):236–49. doi:10.1002/Evan.20283.

17. Okada N, Shedlock AM, Nikaido M. Retroposon mapping in molecular systematics. Mobile genetic elements: protocols and genomic applications. Methods in molecular biology. Totowa, NJ: Humana Press; 2004. p. 189–226.

18. Ray DA, Xing J, Salem A-H, Batzer MA. SINEs of a *nearly* perfect character. Syst Biol. 2006;55:928–35.

19. Shedlock AM, Milinkovitch MC, Okada N. SINE evolution, missing data, and the origin of whales. Syst Biol. 2000;49(4):808–17.

20. Schmitz J, Ohme M, Zischler H. SINE insertions in cladistic analyses and the phylogenetic affiliations of *Tarsius bancanus* to other primates. Genetics. 2001;157(2):777–84.

21. Nikaido M, Piskurek O, Okada N. Toothed whale monophyly reassessed by SINE insertion analysis: the absence of lineage sorting effects suggests a small population of a common ancestral species. Mol Phylogen Evol. 2007;43(1):216–24. doi:10.1016/j.ympev.2006.08.005.

22. van de Lagemaat LN, Gagnier L, Medstrand P, Mager DL. Genomic deletions and precise removal of transposable elements mediated by short identical DNA segments in primates. Genome Research. 2005;15(9):1243–9. doi:10.1101/gr.3910705.

23. Macas J, Neumann P, Navratilova A. Repetitive DNA in the pea (Pisum sativum L.) genome: comprehensive characterization using 454 sequencing and comparison to soybean and Medicago truncatula. BMC Genomics. 2007;8:427. doi:10.1186/1471-2164-8-427.

24. Novak P, Neumann P, Macas J. Graph-based clustering and characterization of repetitive sequences in next-generation sequencing data. BMC Bioinformatics. 2010;11:378. doi:10.1186/1471-2105-11-378.

25. Pagan HJ, Macas J, Novak P, McCulloch ES, Stevens RD, Ray DA. Survey sequencing reveals elevated DNA transposon activity, novel elements, and variation in repetitive landscapes among vesper bats. Genome Biol Evol. 2012;4(4):575–85. doi:10.1093/gbe/evs038.

26. Sun C, Shepard DB, Chong RA, Arriaza JL, Hall K, Castoe TA, et al. LTR retrotransposons contribute to genomic gigantism in plethodontid salamanders. Genome Biol Evol. 2012;4(2):168–83. doi:10.1093/Gbe/Evr139.

27. Kawai K, Nikaido M, Harada M, Matsumura S, Lin LK, Wu Y, et al. Intra- and interfamily relationships of Vespertilionidae inferred by various molecular markers including SINE insertion data. J Mol Evol. 2002;55(3):284–301.

28. Teeling EC. Bats (Chiroptera). In: Hedges SB, Kumar S, editors. The Timetree of Life. Oxford University Press; 2009. p. 499–503.

29. Lack JB, Van Den Bussche RA. Identifying the confounding factors in resolving phylogenetic relationships in Vespertilionidae. J Mammal. 2010;91(6):1435–48.

30. Churakov G, Grundmann N, Kuritzin A, Brosius J, Makalowski W, Schmitz J. A novel web-based TinT application and the chronology of the Primate Alu retroposon activity. BMC evolutionary biology. 2010;10:376. doi:10.1186/1471-2148-10-376.

31. Pritham EJ, Feschotte C. Massive amplification of rolling-circle transposons in the lineage of the bat *Myotis lucifugus*. Proc Natl Acad Sci USA. 2007;17(4):422–32.

32. Witherspoon DJ, Xing J, Zhang Y, Watkins WS, Batzer MA, Jorde LB. Mobile element scanning (ME-scan) by targeted high-throughput sequencing. BMC Genomics. 2010;11:410. doi:10.1186/1471-2164-11-410.

33. Smit AFA, Hubley R, Green P. Repeatmasker at http://repeatmasker.org.

34. Jurka J, Kapitonov VV, Pavlicek A, Klonowski P, Kohany O, Walichiewicz J. Repbase Update, a database of eukaryotic repetitive elements. Cytogenet Genome Res. 2005;110(1–4):462–7.

35. Price AL, Eskin E, Pevzner PA. Whole-genome analysis of Alu repeat elements reveals complex evolutionary history. Genome Res. 2004;14(11):2245–52. doi:10.1101/gr.2693004.

36. *COSEG 0.2.1.* http://www.repeatmasker.org

37. Ronquist F, Huelsenbeck JP. MrBayes 3: Bayesian phylogenetic inference under mixed models. Bioinformatics. 2003;19(12):1572–4.

38. Ichiyanagi K, Nakajima R, Kajikawa M, Okada N. Novel retrotransposon analysis reveals multiple mobility pathways dictated by hosts. Genome Res. 2007;17(1):33–41. doi:10.1101/gr.5542607.

4

The contribution of transposable elements to size variations between four teleost genomes

Bo Gao, Dan Shen, Songlei Xue, Cai Chen, Hengmi Cui[*] and Chengyi Song[*]

Abstract

Background: Teleosts are unique among vertebrates, with a wide range of haploid genome sizes in very close lineages, varying from less than 400 mega base pairs (Mb) for pufferfish to over 3000 Mb for salmon. The cause of the difference in genome size remains largely unexplained.

Results: In this study, we reveal that the differential success of transposable elements (TEs) correlates with the variation of genome size across four representative teleost species (zebrafish, medaka, stickleback, and tetraodon). The larger genomes represent a higher diversity within each clade (superfamily) and family and a greater abundance of TEs compared with the smaller genomes; zebrafish, representing the largest genome, shows the highest diversity and abundance of TEs in its genome, followed by medaka and stickleback; while the tetraodon, representing the most compact genome, displays the lowest diversity and density of TEs in its genome. Both of Class I (retrotransposons) and Class II TEs (DNA transposons) contribute to the difference of TE accumulation of teleost genomes, however, Class II TEs are the major component of the larger teleost genomes analyzed and the most important contributors to genome size variation across teleost lineages. The hAT and Tc1/Mariner superfamilies are the major DNA transposons of all four investigated teleosts. Divergence distribution revealed contrasting proliferation dynamics both between clades of retrotransposons and between species. The TEs within the larger genomes of the zebrafish and medaka represent relatively stronger activity with an extended time period during the evolution history, in contrast with the very young activity in the smaller stickleback genome, or the very low level of activity in the tetraodon genome.

Conclusion: Overall, our data shows that teleosts represent contrasting profiles of mobilomes with a differential density, diversity and activity of TEs. The differences in TE accumulation, dominated by DNA transposons, explain the main size variations of genomes across the investigated teleost species, and the species differences in both diversity and activity of TEs contributed to the variations of TE accumulations across the four teleost species. TEs play major roles in teleost genome evolution.

Keywords: Transposable elements, Teleosts, Genome size evolution, Activity, Diversity

Background

TEs are mobile genetic units and are a major constituent of a cell's "mobilome". They exhibit a broad range of diversity in their structure and transposition mechanisms, and are subdivided into two classes depending on their transposition mode: via RNA for class I retrotransposons and via DNA for class II transposons [1]. Class I retrotransposons include long terminal repeat retrotransposons (LTRs),

long interspersed nuclear elements (LINEs), and short interspersed nuclear elements (SINEs) [2]. Class II transposons can be divided into three major subclasses: cut-and-paste DNA transposons, rolling-circle DNA transposons (Helitrons), and self-synthesising DNA transposons (Polintons/Mavericks) [3]. Cut-and-paste transposons, which are very diverse, have been classified into superfamilies (hAT, Tc1/Mariner, etc.) based on the similarity of their transposases and on shared structural features, including the terminal inverted repeat (TIR) sequence and the length of the target site duplication (TSD) that flanks the TIR and is generated during integration [3]. Due to their unique

---
* Correspondence: hmcui@yzu.edu.cn; chengyilab@hotmail.com
Institute of Epigenetics & Epigenomics, College of Animal Science & Technology, Yangzhou University, Yangzhou, Jiangsu 225009, China

ability to transpose, and because they frequently amplify, TEs are major determinants of genome size [4, 5] and have been highly influential in shaping the structure and evolution of eukaryotic genomes. TEs constitute the largest component of mammalian genomes [6–8]; using the RepeatMasker approach [9, 10] it was predicted that approximately half of the human genome is covered by TEs, while recent annotation by the P-clouds pipeline suggests the TE coverage in human genome may be closer to two-thirds [11]. Most TEs of mammals are belong to class I retrotransposons, and the L1 family of LINEs is still active [6–8, 10].

Teleostean fish constitute the most diverse vertebrate group, and this diversity is also reflected in the diversity of their genome size and structure [12]. Although the available genome sequences for analysis (over 10 species) is minuscule in the huge species diversity of this clade, four representative teleost species, zebrafish (*Danio rerio*, Dr), medaka (*Oryzias latipes*, Ol), stickleback (*Gasterosteus aculeatus*, Ga), and tetraodon (*Tetraodon nigroviridis*, Tn), being of particular interest both experimentally and evolutionarily, have been sequenced as well [13–16]. Medaka, stickleback, and tetraodon belong to the superorder of *Acanthopterygii*, zebrafish belongs to the superorder of *Ostariophysi*; they all arose in the triassic period and are relatively close compared with the other class fishes [17]. However, genome sizes vary across these four teleost species by over four times. The zebrafish genome, with a size of approximately 137.17 Mb, is the largest, followed by the medaka with 869.00 Mb, then the stickleback with 461.53 Mb, while the tetraodon genome, with 358.62 Mb, is the smallest [12]. The variation of genome size between these close lineages remains largely unexplained. Transposable elements (TEs), as a major component of vertebrate genomes, may be a potential source for understanding the fish genome evolution. The initial annotations of four teleost (zebrafish, medaka, stickleback, and tetraodon) genomes have suggested that major differences in TE content exist between lineages [13–16]; and comparisons of TE diversity and evolution have revealed that teleost genomes contain the highest diversity of TE superfamilies in vertebrates [18], however, the TE contents in the early assembles of medaka, stickleback, and tetraodon tent to be underestimated and inaccurate due to the repeat database is far from complete; information on the distribution of TE diversity and density, and the evolution dynamics intra-species of teleosts, and the knowledge of the roles of TEs in teleost genome architecture and evolution is still reduced and fragmented. To better understand the different success rates of TEs and the evolution of genomes within teleosts, in this study we re-annotated the mobilomes of four representative teleost species (zebrafish, medaka, stickleback, and tetraodon) by using multiple de novo repeat prediction pipelines (RepeatModel, MGEScan-non-LTR, LTRharvest, RetroTector) with a combination of known repeat elements from the RepBase database; we identified diverse autonomous families of DNA transposons (hAT and Tc1 superfamilies) and retrotransposons, investigated the evolutionary pattern of TEs and the phylogenetic relationship among various TE clades and superfamilies, and highlighted the differences of TE activity, diversity and abundance within four teleost species. By integrating analyses of these four teleost species, we can perform a comprehensive analysis of mobilomes across the four species and make inferences about the causes of genome size variations within the four teleosts.

Results
Dramatically different expansion of TEs across the four teleost genomes
The joint annotation of teleost mobilomes with the species-specific custom TE libraries, which combined the previously-known elements from RepBase and the elements newly identified by multiple de novo methods as described in the Methods section, revealed a significantly different expansion of TEs within four teleost species (Table 1 and Fig. 1). The largest genome, that of the zebrafish, shows a dramatic accumulation of TEs, and the total interspersed repeats comprise over half of the sequenced genome (56.49 %/773.70 Mb). This is highest of the four investigated teleost species, followed by the medaka (33.70 %/236.28 Mb), and the stickleback (14.21 %/63.48 Mb). In the smallest genome, that of the tetraodon, which also represents the most compact genome described in vertebrates, the repeat content only represents 7.13 % (21.55 Mb) of the genome (Table 1 and Fig. 1a). The variation of genome size correlates with TE contents across the four teleost species (Fig. 1b). Our data clearly shows that differential accumulations of TEs contributed to the size variation of the four teleost genomes.

The greatest difference in TEs between the teleost species lies in the abundance of class II TEs (DNA transposons; Table 1 and Fig. 1a). This class of repeats has a striking amplification in the largest genome of zebrafish, where they contribute over 41.07 % (562.49 Mb) of the sequenced genome. In the second largest genome, the medaka, DNA repeats contribute 11.00 % (77.14 Mb) of the genome (Table 1). However, the proliferation of DNA transposons in the smaller genomes of the stickleback and tetraodon is weak, and this class of TEs only represents 4.47 % (19.96 Mb) and 1.55 % (4.68 Mb) of their sequenced genomes, respectively (Table 1). Retrotransposons (class I transposons), including SINE, LINE and LTR repeats, also display different expansions between teleost species. The overall contents for retrotransposons represent 12.00 % (164.29 Mb) of the zebrafish genome, which is substantially higher than that in the medaka (8.37 %/58.71 Mb), stickleback (6.61 %/

Table 1 TE coverage in teleost genomes[a]

	Zebrafish		Medaka		Stickleback		Tetraodon	
	Count	(%/Mb)	Count	(%/Mb)	Count	(%/Mb)	Count	(%/Mb)
Total retrotransposons	533112	12.00/164.29	215153	8.37/58.71	105276	6.61/29.50	35607	4.00/12.08
SINE	136879	2.24/30.64	30578	0.68/4.79	11523	0.67/2.97	1498	0.09/0.26
LINE	132888	3.85/52.78	112487	4.97/34.86	35604	2.60/11.61	19385	1.97/5.94
LTR	160149	5.90/80.87	72088	2.72/19.05	58159	3.34/14.92	14724	1.95/5.89
DNA	2368307	41.07/562.49	282359	11.00/77.14	73571	4.47/19.96	21901	1.55/4.68
Unclassified	228249	3.43/46.92	397468	14.32/100.42	72717	3.14/14.02	18465	1.58/4.78
Total interspersed repeats		56.49/773.70		33.70/236.28		14.21/63.48		7.13/21.55
Small RNA	13817	0.12/1.65	7223	0.16/1.09	2950	0.10/0.45	784	0.04/0.13
Satellite	75515	1.50/20.61	3046	0.16/1.13	1309	0.09/0.41	560	0.08/0.23
Simple repeats	42321	0.99/13.50	15283	0.29/2.03	8876	0.25/1.12	22873	0.74/2.25
Low complexity	1128	0.03/0.35	149	0.00/0.03	243	0.01/0.04	300	0.02/0.05

[a]The custom library combined with the repeats from RepBase (version 20150807) and de novo repeats was used for the all investigated teleost genomes

29.50 Mb), and tetraodon (4.00 %/12.08 Mb) genomes; the zebrafish represents the highest abundance of both LTR (5.90 %) and SINE (2.24 %) retrotransposons across teleost species; while the medaka shows the highest accumulation of LINEs at 4.97 % of the total sequenced genome (Table 1). Compared with other types of TEs, SINEs represent a relatively weak proliferation in most teleost species except zebrafish (Table 1). The proportion of satellites in the zebrafish genome (1.50 %) is higher than that observed in the medaka (0.16 %), stickleback (0.09 %), and tetraodon (0.08 %) genomes. The proportion of simple repeats in the zebrafish genome (0.99 %) is higher than that in the tetraodon (0.74 %), medaka (0.29 %) and stickleback (0.25 %) genomes (Table 1).

Dramatically different accumulation of DNA transposons across the four teleost genomes

A comparison of the diversity and abundance distributions of DNA TEs across the four teleost genomes revealed striking differences both between superfamilies and between species (Fig. 2, Table 2, and Additional file 1: Table S1). In total, 19 superfamilies of Class II transposons, representing all three main types of DNA transposons (cut-and-paste, rolling-circle, and self-synthesising) were detected in the four teleost genomes (Additional file 1: Table S1), and the results of abundance distribution of DNA repeats are summarized in Table 2. Among the three main types of DNA transposons, both rolling-circle (Helitron) and self-synthesising (Mavericks) DNA transposons were detected within the four teleost genomes with the absence of self-synthesising (Mavericks) DNA transposon in the medaka; these two superfamilies represent much lower abundance within these teleost species, with less than 0.2 % genome coverage, with the exception of Helitron, which has greater expansion in the zebrafish

genome and contributes 1.42 % (19.54 Mb) to the genomic sequence (Table 2). The diversity of repeat types of cut-and-paste DNA transposons observed at the level of the superfamily in the zebrafish (18), medaka (14) and stickleback (13) genomes is broadly similar, while the tetraodon genomes contains reduced superfamilies; several superfamilies of cut-and-paste DNA transposons, including Academ, Kolobok, MULE-MuDR, PIF-ISL2EU, and Sola, observed in the other three teleost species, are absent in the tetraodon genome (Table 2). At the family level, the larger genomes also appear to have many more non-redundant families in each superfamily than the small ones; totally, 1249, 234, 161, 74 non-redundant families were detected in zebrafish, medaka, stickleback, and tetraodon; Typically, over 16 times more non-redundant families were observed in the zebrafish than in the tetraodon (Additional file 1: Table S1). This indicates that the greater TE content in the lineages of large genome compared with the lineages of small ones is based not only on greater numbers of elements but also on greater element diversity at the more fine-scale family level.

These DNA transposons dominate the size variation in teleost genomes; the larger genomes accumulate many more DNA repeats than smaller ones. Typically, over 100 times more genome content (562.49 Mb) derived from DNA transposon amplification was identified in the zebrafish than in the tetraodon (4.68 Mb), and almost all types of DNA repeats appear to occur more frequently in the larger genomes than the smaller ones (Tables 1 and 2). Two dominant families of cut-and-paste DNA transposons in all four teleost species are hAT and Tc1/Mariner (Table 2). Four of the other cut-and-paste DNA superfamilies (CMC-EnSpm, PIF-Harbinger, Kolobok, and PiggyBac) have also amplified to significant numbers (over 1 %) in the zebrafish genome. In addition to hAT and Tc1/Mariner, the PIF-

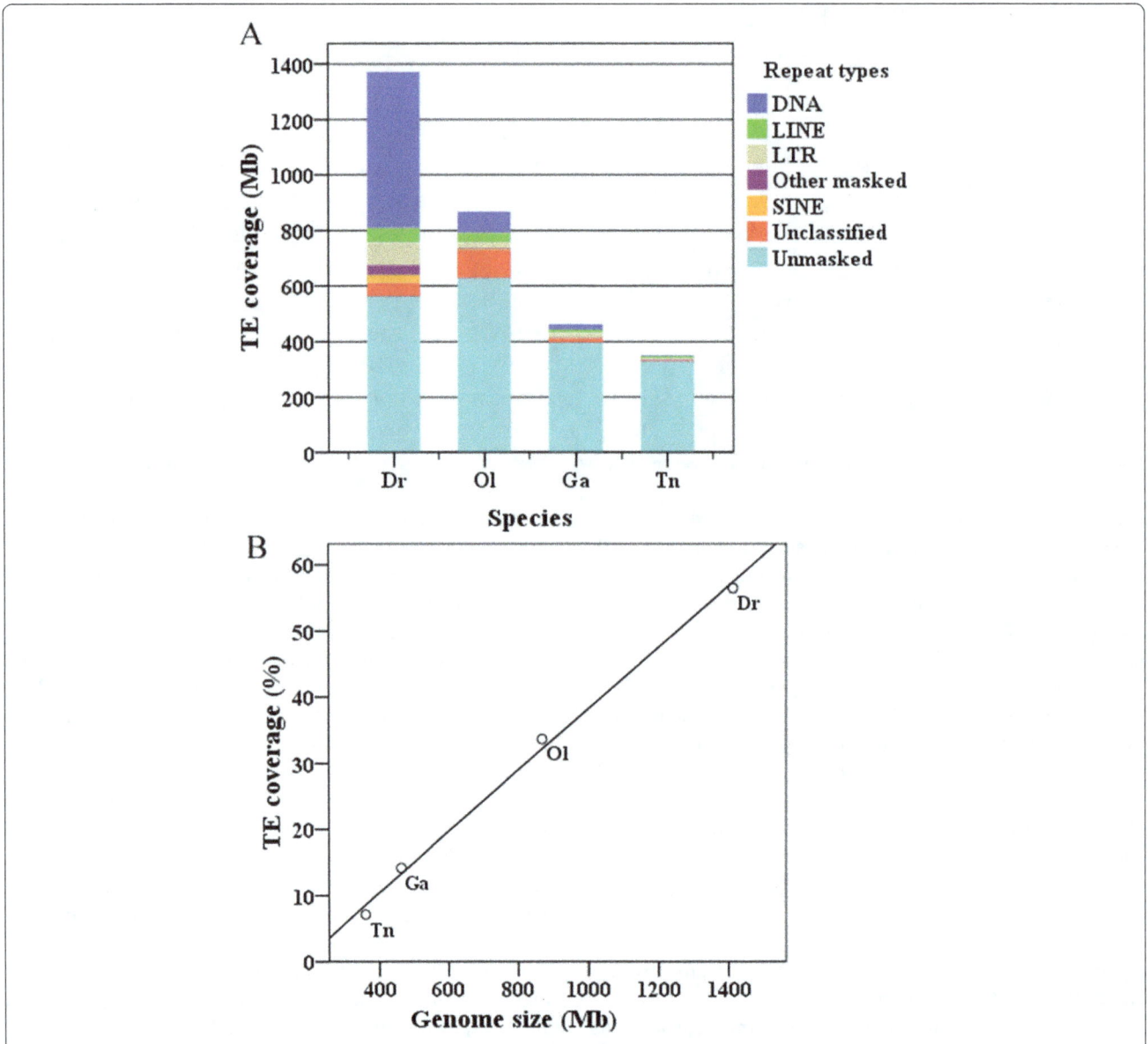

Fig. 1 Genome size and TE coverage in teleost genomes. **a** TE coverage in the four teleost species; (**b**) correlation analysis between the variations in genome size and TE content. Genome sizes were plotted against percentages of TE coverage for four teleost species. The *black line* represents the linear regression of the plot

Harbinger superfamily in the medaka genome has amplified to significant numbers as well, and comprised 1.34 % (11.66 Mb) of the genomic sequences. The other superfamilies did not show significant expansion (<1 %) in the four teleost genomes (Fig. 2 and Table 2).

The hAT is the most abundant and diverse DNA transposon superfamily, represented by multiple families in all four teleost genomes (Ac, Charlie, Tip100, Tol2, hobo etc.) (Additional file 1: Table S1), which contributes 11.73 % (160.97 Mb), 3.61 % (31.36 Mb), 2.11 % (9.73 Mb), and 0.75 % (2.71 Mb) to the zebrafish, medaka, stickleback, and tetraodon genomes, respectively (Fig. 2 and Table 2). Seven, 7, and 5 autonomous

subfamilies of hAT in medaka, stickleback, and tetradodon identified by TBLAST program (Additional file 1: Table S1), were combined with the eight autonomous subfamilies of hAT in zebrafish from RepBase to build the Phylogenetic tree. And the phylogenetic analysis of the hAT autonomous subfamilies with known reference elements revealed that these autonomous hAT subfamilies were classified into the Ac, Charlie, and Tip100 families, and majority of them belong to Ac and Charlie families, only one Tip100 family was detected in zebrafish, medaka, and stickleback, respectively (Fig. 3a). The Tc1/Mariner is the second most abundant DNA transposon superfamily in the teleost genomes, and contains

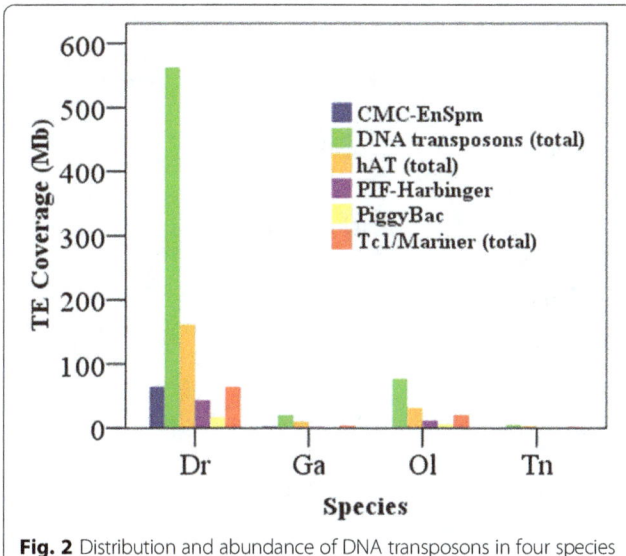

Fig. 2 Distribution and abundance of DNA transposons in four species of teleost

diverse families (Tc1, pogo, ISRm11, Stowaway etc.) (Additional file 1: Table S1), and comprises 4.68 % (64.16 Mb), 2.35 % (20.44 Mb), 0.78 % (3.58 Mb), and 0.39 % (1.40 Mb) of the zebrafish, medaka, stickleback, and tetraodon genomes, respectively (Fig. 2 and Table 2). Five, 14, 7, and 5 new autonomous subfamilies of Tc1/Mariner superfamily were extracted in zebrafish, medaka, stickleback, and tetradodon (Additional file 1: Table S1), respectively, and used for the phylogenetic analysis. Phylogenetic tree revealed all autonomous Tc1/Mariner transposons in teleosts belong to the Tc1 and pogo families, majority of them belong to Tc1 family, and few of them were classified as pogo family, no autonomous Mariner transposon was detected in all four teleosts (Fig. 3b).

Comparison of the sequence divergence distribution of DNA TEs revealed an extraordinary difference of proliferation dynamics across the four teleost genomes (Fig. 4). Overall, the DNA transposons within the largest genome of zebrafish have been active over a longest time period, and exhibited a strongest activity during the evolution history compared with other teleost, as shown by the broadest distribution of divergence ranging from 0 to 35 % and a very sharp peak of divergence at about 10 % (Fig. 4a). In contrast, the accumulation of DNA transposons in both of medaka and stickleback lineages is much weaker than that in zebrafish lineage, and tends to be very recent, with peaks of divergence less than 5 % and striking lacks of ancient proliferation (Fig. 4b and c). While the tetraodon lineage exhibits an extremely low level of activity of DNA TEs (Fig. 4d). In-deep analysis revealed that families in both the hAT and Tc1/Mariner superfamilies display dramatically differential accumulations during their evolutionary histories as well. The dominant families of hAT in teleost genomes are

Charlie and Ac; whereas the dominant families of Tc1/Mariner in teleost genomes are Tc1 and pogo (Table 2). Both Charlie and Ac families in the zebrafish and medaka genomes have undergone one round of substantial accumulation between the divergence of 5 and 15 %, followed by a decrease in recent activity (Fig. 4e and f), while the predominantly recent activities of Charlie and Ac families were observed in the stickleback genome in contrast with the very weak activity of these families in the tetraodon genome (Fig. 4g and h). Tc1 family in zebrafish has undergone one round of sharp burst at the divergence of 8 %, followed by recent decrease in activity and dominates the evolution of Tc1/Mariner superfamilies in this lineage; while both Tc1 and pogo families in the medaka genome have undergone two rounds of weak expansion in the evolution histories. Tc1 in stickleback has undergone one round of recent proliferation; the activities of pogo in stickleback, and Tc1 and pogo families in tetraodon are very low in the whole evolution histories (Fig. 4i, j, k and l). Both hAT and Tc1/Mariner superfamilies in some teleost genomes contain active families as shown by the distribution of many elements with <5 % divergence from the consensus (Fig. 4e-k).

Different distribution of LINE and LTR family diversity within the four teleost genomes

To characterize the family distribution of LINEs in the four teleost species, we applied the MGEScan-non-LTR program [19] to extract the LINE elements. In total, 1324, 436, 188, and 51 'ORF-preserving' LINEs were identified in the genomes of the zebrafish, medaka, stickleback, and tetraodon, respectively. The elements with a long ORF2 (>700aa) and intact RT domain were retained and designated as autonomous LINEs. These newly-identified LINEs were combined with the known autonomous LINEs (ORF2>700aa and intact RT domain) from RepBase, and classified into families based on amino acid sequence similarity (80 %) of ORF2 and the structure of ORFs (Additional file 2: Table S2). A dramatically different distribution of LINE families within species was found: the zebrafish, representing the most diverse lineage, contains 118 LINE families, while the medaka, stickleback, and tetraodon only contain 8, 11, and 2 LINE families, respectively (Table 3).

Phylogenetic analysis of these families revealed 6 clades of LINEs in the teleost species (L1, L2, I, Rex-Babar, RTE, and R2), and these clades differ drastically in family diversity among teleost lineages (Table 3 and Fig. 5). The L1 clade is very diverse in the family structure and was further classified into Swimmer, Tx1-a, and Tx1-b and Tx1-c branches, with each branch containing diverse families. The clades of L2, I, and Rex-Babar were less diverse in family structure compared with L1. The R2 and RTE clades had very little diversity in family

Table 2 Abundance of DNA transposons in teleost genomes

Type/superfamily/family	TE coverage (copy number/base pairs masked/%)			
	Zebrafish	Medaka	Stickleback	Tetraodon
Cut and paste TE				
Academ	987/87534/0.01	791/145168/0.02	334/48465/0.01	
CMC-Chapaev-3		42/21183/0.00		
CMC-EnSpm	382820/64375022/4.69	162/35330/0.00	5353/1984708/0.43	1524/119135/0.03
Crypton	36092/4088708/0.30	8275/1835625/0.21		
Dada	17226/2082682/0.15	549/121682/0.01	118/44503/0.01	936/109835/0.03
Ginger	6053/551259/0.04	236/55011/0.01		
IS3EU	17838/2815133/0.21			
Kolobok	128817/33082060/2.41	14/2093/0.00	583/74420/0.02	
Merlin	76084/8473015/0.62		710/437407/0.09	
MULE-MuDR	4276/2180335/0.16	206/46981/0.01	70/16079/0.00	
MULE-NOF	1159/272049/0.02			
P	7170/1900925/0.14			
PIF-Harbinger	110931/43436661/3.17	32217/11657322/1.34	3030/1513874/0.33	539/106002/0.03
PIF-ISL2EU	2262/959233/0.07	216/157939/0.02	365/213849/0.05	
PiggyBac	45457/17290031/1.26	34982/6161692/0.71	890/125960/0.03	359/119750/0.03
Sola	15251/3243347/0.24	1245/325083/0.04	419/220972/0.05	
Zisupton	6559/1093949/0.08			
Tc1/Mariner (total)	192070/64157597/4.68	65677/20436282/2.35	9784/3576109/0.78	1401857/0.39
Tc1	122464/47220045/3.44	37350/13303937/1.53	6236/2542455/0.55	1578/513677/0.14
pogo	5844/1778814/0.13	27376/6703781/0.77	2439/635035/0.14	2781/832008/0.23
Other families	4632/1895302/0.14	951/428564/0.05	1109/398619/0.09	200/56172/0.02
Unclassified Tc1/Mariner	59130/13263436/0.97			
hAT (total)	708715/160965573/11.73	116937/31365783/3.61	43229/9730722/2.11	12659/2713799/0.75
Ac	222707/57824458/4.22	29754/7436916/0.86	23961/3811137/0.83	1243/339847/0.09
Charlie	133312/26390661/1.92	72169/20423089/2.35	11098/4133497/0.90	7615/1794070/0.50
Other families	97917/19890417/1.45	7404/1871492/0.22	3074/511394/0.11	1211/362560/0.10
Unclassified hAT	254779/56860037/4.15	7610/1634286/0.19	5096/1274694/0.28	2590/217322/0.06
Self-synthesizing TE				
Maverick	11019/1934799/0.14		538/558181/0.12	1243/101677/0.03
Rolling-circle TE				
Helitron	105567/19541897/1.42	3043/304374/0.04	1564/204360/0.04	187/50445/0.01

structure, and only a few families were detected (Table 3 and Fig. 5).

The LTR elements, including ERVs, in the four teleost genomes were extracted using LTRharvest and Retro-Tector pipelines. The LTR elements with a long ORF (>500aa) and intact RT domain were retained and designated as autonomous LTRs. These LTRs were combined with the known autonomous LTRs (ORF >500aa and intact RT domain) from RepBase, and clustered into LTR families based on amino acid sequence similarity (80 %) (Additional file 3: Table S3). A striking difference in family distribution across species was found; the zebrafish lineage shows an extraordinary diversity of LTRs with 261 LTR families, while the medaka, stickleback, and tetraodon contain only 38, 77, and 8 LTR families, respectively (Table 4). Phylogenetic analysis of these families revealed 6 groups of LTRs (BEL/PAO, Copia, DIRS, Ngaro, Gypsy, and ERV) in teleost species, and these groups differ drastically in family diversity between teleost lineages (Figs. 6 and 7). The Gypsy group is incredibly diverse. In total we identified 7 clades of Gypsy within teleost species by RT phylogenetic analysis (Fig. 6a), six of which correspond to known clades (Gmr/Osvaldo, Barthez, CsRn1, V-calde, Mag, and Skipper),

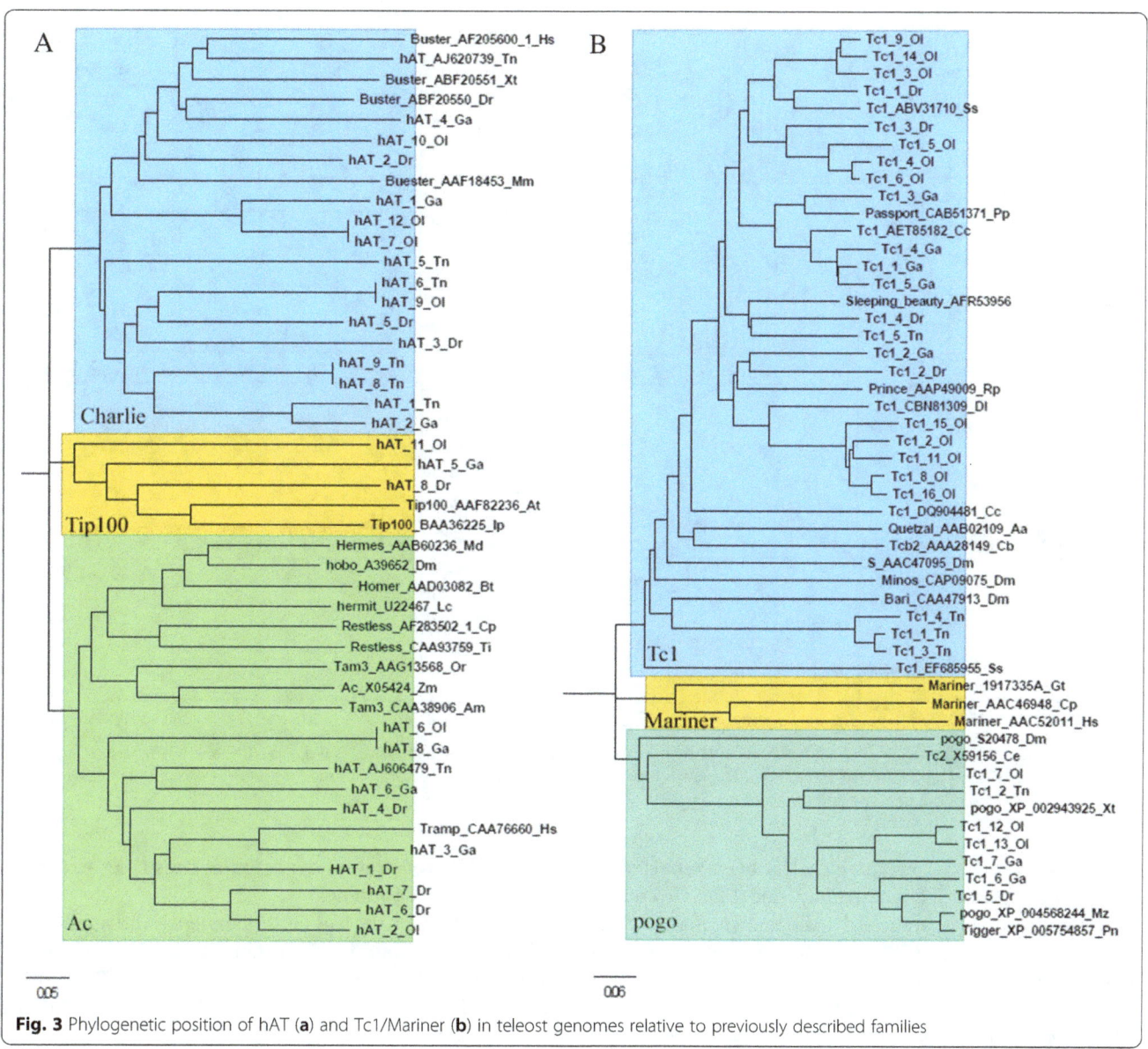

Fig. 3 Phylogenetic position of hAT (**a**) and Tc1/Mariner (**b**) in teleost genomes relative to previously described families

that have been reported in previous reports [20–26]. One new clade (named ReTe1) in teleost species was identified, which doesn't branch from any of the known reference elements; this clade is close to the Skipper and Barthez clades, but forms a distinct branch. ReTe1 clade distributes in zebrafish, medaka, and stickleback and contains diverse families, but it is absent in the tetraodon lineage (Fig. 6a). BEL/PAO is the second most abundant LTR group in teleost species and is represented by three distinct clades (Suzu, PAO, and Sinbad), but this group is absent in the tetraodon genome (Fig. 6b). The Suzu clade, which is homologous with the known reference elements [23], contains several families from the zebrafish, stickleback, and medaka; the PAO clade contains one family from the medaka genome and a number of families from the zebrafish, with a certain degree of structural similarity to the Zebel reference

element identified in previous study [23]; while the Sinbad clade is very diverse, with three distinct branches, and contains many families from the stickleback and zebrafish genomes (Fig. 6b), which have homology to the known Kobel reference identified previously [23]. The Copia, DIRS, and Ngaro groups show very little family structure compared with the BEL/PAO and Gypsy groups (Fig. 6b).

In total, 10, 1, and 16 ERV families were identified in the genomes of the zebrafish, medaka, and stickleback, respectively, and no ERV families were detected in the tetraodon genome (Table 4 and Fig. 7). These ERVs were classified into 2 clades (Eplison retrovirus and Spuma retrovirus) and belong to the Class I and Class III ERV groups by phylogenetic analysis. No ERVs of Class II was detected in teleost species (Fig. 7). The majority of teleost ERVs belong to the known clade of Eplison

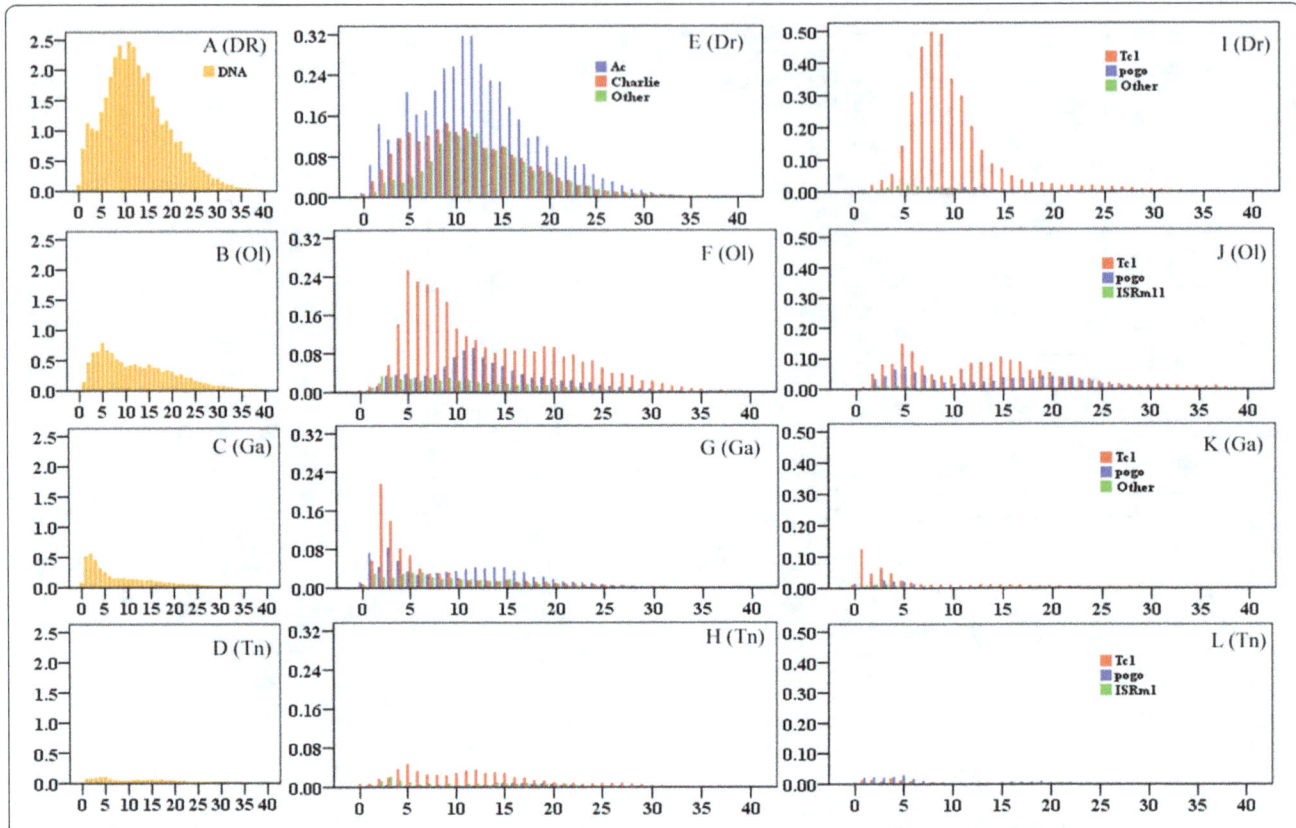

Fig. 4 Divergence distribution of DNA transposons (**a-d**), hAT (**e-h**), and Tc1/Mariner (**i-l**) superfamilies in the zebrafish (**a**, **e**, **i**), medaka (**b**, **f**, **j**), stickleback (**c**, **g**, **k**), and tetradon (**d**, **h**, **l**) genomes. The x-axis represents the substitution rate from consensus sequences (%), and the y-axis represents the percentage of the genome comprised of repeat classes (%)

retroviruses of Class I ERV, which has been reported in fishes and *Xenopus* [27, 28]. Only one ERV, from the zebrafish genome, is branched with known foamy virus proteins from mammals [29], and classified as the Spuma clade of Class III ERV (Fig. 7).

Table 3 Distribution of LINE families in teleost genomes[a]

Clade/Branch	Zebrafish	Medaka	Stickleback	Tetraodon
Total	118	8	11	2
I	9			
L1	82	6	3	2
Swimmer	45			
Tx1-a	14			
Tx1-b	15			2
Tx1-c	8			
L2	11	1	3	
R2	2	1	1	
Rex	9		2	
RTE	5		2	

[a]The newly identified LINEs by MGEScan-non-LTR programme were combined with known LINEs from RepBase, and the family was built up based on the similarity of amino acid sequence of LINE elements (80 %) and the structure of ORFs

Differential proliferation dynamics of class I TEs across the four teleost genomes

A comparison of the age and abundance distributions of TEs across the four teleost genomes revealed contrasting proliferation dynamics both between class I TEs (SINE, LINE, and LTR) and between species (Fig. 8 and Additional file 4: Table S4).

Generally, the retrotransposons within the larger genomes of the zebrafish and medaka have been active over an extended time period, in contrast with the predominantly recent activity in the smaller stickleback genome, or the extremely low level of activity in the tetraodon genome (Fig. 8). Both LTRs and LINEs in the zebrafish and stickleback genomes show evidence of very strong recent activity, in contrast to the recent decrease in activity for most types of retrotransposons in teleost species. Compared with other retrotransposons, SINEs present a very low level of activity in most teleost species, except for the zebrafish, where this repeat type has undergone one round of substantial accumulation between the divergence of 10 and 15 %, followed by a dramatic decrease in recent activity. Current activity is very limited, as shown by the distribution of very few repeats with <5 % divergence from the consensus (Fig. 8).

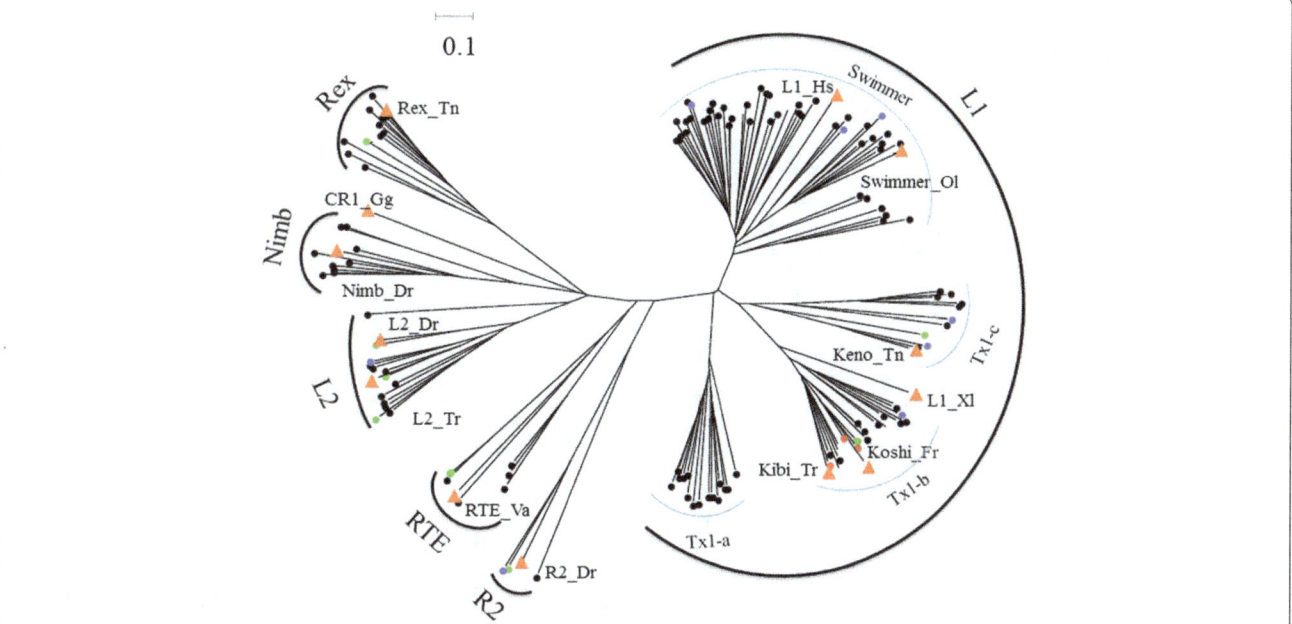

Fig. 5 Phylogenetic relationships among the 6 clades of LINEs in the teleost genomes. The nodes of sequences from zebrafish, medaka, stickleback, and tetraodon are shown as *black, blue, green,* and *red dots,* respectively; and the nodes of reference elements are indicated by *big yellow triangles.* The GenBank accession numbers used for phylogenetic analysis are as follows: Rex_Tn, AJ312227; CR1_Gg, U88211; Nimb_Dr, AL672145; L2_ Dr, AB211150; L2_Tr, AF086712; RTE_Va, AF332697; R2_Dr, AB097126; Kibi_Tr, AB097136; L1_Xl, M26915; Keno_Tn, AB111948; Swimmer_Ol, AF055640; L1_Hs, U93574)

Table 4 Distribution of LTR families in teleost genomes[a]

Group	Clade	Zebrafish	Medaka	Stickleback	Tentradon
Total		261	38	77	8
BEL		54	6	11	
	Suzu	2	1	2	
	Sinbad	32	4	9	
	PAO	20	1		
Copia		4	1	3	
DIRS		3			
ERV		10	1	16	
	Epsilonretrovirus	9	1	16	
	Spumaretrovirus	1			
Ngaro		5			
Gypsy		190	37	61	8
	Osvaldo/Gmr	61	3	22	1
	Barthez	60	11	10	1
	Skipper	4	1	1	1
	CsRN1		1	2	1
	V-clade	29	17	19	
	ReTe1	8	3	2	3
	Mag	28	1	5	1

[a]The newly identified LTRs by LTRHarvest and RetroTector programmes were combined with known LTRs from RepBase, and the family was built up based on the similarity of amino acid sequence of LTR elements (80 %)

The clades of L1, L2, RTE, and Rex-Babar are the major repeat types of LINE in teleost species and have experienced substantial expansion during their evolutionary histories, while the other clades did not get significant amplification (Additional file 4: Table S4). The predominant clade of LINEs in most teleost genomes is L2, which contributes 1.61, 1.57, and 1.20 % to the genomes of the zebrafish, medaka, and stickleback, respectively (Additional file 4: Table S4). An in-depth divergence analysis revealed that the L2 clade has been highly active over an extended time period and shows predominantly recent activity in these teleost species (Additional file 5: Figure S1A, B, and C). The second most abundant clade of LINEs in zebrafish is L1, which represents 1.24 % coverage of the genome, with highly recent activity (Additional file 4: Table S4 and Additional file 5: Figure S1). RTE in medaka and Rex-Babar in stickleback represent the second most abundant clade of LINEs, respectively, Rex-Babar is the major clade of LINE in the tetraodon lineage, whereas the activity of all other clades of LINE within this lineage is very limited (Additional file 4: Table S4 and Additional file 5: Figure S1). The substantial recent expansion of Rex-Babar within the stickleback and tetraodon genomes was in contrast with the weak accumulation of this clade in the lineages of the zebrafish and medaka (Additional file 4: Table S4 and Additional file 5: Figure S1).

The most abundant group of LTRs in all four teleost species is Gypsy, which comprises 2.42, 1.24, 1.85,

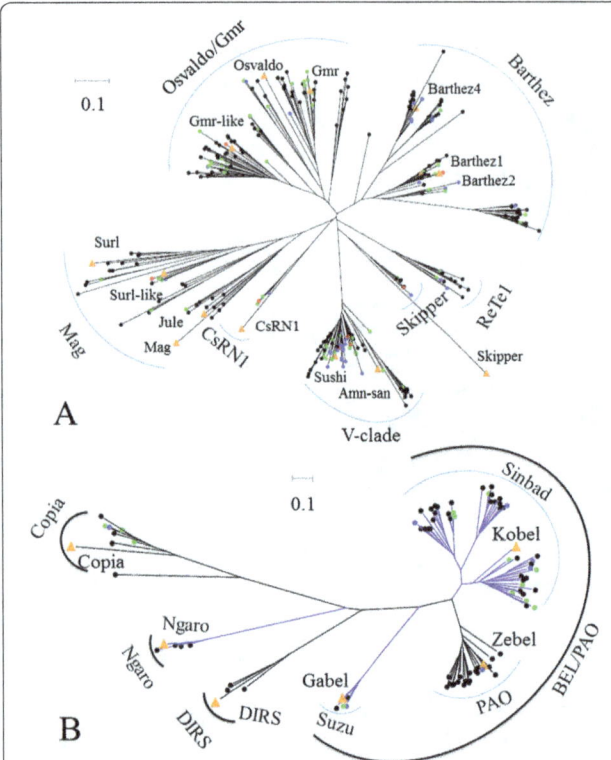

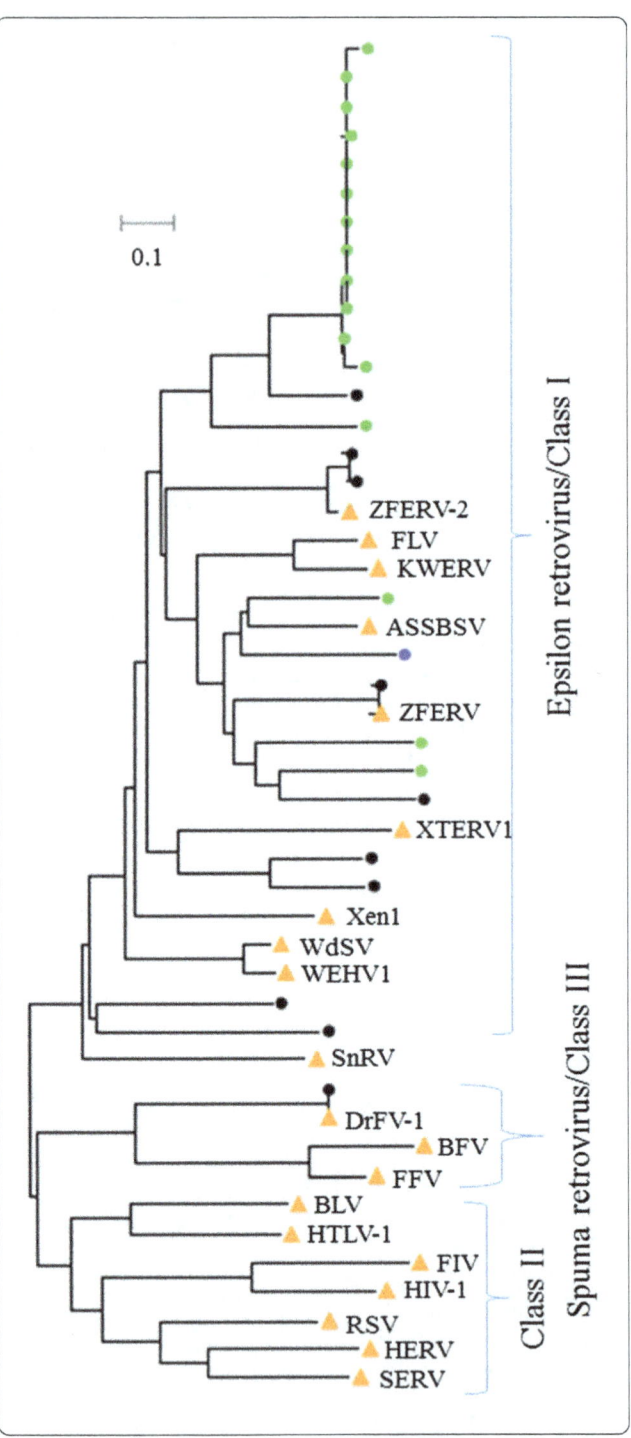

Fig. 6 RT phylogenetic tree of Gypsy (**a**) and other groups (**b**) of LTRs in the teleost genomes. The nodes of sequences from the zebrafish, medaka, stickleback, and tetraodon genomes are shown as *black, blue, green,* and *red dots*, respectively; and the nodes of reference elements are shown with *big yellow triangles*. The GenBank accession numbers used for phylogenetic analysis are as follows: Surl, M75723; Surl-like, AABS01002378; Jule, AY298856; Mag, X17219; CsRn1, AY013571; Sushi, AAC33526; Amn-san, 187466581; Skipper, AF049230; Barthez1, AJ621589; Barthez2, AJ621590; Barthez4, AJ621591; Gmr-like, AJ621595; Gmr, AF104899; Osvaldo, CAB39733; Copia, CAD27357; Ngaro, AAN71721; DIRS, AF442732; Kobel, 154426342; Zebel, 38304119; Gabel, 83921752

and 1.24 % of the zebrafish, medaka, stickleback, and tetraodon genomes, respectively. This group exhibits a distinct mode of evolution with a substantially recent accumulation within the zebrafish and stickleback genomes, in contrast with the relatively old proliferation dynamics within the medaka and tetraodon lineages (Additional file 4: Table S4 and Additional file 6: Figure S2). The DIRS group shows significant proliferation only in the zebrafish lineage (1.06 %) with predominantly recent activity, which is very rare within the other three teleost species. Substantial expansion of ERVs within the zebrafish (0.66 %) and stickleback (0.96) lineages was observed, which is relatively higher than in the medaka (0.08 %) and tetraodon (0.18 %) lineages; while apparent accumulations of Ngaro in the zebrafish (0.89 %) and medaka (0.65) lineages were observed, compared to an extremely low abundance in the stickleback (0.11 %) and tetraodon (0.12 %) lineages (Additional file 4: Table S4 and Additional file 6: Figure S2).

Discussion

TE proliferation and genomic expansion in teleosts

Using species-specific TE libraries, which combine the update RepBase database, and the de novo repeats

Fig. 7 The RT phylogenetic tree of ERVs in teleost genomes. The nodes of sequences from zebrafish, medaka, stickleback, and tetraodon genomes are shown as *green, yellow, blue,* and *red dots,* and the GenBank accession numbers used for phylogenetic analysis are as follows: ZFERV-2 (Zebrafish endogenous retrovirus 2), 162808041; FLV (Feline leukemia virus), NP_047255; KWERV (killer whale endogenous virus), GQ222416; ASSBSV (Atlantic salmon swim bladder sarcoma virus), ABA54982; ZFERV (Zebrafish endogenous retrovirus), AAM34208; XTERV1 (Xenopus tropicalis endogenous virus 1), HM765512; Xen1 (Xenopus laevis endogenous virus 1), AJ506107; WdSV (Walleye dermal sarcoma virus), AAC82611; WEHV1 (Walleye epidermal hyperplasia virus 1), AAD30048; SnRV (Snakehead fish retrovirus), AAC54861; DrFV-1 (Danio rerio Foamy Virus type 1), 85857417; BFV (Bovine foamy virus), NP_044929; FFV (Feline foamy virus), NP_056914; BLV (Bovine leukemia virus), AAC82587; HTLV-1 (Human T-lymphotropic virus 1), AAC82581; FIV, Feline immunodeficiency virus; HIV-1 (Human immunodeficiency virus 1), AAA43076; RSV (Rous sarcoma virus), BAD98246; HERV (human endogenous retrovirus K10), AAA88033; SERV (Simian endogenous retrovirus), AAC97565

extracted by multipiplines, we re-annotated the mobilomes of the four representative teleosts (zebrafish, medaka, stickleback, tetraodon). The estimated fraction of repeats within zebrafish in this study (56.49 %) is similar to the 52.2 % of the previous report [16], and substantially higher than that of most investigated vertebrates, including carp (31.3 %) [30], lizards (34.4 %) [31], western clawed frog (34.5 %) [32], and birds (7–9 %) [33, 34], but comparable to the 45–52 % density in some mammalian genomes [35]. However, the coverage of repeat contents in the genome of the medaka (33.70 %) by this study is much higher (about 16.2 %) than that in the early TE annotation of the medaka genome [15]. This disagreement may be due to a significant original underestimation, since the medaka repeat database is far from complete and dense repeat regions are underrepresented in the previous draft assembly. While the density of interspersed repeats in the tetraodon genome (7.13 %) is clearly higher than the 2.7 % observed in the its close relative, fugu [4], previous size estimations suggested that the tetraodon genome might be more compact than the genome of fugu [36]. The coverage of repeats within the stickleback genome (14.21 %) annotated in the current study is far below the 25.2 % of the previous estimate [14]; the cause of this discrepancy is unclear, since the annotation method in that report is unavailable.

In this study, we confirmed that teleosts are unique among vertebrates in their overall TE composition, which represents an extraordinarily different expansion of TEs (7.13–56.49 %) across four lineages that far exceeds the variation of TEs reported in extant mammals (36–52 %) [8, 35], salamanders (25–48 %) [37], or birds (7–9 %) [33, 34]. The relationship between genome size and TE coverage in different organisms has previously revealed a general positive trend [5, 18, 38, 39]; species with larger genomes have commensurately larger proportions of TE-derived DNA. Our findings confirmed this correlation within the four teleost lineages, and the total TE contents estimated for our four teleost species match very well with the predictions based on genome size, which were well illustrated by the smallest genome of the tetraodon (7.13 % comprised of TEs) and the largest genome of the zebrafish (56.49 % comprised of TEs). Furthermore, this study uncovered that the difference is largely due to the differential expansion of class II TEs (DNA transposons) across the four teleost species. These results suggest that the differential expansion of TEs, particularly DNA transposons, is a major molecular mechanism contributing to the size variation of genomes in the four teleost species. This is similar to that in western clawed frog as an amphibian [32], but contrasts with most mammals and reptiles, where the expansion of the genome is dominated by LTR or non-LTR retrotransposons [7, 8, 10, 31, 37].

Comparison of the diversity and activity of TEs between the four teleost genomes

In the current study, we found that teleost fish genomes represent extremely high diversity of TEs compared with the other vertebrate genomes, which is in agreement with the previous studies [18, 21, 22, 40]; furthermore, we performed a systematic comparative analysis of the intra-lineage diversity and activity of TEs across the four teleosts, and our data suggested that the differences in genome content among taxa are not limited to differences in a specific type of TE accumulation. The differences in both the diversity and activity of TEs contribute to the variances of TEs across teleost lineages. The diversity of TEs at the group level across teleost genomes is broadly similar, but the diversity at the clade (superfamily) and family level shows significant differences, and the smaller genomes have reduced clade (superfamily) and family diversity compared with the larger genomes, which has also been observed in snake lineages [41]. On the other hand, species differences in TE activity may result in changes in TE accumulation as well. In the current study, we found that zebrafish, with a fairly high TE content, represents a long-lasting and higher level of TE activity in its evolutionary history compared with the other three teleost lineages, and many DNA, LTR and LINE families show evidence of recent and ongoing proliferation, while most types of these transposons in the medaka, stickleback, and tetraodon genomes represent either a relatively young expansion and/or a rapid decrease in activity, or extremely low activity during their evolutionary history. Uncovering the reasons of the variation of diversity and activity across these teleost species is a very difficult task, particularly because TEs can also be introduced through horizontal transfer into lineages. The fertilization way,

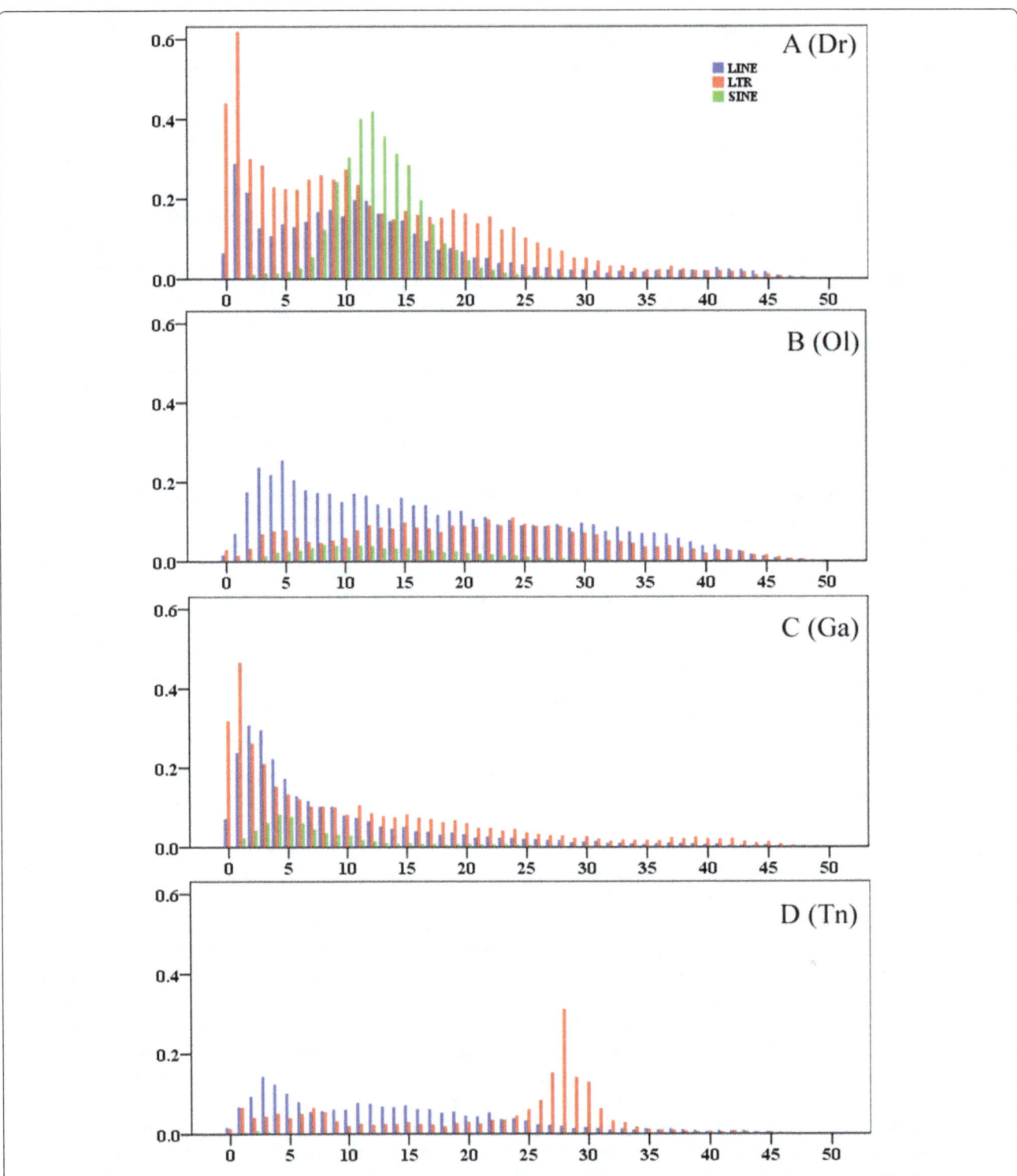

Fig. 8 Divergence distribution of retrotransposon types (LINE, LTR, and SINE) in the zebrafish (**a**), medaka (**b**), stickleback (**c**), and tetradon (**d**) genomes. The x-axis represents the substitution rate from consensus sequences (%), and the y-axis represents the percentage of the genome comprised of repeat classes (%)

body temperature, and host defense mechanisms in opposition to TE activity (or family competition) have been suggested as biological features that may shape susceptibility to TEs in vertebrates [42, 43]. Internal fertilization may minimize exposure of gametes (and embryos) to horizontal transfer of TEs compared with external fertilization, however the four teleost lineages share the same fertilization way, and the body temperature of the four investigated teleosts, varying with the temperature of their surroundings, may also not be the principal determinant. Thus the family competition, the capacity to replicate and compete with other TEs, which is determined by the host defense mechanisms and TE itself, may be the major determinant of TE differences across the four teleost species. At least two host controlling mechanisms of the family competition of TEs: (i) cosuppression usually mediated by small interfering RNA (siRNA) and (ii) methylation, have been proved in *C. elegans* [44] and mice [45], may play roles in the evolution of diversity and activity of TEs in teleost as well. However, tests of these hypotheses and critical reevaluation will be required for further deep understanding of the regulation, mobility, and rates of expansion and extinction of TEs in teleosts.

Evolutionary dynamics of TEs in teleost genomes compared with other vertebrates

Evolutionary dynamics of TEs between vertebrates differ drastically. The genomes of mammals and birds contain few types of TE lineages which are very abundant but relatively inactive [7, 10, 33, 34]. However, our study distinctly shows that the level of class I and class II transposon diversity and activity in teleost genomes is much higher than that seen in either bird or mammalian genomes [16, 39, 46, 47], is similar to that observed in coelacanths [48] and cod [49], and comparable with the prevalence in lizards and western clawed frog [31, 32]. Recently active TEs (with a divergence of less than 5 %) are more common in teleost genomes than in mammals or birds [8, 10, 33, 34].

The estimated fractions of LINEs in teleost genomes (1.97–4.97 %) are substantially lower than in lizards (12.34 %) and mammals (about 20 %) [6, 8, 10, 31], and comparable to that of birds (6 %), coelacanths (6.43 %), cod (3.3 %), and western clawed frog (5.4 %) [32–34, 48, 49]. However, LINEs within teleost genomes represent extremely high diversity with 6 groups. The L1 clade of LINEs contains numerous families and shows signs of recent activity. Some clades of LINEs were observed in teleost genomes, but were absent from western clawed frog, lizards, chickens and humans [10, 31, 32, 34]. Many LINE clades and families within teleost genomes seem to be recent insertions, based on their divergence analysis; this is similar to the proliferation dynamics of LINEs in lizards and western clawed frog [31, 32]. Among these is an unusually high diversity of very young families of L1 retrotransposons in the zebrafish genome, which represents the most diverse group of LINEs, containing four branches (Swimmer, Tx1-a, Tx1-b, and Tx1-c). Each branch yields highly prolific families, yet this group only covers 1.24 % of the zebrafish

genome. This contrasts with observations of both mammalian and bird genomes, where only a single active family of L1 of LINEs has predominated over 10 Mya, with about a 20 % coverage of genome. In birds the most predominant TE elements are CR1 LINEs (about 6 % of the genome) and these have been demonstrated to be degenerated and nonfunctional [7, 10, 34].

Compared to lizards, western clawed frog, mammals, and birds [7, 10, 31, 32, 34], LTR retrotransposons are also very diverse and active in teleost genomes. Representatives of the seven major groups of LTR elements, including endogenous retroviruses (BEL/PAO, Copia, DIRS, ERV, Gypsy, Ngaro), with diverse clades and numerous families were identified. In particular, an unexpectedly high diversity of Gypsy (7 clades) and BEL/PAO (3 clades) were found in teleost genomes, and each clade contains diverse active families. While the Ngaro group is absent in western clawed frog and lizards [31, 32], only ERV may still be active in birds and mammals, and all other LTR groups (BEL/PAO, Copia, DIRS, Gypsy, and Ngaro) are absent or only present as fossils [7, 9, 33, 34]. This high diversity of LTR retrotransposons was already noted within teleost genomes in previous analysis [14, 40]. The estimated fractions of LTRs within the lineages vary from 1.95 % of the tetraodon genome to 5.90 % of the zebrafish, which are substantially higher than in coelacanths (0.86 %), and comparable to that in cod (4.88 %) and western clawed frog (1.75 %) [32, 48, 49].

Teleosts are unique among vertebrates in their proliferation dynamics of DNA transposons; DNA transposons vary dramatically in abundance across teleost species, dominate the variations in genome size, and also represent the highest level of diversity among vertebrates. The coverage of DNA transposons varies across teleost genomes, from 1.55 % in the tetraodon genome to 41.07 % in the zebrafish. The zebrafish genome contains a marked excess of DNA transposons, which is unique among sequenced vertebrate genomes, and is substantially higher than in very close lineages of carp (17.53 %). Indeed, only western clawed frog genome, which is comprised of 25 % DNA transposons, are comparable. The estimated fractions of DNA transposons in the medaka (11.00 %) and stickleback (4.47 %) genomes are substantially higher than in coelacanths (0.20 %) [48], lungfish (1.3 %) [50], birds (less than 1 %) [33, 34] and mammals (less than 3 %) [7, 10], but comparable to that in lizards (8.86 %) [31], salamanders (6.37 %) [37], and cod (6.39 %) [49].

The diversity of teleost DNA transposons, which was already noted previously [18, 30], far exceeds that in other examined vertebrates, including mammals, birds, coelacanths, cod, lizards, and western clawed frog [31, 32, 34, 46, 48]. A particularly high abundance and diversity of hAT and Tc1/Mariner was found in teleost genomes. Nine superfamilies of DNA transposons, including Ginger, Sola, CMC-EnSpm, Crypton, Dada, MULE-MuDR, P, PIF-ISL2EU, and Academ, were observed in teleosts that were absent in lizards, western clawed frog, and coelacanths [31, 32, 48]. In addition, diverse autonomous hAT and Tc1/Mariner subfamilies were identified in teleost genomes, suggesting that the DNA transposons seem to be relatively young and active in teleosts, in contrast to the few recently active DNA transposons found in mammals and birds [7, 10, 33, 34]. Overall, teleosts have an extremely wide diversity and high level of activity of TEs, but represent a significantly different success of TEs across lineages, while mammalian genomes are enriched with L1 elements but a low level of diversity and have a high degree of TE expansion, and bird genomes exhibit low TE density with very little mobile element activity.

Conclusion

In this study, we investigated the diversity, activity, and abundance distribution of TEs among four closely related teleost species. In contrast to other vertebrates, teleosts display contrasting profiles of mobilomes across the four investigated lineages. The larger genomes represent a higher diversity and activity within each family and a greater abundance of TEs compared with the smaller genomes. The differences in TE expansion, dominated by DNA transposons, explain the main size variation in the four teleost genomes, and the species differences in both the diversity and activity of TEs contribute to the variations in TE accumulations. TEs play pivotal roles in teleost genome evolution.

Methods
Computational identification of interspersed repeats
The zebrafish (GRCz10), medaka (MEDAKA1), stickleback (BROADS1), and tetraodon (TETRAODON8) genomes were downloaded from the Ensembl database (http://asia.ensembl.org/index.html). The repeat contents of the zebrafish, medaka, stickleback and tetraodon genomes were assessed using RepeatMasker (http://www.repeatmasker.org/), RepeatModeler (http://repeatmasker.org/RepeatModeler.html) and ab initio repeat prediction programmes. The RepBase (http://www.girinst.org/) of consensus repeat sequences [51] was used to identify repeats in the genome derived from known classes of elements. RepeatModeler was used to build de novo repeats. The autonomous hAT and Tc1/Mariner DNA transposons were queried using TBLASTN to detect the presence of coding sequences related to all known DNA transposon superfamilies in RepBase [51]. The top 10–40 non-overlapping hits (generally Evalue $<10^{-5}$) were extracted, along with 500 bp of flanking sequence, aligned using a local installation of MUSCLE [52], and used to construct consensus sequences.

For each consensus, coding sequences were predicted by using Open Reading Frame (ORF) Finder (http://www.ncbi.nlm.nih.gov/projects/gorf/). The non-LTR retrotransposons were identified by MGEScan-non-LTR [19], and the LTR retrotransposons, including endogenous retroviruses (ERVs), were identified by LTRharvest [47] and RetroTector [53]. The autonomous LTRs were classified into families based on amino acid sequence similarity (80 %) of the ORF containing RT domain; while the autonomous LINEs were classified into families based on the structure of ORFs and amino acid sequence similarity (80 %) of the ORF2.

Repeats characterized as putative TEs by the previous approach were joined to the RepBase database of TEs (update 20150807), and the redundancies were filtered out to create a custom library for comparison to find the distribution and coverage of TEs in the genome using RepeatMasker (RepeatMasker -open-4.0.5). The redundant repeats were removed based on the 80-80 rule, which considers two sequences as belonging to same TE family if they can be aligned over more than 80 % of their length, with over 80 % identity. The new non-redundant repeats of the four teleost species were given in fasta file format in Additional files (Additional files 7, 8, 9 and 10).

Phylogenetic analysis
Bootstrapped (1000 replicates) neighbour-joining (NJ) phylogenetic trees were generated using MEGA5 [54] based on a muscle multiple protein alignment with the conserved domain of the DNA transposases or RT (reverse transcription) domain of retrotransposons. For the hAT superfamily, we used a conserved 39 aa-long region of hAT transposase [55] to build the alignment, and then deduced the NJ tree. For the Tc1/Mariner superfamily, the NJ tree was generated by using a multiple sequence alignment with the most conserved domain of the Tc1/Mariner transposase (about 150 aa) corresponding to the catalytic "DDE" domain, as in [56]. For retrotransposons (LINEs, LTRs and ERVs), the NJ tree was generated by using an amino acid multiple alignment of the conserved RT domain from retrotransposons and reference elements. All these alignment are available upon request.

Divergence distribution of interspersed repeats
The average number of substitutions per site (K) for each fragment was estimated according to the divergence levels reported by RepeatMasker, using the one-parameter Jukes-Cantor formula $K = -300/4 \times Ln(1 - D \times 4/300)$ as in [7], where D represents the proportion of sites that differ between the fragmented repeat and the consensus sequence.

Additional files

Additional file 1: Table S1. Numbers of families of DNA repeats from RepBase and the de novo repeats identified by TBLASTN. (PDF 91 kb)

Additional file 2: Table S2. Characteristics of Non-LTR retrotransposons present in teleost genomes, separated by family. (PDF 158 kb)

Additional file 3: Table S3. Characteristics of LTR retrotransposons present in teleost genomes, separated by family. (PDF 371 kb)

Additional file 4: Table S4. Abundance of retrotransposons in teleost genomes. (PDF 181 kb)

Additional file 5: Figure S1. Divergence distribution of the major clades of LINEs in the zebrafish (A), medaka (B), stickleback (C), and tetraodon (D) genomes. The x-axis represents the substitution rate from consensus sequences (%), and the y-axis represents the percentage of the genome comprised of repeat classes (%). (JPG 285 kb)

Additional file 6: Figure S2. Divergence distribution of the major groups of LTRs in the zebrafish (A), medaka (B), stickleback (C), and tetraodon (D) genomes. The x-axis represents the substitution rate from consensus sequences (%), and the y-axis represents the percentage of the genome comprised of repeat classes (%). (JPG 312 kb)

Additional file 7: TE sequence 1: new repeats identified in zebrafish. (FAS 390 kb)

Additional file 8: TE sequence 2: new repeats identified in medaka. (FAS 376 kb)

Additional file 9: TE sequence 3: new repeats identified in stickleback. (FAS 357 kb)

Additional file 10: TE sequence 4: new repeats identified in tetraodon. (FAS 97 kb)

Abbreviations
Dr: danio rerio; ERV: endogenous retroviruse; Ga: gasterosteus aculeatus; LINE: long interspersed nuclear elements; LTR: long terminal repeats; Mb: mega base pairs; NJ: neighbour-joining; ORF: open reading frame; Ol: oryzias latipes; RT: reverse transcription; SINE: short interspersed nuclear elements; TSD: target site duplication; TIR: terminal inverted repeat; Tn: Tetraodon nigroviridis; TEs: Transposable elements.

Competing interests
The authors declare that they have no competing interests.

Authors' contributions
CS and HC conceived the study, BG participated in its design. CC, SX, DS performed the analyses. BG, CS, and HC wrote the manuscript. All authors read and approved the final manuscript.

Acknowledgements
This work was funded by the Natural Science Foundation of China (NSFC) (31200920), by the National Major Transgenic Project of China (2014ZX08006005-008), and by the Priority Academic Program Development of Jiangsu Higher Education Institutions.

References
1. Finnegan DJ. Eukaryotic transposable elements and genome evolution. Trends Genet. 1989;5(4):103–7.
2. Xiong Y, Eickbush TH. Origin and evolution of retroelements based upon their reverse transcriptase sequences. EMBO J. 1990;9(10):3353–62.
3. Kapitonov VV, Jurka J. A universal classification of eukaryotic transposable elements implemented in Repbase. Nat Rev Genet. 2008;9(5):411–2. author reply 414.
4. Petrov DA. Evolution of genome size: new approaches to an old problem. Trends Genet. 2001;17(1):23–8.
5. Piegu B, Guyot R, Picault N, Roulin A, Sanyal A, Kim H, et al. Doubling genome size without polyploidization: dynamics of retrotransposition-driven genomic expansions in Oryza australiensis, a wild relative of rice. Genome Res. 2006;16(10):1262–9.
6. Groenen MA, Archibald AL, Uenishi H, Tuggle CK, Takeuchi Y, Rothschild MF, et al. Analyses of pig genomes provide insight into porcine demography and evolution. Nature. 2012;491(7424):393–8.
7. Waterston RH, Lindblad-Toh K, Birney E, Rogers J, Abril JF, Agarwal P, et al. Initial sequencing and comparative analysis of the mouse genome. Nature. 2002;420(6915):520–62.
8. Li R, Fan W, Tian G, Zhu H, He L, Cai J, et al. The sequence and de novo assembly of the giant panda genome. Nature. 2010;463(7279):311–7.
9. Smit AF. The origin of interspersed repeats in the human genome. Curr Opin Genet Dev. 1996;6(6):743–8.
10. Lander ES, Linton LM, Birren B, Nusbaum C, Zody MC, Baldwin J, et al. Initial sequencing and analysis of the human genome. Nature. 2001; 409(6822):860–921.
11. de Koning AP, Gu W, Castoe TA, Batzer MA, Pollock DD. Repetitive elements may comprise over two-thirds of the human genome. PLoS Genet. 2011; 7(12), e1002384.
12. Volff JN. Genome evolution and biodiversity in teleost fish. Heredity (Edinb). 2005;94(3):280–94.
13. Jaillon O, Aury JM, Brunet F, Petit JL, Stange-Thomann N, Mauceli E, et al. Genome duplication in the teleost fish Tetraodon nigroviridis reveals the early vertebrate proto-karyotype. Nature. 2004;431(7011):946–57.
14. Jones FC, Grabherr MG, Chan YF, Russell P, Mauceli E, Johnson J, et al. The genomic basis of adaptive evolution in threespine sticklebacks. Nature. 2012;484(7392):55–61.
15. Kasahara M, Naruse K, Sasaki S, Nakatani Y, Qu W, Ahsan B, et al. The medaka draft genome and insights into vertebrate genome evolution. Nature. 2007;447(7145):714–9.
16. Howe K, Clark MD, Torroja CF, Torrance J, Berthelot C, Muffato M, et al. The zebrafish reference genome sequence and its relationship to the human genome. Nature. 2013;496(7446):498–503.
17. Hedges SB, Marin J, Suleski M, Paymer M, Kumar S. Tree of life reveals clock-like speciation and diversification. Mol Biol Evol. 2015;32(4):835–45.
18. Chalopin D, Naville M, Plard F, Galiana D, Volff JN. Comparative analysis of transposable elements highlights mobilome diversity and evolution in vertebrates. Genome Biol Evol. 2015;7(2):567–80.
19. Rho M, Tang H. MGEScan-non-LTR: computational identification and classification of autonomous non-LTR retrotransposons in eukaryotic genomes. Nucleic Acids Res. 2009;37(21), e143.
20. Volff JN, Korting C, Altschmied J, Duschl J, Sweeney K, Wichert K, et al. Jule from the fish Xiphophorus is the first complete vertebrate Ty3/Gypsy retrotransposon from the Mag family. Mol Biol Evol. 2001;18(2):101–11.
21. Volff JN, Bouneau L, Ozouf-Costaz C, Fischer C. Diversity of retrotransposable elements in compact pufferfish genomes. Trends Genet. 2003;19(12):674–8.
22. Fischer C, Bouneau L, Coutanceau JP, Weissenbach J, Ozouf-Costaz C, Volff JN. Diversity and clustered distribution of retrotransposable elements in the compact genome of the pufferfish Tetraodon nigroviridis. Cytogenet Genome Res. 2005;110(1–4):522–36.
23. Llorens C, Munoz-Pomer A, Bernad L, Botella H, Moya A. Network dynamics of eukaryotic LTR retroelements beyond phylogenetic trees. Biol Direct. 2009;4:41.
24. Goodwin TJ, Poulter RT. A group of deuterostome Ty3/ gypsy-like retrotransposons with Ty1/ copia-like pol-domain orders. Mol Genet Genomics. 2002;267(4):481–91.
25. Bae YA, Moon SY, Kong Y, Cho SY, Rhyu MG. CsRn1, a novel active retrotransposon in a parasitic trematode, Clonorchis sinensis, discloses a new phylogenetic clade of Ty3/gypsy-like LTR retrotransposons. Mol Biol Evol. 2001;18(8):1474–83.
26. Marin I, Llorens C. Ty3/Gypsy retrotransposons: description of new Arabidopsis thaliana elements and evolutionary perspectives derived from comparative genomic data. Mol Biol Evol. 2000;17(7):1040–9.
27. Martineau D, Bowser PR, Renshaw RR, Casey JW. Molecular characterization of a unique retrovirus associated with a fish tumor. J Virol. 1992;66(1):596–9.
28. Sinzelle L, Carradec Q, Paillard E, Bronchain OJ, Pollet N. Characterization of a Xenopus tropicalis endogenous retrovirus with developmental and stress-dependent expression. J Virol. 2011;85(5):2167–79.
29. Jern P, Sperber GO, Blomberg J. Use of endogenous retroviral sequences (ERVs) and structural markers for retroviral phylogenetic inference and taxonomy. Retrovirology. 2005;2:50.

30. Xu P, Zhang X, Wang X, Li J, Liu G, Kuang Y, et al. Genome sequence and genetic diversity of the common carp. Cyprinus Carpio Nat Genet. 2014; 46(11):1212–9.

31. Alfoldi J, Di Palma F, Grabherr M, Williams C, Kong L, Mauceli E, et al. The genome of the green anole lizard and a comparative analysis with birds and mammals. Nature. 2011;477(7366):587–91.

32. Hellsten U, Harland RM, Gilchrist MJ, Hendrix D, Jurka J, Kapitonov V, et al. The genome of the Western clawed frog Xenopus tropicalis. Science. 2010; 328(5978):633–6.

33. International Chicken Genome Sequencing Consortium. Sequence and comparative analysis of the chicken genome provide unique perspectives on vertebrate evolution. Nature. 2004;432(7018):695–716.

34. Warren WC, Clayton DF, Ellegren H, Arnold AP, Hillier LW, Kunstner A, et al. The genome of a songbird. Nature. 2010;464(7289):757–62.

35. Renfree MB, Papenfuss AT, Deakin JE, Lindsay J, Heider T, Belov K, et al. Genome sequence of an Australian kangaroo, Macropus eugenii, provides insight into the evolution of mammalian reproduction and development. Genome Biol. 2011;12(8):R81.

36. Neafsey DE, Palumbi SR. Genome size evolution in pufferfish: a comparative analysis of diodontid and tetraodontid pufferfish genomes. Genome Res. 2003;13(5):821–30.

37. Sun C, Shepard DB, Chong RA, Lopez Arriaza J, Hall K, Castoe TA, et al. LTR retrotransposons contribute to genomic gigantism in plethodontid salamanders. Genome Biol Evol. 2012;4(2):168–83.

38. Vitte C, Panaud O. LTR retrotransposons and flowering plant genome size: emergence of the increase/decrease model. Cytogenet Genome Res. 2005; 110(1–4):91–107.

39. Hawkins JS, Kim H, Nason JD, Wing RA, Wendel JF. Differential lineage-specific amplification of transposable elements is responsible for genome size variation in Gossypium. Genome Res. 2006;16(10):1252–61.

40. Basta HA, Buzak AJ, McClure MA. Identification of novel retroid agents in Danio rerio, Oryzias latipes, Gasterosteus aculeatus and Tetraodon nigroviridis. Evol Bioinform Online. 2007;3:179–95.

41. Castoe TA, Hall KT, Mboulas MLG, Gu WJ, de Koning APJ, Fox SE, et al. Discovery of highly divergent repeat landscapes in snake genomes using high-throughput sequencing. Genome Biol Evol. 2011;3:641–53.

42. Abrusan G, Krambeck HJ. Competition may determine the diversity of transposable elements. Theor Popul Biol. 2006;70(3):364–75.

43. Huang CR, Burns KH, Boeke JD. Active transposition in genomes. Annu Rev Genet. 2012;46:651–75.

44. Sijen T, Plasterk RH. Transposon silencing in the Caenorhabditis elegans germ line by natural RNAi. Nature. 2003;426(6964):310–4.

45. Walsh CP, Chaillet JR, Bestor TH. Transcription of IAP endogenous retroviruses is constrained by cytosine methylation. Nat Genet. 1998; 20(2):116–7.

46. Yuan YW, Wessler SR. The catalytic domain of all eukaryotic cut-and-paste transposase superfamilies. Proc Natl Acad Sci U S A. 2011;108(19):7884–9.

47. Ellinghaus D, Kurtz S, Willhoeft U. LTRharvest, an efficient and flexible software for de novo detection of LTR retrotransposons. BMC Bioinformatics. 2008;9:18.

48. Amemiya CT, Alfoldi J, Lee AP, Fan S, Philippe H, Maccallum I, et al. The African coelacanth genome provides insights into tetrapod evolution. Nature. 2013;496(7445):311–6.

49. Star B, Nederbragt AJ, Jentoft S, Grimholt U, Malmstrom M, Gregers TF, et al. The genome sequence of Atlantic cod reveals a unique immune system. Nature. 2011;477(7363):207–10.

50. Metcalfe CJ, Filee J, Germon I, Joss J, Casane D. Evolution of the Australian lungfish (Neoceratodus forsteri) genome: a major role for CR1 and L2 LINE elements. Mol Biol Evol. 2012;29(11):3529–39.

51. Jurka J, Kapitonov VV, Pavlicek A, Klonowski P, Kohany O, Walichiewicz J. Repbase Update, a database of eukaryotic repetitive elements. Cytogenet Genome Res. 2005;110(1–4):462–7.

52. Edgar RC. MUSCLE: multiple sequence alignment with high accuracy and high throughput. Nucleic Acids Res. 2004;32(5):1792–7.

53. Sperber GO, Airola T, Jern P, Blomberg J. Automated recognition of retroviral sequences in genomic data–RetroTector. Nucleic Acids Res. 2007; 35(15):4964–76.

54. Tamura K, Peterson D, Peterson N, Stecher G, Nei M, Kumar S. MEGA5: Molecular Evolutionary Genetics Analysis using maximum likelihood, evolutionary distance, and maximum parsimony methods. Mol Biol Evol. 2011;28(10):2731–9.

55. Kempken F, Windhofer F. The hAT family: a versatile transposon group common to plants, fungi, animals, and man. Chromosoma. 2001;110(1):1–9.

56. Pritham EJ, Feschotte C, Wessler SR. Unexpected diversity and differential success of DNA transposons in four species of entamoeba protozoans. Mol Biol Evol. 2005;22(9):1751–63.

5

Genomic analysis of mouse VL30 retrotransposons

Georgios Markopoulos[1,2], Dimitrios Noutsopoulos[3], Stefania Mantziou[1], Demetrios Gerogiannis[4], Soteroula Thrasyvoulou[1], Georgios Vartholomatos[5], Evangelos Kolettas[1,2] and Theodore Tzavaras[1*]

Abstract

Background: Retrotransposons are mobile elements that have a high impact on shaping the mammalian genomes. Since the availability of whole genomes, genomic analyses have provided novel insights into retrotransposon biology. However, many retrotransposon families and their possible genomic impact have not yet been analysed.

Results: Here, we analysed the structural features, the genomic distribution and the evolutionary history of mouse VL30 LTR-retrotransposons. In total, we identified 372 VL30 sequences categorized as 86 full-length and 49 truncated copies as well as 237 solo LTRs, with non-random chromosomal distribution. Full-length VL30s were highly conserved elements with intact retroviral replication signals, but with no protein-coding capacity. Analysis of LTRs revealed a high number of common transcription factor binding sites, possibly explaining the known inducible and tissue-specific expression of individual elements. The overwhelming majority of full-length and truncated elements (82/86 and 40/49, respectively) contained one or two specific motifs required for binding of the VL30 RNA to the poly-pyrimidine tract-binding protein-associated splicing factor (PSF). Phylogenetic analysis revealed three VL30 groups with the oldest emerging ~17.5 Myrs ago, while the other two were characterized mostly by new genomic integrations. Most VL30 sequences were found integrated either near, adjacent or inside transcription start sites, or into introns or at the 3' end of genes. In addition, a significant number of VL30s were found near Krueppel-associated box (KRAB) genes functioning as potent transcriptional repressors.

Conclusion: Collectively, our study provides data on VL30s related to their: (a) number and structural features involved in their transcription that play a role in steroidogenesis and oncogenesis; (b) evolutionary history and potential for retrotransposition; and (c) unique genomic distribution and impact on gene expression.

Keywords: VL30s, Retrotransposon, PSF, Mouse genome, Exaptation

Background

Mobile DNA sequences occupy almost half of many mammalian genomes [1, 2], and in the past they have been considered as solely selfish or junk DNA sequences. Recent studies, however, have provided evidence that mobile DNA can be harnessed for host functions, and it is now accepted that mobile DNA-host interactions play a major role in organism physiology, pathology and evolution [3]. Mobile DNA elements are divided into two major groups, DNA transposons and retrotransposons. While DNA transposons mobilize through a conservative mechanism, comprising their

excision and subsequent integration into a new site, retrotransposon numbers increase in the genome through an RNA intermediate via a mechanism known as retrotransposition [3] and, based on their genomic structure, they are further subdivided into Long Terminal Repeat (LTR)- and non-LTR retrotransposons.

Viral-like 30 elements (VL30s) are a family of 5–6 Kb retrovirus-like DNA sequences present in mouse and rat genomes, and are classified as LTR retrotransposons [4]. VL30 LTRs contain a large repertoire of transcription factor binding sites that provide a remarkable plasticity for VL30 RNA expression [4, 5]. VL30 transcription is tissue-specific [6], up-regulated by histone phosphorylation, acetylation and DNA demethylation [7], and can be induced by several different factors including C2 ceramide, a dominant negative p53 oncoprotein [8] or in

* Correspondence: thtzavar@cc.uoi.gr
[1]Laboratory of General Biology, Faculty of Medicine, School of Health Sciences, University of Ioannina, Ioannina 45110, Greece
Full list of author information is available at the end of the article

cerebral ischemia [9]. VL30s are considered as immediate early response genes as their RNA expression is rapidly induced following transient inhibition of protein synthesis or mitogenic stimulation [10, 11]. Another important feature of VL30s is that their RNA is packaged into C-type retroviral particles of murine leukemia viruses and can be transmitted to heterologous cells [12, 13], making them an important potential tool for gene transfer. VL30 transcripts play functional roles as inducible VL30 transcription regulates expression of neighboring genes, creating a transcription network [14]. Induced VL30 RNA seems to be a critical factor involved in oncogenesis and steroidogenesis [15, 16], as it contains specific motifs for binding to spliceosome factor (PSF), leading to RNA induction of PSF-repressed genes.

Sequence analysis of two VL30 members, NVL-3 [17] and BVL-1 [18], has revealed no coding open reading frames due to multiple stop codons for *gag* and *pol*, and no evidence for *env* retroviral genes, classifying them as non-autonomous LTR retrotransposons. Specifically, the NVL-3 member is retrotransposition-competent [19, 20] and its retrotransposition is induced by the large T antigen of Simian virus 40 [19] or oxidative stress agents such as hydrogen peroxide [21], vanadium [22] and arsenic [23]. Finally, induced VL30 retrotransposition activates a caspase-independent and p53-dependent cell death pathway associated with mitochondrial and lysosomal damage [24].

The availability of bioinformatics tools allows the analysis of genomic DNA sequences related to their genomic distribution, evolution and prediction of functional properties. Such analyses have revealed that more than 2/3 of mammal genomes may be comprised of mobile DNA [25], a much higher percentage than the ~50 % initially calculated [1]. They have also provided novel insights on the features and biology of LTR retrotransposons such as human HERV-Ks [26] and many mouse LTR retrotransposons [27, 28].

Data on mouse VL30s comes from studies based on conventional methods, such as hybridization analysis and sequencing [4, 5], and gross estimations [29], prior to publication of the mouse genome, but a detailed genomic analysis of mouse VL30s is not yet available. In this study, we systematically annotated VL30 sequences and analyzed their number, structure, phylogeny and genomic distribution providing new insights for their potential role in the mouse genome.

Results
Copy number, structural features and variation of VL30 sequences
As the only available genomic data for VL30s come from sequencing of a few individual elements [4], prior to publication of the mouse genome, we attempted to measure their total number, analyse their structural features and predict functional traits of particular VL30 sequences.

To search for VL30 sequences in the UCSC mouse Database [30, 31], we assumed that: (i) an intact VL30 element, independent of its length, should contain near identical 5′ and 3′ LTRs, an intact retroviral Primer Binding Site (PBS) required for binding of a primer tRNA to initiate reverse transcription, and a polypurine Tract (PPT) responsible for initiating (+) strand synthesis during reverse transcription; (ii) a truncated VL30 should contain an internal sequence but not necessarily two intact LTRs; and (iii) solo LTRs, of varying sequence length, should only contain LTR sequences as defined by the Repeat masker tool [32]. Using a combination of Repeat Masker analysis (http://www.repeatmasker.org) for VL30 internal sequences [32], BLAT querying [33] for known VL30 consensus sequences from Repbase [34] and *in-silico* PCR for specific VL30 primers, corresponding to a conserved internal region [35], our genomic search initially revealed 372 total VL30 sequences. All sequence data were extracted along with 2000 bp of 5′ and 3′ flanking genomic regions.

Using the LTR-finder tool [36] in the above extracted VL30 sequence data we distinguished full-length from truncated and solo LTR sequences and found that the mouse genome contains 86 full-length elements in a sequence length ranging from 4075 to 6321 bp. In addition, manual annotation of the remaining non-intact VL30 sequence data revealed 49 truncated elements and 237 relative solo LTRs. The particular genomic coordinates of full-length, truncated and solo LTR VL30s as well as their sequence features are provided in Additional file 1.

Next, we examined the full-length VL30s for variations in their LTR sequences and retroviral replication signals. It was found that LTRs were quite heterogeneous in sequence length ranging from 436- to 681 bp (Fig. 1). In keeping with this, an LTR-related structural feature of retroviruses are Target Site Duplications (TSDs), a few bp long, generated following integration of their LTRs into the genome. Hence, we attempted to analyze the TSDs of full-length VL30 sequences found, using the LTR-finder tool [36]. The examination of mouse genomic sequences, immediately adjacent to VL30 LTRs, showed TSDs in 76 out of 86 full-length elements with a random 4-bp long motif (Additional file 1). The 86 full-length elements were distinguished by heterogeneous 15–20 bp long PBS signals, allowing binding of various tRNAs. Specifically, 57 elements contained PBS signals corresponding to Gly tRNA species, while a distinctively smaller number of 17 and 10 elements contained PBS signals corresponding to Pro and Gln tRNA species, respectively. Only two elements contained Met or Thr

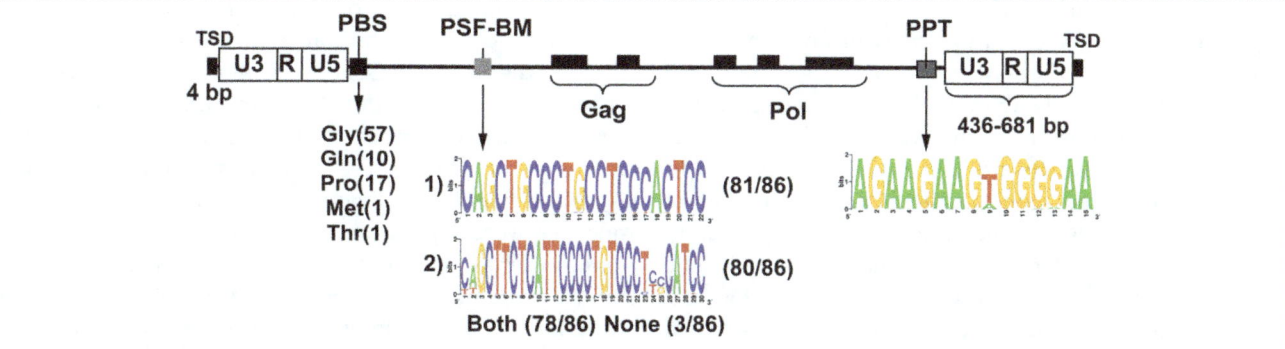

Fig. 1 Structural properties of full-length VL30 elements. A representative VL30 element is depicted, using sequence data from UCSC genome browser, LTR-finder and additional BLAST searches. Depicted from 5′ to 3′: LTRs: Long Terminal Repeats, PBS: Primer Binding Site; PSF-BM: PSF Binding Motif; Gag, Pol: regions related to retroviral genes (but contain multiple stop-codons); PPT: Poly-Purine Tract; TSD: Target Site Duplication. Gly, Gln, Pro, Met and Thr indicate tRNA species complementary to PBS signals of VL30 elements, while in parentheses is shown the number of elements that use each tRNA. Below PSF-BM are presented the sequences of the two specific motifs of PSF binding, while below PPT region the 15-bp consensus sequence

tRNA species. Further analysis revealed that all full-length elements had a PPT region, which was highly conserved in 68 elements containing the 15-bp consensus sequence AGAAGAAGTGGGGAA, while the rest 18 elements had one or two A/T substitutions (Fig. 1 and Additional file 1).

Early studies have shown that mouse and rat VL30 RNAs can be efficiently packaged into MoMLV virions [37, 38] through a retroviral Ψ packaging-sequence signal. Given that a mouse VL30 Ψ consensus sequence is not known, we attempted to identify such a sequence comparing the full-length sequences with the known 68-bp purine-rich sequence GGCAAGCCGGCCGGCG [38], downstream of the rat 5′ LTR VL30, which is critical for RNA dimerization and packaging. Examining the entire internal sequence of all 86 full-length sequences by the Blast algorithm, between 5′ and 3′ LTRs as well as downstream of U5 and upstream of U3 regions, neither a conserved nor a rat-related Ψ signal motif was found.

Finally, based on the property of VL30 RNA to bind PSF [39], we analyzed, using Blast alignment, all full-length VL30 sequences for the two known specific motifs of PSF binding: CAGCTGCCCTGCCTCCCACTCC and CAGCTTCTCATTCCCCTGTCCCTCCCATCC [15]. We found that in contrast to three PSF-negative elements, five contained only one while all the rest of the 78 sequences contained both motifs (Fig. 1 and Additional file 1). Next, we asked whether truncated elements also had PSF binding motifs. A similar analysis showed that 9 out of 49 truncated elements had no PSF signal motifs. In contrast, 6 elements contained one while 34 contained both PSF-binding motifs (Additional file 1).

VL30s transcriptional and protein coding potential

While both LTR sequences participate in the integration of a retrovirus into the host genome, the 5′ LTR serves as a promoter regulating the entire retroviral RNA expression accomplished through cellular transcription factors which interact with specific DNA motifs located in this region.

We attempted to identify transcription factor binding sites (TFBS) over the entire 5′ LTR sequences of full-length VL30s using the Matinspector tool [40] in the Genomatix database (see Methods). Specifically, we analyzed the group of 5′ LTRs of all 86 full-length VL30s found along with a group of 5′ LTRs of 7 known VL30 elements (such as NVL1/2, NVL3, BVL1, VL3, VM1, VL11 and B10) for common or unique TFBS. Our analysis revealed 10,371 TFBS of 53 distinct transcription factors (TFs) to exist in 70 % of both element groups and a total number of 15,039 (13,912 in 86 full-length elements, denoted in parenthesis thereof) TFBS of 197 TFs in at least one element (Additional file 2). The number of TF binding sites largely varied depending on the particular VL30 LTR and chromosome. For example, the binding sites of HOMF (homeobox-domain factor) and VTBP (Vertebrate TATA Binding Protein) were present in all 93 (86) (Additional file 2) analyzed LTRs and in all chromosomes with a high number of 550 and 370 sites (516 and 346) ranging between 1–12 and 1–7 per chromosome, respectively. To a lesser extent, the glucocorticoid responsive/related elements (GREF) and nuclear factor kappa-light chain enhancer of activated B cells (NF-κB) TF binding sites were present in 87 and 85 LTRs, with 265 and 131 binding sites, respectively, ranging 1–7 and 1–4 sites per chromosome. Finally, some TF binding sites were extremely infrequent with 1–2 sites restricted either to only one or two LTRs in a particular chromosome, such as those of HASF and MEF3 with two binding sites on 2 different chromosomes, or CABL and HNFP with 1 binding site in different chromosome, respectively (Additional file 2).

In addition, we analysed the truncated and solo LTR sequences since they can also act as promoters or enhancers, driving the expression of adjacent genomic loci. We scored 8558 and 44,428 total TFBS in truncated and solo LTR sequences, respectively. In analogy to full-length elements, HOMF and VTBP TFBSs we present in almost all truncated sequences with a high number of 238 (in 47 out of 49 sequences) and 122 (in 48 out of 49 sequences) sites, respectively. As regards solo LTRs the respective TFBSs for HOMF and VTBP were 1266 and 643 in 234 out of 237 sequences, respectively (Additional file 2).

Next, we extended our analysis searching for VL30-associated ESTs in the mouse section of dbEST. We retrieved 592 total ESTs starting or terminating within 99 distinct VL30 sequences. In particular, 414 different ESTs were identified in two-thirds of the full-length VL30 sequences (57 out of 86). Moreover, we found 85 and 93 ESTs in 14 truncated and 28 solo LTR sequences, respectively (Additional file 3).

In a final step, we searched whether VL30s have protein coding potential, using the ORF-finder [41] in the sequences of full-length elements. The analysis of their internal sequences between 5′ and 3′ LTRs, revealed multiple stop codons and no intact full-open reading frames corresponding to typical *gag*, *pol* and *env* retroviral genes (data not shown).

The genomic distribution of VL30s

Taking into account the total number of 372 VL30 sequences found, we asked whether there are random or "hot spot" integrations of VL30s in the mouse genome, as their genomic distribution is unknown.

Using the Ensembl Genome Browser tool [42], we created a map representing the distribution of full-length, truncated as well as solo LTR VL30 sequences in the mouse chromosomes (Fig. 2). By this analysis we primarily found that all full-length, truncated and solo LTR sequences were represented in all 21 autosomal and sex chromosomes. Moreover, assuming a random insertion model based on chromosome size, we analyzed their distribution at the chromosomal level comparing the observed with the relative expected number of VL30 integrations. Statistical analysis was revealed a significant difference between the two distributions (chi-square = 43.66, df = 20, p <0.002). We found that VL30s density was much higher than expected on chromosomes 3, 7, 12, 13, 17 and X. In contrast, chromosomes 2, 11, 15, 19 and Y had fewer elements than expected (Fig. 3). In addition, we observed the presence of 8 almost identical truncated VL30s integrated in a particular genomic domain of 1.5 Mb at XqF2 and XqF3 regions of chromosome X (Fig. 4a). In a further step, we examined the GC content of VL30 integration sites. By analyzing the GC content of 400 bp genomic DNA flanking full-length VL30 sequences, we found that VL30s were integrated in genomic regions characterized by a ~42 % GC content (Additional file 4).

Next, uploading the VL30 genomic coordinates (Additional file 1) into the GREAT annotation tool [43], we analyzed the distribution of VL30 sequences in relation to mouse gene regions. Out of 372 total VL30 sequences, 317 sequences were integrated up to 500

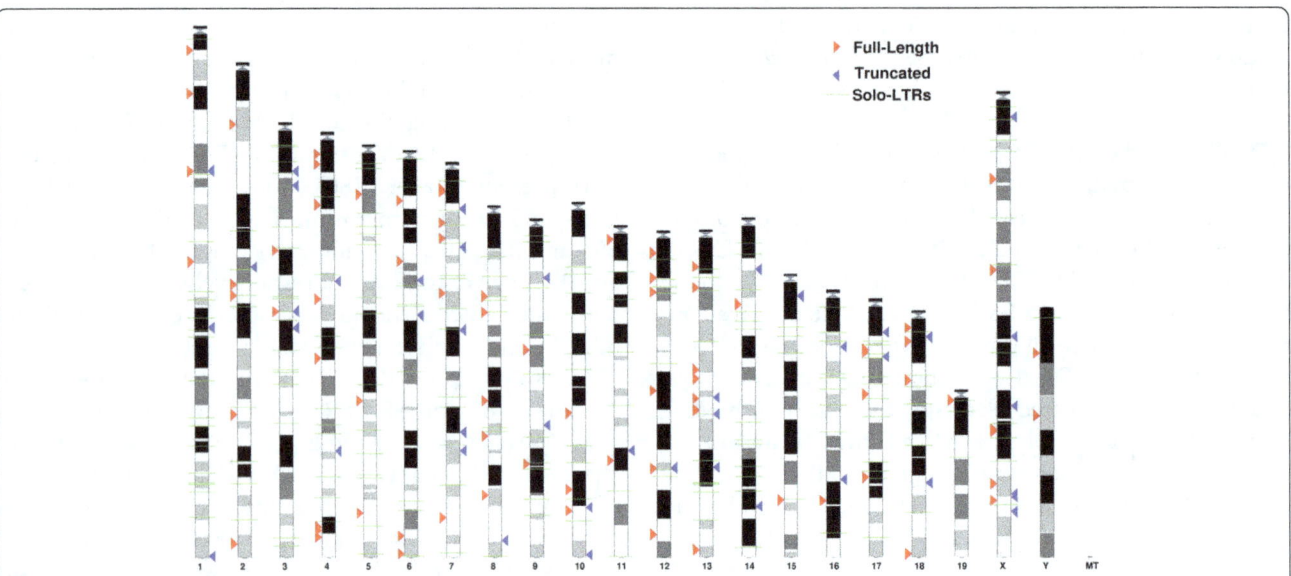

Fig. 2 Chromosomal distribution of VL30 elements. Full-length VL30s, truncated VL30s and solo LTRs have been annotated to their respective genomic positions and projected to mouse chromosome ideograms, using the Ensembl Genome Browser. Red arrows denote full-length VL30s, blue arrows truncated elements and green horizontal lines solo LTRs

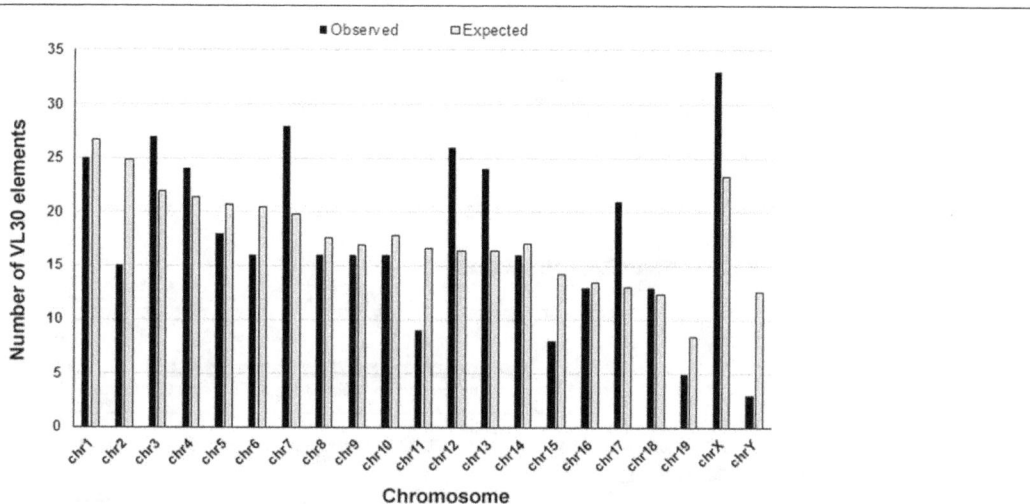

Fig. 3 Mouse genomic distribution of VL30 elements. Histogram bars indicate the observed (black) and expected (light grey) number of VL30 elements from each mouse chromosome. The expected number of VL30 elements on each chromosome was calculated following multiplication of the chromosomal length with average density of VL30s in the mouse genome

kilobases upstream of transcription start sites (TSS). Specifically, 33 full-length, 21 truncated and 87 solo LTR sequences were found at a distance between −500 and −50 kilobases upstream of TSS; 29 full-length, 10 truncated and 84 solo LTR sequences at a distance between −50 and −5, while 5 full-length, 1 truncated and 15 solo LTR sequences were the nearest integrated sequences upstream of TSSs. Moreover, 107 total sequences: full-length, truncated and solo LTRs were integrated downstream of TSSs, at a distance less than 50 kb (Fig. 5 and Additional file 5). In keeping with this, analysis of their position in relation to TSSs revealed 30 sequences adjacent to and 4 ones residing within TSSs, as exemplified in Fig. 4a-f, and 22 sequences were located within introns while 2 ones at the 3′ end of genes (Table 1 and example in Fig. 4). Evaluating their integration distance from known genes, we found a statistically significant association of VL30 sequence integrations in the vicinity of genes encoding zinc finger proteins (ZFPs) (binomial raw *p*-value, $p = 1.23 \times 10^{-10}$), which contain the Krueppel-associated box (KRAB) (Additional file 6). Finally, we also found sequences located near regulatory genes such as PSF (*Sfpq*) (Fig. 4e), and several transcription factors (*Gtf2e2, Ap1m1, Crebzf*), kinases (*Cks2, Camkk2*), receptors (*Gpr113, Olfr1033, TCR-beta chain*), as well as genes involved in cell differentiation (*Tnks, Zfp568, Tdrd9, α*) (Table 1).

Phylogeny of VL30 and estimation of their integration time in the mouse genome

There is a relatively limited data on VL30 phylogeny concerning only a small subset of 18 sequences, which was obtained prior to mouse genome sequencing [5].

To obtain information on VL30 phylogeny, we attempted a relative analysis of VL30 sequences based on differences of their 5′ LTR, which mostly define the functional properties of each element. We avoided analysing truncated copies since they show extensive divergence in LTR length (truncated LTRs) and even absence of entire LTR regions. We used ClustalW2 algorithm [44] to perform multiple alignment between the 86 5′LTR of full-length elements and as reference the LTR sequences of seven known VL30s (VM1, VL11, BVL1, B10, NVL1/2, NVL3 and VL3 elements) [4, 5]. We generated a phylogenetic tree using the Maximun Likelihood method and Tamura-Nei model, for correction of multiple hits by taking into account the differences in substitution rate between nucleotides and the inequality of nucleotide frequencies. Based on the phylogenetic tree obtained (Fig. 6), we found that VL30 elements have diverged into three distinct groups. Group I was the oldest group comprising 16 elements, which were not closely related to the aforementioned reference sequences, forming a separate monophyletic group to the elements of group II and III. Group II contained 24 elements related to NVL3 and VL3 in two sub-groups; one highly related to NVL3 and a divergent one more related to VL3. Finally, Group III contained 46 elements divided in four sub-groups: III-A, III-B and III-C, which were highly similar while that of III-D was more divergent. This phylogenetic grouping can be confirmed by the tRNA species used by each element. Specifically, Group I is quite heterogeneous as 10, 5 and 1 members use Gln, Pro or Thr tRNA species, respectively. In contrast, Group II is divided into two distinct sub-groups containing either exclusively Gly (sub-group related to NVL3) or mainly Pro tRNAs

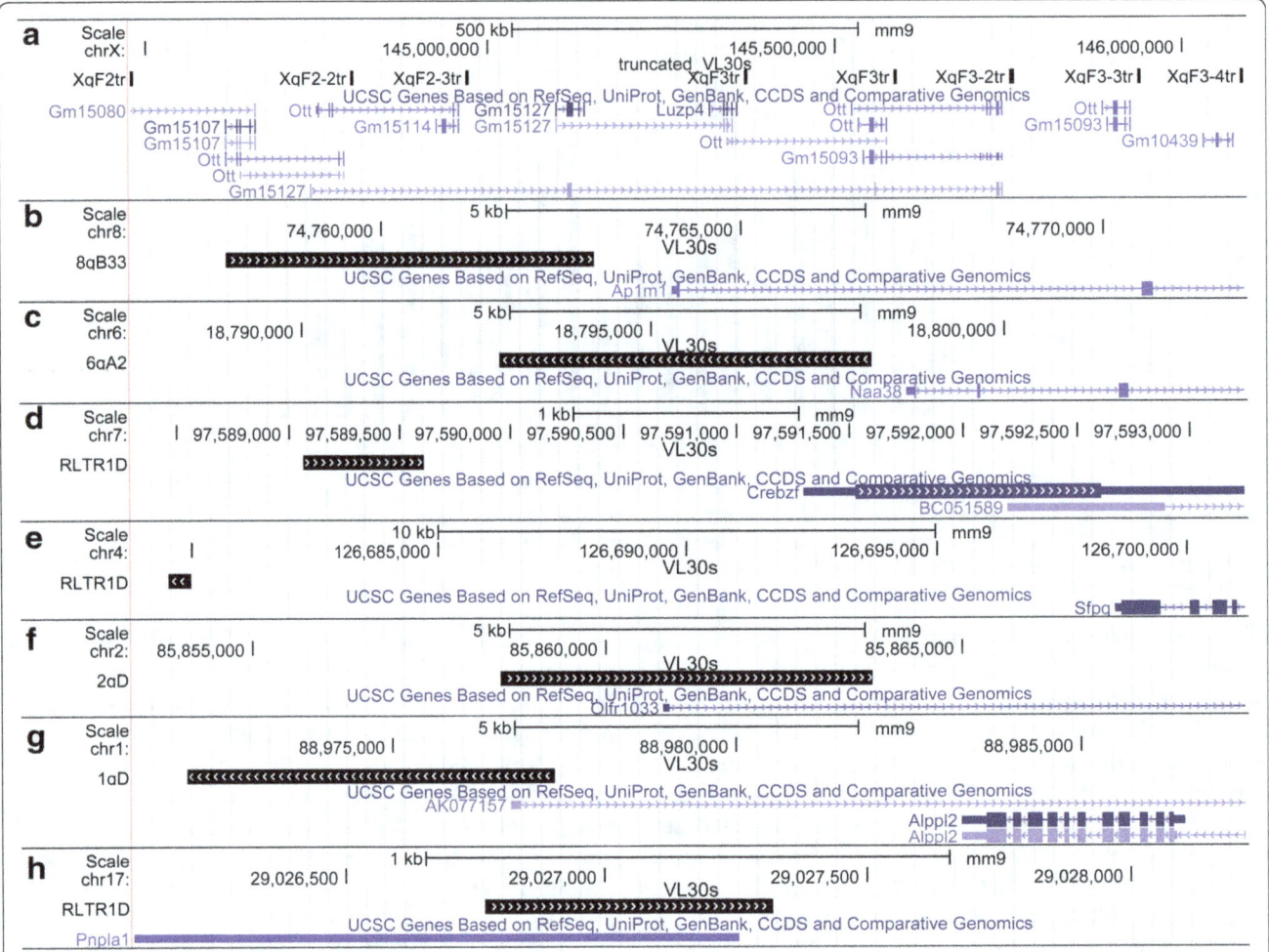

Fig. 4 Examples of VL30 integration events with possible significance for the mouse genome. Genomic regions (panels **a-h**) containing integration events with a possible significance in the mouse genome were found after GREAT analysis and manually confirmed in UCSC Genome Browser. Genome graphs were extracted from the UCSC Genome Browser. The chromosome number, scale, VL30 element name and associated genes name in each region are shown

(sub-group related to VL3). As regards Group III, with exception of one element, all others contain Gly tRNA (Fig. 6).

We next calculated the sequence divergence between paired 5′ and 3′ LTRs as previously described [45], in order to obtain evolutionary data for each individual element. We found that most VL30s shared highly related 5′ and 3′ LTRs pairs. In addition, using a mean nucleotide substitution rate of 4.6×10^{-9} [2], we also estimated LTR integration time [46]. We found that VL30s (Group I) emerged about 17.5 Myrs ago (Fig. 7) and most integrations date from 0.45 Myrs ago till today. Finally, 42 full-length elements shared identical 5′ and 3′ LTRs.

To confirm the evolutionary relationships analyzed, we combined the data of the two aforementioned analyses by highlighting with rhombus symbols the elements with

identical 5′ and 3′ LTRs in the evolutionary tree obtained (Fig. 6). Indeed, Group I was the oldest one as its members shared no identical LTRs, and 7qA1, having the most divergent 5′ and 3′ LTRs, was the "oldest" VL30 element. Group II was the immediately younger subgroup comparing to Group I and contained 13 sequences (out of 21 total) highly similar to VL3 that shared identical 5′ and 3′ LTRs. Finally, Group III elements, although consisting of three highly related and one divergent subgroup, were the overall youngest group as their high majority, 34 out of 46 elements, had identical LTRs (Fig. 6). Collectively, based on the above dating analysis, Group I contains elements that are 17.5-3.1 Myrs old (from 7qA1 to 2qF1), Group II of 9.3 Myrs old up to date (from 13qB3.2 to 8 elements with identical LTRs) and the youngest Group III of 6.2 Myrs old up to date (from 6qA2 to 34 elements with identical LTRs).

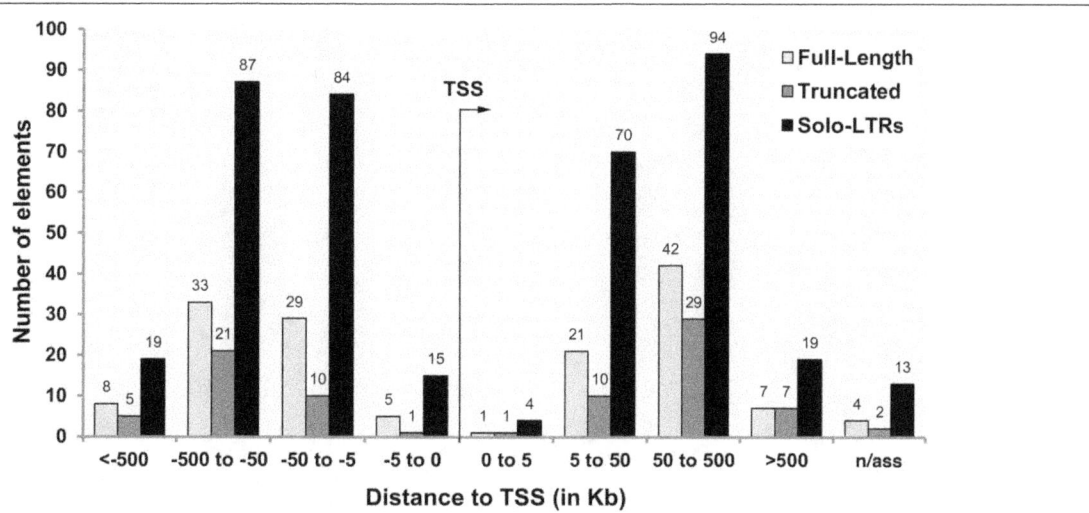

Fig. 5 Distribution of VL30s relative to transcription start sites of mouse genes. Analysis with GREAT tool under default settings was performed for full-length, truncated VL30s and solo LTRs. The chart shows the relative distance (0–5, 5–50, 50–500, 500–1000 Kb) of full-length (light grey), truncated (dark grey) or solo LTRs (black) to Transcription Start Site (TSS). n/ass denotes VL30s not associated to any genes

Discussion

The present study, based on the complete sequencing of the mouse genome and utilizing advanced bioinformatic tools, presents an *in silico* analysis of VL30s providing information about their structural characteristics, chromosomal distribution and evolution.

Features of VL30 elements

Prior to this study, it was generally assumed that the mouse genome contained ~150–200 VL30 full-length elements [4], while our analysis revealed 372 total VL30 sequences. Specifically, we found 86 full-length elements, 49 truncated elements, and 237 solo LTR sequences. Accordingly, the total number of truncated and solo LTR sequences is 3.3-times more than those of full-length elements. This implies that both types of non-intact VL30 sequences rather derived after several mutation/recombination events during evolution. Hence, in contrast to previous assumptions, the mouse genome contains 86 full-length elements comprising the 23.1 % of the total VL30 sequences. 76 out of 86 full-length elements analysed had a hallmark of their genome integration, a conserved 4-bp TSD [4, 20]. LTRs were found to be quite heterogeneous, ranging from 436 to 681 bp. Thus, as VL30s are retrotransposition-competent [19], this finding might be explained by the occurrence of mutations during transcription and/or reverse transcription during retrotransposition, or by recombination events during evolution. Regarding the PBS signals, the large majority had Gly tRNA species specificity, while 17 and 10 elements had a Pro and Gln tRNA species (Fig. 1), respectively. Based on this, we may categorize full-length elements into three distinct groups according to the replication tRNA species:

the prevalent group of Gly tRNA-positive elements and those of less frequent groups Pro- and Gln tRNA-positive elements. In addition, the different groups of tRNA species in association to conserved PPTs found (Fig. 1), shows the versatile nature of VL30s replication. Importantly, all internal sequences were characterized by the absence of an *env* gene while *gag* and *pol* genes had multiple stop codons, documenting that VL30s have no intact ORFs or code for very small peptides. Although VL30s are retrotransposition-competent these data emphasize that none of the full-length VL30s is an autonomous-LTR retrotransposon and their retrotransposition is mediated *in trans*-complementation, probably, by an endogenous MoMLV reverse transcriptase [19]. Strikingly, out of 135 total VL30 sequences, 86 full-length and 49 truncated, only 12 sequences - 3 full length and 9 truncated ones - had no PSF-binding motifs (Fig. 1 & Additional file 1). This large majority of PSF-positive sequences, irrespective of their full length or truncated state, document an almost universal property of VL30s, unique among retroelements, involved in mouse steroidogenesis and oncogenesis [39]. Finally, we were unable to locate or determine a specific or consensus Ψ signal for mouse VL30s, following comparison with that of rat VL30s. Though mouse VL30 elements undoubtedly have a strong but unknown yet Ψ signal, as their RNA can be encapsidated in C-type retroviruses [12]. Thus, we believe that the different Ψ signal in rat and mouse VL30 elements distinguishes evolutionarily these murine species.

VL30s-related transcription factors and ESTs

It is well known that cellular transcription factors (TFs) interacting with specific DNA motifs located in 5′ LTRs

Table 1 VL30 elements closely located to mouse genes

Region	VL30	Associated gene, (distance to TSS)	Orientation[a]	Possible role
Adjacent to TSS	4qA1-2	Rbm12b (−4469)	+	Alternative promoter, Enhancer Or Effects in gene expression
	4qE2	Ube4b (−9958)	−	
	5qG2	Rhbdd2 (−6349)	+	
	6qA2	Naa38 (−3143)	+	
	7qA3-2	Eid2 (−8861)	+	
	8qA4	Gtf2e2 (−15672)	+	
	8qB33	Ap1m1 (−3642)	+	
	8qC3	Gm6531 (−12209)	+	
	8qD3	Tmco7 (−4073)	+	
	12qC1	Fbxo33 (−5398)	−	
	13qA5	Cks2 (−8731)	+	
	15qE1	Ndufa6 (−10978)	−	
	18qA1-2	Rbbp8 (−12595)	+	
	8qE1tr	Afg3l1 (−3684)	−	
	RLTR1C-chr6-1	Tmem168 (−3944)	−	
	RLTR1D-chr5-3	Camkk2 (−2560)	+	
	RLTR1C-chr4-1	Zmym1 (−2767)	+	
	RLTR1D-chr8-2	Tnks (−5185)	+	
	RLTR1D-chr7-5	Crebzf (−1957)	+	
	RLTR1D-chr4-6	Sfpq (−18784)	−	
	RLTR6_Mm-chr5-4	Gpr113 (−865)	−	
	RLTR6_Mm-chr6-2	Avl9 (−4915)	−	
	RLTR1C-chr4-2	Fam54b (−3954)	+	
	RLTR6_Mm-chr5-6	Mrfap1 (−2311)	+	
	RLTR1D-chr7-2	Ceacam5 (−4916)	−	
	RLTR6_Mm-chr7-2	Zfp568 (−4304)	+	
	RLTR6_Mm-chr11-2	Rpl23 (−2735)	+	
	RLTR1C-chr13-1	Mrpl32 (−4663)	−	
	RLTR1D-chr13-2	Ankdd1b (−3400)	−	
TSS within VL30	1qD	AK077157 (0)	−	Promoter
	2qD	Olfr1033 (0)	+	
	4qB3tr	Gm12505 (0)	−	
	18qA2	AK076984 (0)	−	
Intronic	1qA2	Sntg1 (+62021)	−	Alternative splicing region or disruption of transcription
	2qA3	Abi1 (+27853)	−	
	6qB1	TCR-beta chain (+20882)	+	
	6qG2	Slco1a6	−	
	7qA3	gm6902	−	
	12qA11	Slc7a15	−	
	12qF1	Tdrd9 (+47580)	+	
	12qa13	Gm4983	+	
	7qc3	A330076H08Rik	−	
	7qE2tr	Gm1966 (+26035)	−	
	7qE2-2tr	Neu3 (+9192)	−	

Table 1 VL30 elements closely located to mouse genes *(Continued)*

	RLTR1D-chr14-2	Dydc2 (+5144)	−	
	RLTR1D-chr1-12	Parp1 (+5484)	−	
	RLTR6_Mm-chr2-2	Ap4e1 (+13898)	+	
	RLTR6_Mm-chr2-3	Plk1s1 (+34325)	+	
	RLTR6_Mm-chr2-4	a (+15736)	+	
	RLTR1C-chr3-1	Slc7a12 (+4176)	−	
	RLTR6_Mm-chr5-9	N4bp2l2 (+1602)	+	
	RLTR6_Mm-chr8-2	Gsr (+11508)	+	
	RLTR1C-chr9-2	Rasgrf1 (+76260)	−	
	RLTR1D-chrX-2	Otc (+44070)	−	
	RLTR1D-chr14-2	Dydc2 (+5144)	−	
End of gene	RLTR6_Mm-chr11-1	Wdr92 (+22760)	+	Transcription termination
	RLTR1D-chr6-3	BC064078	+	

[a]+ same orientation to that of gene, − opposite orientation

regulate retroviral transcription [47]. We found that although the number of TF binding sites (TFBS) largely varied depending on the particular VL30 LTR, we identified 10,371 TFBS allocated to 53 distinct TFs, common to 70 % of 86 new full-length elements, and a total number of 14,880 TFBS from 197 TFs in all 5′ LTRs analyzed (Additional file 2). Notably, while some TFBS were present in almost all full-length elements such as those of HOMF and VTBP, others were extremely infrequent with 1–2 TFBS in only one or both LTRs (Additional file 2). In relation to the glucocorticoid responsive/related TFBS of GREF, they were present in all LTRs of full-length elements with 264 total binding sites, ranging from 1 to 7 sites per element. This particular finding justifies our previous observation on VL30 RNA induction through nuclear steroid/receptor complexes binding on GREF following estradiol, diethylstilbestrol, progesterone or dexamethasone treatment [8]. Significantly, we identified 52,986 total TFBSs in both truncated as well as solo LTR sequences that may be similarly involved in gene expression, as full-length VL30 LTRs. In support of this comes the finding that 99 distinct VL30 sequences had 592 VL30-associated ESTs (Additional file 3), implying their involvement in initiation or termination of RNA expression. In conclusion, VL30s: (a) bear a large number of common or unique TFBS, explaining the versatile VL30 RNA expression, which can be tissue-specific or up-regulated by various or pleiotropic stimuli, and (b) provide transcriptional initiation or termination sites, which may affect cellular gene expression.

VL30s genomic distribution

The analysis of chromosomal distribution of VL30s shows that they are ubiquitous in mouse autosomal and sex chromosomes (Fig. 2) and integrated in regions of ~42 % GC content (Additional file 4), which is the exact GC average of the mouse genome [2]. Nevertheless, two lines of evidence support their preferential chromosomal integration documented by: (a) chromosomes 3, 7, 12, 13, 17 and X where the number of integrations is higher that than those expected (Fig. 3), and (b) a possible integration "hot-spot" mapped in 1.5 Mb of XqF2-F3, where 8 truncated VL30 elements were inserted (Fig. 4) probably as a result of serial unequal recombination events between loci of the X chromosome.

As concerns VL30 genomic distribution with respect to gene regions, we found that the large majority of 372 VL30 sequences were preferentially integrated near or in gene regions. In particular, 285 or 272 integrations were located upstream or downstream of TSSs, respectively, in a distance less than 500 Kb (Fig. 5). The significance of such integrations is that they may play an important role in the epigenetic regulation of gene expression. Specifically, first, VL30s affect nearby gene expression acting in a distance of up to 220 Kb [14], and contain many TFBS. Based on this, VL30 integrations at about of <20 Kb distance upstream of 29 known genes (Table 1) may play a *cis*-acting regulatory role in modulating their expression. Thus, VL30 LTRs integrated adjacent, upstream or downstream of gene Transcriptional Start Sites (TSS) (Fig. 5) may act as promoters or enhancers for a large number of genes (Table 1). Notably, VL30 members are retrotransposition-competent and their retrotransposition is induced up to ~90,000- or 420,000-times higher than the natural retrotransposition frequency by arsenic [23] or H_2O_2 [21], respectively. Conceivably, in these cases, the number of new integrations might lead to a dramatic change of the whole genome expression profile. Secondly, we found three full-length and one truncated VL30 element containing a TSS (Table 1) and their LTRs may serve as primary promoters for these genes. Thirdly, 22 integrations found

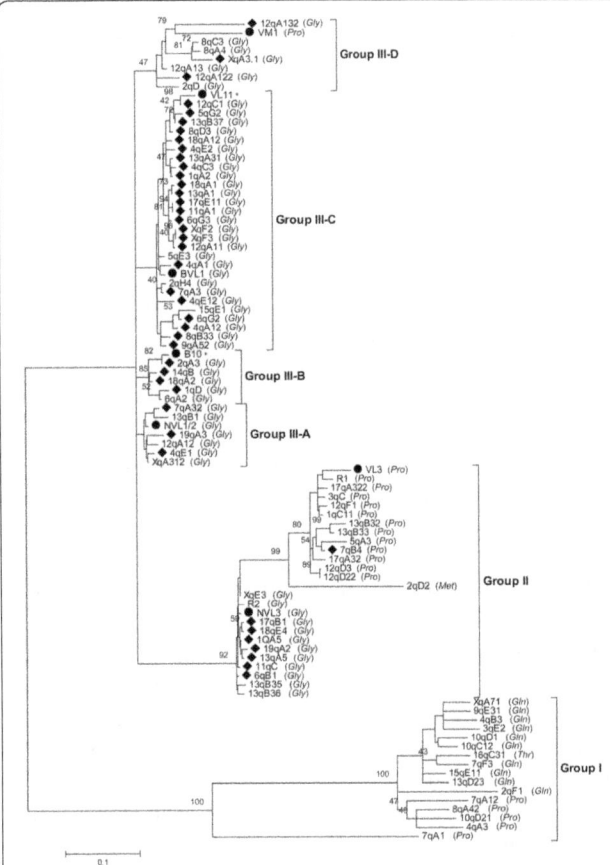

Fig. 6 Molecular phylogenetic analysis of full-length VL30 LTRs by the Maximum Likelihood method. The bootstrap consensus tree representing the phylogeny of LTRs from 86 full-length and 6 known VL30s is presented (500 tests). The percentage of replicate trees in which the associated taxa clustered together is shown above the branches. The tree is drawn to scale, with branch lengths measured in the number of substitutions per site. Bootstrap replicates denote known elements or elements with identical 5′ and 3′ LTRs. The respective tRNA species used by each VL30 element is shown in parenthesis. Asterisk denotes solo LTR sequences

inside introns and 2 ones at the 3′ end of genes (Table 1). Intronic integrations may affect mRNA splicing, as the presence of splicing donor sequences in some VL30s [4] is consistent with this concept. In reference to integrations at the 3′ end of genes (Table 1) we believe that they possibly act as transcription termination signals through VL30 polyA sequences since retrotransposon integrations lead to premature transcriptional termination even at a >12.5 Kb distance [48].

Finally, we identified a significant number of integrations near genes of the KRAB (Krueppel-associated Box containing) family (Additional file 6), known to participate in epigenetic silencing of endogenous retroviruses [49, 50]. Given that: (i) retrotransposons participate in or control regulatory networks in embryonic stem cells [51] or more premature developmental stages [52]; (ii)

VL30s contain a large repertoire of TFBS (Additional file 2) and are competent of inducible [5, 8] and cell type-specific expression [6]; and (iii) Mouse VL30s inserted near genes affect their expression [14]; we suggest that the unique genomic distribution of VL30s near genes, such as those of the KRAB gene family, might ultimately lead to their selection as functional elements influencing epigenetic gene regulation and participating in regulatory networks.

VL30 phylogeny

Based on the evolutionary tree created in the present study (Fig. 6) and on the calculation of the integration time for individual VL30s phylogeny (Fig. 7), we found that VL30s are divided into three evolutionary groups.

Our dating analysis is based on nucleotide substitution rate that creates divergence between 5′ and 3′ LTRs of individual elements. Even though we cannot exclude possible biases in date estimates (deamination biases for instance), we followed that method since it is the most widely accepted for LTR elements, has been used for HERV elements [45] and our data from intra-element data analysis come in agreement with their position in the phylogenetic tree. We document that Group I is the oldest one emerged ~17.5 Myrs ago, after the calculated rodent-primate divergence of ~41 Myrs ago [53] and this may explain why VL30s are rodent-specific, not existing in primates. Moreover, the element 7qA1 seems to be the archetype or "the VL30 eve" (Fig. 7). Group II and Group III elements comprised mostly of full-length elements, emerged 9.3 and 6.2 Myrs ago, respectively and are characterized by highly conserved LTRs (Fig. 6) and replication signals (Fig. 1). The elements of both groups are active or retrotransposition-competent, as 42 out of 86 full-length elements share identical 5′ and 3′ LTRs representing present-day integrations, and being possible candidates for the evolutionary youngest elements. VL30 grouping is further strengthened by the fact that while Group I elements are characterized by heterogeneity using 3 different tRNA species, Group II and Group III contained elements using Gly or Pro and Gly tRNA species, respectively (Fig. 6 and Additional file 1). We believe that the oldest elements probably diverged due to recombination/mutation events and upon evolutionary pressure underwent selection and fixation in the mouse genome. We cannot also exclude that these different PBS signals may indicate independently derived mutations from cross-species transmission events. Our data are also supported by previous findings that: (a) VL30s are highly polymorphic (at ~40 % of integrations) across 18 mouse strains [54]; (b) VL30 RNA is the only one (among LINEs, SINEs and IAP retroelements and DNA transposons) up-regulated by both histone hyperacetylation and DNA demethylation agents [7]; (c)

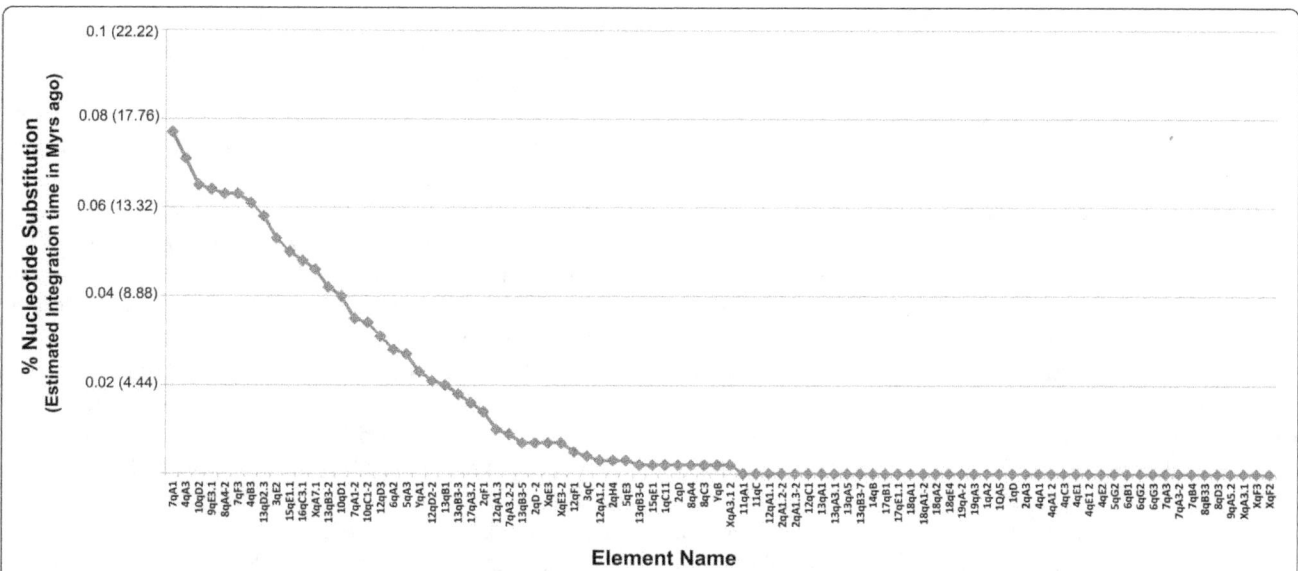

Fig. 7 Estimation of divergence and time of integration for full-length VL30s. The graph represents the percentage (%) divergence of 5' and 3' LTRs, as found by BLAST querying, and the respective calculated time of integration, taking into account the mean mutation rate of the mouse genome (4.6×10^{-9} mutations per site per generation)

overexpression of VL30 elements following DNA demethylation is associated with abnormal placental formation of *Mus musculus* × *Muscaroli* hybrids [55]; and (d) hydrogen peroxide induces VL30 retrotransposition of NVL3 element at ~42 % [21], the highest frequency ever measured in cultured mammalian cells. Taken together, our evolutionary data support that VL30 elements are highly active and rapidly expanding in the mouse genome.

Moreover, taking into account: (i) the recent expansion of VL30s in the mouse lineage (Fig. 2); (ii) the universal feature of VL30 RNAs to bind PSF (Fig. 1), leading to oncogenesis and steroidogenesis [15, 16]; and (iii) their possible role as insertion mutagens and inducers of cell death [24], we believe that PSF binding may act in some cases as a positive selection mechanism, leading to "tolerance" of VL30 integrations, since the deleterious effect of new retrotranspositions may be compensated for by the physiological action of VL30 RNA in steroidogenesis.

Conclusion

This is the first extensive study on the LTR retrotransposon VL30s, which reveals that the mouse genome harbors precisely 372 total VL30 sequences characterized as 86 full-length, 49 truncated and 237 solo LTRs. The full-length VL30s are categorized into three distinct evolutionary groups. The finding that 42 out of 86 ones share identical LTRs and are marked by various genomic TSDs implies their potential to mobilize as non-autonomous retrotransposons, leading thus to a significant genome reorganization. A hallmark of VL30 genomic distribution

is their preferential integration in 6 out 21 mouse chromosomes. At the transcriptional level, they can be diversely induced by various stimuli as they bear a very large number of TFBS also related to tissue-specificity. The most striking functional mark of VL30s, among other retroelements, is their involvement in steroidogenesis and oncogenesis as 91.2 % of both full length and truncated ones are PSF-positive. Finally, the large majority of the 372 sequences bearing LTR integrated near or in gene regions denotes their role in gene regulation, serving either as promoters/enhancers for many cellular genes or transcription termination signals through their polyA-signals. Retrotransposons, resembling the Janus double-faced Roman god, have a dual impact on the genome. Specifically, while new integrations of the retrotransposon copies might be mutagenic or deleterious leading to cell death their LTRs could be exploited by the cell for gene regulation, adaptation and general homeostasis upon environmental changes. Conclusively, VL30s, fulfilling both effects of retrotransposon action, might have been evolved as important mouse "LTR controlling elements" in parallel to those that Barbara McClintock stated for maize transposons [55].

Methods

Bioinformatics mining tools for identification of VL30 sequences

The UCSC genome browser (http://genome.ucsc.edu/) [30, 31] bioinformatic tools were used to screen manually the mouse genome (version mm9) for VL30 sequences. A combination of Repeat Masker analysis (http://www.repeatmasker.org) [32], BLAT querying [33] for VL30 consensus sequences from Repbase [34] and *in-silico* PCR for specific VL30s primers [35] was used to annotate VL30 sequences in their respective genomic regions. The LTR-finder tool (http://tlife.fudan.edu.cn/ltr_finder/) [36] was utilized to identify full-length VL30s. Sequences not corresponding to full-length VL30s were manually annotated as truncated VL30 elements or solo LTRs. The transfer of VL30 genomic coordinates between mouse genome version mm9 and the current one of mm10 was performed by the LiftOver tool of UCSC database.

The BLAST algorithm [56], in the NCBI database (http://blast.ncbi.nlm.nih.gov/Blast.cgi), was used to locate PSF binding motifs, (based on the two consensus sequences described in [15]) and the open reading frame (ORF)-finder [41] to search for potential ORFs in full-length VL30s. The analysis of full-length VL30 LTRs for transcription factor binding sites was performed using the Matinspector tool [40] in the Genomatix database (www.genomatix.de/matinspector.html). Retroviral Poly-Purine Tract (PPT) and PSF-binding motif consensus sequences for full-length elements were generated using the Weblogo software [57]. The GC-content of genomic regions [58], adjacent to full-length VL30s, was calculated in the Galaxy platform [59].

Phylogenetic and nucleotide substitution analysis

The multiple alignment of VL30 5′ LTR regions was performed with the ClustalW2 algorithm [44] (http://www.ebi.ac.uk/Tools/msa/clustalw2/). The phylogeny of VL30s was inferred using the Maximum Likelihood method, based on the Tamura-Nei nucleotide substitution model [60], including in our analysis the LTRs of seven known elements (NVL1/2, NVL3, BVL1, VL3, VM1, VL11 and B10). The bootstrap test (500 replicates) was applied to validate phylogeny [61] and all evolutionary analyses were conducted by the molecular evolutionary genetics analysis 5 (MEGA5) software [62].

Nucleotide substitution analysis was performed according to a previous study [63] to calculate the divergence between paired 5′ and 3′ LTRs, as described for HERV retrotransposons [45]. Paired 5′ and 3′ LTR sequences were compared by BLAST alignment [56] to find differences in nucleotide sequence, as product of spontaneous mutation events during evolution. Based on the number of nucleotide substitutions between paired LTRs we calculated VL30s integration time according to an estimate of the mean mutation rate in the mouse genome, 4.6×10^{-9} mutations per base per year [2].

Analysis of VL30s genomic distribution

The distribution of VL30s was depicted using the Ensembl Genome Browser [42] uploading their genomic coordinates. To study VL30 chromosomal distribution, chromosome sizes were obtained from summary tables in UCSC website and the fraction of each chromosome was calculated (by dividing the chromosome size to the total genome size). The expected number of element integrations per chromosome was calculated by multiplying the total number of elements with chromosome fraction, as done in a previous study for SVA elements [64]. The chi-squared test was used to analyze the chromosomal distribution of VL30 elements by comparing the observed with the expected number of VL30 elements, assuming a random insertion model.

The relative distribution of VL30s to Transcription Start Sites (TSS) of mouse genes was analyzed by the Genomic Regions Enrichment of Annotations Tool (GREAT) (great.stanford.edu) [43], with VL30 coordinates uploaded to UCSC Table Browser [65]. Briefly, GREAT algorithm determines the distance of a given genomic region to the nearest TSS found in a distance less than 1000 Kb. Then it categorizes elements based on the respective distance in groups (0–5 Kb, 5–50 Kb, 50–500 Kb and >500 Kb). Finally, it calculates of association between elements and previously annotated gene groups. The obtained association data were manually confirmed and further examined in the UCSC Genome Browser.

Statistical analysis

STATISTICA (version 13) was used for statistical analysis. The chi-squared test was used to analyze chromosomal distribution of VL30 elements, based on the expected values from a random distribution model. VL30s and gene TSS association statistics were calculated in the GREAT platform [43]. GREAT performs a binomial test over genomic regions and the hypergeometric test over genes, so that a possible bias from the one test is compensated by the other test. Only the output regions with significant associations from both independent statistical tests were obtained.

Additional files

Additional file 1: Genomic coordinates and structural features of full-length, truncated and solo-LTR VL30 sequences. The structural features of each individual full-length element after LTR-finder analysis are shown using the genome version mm10. The presence of PSF-binding motifs, with the percentage (%) similarity for each motif to consensus sequence is also shown for full-length and truncated elements. (XLS 79 kb)

Additional file 2: Common and unique transcription factor binding sites in LTRs of full-length, truncated and solo LTR VL30 sequences. The Microsoft Excel file contains three sheets each one presenting all transcription factor binding sites of full-length, truncated and solo LTR VL30 sequences, respectively. (XLS 554 kb)

Additional file 3: ESTs associated with VL30 elements. The Microsoft Excel file contains detailed information about the ESTs starting or terminating within distinct VL30 sequences. (XLS 84 kb)

Additional file 4: GC-content adjacent to full-length VL30s integrations. The GC-content of 400 bp upstream and downstream of full-length VL30s was analyzed in the Galaxy platform. The graph shows the average GC-content in each chromosome. Error bars represent 95 % confidence intervals (+/- the standard deviation). (JPG 483 kb)

Additional file 5: VL30 elements integrated nearby mouse genes. The table provides information about all VL30 elements integrated in the vicinity of mouse genes and their relative distance to transcription start sites (TSS). (PDF 254 kb)

Additional file 6: VL30 elements associated with Krueppel-associated box (KRAB) zinc finger proteins. The table provides information about all VL30 elements associated with KRAB zinc finger proteins and their relative distance to TSS. (PDF 10 kb)

Abbreviations

EST: expressed sequence tag; KRAB: Krueppel-associated box; LTR: Long Terminal Repeat; PSF: poly-pyrimidine tract-binding protein-associated splicing factor; TFBS: transcription factor binding site; TSD: Target Site Duplication; TSS: transcription start site; VL30s: viral like 30 elements.

Competing interests

The authors declare that they have no competing interests.

Authors' contributions

GM analysed data for structural features, motif-recognition and transcription-factor binding sites, performed nucleotide substitution analysis and contributed in the study design. DN analysed the VL30-gene associations. SM analysed phylogenetic data. DG performed VL30s data mining. ST performed GC content analysis. GV and EK contributed in the study design and manuscript preparation. TT conceived the study and wrote the final manuscript. All authors read and approved the final manuscript.

Funding

This work was supported by a fund from Empirikeion Foundation (6351/2004), Athens, Greece and small institutional funds from University of Ioannina.

Author details

[1]Laboratory of General Biology, Faculty of Medicine, School of Health Sciences, University of Ioannina, Ioannina 45110, Greece. [2]Biomedical Research Division, Institute of Molecular Biology and Biotechnology, Foundation of Research and Technology (IMBB-FORTH), University Campus, Ioannina 45110, Greece. [3]Laboratory of Molecular Biology and Genetics, Department of Biological Applications and Technology, School of Health Sciences, University of Ioannina, Ioannina 45110, Greece. [4]Department of Computer Science, School of Sciences, University of Ioannina, Ioannina 45110, Greece. [5]Hematology Laboratory, Unit of Molecular Biology, University Hospital of Ioannina, Ioannina 45110, Greece.

References

1. Lander ES, Linton LM, Birren B, Nusbaum C, Zody MC, Baldwin J, et al. Initial sequencing and analysis of the human genome. Nature. 2001;409:860–921.
2. Waterston RH, Lindblad-Toh K, Birney E, Rogers J, Abril JF, Agarwal P, et al. Initial sequencing and comparative analysis of the mouse genome. Nature. 2002;420:520–62.
3. Goodier JL, Kazazian Jr HH. Retrotransposons revisited: the restraint and rehabilitation of parasites. Cell. 2008;135:23–35.
4. French NS, Norton JD. Structure and functional properties of mouse VL30 retrotransposons. Biochim Biophys Acta. 1997;1352:33–47.
5. Nilsson M, Bohm S. Inducible and cell type-specific expression of VL30 U3 subgroups correlate with their enhancer design. J Virol. 1994;68:276–88.
6. Faulkner GJ, Kimura Y, Daub CO, Wani S, Plessy C, Irvine KM, et al. The regulated retrotransposon transcriptome of mammalian cells. Nat Genet. 2009;41:563–71.
7. Brunmeir R, Lagger S, Simboeck E, Sawicka A, Egger G, Hagelkruys A, et al. Epigenetic regulation of a murine retrotransposon by a dual histone modification mark. PLoS Genet. 2010;6:e1000927.
8. Tzavaras T, Eftaxia S, Tavoulari S, Hatzi P, Angelidis C. Factors influencing the expression of endogenous reverse transcriptases and viral-like 30 elements in mouse NIH3T3 cells. Int J Oncol. 2003;23:1237–43.
9. Costain WJ, Rasquinha I, Graber T, Luebbert C, Preston E, Slinn J, et al. Cerebral ischemia induces neuronal expression of novel VL30 mouse retrotransposons bound to polyribosomes. Brain Res. 2006;1094:24–37.
10. Magun BE, Rodland KD. Transient inhibition of protein synthesis induces the immediate early gene VL30: alternative mechanism for thapsigargin-induced gene expression. Cell Growth Differ. 1995;6:891–7.
11. Singh K, Saragosti S, Botchan M. Isolation of cellular genes differentially expressed in mouse NIH 3 T3 cells and a simian virus 40-transformed derivative: growth-specific expression of VL30 genes. Mol Cell Biol. 1985;5:2590–8.
12. Scolnick EM, Vass WC, Howk RS, Duesberg PH. Defective retrovirus-like 30S RNA species of rat and mouse cells are infectious if packaged by type C helper virus. J Virol. 1979;29:964–72.
13. Dolberg D, Fan H. Further characterization of virus-like 30S (VL30) RNA of mice: initiation of reverse transcription and intracellular synthesis. J Gen Virol. 1981;54:281–91.
14. Herquel B, Ouararhni K, Martianov I, Le Gras S, Ye T, Keime C, et al. Trim24-repressed VL30 retrotransposons regulate gene expression by producing noncoding RNA. Nat Struct Mol Biol. 2013;20:339–46.
15. Song X, Sui A, Garen A. Binding of mouse VL30 retrotransposon RNA to PSF protein induces genes repressed by PSF: effects on steroidogenesis and oncogenesis. Proc Natl Acad Sci U S A. 2004;101:621–6.
16. Wang G, Cui Y, Zhang G, Garen A, Song X. Regulation of proto-oncogene transcription, cell proliferation, and tumorigenesis in mice by PSF protein and a VL30 noncoding RNA. Proc Natl Acad Sci U S A. 2009;106:16794–8.
17. Adams SE, Rathjen PD, Stanway CA, Fulton SM, Malim MH, Wilson W, et al. Complete nucleotide sequence of a mouse VL30 retro-element. Mol Cell Biol. 1988;8:2989–98.
18. Hodgson CP, Fisk RZ, Arora P, Chotani M. Nucleotide sequence of mouse virus-like (VL30) retrotransposon BVL-1. Nucleic Acids Res. 1990;18:673.
19. Noutsopoulos D, Vartholomatos G, Kolaitis N, Tzavaras T. SV40 large T antigen up-regulates the retrotransposition frequency of viral-like 30 elements. J Mol Biol. 2006;361:450–61.
20. Tzavaras T, Kalogera C, Eftaxia S, Saragosti S, Pagoulatos GN. Clone-specific high-frequency retrotransposition of a recombinant virus containing a VL30 promoter in SV40-transformed NIH3T3 cells. Biochim Biophys Acta. 1998;1442:186–98.
21. Konisti S, Mantziou S, Markopoulos G, Thrasyvoulou S, Vartholomatos G, Sainis I, et al. H (2) O (2) signals via iron induction of VL30 retrotransposition correlated with cytotoxicity. Free Radic Biol Med. 2012;52:2072–81.
22. Noutsopoulos D, Markopoulos G, Koliou M, Dova L, Vartholomatos G, Kolettas E, et al. Vanadium induces VL30 retrotransposition at an unusually high level: a possible carcinogenesis mechanism. J Mol Biol. 2007;374:80–90.
23. Markopoulos G, Noutsopoulos D, Mantziou S, Vartholomatos G, Monokrousos N, Angelidis C, et al. Arsenic induces VL30 retrotransposition: the involvement of oxidative stress and heat-shock protein 70. Toxicol Sci. 2013;134:312–22.
24. Noutsopoulos D, Markopoulos G, Vartholomatos G, Kolettas E, Kolaitis N, Tzavaras T. VL30 retrotransposition signals activation of a caspase-independent and p53-dependent death pathway associated with mitochondrial and lysosomal damage. Cell Res. 2010;20:553–62.
25. de Koning AP, Gu W, Castoe TA, Batzer MA, Pollock DD. Repetitive elements may comprise over two-thirds of the human genome. PLoS Genet. 2011;7:e1002384.
26. Belshaw R, Dawson AL, Woolven-Allen J, Redding J, Burt A, Tristem M. Genomewide screening reveals high levels of insertional polymorphism in the human endogenous retrovirus family HERV-K (HML2): implications for present-day activity. J Virol. 2005;79:12507–14.
27. DeBarry JD, Ganko EW, McCarthy EM, McDonald JF. The contribution of LTR retrotransposon sequences to gene evolution in Mus musculus. Mol Biol Evol. 2006;23:479–81.
28. McCarthy EM, McDonald JF. Long terminal repeat retrotransposons of Mus musculus. Genome Biol. 2004;5:R14.
29. French NS, Norton JD. Analysis of retrotransposon families in genomic DNA by two-dimensional restriction mapping: detection of VL30 insertions in mouse thymic lymphoma. Biochim Biophys Acta. 1994;1219:484–92.
30. Kent WJ, Sugnet CW, Furey TS, Roskin KM, Pringle TH, Zahler AM, et al. The human genome browser at UCSC. Genome Res. 2002;12:996–1006.

31. Fujita PA, Rhead B, Zweig AS, Hinrichs AS, Karolchik D, Cline MS, et al. The UCSC genome browser database: update 2011. Nucleic Acids Res. 2011;39:D876–82.

32. Huda A, Jordan IK. Analysis of Transposable Element Sequences Using CENSOR and RepeatMasker Bioinformatics for DNA Sequence Analysis. In: Posada D, editor. Humana Press; 2009. p. 323–36. doi:10.1007/978-1-59745-251-9_16.

33. Kent WJ. BLAT–the BLAST-like alignment tool. Genome Res. 2002;12:656–64.

34. Jurka J, Kapitonov VV, Pavlicek A, Klonowski P, Kohany O, Walichiewicz J. Repbase update, a database of eukaryotic repetitive elements. Cytogenet Genome Res. 2005;110:462–7.

35. Puschendorf M, Stein P, Oakeley EJ, Schultz RM, Peters AH, Svoboda P. Abundant transcripts from retrotransposons are unstable in fully grown mouse oocytes. Biochem Biophys Res Commun. 2006;347:36–43.

36. Xu Z, Wang H. LTR_FINDER: an efficient tool for the prediction of full-length LTR retrotransposons. Nucleic Acids Res. 2007;35:W265–8.

37. Besmer P, Olshevsky U, Baltimore D, Dolberg D, Fan H. Virus-like 30S RNA in mouse cells. J Virol. 1979;29:1168–76.

38. Torrent C, Gabus C, Darlix JL. A small and efficient dimerization/packaging signal of rat VL30 RNA and its use in murine leukemia virus-VL30-derived vectors for gene transfer. J Virol. 1994;68:661–7.

39. Song X, Sun Y, Garen A. Roles of PSF protein and VL30 RNA in reversible gene regulation. Proc Natl Acad Sci U S A. 2005;102:12189–93.

40. Cartharius K, Frech K, Grote K, Klocke B, Haltmeier M, Klingenhoff A, et al. MatInspector and beyond: promoter analysis based on transcription factor binding sites. Bioinformatics. 2005;21:2933–42.

41. Asch BB, Asch HL. Expression of the retrotransposons, intracisternal A-particles, during neoplastic progression of mouse mammary epithelium analyzed with a monoclonal antibody. Cancer Res. 1990;50:2404–10.

42. Heberlein C, Kawai M, Franz MJ, Beck-Engeser G, Daniel CP, Ostertag W, et al. Retrotransposons as mutagens in the induction of growth autonomy in hematopoietic cells. Oncogene. 1990;5:1799–807.

43. McLean CY, Bristor D, Hiller M, Clarke SL, Schaar BT, Lowe CB, et al. GREAT improves functional interpretation of cis-regulatory regions. Nat Biotech. 2010;28:495–501.

44. Davis CM, Constantinides PG, van der Riet F, van Schalkwyk L, Gevers W, Parker MI. Activation and demethylation of the intracisternal A particle genes by 5-azacytidine. Cell Differ Dev. 1989;27:83–93.

45. Sverdlov ED. Retroviruses and primate genome evolution. Georgetown: Landes Bioscience; 2005.

46. Takahata N, Kimura M. A model of evolutionary base substitutions and its application with special reference to rapid change of pseudogenes. Genetics. 1981;98:641–57.

47. Robbez-Masson L, Rowe HM. Retrotransposons shape species-specific embryonic stem cell gene expression. Retrovirology. 2015;12:45.

48. Li J, Akagi K, Hu Y, Trivett AL, Hlynialuk CJ, Swing DA, et al. Mouse endogenous retroviruses can trigger premature transcriptional termination at a distance. Genome Res. 2012;22:870–84.

49. Matsui T, Leung D, Miyashita H, Maksakova IA, Miyachi H, Kimura H, et al. Proviral silencing in embryonic stem cells requires the histone methyltransferase ESET. Nature. 2010;464:927–31.

50. Rowe HM, Jakobsson J, Mesnard D, Rougemont J, Reynard S, Aktas T, et al. KAP1 controls endogenous retroviruses in embryonic stem cells. Nature. 2010;463:237–40.

51. Kunarso G, Chia N-Y, Jeyakani J, Hwang C, Lu X, Chan Y-S, et al. Transposable elements have rewired the core regulatory network of human embryonic stem cells. Nat Genet. 2010;42:631–4.

52. Smith ZD, Chan MM, Mikkelsen TS, Gu H, Gnirke A, Regev A, et al. A unique regulatory phase of DNA methylation in the early mammalian embryo. Nature. 2012;484:339–44.

53. Kumar S, Hedges SB. A molecular timescale for vertebrate evolution. Nature. 1998;392:917–20.

54. Nellaker C, Keane TM, Yalcin B, Wong K, Agam A, Belgard TG, et al. The genomic landscape shaped by selection on transposable elements across 18 mouse strains. Genome Biol. 2012;13:R45.

55. Brown JD, Piccuillo V, O'Neill RJ. Retroelement demethylation associated with abnormal placentation in Mus musculus x Mus caroli hybrids. Biol Reprod. 2012;86:88.

56. Altschul SF, Gish W, Miller W, Myers EW, Lipman DJ. Basic local alignment search tool. J Mol Biol. 1990;215:403–10.

57. Crooks GE, Hon G, Chandonia JM, Brenner SE. WebLogo: a sequence logo generator. Genome Res. 2004;14:1188–90.

58. Rice P, Longden I, Bleasby A. EMBOSS: the European Molecular Biology Open Software Suite. Trends Genet. 2000;16:276–7.

59. Goecks J, Nekrutenko A, Taylor J, Team TG. Galaxy: a comprehensive approach for supporting accessible, reproducible, and transparent computational research in the life sciences. Genome Biol. 2010;11:R86.

60. Tamura K, Nei M. Estimation of the number of nucleotide substitutions in the control region of mitochondrial DNA in humans and chimpanzees. Mol Biol Evol. 1993;10:512–26.

61. Felsenstein J. Confidence-limits on phylogenies - an approach using the bootstrap. Evolution. 1985;39:783–91.

62. Tamura K, Peterson D, Peterson N, Stecher G, Nei M, Kumar S. MEGA5: Molecular Evolutionary Genetics Analysis using maximum likelihood, evolutionary distance, and maximum parsimony methods. Mol Biol Evol. 2011;28:2731–9.

63. Tristem M. Identification and characterization of novel human endogenous retrovirus families by phylogenetic screening of the human genome mapping project database. J Virol. 2000;74:3715–30.

64. Wang H, Xing J, Grover D, Hedges DJ, Han K, Walker JA, et al. SVA elements: a hominid-specific retroposon family. J Mol Biol. 2005;354:994–1007.

65. Karolchik D, Hinrichs AS, Furey TS, Roskin KM, Sugnet CW, Haussler D, et al. The UCSC table browser data retrieval tool. Nucleic Acids Res. 2004;32:D493–6.

6

Distribution of the DNA transposon family, *Pokey* in the *Daphnia pulex* species complex

Shannon H. C. Eagle and Teresa J. Crease[*]

Abstract

Background: The *Pokey* family of DNA transposons consists of two putatively autonomous groups, *Pokey*A and *Pokey*B, and two groups of Miniature Inverted-repeat Transposable Elements (MITEs), m*Pok*1 and m*Pok*2. This TE family is unusual as it inserts into a specific site in ribosomal (r)DNA, as well as other locations in *Daphnia* genomes. The goals of this study were to determine the distribution of the *Pokey* family in lineages of the *Daphnia pulex* species complex, and to test the hypothesis that unusally high *Pokey*A number in some isolates of *Daphnia pulicaria* is the result of recent transposition. To do this, we estimated the haploid number of *Pokey*, m*Pok*, and rRNA genes in 45 isolates from five *Daphnia* lineages using quantitative PCR. We also cloned and sequenced partial copies of *Pokey*A from four isolates of *D. pulicaria*.

Results: Haploid *Pokey*A and *Pokey*B number is generally less than 20 and tends to be higher outside rDNA in four lineages. Conversely, the number of both groups is much higher outside rDNA (~120) in *D. arenata*, and *Pokey*B is also somewhat higher inside rDNA. m*Pok*1 was only detected in *D. arenata*. m*Pok*2 occurs both outside (~30) and inside rDNA (~6) in *D. arenata*, but was rare (≤2) outside rDNA in the other four lineages. There is no correlation between *Pokey* and rRNA gene number (mean = 240 across lineages) in any lineage. Variation among cloned partial *Pokey*A sequences is significantly higher in isolates with high number compared to isolates with an average number.

Conclusions: The high *Pokey* number outside rDNA in *D. arenata* and inside rDNA in some *D. pulicaria* isolates is consistent with a recent increase in transposition rate. The *D. pulicaria* increase may have been triggered by insertion of *Pokey*A into a region of transcriptionally active rDNA. The expansion in *D. arenata* (thought to be of hybrid origin) may be a consequence of release from epigenetic repression following hybridization. Previous work found *D. obtusa* to be very different from the *D. pulex* complex; mean *Pokey*A is higher in rDNA (~75), rDNA array size is nearly twice as large (415), and the two are positively correlated. The predominance of *Pokey* in only one location could be explained by purifying selection against ectopic recombination between elements inside and outside rDNA.

Keywords: *Daphnia*, Transposon, *Pokey*, Ribosomal DNA, MITE

Background

Transposable elements (TEs) are segments of DNA that can move around the genome. TEs are often detrimental to the host because they can interrupt gene function, their transposition has energy costs (e.g. the repair of double strand breaks), they can cause ectopic recombination [1], and the epigenetic mechanisms used to control them can shut down surrounding genes [2]. Usually, DNA transposons must undergo horizontal transfer to remain active

[3]. However, *Pokey*, a class II cut-and-paste DNA transposon in the *piggyBac* superfamily [4] has been vertically inherited in the subgenus, *Daphnia* [5]. *Pokey* inserts at TTAA sites and creates a target site duplication (TSD). It ranges in size from 4.5 to 10 kilobase pairs (kb); the length variation is due to repeat sequences at the 5′ end derived from the ribosomal intergenic spacer (IGS) [4] and/or the ribosomal internal transcribed spacer (ITS) [6]. *Pokey* is the only DNA transposon known to insert in a specific location in ribosomal DNA (rDNA); other rDNA elements, such as R1 and R2, are non-Long Terminal Repeat (LTR) retrotransposons [7]. Nevertheless, Kojima and Jurka [8] recently reported the discovery of class II *Dada* TEs in

* Correspondence: tcrease@uoguelph.ca
Department of Integrative Biology, University of Guelph, Guelph, ON N1G 2 W1, Canada

organisms as taxonomically diverse as fish, molluscs, annelids, and cladoceran crustaceans (*Daphnia*). Like *Pokey*, they show target-site specificity for particular multicopy RNA genes (small nuclear RNA, transfer RNA), encode a putatively cut-and-paste transposase, and create TSDs on insertion. However, unlike other elements that target multicopy RNA genes, *Pokey* also inserts into TTAA sites outside rDNA in the genomes of North American (NA) *D. pulex* [9] and NA *D. pulicaria* [10].

In eukaryotes, rDNA is a multigene family arranged in tandem arrays of units each consisting of the 18S, 5.8S, and 28S rRNA genes plus the ITS, the external transcribed spacer (ETS), and the IGS. The number of rDNA units per eukaryotic genome varies from less than 50 to over 25,000 [11] and is usually much higher than the number required for viability [7]. rDNA typically displays the phenomenon of concerted evolution, which means that intraspecific sequence divergence between rDNA copies is generally much lower than interspecific divergence [7]. The mechanisms of concerted evolution are thought to be gene conversion and unequal crossing over [7], which can cause the number of rDNA units to fluctuate. Due to its multicopy nature and the presence of copies above the number required for viability, some TEs, such as R1, R2, and *Pokey*, have persisted in rDNA for very long periods of time [7] even though the homogenizing mechanisms of concerted evolution are expected to continually remove them. In addition, TEs inserted in rRNA genes render the genes non-functional, so selection is expected to favor loss of these elements [7].

Daphnia are freshwater cladoceran crustaceans with a worldwide distribution [12]. Based on molecular analyses, the *D. pulex* species complex (subgenus *Daphnia*) is composed of eight lineages between which hybridization frequently occurs: North American (NA) *D. pulex*, *D. melanica*, *D. middendorffiana*, North American (NA) *D. pulicaria*, *D. tenebrosa*, European (EU) *D. pulex*, European (EU) *D. pulicaria* [13], and *D. arenata* [14]. Although these analyses suggest that the North American and European populations of *D. pulex* and *D. pulicaria* should be classified as separate species, they have not been officially described as such [13]. While most cladocerans reproduce by cyclical parthenogenesis (diapausing eggs are produced sexually and direct-developing eggs are produced apomictically), populations of obligate parthenogens (diapausing eggs are also produced apomictically) also occur in some lineages of this species complex [13].

Eagle and Crease [10] found that the total haploid number of *Pokey* in NA *D. pulex* and NA *D. pulicaria* is usually less than 20 copies, most of which are found outside rDNA. In addition, *Pokey* number in rDNA is not correlated with rRNA gene number. In contrast, LeRiche et al. [15] found *Pokey* number in another species in the same subgenus, *Daphnia obtusa*, to be as high as 154 (mean = 75), almost

all of which are located inside rDNA. Moreover, *Pokey* number in rDNA is strongly correlated with rRNA gene number.

Both autonomous (encoding a functional transposase) and non-autonomous (the transposase gene is degraded or partially deleted) copies of *Pokey* have been found [6]. Further, Miniature Inverted Repeat Transposable Elements (MITEs), which are very short (usually less than 600 nucleotides) [16], non-autonomous TEs have been observed in *D. arenata* [6]. Overall, Elliott et al. [6] discovered four TEs in the *D. arenata Pokey* family: the original *Pokey* group discovered in *D. pulex*, now called *Pokey*A; a variant containing a transposase gene (complete or partial) that was initially seen in *D. obtusa* [5], now called *Pokey*B; a MITE with terminal inverted repeats (TIRs) similar to those of *Pokey*A, called m*Pok*1, and a MITE with TIRs similar to those of *Pokey*B, called m*Pok*2. The m*Pok* elements average 760 base pairs (bp) in length [6]. *Pokey*A and m*Pok*1 have 16 bp imperfect TIRs, and *Pokey*B and m*Pok*2 have 12 bp imperfect TIRs [6]. Like *Pokey*A, *Pokey*B also contains an open reading frame (ORF) encoding a putative transposase and both ORFs contain an intron [6]. Mean sequence divergence between copies of *Pokey*A and *Pokey*B containing a complete or partial transposase gene from *D. arenata* is 40 %, suggesting that the two TE groups did not diverge from one another recently.

Three additional types of TEs have now been discovered in the *D. arenata Pokey* family, but we do not know if they occur in other lineages in the *D. pulex* complex. Our first goal was to determine the distribution of *Pokey* and m*Pok* in the *D. pulex* complex. We did this using quantitative polymerase chain reaction (qPCR) to estimate gene number in 45 cyclically parthenogenetic isolates representing five lineages: NA *D. pulex* ($N = 26$), NA *D. pulicaria* ($N = 5$), *D. arenata* ($N = 4$), EU *D. pulex* ($N = 8$), and EU *D. pulicaria* ($N = 2$) (Fig. 1, Additional file 1: Table S1). Our second goal was to test the hypothesis of recent transposition of *Pokey*A in NA *D. pulicaria*, suggested by Eagle and Crease [10], by estimating divergence among partial *Pokey*A sequences from four isolates; two with unusually high *Pokey*A number and two with numbers close to the average for this species.

Results
Pokey and m*Pok* number in the *D. pulex* complex
We used qPCR (Additional file 1: Table S2) to estimate the haploid number of the four TEs in the *Pokey* family (*Pokey*A, *Pokey*B, m*Pok*1, m*Pok*2) in the entire genome (t*Pokey*, tm*Pok*) and in rDNA (r*Pokey*, rm*Pok*) by comparing their PCR amplification rate to the PCR amplification rate of two single copy genes, *Gtp* (a member of the RAB subfamily of small GTPases) and *Tif* (a transcription initiation factor) [10], [17]. We calculated the number of *Pokey* and m*Pok* outside rDNA (g*Pokey*, gm*Pok*) by subtracting

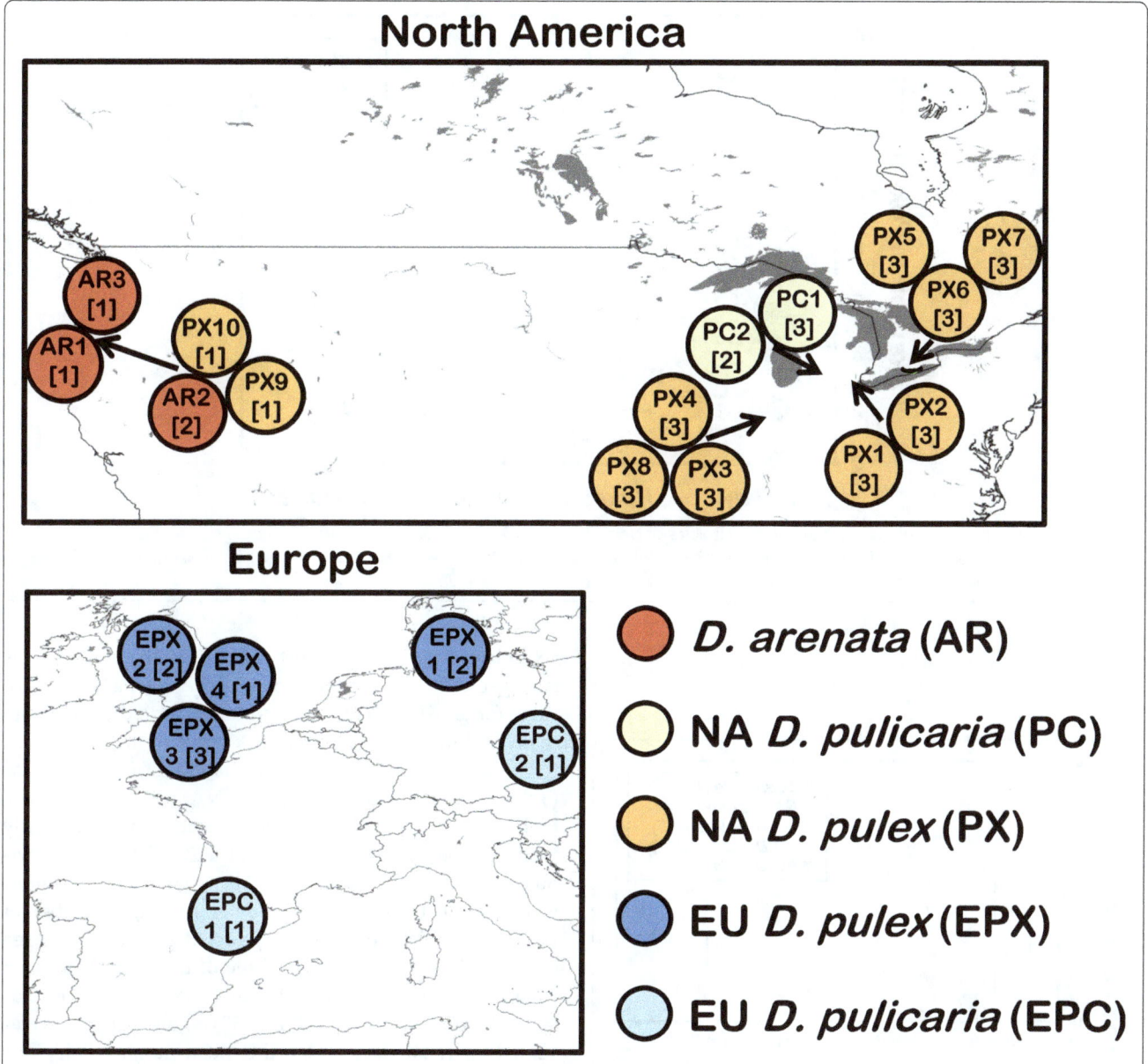

Fig. 1 Location of *Daphnia* populations sampled. The numbers in square brackets indicate the number of isolates sampled from that population. NA = North America, EU = Europe

rPokey or rmPok from tPokey or tmPok. We also estimated the number of 18S and 28S rRNA genes in each isolate.

The mean haploid number of all PokeyA and PokeyB in NA *D. pulex* is 11.4 and never exceeds 19 copies per isolate (Additional file 1: Table S3). Mean gPokeyA plus gPokeyB number (8.6) exceeds mean rPokeyA plus rPokeyB number (2.8) and this is true for both PokeyA (5.3 vs 2.8) and PokeyB (3.6 vs 0.1, Fig. 2), individually. PokeyA number is higher than PokeyB number in NA *D. pulex*, inside and outside rDNA (Fig. 2). These patterns are similar in each of the 26 NA *D. pulex* isolates (Additional file 1: Table S3).

The mean total number of PokeyA plus PokeyB per haploid genome in three of the other four lineages (NA *D. pulicaria*, EU *D. pulex*, EU *D. pulicaria*) is similar to NA *D. pulex*, with means ranging from 11.4 to 29.4 (Fig. 3). In contrast, the mean total number in *D. arenata* is substantially higher (125.4) (Additional file 1: Table S3), and the mean number of gPokeyA plus gPokeyB is highest in *D. arenata* (Table 1, Fig. 3). Although the mean is lower compared to *D. arenata*, we still detected gPokeyB in all of the other 41 isolates, and gPokeyA in all but six of them (one NA *D. pulex*, one NA *D. pulicaria* and four EU *D. pulex*; Additional file 1: Table S3). The gPokeyA estimate in these

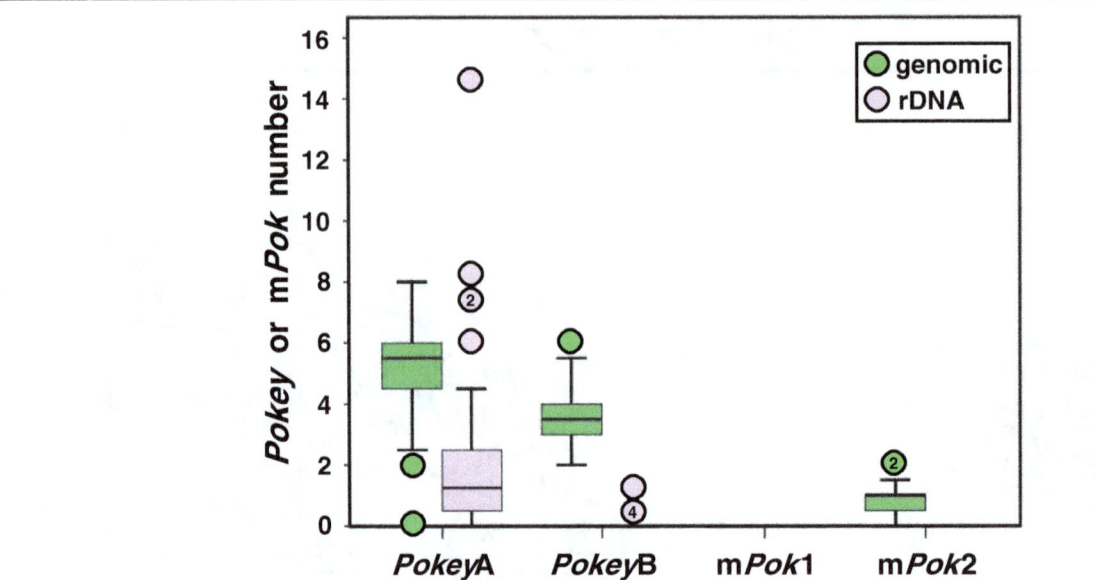

Fig. 2 Box plot of haploid *Pokey* family number in 26 isolates of North American *Daphnia pulex*. *Pokey* or m*Pok* numbers greater or less than 1.5 times the box length are indicated with circles. The number within the circle indicates how many isolates have that particular number of *Pokey* or m*Pok*. Genomic = *Pokey* or m*Pok* inserted outside 28S, rDNA = *Pokey* or m*Pok* inserted in 28S

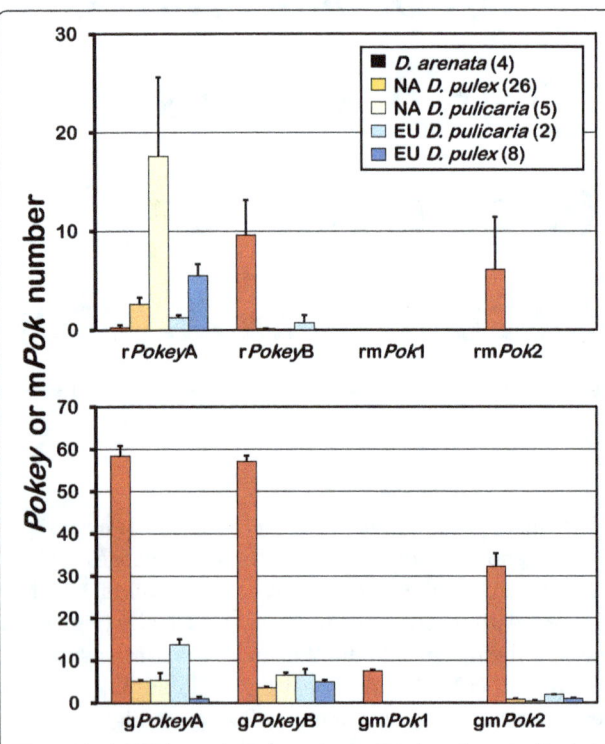

Fig. 3 Haploid *Pokey* and m*Pok* number in 45 isolates from the *Daphnia pulex* complex. Black vertical bars are standard errors. NA = North America, EU = Europe, r*Pokey* = *Pokey* inserted in 28S, rm*Pok* = m*Pok*1 inserted in 28S, g*Pokey* = *Pokey* inserted outside 28S, gm*Pok* = m*Pok* inserted outside 28S

six isolates was set to zero because the estimate of r*Pokey*A is larger than the estimate of t*Pokey*A giving a negative value for g*Pokey*A. Thus, we cannot exclude the possibility that very low numbers of g*Pokey*A occur in these isolates.

The mean number of r*Pokey*A is lowest in *D. arenata* (Table 1, Fig. 3) and highest in NA *D. pulicaria*. The high mean in NA *D. pulicaria* is due to the unusually high number in two isolates (PC1.2, PC2.1), as previously described by Eagle and Crease [10]. These isolates were chosen to determine whether the high number of r*Pokey*A could be explained by recent transposition (see below). r*Pokey*A is absent in three of the four *D. arenata* isolates, but is only absent in three of the other 41 isolates (Additional file 1: Table S3). This pattern is reversed for r*Pokey*B in which the mean number is highest in *D. arenata* (Table 1, Fig. 3), but the element is absent from rDNA in 35 of the other 41 isolates (Additional file 1: Table S3).

The MITEs, m*Pok*1 and m*Pok*2 are primarily found in *D. arenata* (Fig. 3). Indeed, the only MITE we detected in the other four lineages is gm*Pok*2, but it is present in no more than 2 copies per haploid genome (Table 1), and absent in most isolates (Additional file 1: Table S3). m*Pok*2 was detected both inside (mean = 6) and outside (mean = 32) rDNA in *D. arenata* (Table 1, Fig. 3), but m*Pok*1 was only detected outside rDNA (mean = 7.5) (Table 1, Fig. 3).

rRNA gene number in the *D. pulex* complex
Haploid 18S number ranges from 98 to 460 and 28S number ranges from 88 to 577 in the 45 *Daphnia*

Table 1 Haploid number of *Pokey*, m*Pok*, and rRNA genes in isolates from the *Daphnia pulex* complex

Gene[a]	N[b]	NA[c] *D. pulex*	NA[c] *D. pulicaria*	*D. arenata*	EU[c] *D. pulex*	EU[c] *D. pulicaria*
		26	5	4	8	2
18S	Mean	216.3	310.4	303.8	194.5	366.3
	Std Dev[b]	85.3	112.7	38.3	93.2	35.7
	Range	112.5 to 440.5	176.0 to 460.0	248.0 to 332.5	97.5 to 339.5	341.0 to 391.5
28S	Mean	230.9	405.9	309.3	201.5	347.8
	Std Dev	81.1	128.3	30.3	124.3	9.5
	Range	119.5 to 452.5	234.5 to 576.5	264.5 to 330.5	87.5 to 418.0	341.0 to 354.5
r*Pokey*A	Mean	2.6	17.6	0.3	5.5	1.3
	Std Dev	3.4	17.9	0.5	3.4	0.4
	Range	0 to 14.5	3.5 to 48.0	0 to 1.0	1.0 to 10.5	1.0 to 1.5
r*Pokey*B	Mean	0.1	0	9.6	0	0.8
	Std Dev	0.3		7.0		1.1
	Range	0 to 1.0		5.0 to 20.0		0 to 1.5
rm*Pok*1	Mean	0	0	0	0	0
	Std Dev					
	Range					
rm*Pok*2	Mean	0	0	6.1	0	0
	Std Dev			10.6		
	Range			0 to 22.0		
g*Pokey*A	Mean	5.1	5.3	58.4	1.0	13.8
	Std Dev	1.7	4.0	4.9	1.3	1.8
	Range	0.0 to 8.0	0.0 to 10.5	52.5 to 63.5	0.0 to 3.0	12.5 to 15.0
g*Pokey*B	Mean	3.6	6.5	57.1	4.9	6.5
	Std Dev	1.1	1.5	2.8	1.2	2.1
	Range	2.0 to 6.0	5.0 to 9.0	54.5 to 61.0	3.5 to 7.0	5.0 to 8.0
gm*Pok*1	Mean	0	0	7.5	0	0
	Std Dev			0.7		
	Range			7.0 to 8.5		
gm*Pok*2	Mean	0.8	0.4	32.3	1.0	2.0
	Std Dev	0.5	0.4	6.1	0.4	0.0
	Range	0 to 2.0	0 to 1.0	25.0 to 38.5	0.5 to 1.5	2.0 to 2.0

[a]18S = 18S rRNA genes, 28S = 28S rRNA genes, r*Pokey* = *Pokey* inserted in 28S, g*Pokey* = *Pokey* inserted outside 28S, rm*Pok* = m*Pok* inserted in 28S, gm*Pok* = m*Pok* inserted outside 28S

[b]N = number of isolates. Std Dev = Standard Deviation

[c]NA = North American, EU = European

isolates (Table 1). 28S number is larger than 18S number by 2.5 to 138 copies in 33 of the 45 isolates, and the difference is significant in 23 of these (Fig. 4, Additional file 1: Table S3). Despite the differences between the two genes, their numbers are significantly correlated in NA *D. pulex* with a slope of 0.92 and an R^2 of 0.937 ($p < 10^{-15}$, Table 2, Additional file 2: Figure S1). The slope of the correlation based on the other 19 isolates is 1.19 with an R^2 of 0.873 ($p = <10^{-9}$, Table 2, Additional file 2: Figure S1). There is no correlation between 28S and total r*Pokey* plus rm*Pok* number in the 26 NA *D. pulex*

isolates ($R^2 = 0.04$, Table 2, Additional file 2: Figure S2). Similarly, there is no correlation between 28S and r*Pokey* plus rm*Pok* number in the 19 isolates from the other four lineages ($R^2 = 0.03$, Table 2, Additional file 2: Figure S2).

Variation in partial *Pokey*A sequences in North American *D. pulicaria*

An approximately 1500 bp segment, including 519 bp of the 5′ end of the *Pokey*A transposase and the length-variable region upstream was PCR-amplified, cloned, and

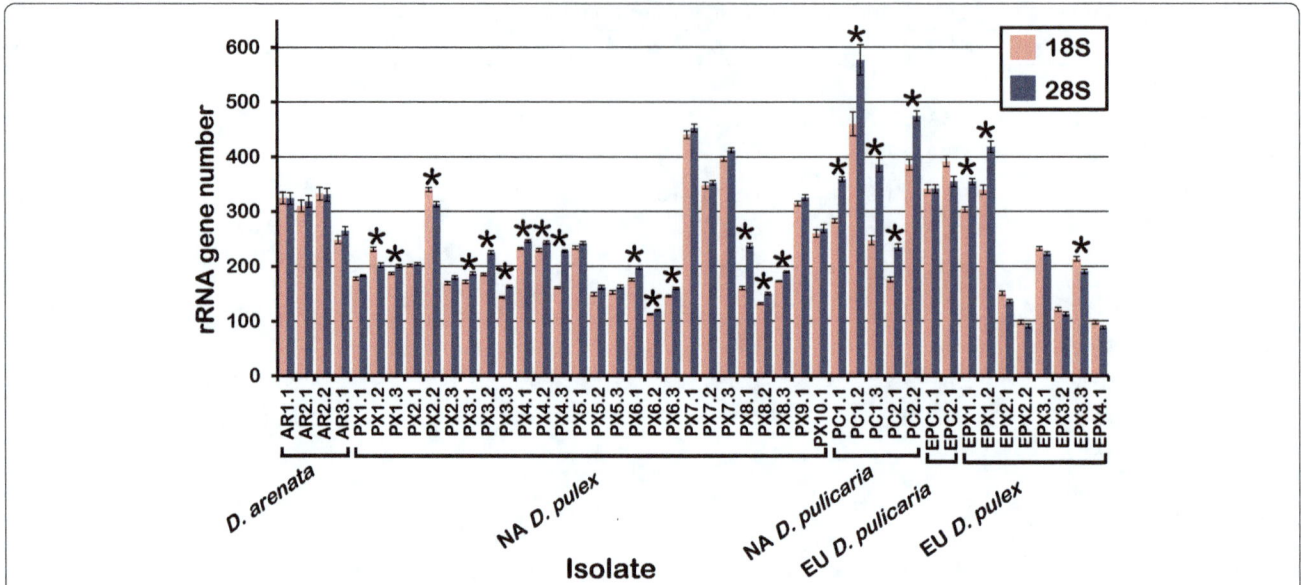

Fig. 4 Haploid 18S and 28S rRNA gene number in isolates from the *Daphnia pulex* complex. * means are significantly different at the 5 % level after sequential Bonferroni correction. 18S = 18S rRNA genes, 28S = 28S rRNA genes. Each isolate is identified by a 2 or 3-letter lineage code followed by x.y, where x is the population and y is the isolate. NA = North America, EU = Europe

sequenced from four isolates of NA *D. pulicaria*. We predicted that sequence divergence would be lower among copies from isolates with high *Pokey*A number if the increase was a consequence of recent transposition. Using data from Eagle and Crease [10], we chose two isolates, one with high and one with average r*Pokey*A number, from each of two populations. The number of other *Pokey* family elements is similar in all four isolates (Table 3). The number of 28S genes in these isolates ranges from 235 to 577 and the one with the lowest number (PC2.1) has the highest number of r*Pokey*A (48, Table 3).

On average, the mean number of differences between each cloned sequence and the isolate consensus sequence is lower in the isolate with a high *Pokey*A number compared to the isolate with an average number from the same population. Moreover, the difference between means is significant in both populations (Table 3). The Neighbor-joining dendrogram of these sequences (Fig. 5) shows two major clusters; cluster 1 contains 11 of 12 sequences from isolate High 1 (PC1.2), as well as sequences from both

Average isolates. Cluster 2 contains all 12 sequences from isolate High 2 (PC2.1), as well as sequences from both Average isolates. All but two of the sequences encode a putatively functional transposase (no insertions, deletions, or stop codons). The exceptions are High 1–19, which contains several stop codons and numerous nucleotide substitutions throughout, and Average 2-32, in which the first 29 bp of the coding sequence are deleted (gaps are not considered in calculations of sequence divergence so this sequence does not appear to be divergent from other sequences in cluster 2).

Discussion

Pokey family distribution in NA and EU *D. pulex* and *D. pulicaria*

In general, *Pokey* number is similar in four of the five lineages we examined; NA and EU *D. pulex* and NA and EU *D. pulicaria*. The mean number of *Pokey*A plus *Pokey*B is less than 20 and both TEs are primarily located outside of rDNA. Our *Pokey*A numbers (Table 1) are similar to

Table 2 Correlation between haploid r*Pokey* family and rRNA gene number in isolates from five lineages of the *Daphnia pulex* complex

lineage	X-axis[c]	Y-axis[c]	slope	y-intercept	R^2	p-value*	Figure
NA *D. pulex*[a]	18S	28S	0.920	31.88	0.937	**6.3×10^{-16}**	Additional file 2: Figure S1
four lineages[b]	18S	28S	1.19	23.39	0.873	**4.9×10^{-9}**	Additional file 2: Figure S1
NA *D. pulex*	28S	r*Pokey* family	−0.008	4.60	0.035	0.361	Additional file 2: Figure S2
four lineages	28S	r*Pokey* family	0.014	6.44	0.026	0.51	Additional file 2: Figure S2

[a]This analysis is based on 26 isolates
[b]This analysis is based on 19 isolates including *D. arenata* (4), EU *D. pulex* (8), and NA (5) and EU (2) *D. pulicaria*
[c]18S = 18S rRNA genes, 28S = 28S rRNA genes, r*Pokey* family = *Pokey* plus m*Pok* inserted in 28S = [r*Pokey*A + r*Pokey*B + rm*Pok*1 + rm*Pok*2]
*p-values < 0.05 are indicated in bold font

Table 3 Haploid number of *Pokey*, m*Pok*, and rRNA genes in four isolates of North American *Daphnia pulicaria*

gene[a]	PC1.1 Average 1[b]	PC1.2 High 1	PC2.1 High 2	PC2.2 Average 2
18S	283	460	176	385.5
28S	358.5	576.5	234.5	474.5
r*Pokey*A	6	17	48	3.5
g*Pokey*A	3.5	7.5	0	5
g*Pokey*B	5	6.5	9	6
gm*Pok*2	1	0	0.5	0.5
mean number of differences from consensus[c]	49.9	10.8	1.6	45.8
standard deviation	43.3	36.9	3.4	32.5
range	0 to 97	0 to 128[d]	0 to 6	12 to 78
p-value[e]		0.027		0.010

[a]18S = 18S rRNA genes. 28S = 28S rRNA genes. r*Pokey*B, rm*Pok*2 and m*Pok*1 were not detected in these four isolates

[b]Average and High correspond to the sequence labels in Fig. 4

[c]The number of differences between the consensus and each of 12 cloned 1500 bp sequences was determined for each isolate. Gaps were included in the analysis. Consecutive gaps were counted as one variant nucleotide position

[d]All other values were 0 or 1 for this isolate

[e]ANOVA was used to compare the mean number of differences across all four isolates. This test was significant ($F = 6.62$, d.f. = 3, $p = 0.0009$). A *Post hoc* Tukey's HSD test was used to determine if means in the Average and High isolates from the same population were significantly different. They p-values refer to the Tukey's tests

estimates from Eagle and Crease [10] for NA *D. pulex* (mean r*Pokey*A = 2.1 and mean g*Pokey*A = 9.6 based on 43 isolates) and NA *D. pulicaria* (mean r*Pokey*A = 6.6 and mean g*Pokey*A = 9.5 based on 26 isolates). However, the NA *D. pulicaria* r*Pokey*A mean in the current study (17.6) is higher than the previous estimate because we only included five isolates, two of which have unusually high r*Pokey*A number, and were chosen for that reason.

Both the current and previous estimates of g*Pokey*A and g*Pokey*B are within the range of other DNA transposons found in the *Daphnia* genome sequence [18]. For example, there are approximately 31 copies per *Tc1/mariner* family (217 copies from 7 families) and approximately 6 copies per *hAT* family (33 copies from 6 families). In contrast, the mean numbers of r*Pokey*A and r*Pokey*B (Table 1) tend to be low compared to the mean number of rDNA-specific non-LTR retroelements, R1 (34) and R2 (12) in *D. melanogaster* [19].

Overall, *Pokey*A outnumbers *Pokey*B, which could be due to a higher transposition rate, weaker purifying selection at the level of the host, and/or lower rates of deletion of *Pokey*A. Due to the fact that they both insert into the same TTAA site in rDNA, it seems unlikely that the strength of purifying selection against these two TEs at the level of the host, or their rate of deletion differs. Therefore, we suggest that the higher number of *Pokey*A may be due to a higher transposition rate, which could be tested using yeast excision assays.

Although *Pokey* number and distribution is similar in the four *D. pulex* and *D. pulicaria* lineages, it was necessary to set g*Pokey*A to 0 in six isolates because r*Pokey*A exceeds t*Pokey*A. The t*Pokey*A primer pair is located at

the 5′ end of the transposase gene, and can only amplify autonomous *Pokey*A. Conversely, it was necessary to locate the r*Pokey*A forward primer near the 3′ end of the gene, even though it was not possible to design primers in that region to target only r*Pokey*A or rm*Pok*1. Consequently, the r*Pokey*A forward primer amplifies both autonomous and non-autonomous r*Pokey*A, as well as rm*Pok*1. However, the higher number of r*Pokey*A compared to t*Pokey*A cannot be due to the presence of rm*Pok*1 because no copies of m*Pok*1 were detected in these isolates by end-point PCR using internal, m*Pok*1-specific primers. Thus, the higher number of r*Pokey*A compared to t*Pokey*A in some isolates is most likely explained by copies of r*Pokey*A that did not amplify with the t*Pokey*A primers (Table 1, Fig. 3), but further study would be required to confirm this.

The number of m*Pok* we observed in NA and EU *D. pulex* and NA and EU *D. pulicaria* is very low (0 to 2 haploid copies per isolate) compared to other *piggyBac* MITEs, for which numbers range from 50 to 2×10^4 in metazoans including nematodes, insects, and vertebrates [20]. Indeed, the absence of any m*Pok*1 in the 41 isolates we examined suggests that this MITE may not occur in these four lineages. In addition, the absence of m*Pok*2 in rDNA and the low number outside rDNA (maximum of two copies) suggests that the activity of m*Pok*2 is very low in these lineages.

Pokey family distribution in *D. arenata*

While m*Pok*1 and m*Pok*2 were both detected in *D. arenata*, the former was only detected outside rDNA and in low numbers suggesting its activity is low. The mean p-distance

Fig. 5 (See legend on next page.)

between copies of m*Pok*1 and the m*Pok*1 consensus sequence is only 2 % in *D. arenata* isolate AR1.1 (Additional file 2: Figure S3). This, along with the fact that it does not occur in the other lineages, suggests that m*Pok*1 could be of recent origin. In contrast, m*Pok*2 occurred both inside and outside rDNA, with a mean haploid number of genomic copies greater than 30. This is at the top of the range for g*Pokey* in the other four lineages (Fig. 3), suggesting that m*Pok*2 may be active in *D. arenata*. Recent activity is also consistent with a p-distance less than 3 % between the m*Pok*2 consensus sequence and 75 % of the m*Pok*2 in isolate AR1.1 (Additional file 2: Figure S3). However, the p-distance ranges from 4 to 22 % for the other copies of m*Pok*2 (Additional file 2: Figure S3). This, and the fact that m*Pok*2 is found in other lineages in the *D. pulex* complex, suggests that this MITE is not of recent origin [6]. Further work is required to determine the distribution of m*Pok* in the *D. pulex* complex, and in other species in the subgenus *Daphnia*.

The number of *Pokey* and m*Pok*, except r*PokeyA*, is substantially higher in *D. arenata* than the other four lineages, resulting in a very different distribution (Fig. 3). *Daphnia arenata* is endemic to western Oregon [13] and may have diverged from NA *D. pulex* during the Pleistocene glaciation, during which it was isolated from other refugial populations south of the glaciers by the Cordilleran ice sheet [21]. This restriction to a small geographic region could have resulted in a smaller effective population size, such that the fate of slightly deleterious mutations (such as some TE insertions) could have been primarily determined by genetic drift instead of purifying selection [22].

Another explanation for the high *Pokey* number in *D. arenata* is a much higher transposition rate compared to the other lineages. Such an increase could be the result of changes in *Pokey*, such as mutations in the transposase that increase transposition rate, or mutations in the terminal repeats that increase recognition by the transposase. These hypotheses could be tested using yeast excision, yeast one-hybrid, and/or electrophoretic mobility shift assays.

A third explanation for high *Pokey* number in *D. arenata* is a loss of epigenetic regulation by the host. The most well-known example of such a loss is the phenomenon of P-M hybrid dysgenesis in *Drosophila*, which occurs when a male from a strain with *P*-elements is mated to a female from a strain without *P*-elements [23]. The female usually provides the epigenetic regulation, but if she does not have *P*-elements, then the paternal *P*-elements are able to

transpose without regulation in the offspring [24], leading to effects such as male recombination, sterility, mutation, chromosomal aberrations, and nondisjunction [23].

Interspecific hybridization is also known to stimulate increased TE activity due to the disruption of host defenses, although this is not always the case (see Ungerer and Kawakami [25] for an example). Labrador et al. [26] found that the *Osvaldo* LTR retrotransposon has a higher transposition rate in hybrids of *Drosophila buzzatii* and *Drosophila koepferae* compared to non-hybrids. Similar increases in transposition rate in interspecific hybrids have been seen for other elements in rice and *Drosophila* (e.g. [27–31]).

Vergilino et al. [32] suggested that *D. arenata* may be a hybrid between NA *D. pulex* and *D. tenebrosa* or EU *D. pulicaria* because the position of *D. arenata* varies on phylogenetic trees generated from different nuclear loci [32–34]. NA *D. pulex* is likely the maternal parent based on analysis of mitochondrial DNA [13, 21]. If *D. arenata* does have a hybrid origin, release from host regulation could explain the relatively high number of *Pokey* and m*Pok*2 and possibly the occurrence of m*Pok*1. Further work is required to establish the distribution of m*Pok* in other *Daphnia* species, determine the origin of m*Pok*1, confirm the hybrid origin of *D. arenata*, and test the hypothesis that the high *Pokey* load in *D. arenata* is a consequence of increased transposition in a hybrid lineage. If this is the case, we would also expect the number of active TEs from other superfamilies to be higher in *D. arenata* compared to other lineages in the *D. pulex* complex, but this remains to be tested.

Distribution of *Pokey* in the subgenus, *Daphnia*

Our results are very different from those obtained for *D. obtusa* in which r*Pokey* number is much higher (mean = 75 for *PokeyA* and 9 for *PokeyB*) and g*Pokey* number is much lower (mean = 0.5 for *PokeyA*) [15]. Moreover, r*PokeyA* is significantly positively correlated with 28S number in *D. obtusa* ($R^2 = 0.54$). The mean haploid rRNA gene number in lineages of the *D. pulex* complex is 240, with values ranging from 90 to nearly 580. Conversely, the mean haploid number in 21 isolates of *D. obtusa* is 415, with values ranging from 180 to 960. LeRiche et al. [15] suggested that r*PokeyA* number could be higher in *D. obtusa* due to the larger size of its rDNA array, which may be a consequence of higher rates of sister chromatid exchange [15]. A large rDNA array provides more *Pokey* insertion sites, and a high rate of sister chromatid exchange provides more opportunity for the element to increase via recombination, even if the transposition rate is low.

Although the number of *Daphnia* species whose *Pokey* distribution has been studied is modest, the emerging pattern is that these elements tend to be concentrated in one genomic location (inside or outside of rDNA) and occur in low number in the other. In the *D. pulex* complex, *Pokey* is concentrated outside rDNA and in *D. obtusa* it is concentrated inside rDNA. Moreover, with the exception of *D. arenata*, g*Pokey* in the *D. pulex* complex (less than 30 copies) tends to be much lower than r*Pokey* in *D. obtusa* (up to 154 copies). Even so, the lower number of g*Pokey* in the *D. pulex* complex is consistent with the number of other DNA elements observed in the *Daphnia* genome sequence as mentioned above [18].

One explanation for the tendency of *Pokey* to be concentrated in only one genomic location is purifying selection against chromosomal rearrangements caused by ectopic recombination between copies inside and outside rDNA. The ectopic recombination model [35] suggests that TEs should accumulate in genomic regions of low recombination due to the deleterious effects of exchange between elements at non-homologous sites. Although we do not know the location of g*Pokey* in *Daphnia* genomes, their relatively low number suggests that their accumulation is resisted by selection. In contrast, Eickbush [36] suggested that ectopic recombination between TEs inserted in rDNA may not have severe effects because the outcome is similar to the effects of concerted evolution. On the other hand, ectopic recombination between g*Pokey* and r*Pokey* could severely impact genome organization and result in strong purifying selection, such that *Pokey* is generally not able to persist in substantial numbers in both locations in the same genome. *Daphnia arenata* is a notable exception to this general pattern as *Pokey*B and m*Pok*2 numbers are unusually high both inside and outside rDNA. If *Pokey* has been recently transposing in *D. arenata*, as suggested above, then there may not have been sufficient time for purifying selection to limit the expansion of *Pokey* in one or the other genomic location.

Variation among partial *Pokey*A sequences in North American *D. pulicaria*

Eagle and Crease [10] suggested that the transposition rate of r*Pokey*A may have recently increased in some populations of NA *D. pulicaria* because its number is unusually high in some isolates from populations PC1 and PC2. Sequence divergence between TE copies that have recently transposed is expected to be low as there has been little time for them to accumulate differences from their parent copy. Our results are consistent with this hypothesis as the sequences from isolates with high *Pokey*A number (PC1.2 and PC2.1) show significantly less deviation from their consensus sequence than do

sequences from isolates with average *Pokey*A numbers in the same population (PC1.1 and PC2.2, Table 3). Moreover, all but two of the sequences encode a potentially functional transposase. However, we cannot rule out the possibility that recent recombination within the rDNA array has also played a role in this r*Pokey*A expansion.

A recent increase in transposition rate is often a consequence of an element's ability to evade host silencing, which has been suggested to depend on the distribution of elements that occur in rDNA. For example, Eickbush et al. [37] found that rates of R2 transcription are higher when the elements are spread throughout the rDNA array than when they are clustered, and indirect evidence suggests that *Pokey* insertions are clustered in the rDNA of NA *D. pulex* [38]. Eickbush et al. [37] concluded that the largest continuous block of uninserted rDNA units tend to be transcribed, while other units remain silent. However, when R2 insertions are spread across the rDNA array, or have recently transposed into the midst of a large continuous block of uninserted units, they are more likely to be transcribed. Similarly, it is possible that recent insertion of *Pokey*A into a large continuous block of uninserted rDNA units in a NA *D. pulicaria* individual (either by recombination or transposition) triggered an increase in *Pokey*A transcription. The hypothesis that transcription rate is higher in individuals with high *Pokey*A number could be tested using nuclear run-on transcription assays and Reverse Transcription-PCR [37]. In addition, fiber-FISH assays could be used to determine the arrangement of *Pokey* within rDNA arrays. This technique has been used to show that *Pokey* is dispersed in isolate AR1.1 (Figure S11 in [18]).

Conclusions

The goals of this study were to determine the distribution of the *Pokey* family (*Pokey*A, *Pokey*B, m*Pok*1 and m*Pok*2) inside and outside rDNA in lineages of the *D. pulex* complex, and to test the hypothesis that r*Pokey*A expansion in NA *D. pulicaria* is the result of recent transposition. We found the distribution to be similar in four of the lineages; *D. pulex* and *D. pulicaria* from North America (NA) and Europe (EU) and in general, *Pokey* is more common outside than inside rDNA. *Pokey*A expansion in NA *D. pulicaria* rDNA appears to be recent and we suggest it could have been triggered by a change in rDNA distribution that reduced the host's ability to regulate *Pokey* transcription.

The *Pokey* family distribution in *D. arenata* is very different from the other four lineages. In particular, the mean number of both *Pokey*A and *Pokey*B outside rDNA is five to six times higher in *D. arenata* and the two types of m*Pok* are nearly exclusive to this lineage. We suggest that the proliferation of *Pokey* and m*Pok* in *D. arenata* may be a consequence of release from epigenetic repression as a result of interspecific hybridization, although a hybrid origin for this lineage requires confirmation.

The mean number of rRNA genes and rPokeyA is much larger in *D. obtusa* compared to the five *D. pulex* lineages. However, both *Pokey*A and *Pokey*B are absent or nearly so outside *D. obtusa* rDNA. Overall, these results suggest that *Pokey* primarily occupies only one or the other genomic location within a species. We suggest that this may be a consequence of purifying selection against ectopic recombination between elements in different genomic locations, which could cause severe genomic rearrangements.

Elliott et al. [6] suggested that *Pokey* originated as a genomic element and subsequently invaded *Daphnia* rDNA, which makes it unique compared to other DNA elements. Moreover, *Pokey* has been vertically inherited in the subgenus *Daphnia* since it diverged from the other *Daphnia* subgenera, *Ctenodaphnia* and *Hyalodaphnia* [10], which may have occurred as long ago as 100 million years [39]. It seems likely that the persistence of *Pokey* in *Daphnia* over such a long period of time is at least partially due to the potential for copies from one genomic location to reinvade the other if it is lost [6]. However, the factors that determine whether *Pokey* is primarily a genomic or an rDNA element within each species, and whether this configuration is stable over time are unknown and warrant further research.

Methods
Daphnia isolates and DNA extractions
We analyzed a total of 45 cyclically parthenogenetic *Daphnia* isolates from five lineages in the *D. pulex* complex (Fig. 1, Additional file 1: Table S1). Twenty-six of these isolates were NA *D. pulex*. One to three isolates were sampled from ten populations in the Midwest United States and southern Ontario. In addition, 19 isolates from the other lineages were included: four *D. arenata*, five NA *D. pulicaria*, eight EU *D. pulex*, and two EU *D. pulicaria*. *Daphnia* individuals were collected and clonally propagated (isolates) as described in Eagle and Crease [10]. DNA was extracted from multiple individuals from each isolate using phenol:chloroform or the Aquagenomics kit (MultiTarget Pharmaceuticals LLC, Salt Lake City, Utah, USA) as in Eagle and Crease [10]. DNA concentrations, which ranged from 13 to 4400 ng/μL (Additional file 1: Table S1), were measured using a NanoDrop® ND-8000 spectrophotometer (ThermoScientific).

Estimation of *Pokey* family and rRNA gene number
We used the SYBR green real-time qPCR and ΔC_T relative quantification method as described in Eagle and Crease [10] to estimate the haploid number of each type of gene. C_T, or cycle threshold, is the point at which the amplification curve crosses a set threshold. We estimated the number of the four TEs in the *Pokey* family (*Pokey*A, *Pokey*B, m*Pok*1, and m*Pok*2) in the entire genome (t*Pokey*,

tm*Pok*) and in rDNA (r*Pokey*, rm*Pok*), as well as the number of 18S and 28S rRNA genes in all 45 *Daphnia* isolates.

A total of 11 primer pairs were used (Table 4). The forward primers for r*Pokey*/rm*Pok* are located near the 3' end of the element and the reverse primer is located in the 28S rRNA gene downstream of the *Pokey* family insertion site. Standard curves were run to determine the percent amplification efficiency (PAE) for each primer pair (Table 4, Additional file 3).

qPCRs had a final volume of 20 μL and contained 1X PerfeCTa SYBR Green FastMix, ROX (Quanta BioSciences) and 0.25 μM of each primer. Reactions were run on a StepOne Plus instrument (Applied Biosystems) using the following protocol: 95 °C for 10 min; followed by 40 cycles of 95 °C for 5 s and 60 °C for 30 s. Melt curves were run by heating the amplicons at 95 °C for 15 s, decreasing the temperature to 60 °C for 1 min, then heating to 95 °C in 0.3 °C increments. The DNA template was either serial dilutions of amplicons generated from plasmid DNA (standard curve plates) or 10 ng of genomic DNA (experimental plates). Samples were run in triplicate with the exception of the tm*Pok*1 and rm*Pok*2 primer pairs for samples in which previous endpoint PCR produced no amplicon (Additional file 3). In such cases, only a single reaction was done. Negative control reactions were run for each primer pair on every standard curve plate. Negative control reactions were also run for each primer pair on 15 out of the 22 experimental plates (Additional file 3). The StepOne Software (Applied Biosystems) set the baseline for each reaction. The threshold was set based on amplicon size as in Eagle and Crease [10] (Table 4). A threshold of 0.2 was used for 50 bp amplicons, and the threshold for larger amplicons was determined using the formula, $0.2 \times 2^{[1-(50/length\ in\ bp)]}$.

C_T values were used to estimate gene number, using the formula: $2^{-\Delta CT}$ where ΔC_T is ([C_T x PAE $_{multicopy\ gene}$] - [C_T x PAE $_{single-copy\ gene}$]), as in Eagle and Crease [10]. If the standard deviation for mean C_T for a gene was greater than 0.2, then the most extreme value was omitted (Additional file 1: Table S2) from further analysis. If there was no clear outlier among the three C_T values, then all three values were used. The two or three C_T values for each multicopy gene were compared to the two or three C_T values for both single copy genes producing 8 to 18 estimates of gene number. The *Tif-Gtp* ratio was 1.66 in isolate PX1.2 and 1.42 in isolate PX2.2; these two isolates likely have three copies of the *Tif* gene instead of the expected two. Therefore, the estimates of gene number using *Tif* as the single copy gene were multiplied by 1.5 in these isolates. In addition, both tm*Pok*1 and r*Pokey*A + rm*Pok*1 amplified in one isolate, AR3.1. To determine if the estimate of r*Pokey*A + rm*Pok*1 included r*Pokey*A, rm*Pok*1 or both, this isolate

Table 4 Primer pairs used for qPCR of the *Pokey* family and rRNA genes in the *Daphnia pulex* complex

Gene[a]	Primer	Primer sequence	Amplicon Size (bp)	Source[b]	Threshold[c]	PAE[d]
18S	18S 1864 F	5'-ccg cgt gac agt gag caa ta	50	[17]	0.200	0.947
	18S 1913 R	5'-ccc agg aca tct aag ggc atc		[17]		
28S	28S 2508 F	5'-gcc tgc tcg tac cga tat cc	50		0.200	0.937
	28S 2558 R	5'-cta gag gct gtt cac ctt gga ga				
t*Pokey*A	Pok 3720 F	5'-cag ttc aaa gag tgg ctc ctc c	50		0.200	0.868
	Pok 3770 R	5'-cgg gtc tga ctt ctg gtt cg				
t*Pokey*B	PokB 1375 F	5'-aaa gag gag aag aat gac ccg g	50		0.200	0.897
	PokB 1425 R	5'-tca gaa gag cac cct acc ttg g				
tm*Pok*1	mPok1 524 F	5'-gga cac cta tgg cgg gat t	50		0.200	0.874
	mPok1 574 R	5'-cgc tga ggt ctg tcg gga				
tm*Pok*2	mPok2 714 F	5'-ggt cag ttg gct ccg aca a	45		0.185	0.911
	mPok2 759 R	5'-aa ccc ttt atc gac gcg aag a		T. Elliott, unpublished		
r*Pokey*A + rm*Pok*1	Pok 6561 F	5'-caa tcg aat ccg acc atc g	66		0.237	0.914
	28S 3073 R	5'-tga cga ggc att tgg cta cc				
r*Pokey*B	PokB 4283 F	5'-aat ttc agt caa gca cgg cc	70		0.244	0.903
	28S 3073 R	5'-tga cga ggc att tgg cta cc				
rm*Pok*2	mPok 714 F	5'-ggt cag ttg gct ccg aca a	68		0.240	0.900
	28S 3073 R	5'-tga cga ggc att tgg cta cc				
Tif	TIF392F	5'-gac atc atc ctg gtt ggc ct	50	[17]	0.200	0.936
	TIF442R	5'-aac gtc agc ctt ggc atc tt		[17]		
Gtp	GTP385R	5'-tat tca gca tgg aga gac ggc	50	[17]	0.200	0.928
	GTP435R	5'-gat gtc gac tga cgc tgg aa		[17]		

[a]18S = 18S rRNA genes, 28S = total 28S rRNA genes, t*Pokey* = total *Pokey*, tm*Pok* = total m*Pok*, r*Pokey* = *Pokey* inserted in 28S, rm*Pok* = m*Pok*1 inserted in 28S
[b]Unless otherwise indicated, primers were designed for this study
[c]A threshold of 0.2 was used for 50 bp amplicons. The threshold for larger amplicons was determined using the formula, 0.2 x 2^[1-(50/length in bp)] as in Eagle and Crease [10]
[d]PAE = primer amplification efficiency

was screened for rm*Pok*1 with end-point PCR (Additional file 3), but no amplicon was detected.

g*Pokey*/gm*Pok* number was calculated as [t*Pokey* − r*Pokey* or tm*Pok* − rm*Pok*]. In six isolates (EPX1.1, EPX1.2, EPX2.1, EPX2.2, PC2.1, and PX6.3), the calculation of g*Pokey*A produced a negative number in which case g*Pokey*A was set to zero, and t*Pokey*A was assumed to be the same as r*Pokey*A.

We used paired t-tests for means, two-sample t-tests assuming unequal variances with the sequential Bonferroni correction [40], and linear regressions (Microsoft Excel, Richmond, Washington, USA) to examine the relationship between 18S and 28S rRNA gene number. We also estimated the correlation between rRNA genes and *Pokey* family number using regression analysis.

Sequencing partial *Pokey*A copies from North American *D. pulicaria*

We cloned and sequenced partial *Pokey*A elements from four NA *D. pulicaria* isolates from two populations. From each population, an isolate with high number of *Pokey*A

(PC1.2, PC2.1) and an isolate with average number of *Pokey*A (PC1.1, PC2.2) were selected (Table 3). The average number of *Pokey*A for NA *D. pulicaria* was based on the results of Eagle and Crease [10]. Partial *Pokey*A sequences were amplified from approximately 50 ng of genomic DNA in a 25 µL reaction, which contained 1X Phusion HF Buffer (NEB), 0.4 mM dNTPs, 0.08 µM each of the Pok-2904 F (5' ggg aca tag gtg tcc cgg) and Pok-6178R (5' tcg acc agg ggt ctt tcc agt c) primers, and 1 units of Phusion DNA Polymerase. Reactions were run on either a PTC-100 Thermocycler (MJ Research) or a T100 Thermocycler (Bio-Rad Laboratories, Inc.) using the following protocol: 3 min initial denaturation at 98 °C; 35 cycles of 10 s denaturation at 98 °C, 30 s annealing at 55 °C, and 2 min elongation at 72 °C; followed by a final elongation at 72 °C for 10 min. The 3.3 kb amplicons were verified by electrophoresis on a 1 % TAE agarose gel, stained with GelRed™ Nucleic Acid Gel Stain (Biotium, Inc.) and visualized under UV light. The amplicons were then cloned into the plasmid, pSC-B-amp/kan using the StrataClone Blunt PCR Cloning Kit (Agilent Technologies) as per the manufacturer's protocol with a few modifications. The modifications were as follows: 2.5 µL of PCR product and 0.5 µL of StrataClone Blunt Vector Mix were used in the ligation; 25-50 µL of StrataClone SoloPack cells was used for the transformation; an incubation of 30–45 min was used because the insert was large (3.3 kb); 500 µL of Terrific Broth was used instead of 250 µL of LB medium; and 50 µL and 400 µL were plated instead of 5 µL and 100 µL.

Positive colonies (white) were triple-streaked onto new plates and grown overnight. One of the three streaks was added with a toothpick to 10 µL of water and incubated at 99.9 °C for 3 min. One microliter was amplified in a 25 µL reaction with two sets of primers: Pok-2904 F with Pok-3811R (5' ccg tgt tac ttc acc atc gg) to generate a 910 bp fragment; and Pok-3720 F (5' cag ttc aaa gag tgg ctc c) with Pok-4488R (5' gaa tcg ctc gcg agt cat gg) to generate a 770 bp fragment. Reactions contained 1X GenScript buffer (GenScript USA Inc.), 0.04 mM dNTPs, 0.04 µM of each primer, and 0.5 units of GenScript DNA polymerase (GenScript USA Inc.). Reactions were run on either a PTC-100 Thermocycler (MJ Research) or a T100 Thermocycler (Bio-Rad Laboratories, Inc.) using the following protocol: 2 min initial denaturation at 94 °C; 35 cycles of 30 s denaturation at 94 °C, 30 s annealing at 55 °C, and 1 min extension at 72 °C; followed by a final elongation at 72 °C for 5 min. Amplicons from 12 clones per isolate were sequenced with the primers Pok-2904 F and Pok-3720 F in 12 µL reactions. The reactions contained 0.3 µL BigDye® Terminator v3.1 Ready Reaction Mix (Applied Biosystems), 0.4X Sequencing Buffer (Applied Biosystems), 0.83 µM of primer, and 2 µL of amplicon. Sequencing reactions were run for 1 min at 96 °C followed by 30 cycles of a 20 s denaturation at 96 °C, a 20 s annealing at 55 °C, and a 4 min extension at 60 °C. Reactions were resolved on an ABI 3730 DNA Analyzer (Applied Biosystems) by the Genomics Facility at the University of Guelph. Sequences were analyzed using CLC Main Workbench software (CLC Bio). We used MEGA 5.0 [41] to generate a 1608 bp alignment containing 48 sequences from the four NA *D. pulicaria* isolates. The maximum composite likelihood model was used to estimate a matrix of pairwise sequence divergence from which a Neighbor-Joining dendrogram was generated. Gaps were excluded from this analysis using the pairwise deletion option. In addition, a consensus sequence was generated for each isolate from the 12 sequences. The number of differences between each sequence and its consensus was determined in MEGA. Gaps were included in this analysis, but consecutive gaps were coded as single nucleotide changes. ANOVA was used to test for heterogeneity among the mean number of differences in each of the four isolates. A *post hoc* Tukey's HSD test was used to compare the means in the average and high isolates from the same population.

Additional files

Additional file 1: **Table S1.** Details of *Daphnia* isolates analyzed in this study. **Table S2.** C_T values from qPCR analysis of *Pokey*, m*Pok*, and rRNA gene number in 45 isolates from the *Daphnia pulex* complex. **Table S3.** Estimates of *Pokey*, m*Pok*, and rRNA gene number in 45 isolates of the *Daphnia pulex* complex. (XLSX 51 kb)

Additional file 2: **Figure S1.** Correlation between 18S and 28S rRNA gene number *D. pulex* lineages. **Figure S2.** Correlation between 28S rRNA gene and r*Pokey* family number in *D. pulex* lineages. **Figure S3.** Repeat landscape for m*Pok* from *Daphnia arenata* isolate AR1.1. (PDF 34 kb)

Additional file 3: Details of qPCR methods. (PDF 232 kb)

Abbreviations

18S, 18S rRNA gene; 28S, 28S rRNA gene; bp, base pair; C_T, threshold cycle; ETS, external transcribed spacer; EU, European; gm*Pok1*, m*Pok1* found outside rDNA; gm*Pok2*, m*Pok2* found outside rDNA; g*Pokey*, *Pokey* elements found outside rDNA; g*Pokey*A, m*Pok1* found outside rDNA; g*Pokey*B, m*Pok2* found outside rDNA; IGS, intergenic spacer; ITS, internal transcribed spacer; kb, kilo base pairs; LTR, long terminal repeats; MITE, Miniature Inverted-repeat Transposable Element; NA, North American; PAE, percent amplification efficiency; qPCR, quantitative Polymerase Chain Reaction; rDNA, ribosomal DNA; rm*Pok1*, m*Pok1* in the 28S gene; rm*Pok2*, m*Pok2* in the 28S gene; r*Pokey*, *Pokey* elements found in the 28S gene; r*Pokey*A, m*Pok1* in the 28S rRNA gene; r*Pokey*B, m*Pok2* in the 28S rRNA gene; rRNA, ribosomal RNA; TE, transposable element; TIR, terminal inverted repeat; tm*Pok1*, m*Pok1* in the genome; tm*Pok2*, m*Pok2* in the genome; t*Pokey*, *Pokey* elements found in the genome; t*Pokey*A, *Pokey*A in the genome; t*Pokey*B, *Pokey*B in the genome; TSD, target site duplication.

Acknowledgements

We thank Andrew Beckerman, Melania Cristescu, France Dufresne, Michael Lynch, Seanna McTaggart and Julia Reger for providing some of the *Daphnia* samples, and the Genomics Facility at the University of Guelph for assistance with the qPCR and sequencing of plasmid clones. Comments from Tyler Elliott and two anonymous reviewers greatly improved the manuscript.

Funding

This research was funded by a Discovery Grant from the Natural Sciences and Engineering Research Council of Canada to TJC. SHCE was partially supported by an Ontario Graduate Studies Science and Technology Scholarship. The funding agencies had no involvement in the design of the study, the collection, analysis, and interpretation of data, or writing the manuscript.

Authors' contributions

SHCE participated in the design of the study, carried out the molecular genetic analyses, analyzed the data, and drafted the manuscript. TJC conceived of the study and participated in its design, and helped to draft the manuscript. Both authors read and approved the final manuscript.

Competing interests

The authors declare that they have no competing interests.

References

1. Nuzhdin SV. Sure facts, speculations, and open questions about the evolution of transposable element copy number. Genetica. 1999;107:129–37.
2. Slotkin RK, Martienssen R. Transposable elements and the epigenetic regulation of the genome. Nat Rev Genet. 2007;8:272–85.
3. Schaack S, Gilbert C, Feschotte C. Promiscuous DNA: horizontal transfer of transposable elements and why it matters for eukaryotic evolution. Trends Ecol Evol. 2010;25:537–46.
4. Penton EH, Sullender BW, Crease TJ. *Pokey*, a new DNA transposon in *Daphnia* (Cladocera: Crustacea). J Mol Evol. 2002;55:664–73.

5. Penton EH, Crease TJ. Evolution of the transposable element *Pokey* in the ribosomal DNA of species in the subgenus *Daphnia* (Crustacea: Cladocera). Mol Biol Evol. 2004;21:1727–39.

6. Elliott TA, Stage DE, Crease TJ, Eickbush TH. In and out of the rRNA genes: characterization of *Pokey* elements in the sequenced *Daphnia* genome. Mobile DNA. 2013;4:20.

7. Eickbush TH, Eickbush DG. Finely orchestrated movements: evolution of the ribosomal RNA genes. Genetics. 2007;175:477–85.

8. Kojima KK, Jurka J. A superfamily of DNA Transposons targeting multicopy small RNA genes. PLoS ONE. 2013;8:68260.

9. Valizadeh P, Crease TJ. The association between breeding system and transposable element dynamics in *Daphnia pulex*. J Mol Evol. 2008;66:643–54.

10. Eagle SHC, Crease TJ. Copy number variation of ribosomal DNA and *Pokey* transposons in natural populations of *Daphnia*. Mobile DNA. 2012;3:4.

11. Prokopowich CD, Gregory TR, Crease TJ. The correlation between rDNA copy number and genome size in eukaryotes. Genome. 2003;46:48–50.

12. Hebert P. The population biology of *Daphnia* (Crustacea, Daphnidae). Biol Rev. 1978;53:387–426.

13. Colbourne J, Crease T, Weider L, Hebert P, Dufresne F, Hobaek A. Phylogenetics and evolution of a circumarctic species complex (Cladocera: *Daphnia pulex*). Biol J Linn Soc. 1998;65:347–65.

14. Hebert P. The *Daphnia* of North America: an illustrated fauna. CD-ROM. 1995. University of Guelph, Guelph, ON

15. LeRiche K, Eagle SHC, Crease TJ. Copy number of the transposon, *Pokey*, in rDNA is positively correlated with rDNA copy number in *Daphnia obtusa*. PLoS ONE. 2014;9:114773.

16. Feschotte C, Pritham EJ. DNA transposons and the evolution of eukaryotic genomes. Ann Rev Genet. 2007;41:331–68.

17. McTaggart SJ, Dudycha JL, Omilian A, Crease TJ. Rates of recombination in the ribosomal DNA of apomictically propagated *Daphnia obtusa* lines. Genetics. 2007;175:311–20.

18. Colbourne JK, Pfrender ME, Gilbert DG, Thomas WK, Tucker A, Oakley TH, et al. The ecoresponsive genome of *Daphnia pulex*. Science. 2011;331:555–61.

19. Averbeck KT, Eickbush TH. Monitoring the mode and tempo of concerted evolution in the *Drosophila melanogaster* rDNA locus. Genetics. 2005;171:1837–46.

20. Feschotte C, Zhang X, Wessler S. Miniature inverted-repeat transposable elements (MITEs) and their relationship with established DNA transposons. In: Craig NL, Craigie R, Gellert M, Lambowitz AM, editors. Mobile DNA II. Washington DC: ASM Press; 2002. p. 1147–58.

21. Crease T, Lee S-K, Yu S-L, Spitze K, Lehman N, Lynch M. Allozyme and mtDNA variation in populations of the *Daphnia pulex* complex from both sides of the Rocky Mountains. Heredity. 1997;79:242–51.

22. Lynch M, Conery JS. The origins of genome complexity. Science. 2003; 302:1401–4.

23. Kidwell MG, Kidwell JF, Sved JA. Hybrid dysgenesis in *Drosophila melanogaster*: a syndrome of aberrant traits including mutation, sterility and male recombination. Genetics. 1977;86:813–33.

24. Malone CD, Hannon GJ. Small RNAs as guardians of the genome. Cell. 2009; 136:656–68.

25. Ungerer MC, Kawakami T. Transcriptional dynamics of LTR retrotransposons in early generation and ancient sunflower hybrids. Gen Biol Evol. 2013;5:329–37.

26. Labrador M, Farré M, Utzet F. Fontdevila A 1999 Interspecific hybridization increases transposition rates of *Osvaldo*. Mol Biol Evol. 1999;16:931–7.

27. Shan X, Liu Z, Dong Z, Wang Y, Chen Y, Lin X, Long L, Han F, Dong Y, Liu B. Mobilization of the active MITE transposons mPing and Pong in rice by introgression from wild rice (*Zizania latifolia* Griseb). Mol Biol Evol. 2005;22: 976–90.

28. Wang H-Y, Tian Q, Ma Y-Q, Wu Y, Miao G-J, Ma Y, Cao D-H, Wang X-L, Lin C, Pang J, Liu B. Transpositional reactivation of two LTR retrotransposons in rice-*Zizania* recombinant inbred lines (RILs). Hereditas. 2010;147:264–77.

29. Wang N, Wang H, Wang H, Zhang D, Wu Y, Ou X, Liu S, Dong Z, Liu B. Transpositional reactivation of the *Dart* transposon family in rice lines derived from introgressive hybridization with *Zizania latifolia*. BMC Plant Biol. 2010;10:190.

30. Vela D, Fontdevila A, Vieira C, García Guerreiro MP. A genome-wide survey of genetic instability by transposition in *Drosophila* hybrids. PLoS ONE. 2014;9:88992.

31. Garcia Guerreiro MP. Interspecific hybridization as a genomic stressor inducing mobilization of transposable elements in *Drosophila*. Mob Genet Elements. 2014;4:34394.

32. Vergilino R, Markova S, Ventura M, Manca M, Dufresne F. Reticulate

evolution of the *Daphnia pulex* complex as revealed by nuclear markers. Mol Ecol. 2011;20:1191–207.

33. Omilian AR, Lynch M. Patterns of intraspecific DNA variation in the *Daphnia* nuclear genome. Genetics. 2009;182:325–36.

34. Crease TJ, Floyd R, Cristescu ME, Innes D. Evolutionary factors affecting *Lactate dehydrogenase* A and B variation in the *Daphnia pulex* species complex. BMC Evol Biol. 2011;11:212.

35. Montgomery E, Charlesworth B, Langley CH. A test for the role of natural selection in the stabilization of transposable element copy number in a population of Drosophila melanogaster. Genet Res. 1987;46:31–41.

36. Eickbush TH. R2 and related site-specific non-long terminal repeat retrotransposons. In: Craig NL, Craigie R, Gellert M, Lambowitz AM, editors. Mobile DNA II. Washington DC: ASM Press; 2002. p. 813–35.

37. Eickbush DG, Ye J, Zhang X, Burke WD, Eickbush TH. Epigenetic regulation of retrotransposons within the nucleolus of *Drosophila*. Mol Cell Biol. 2008;28:6452–61.

38. Glass S, Moszczynska A, Crease TJ. The effect of transposon *Pokey* insertions on sequence variation in the 28S rRNA gene of *Daphnia pulex*. Genome. 2008;51:988–1000.

39. Kotov AA, Taylor DJ. Mesozoic fossils (>145 Mya) suggest the antiquity of the subgenera of Daphnia and their coevolution with chaoborid predators. BMC Evol Biol. 2011;11:129.

40. Rice W. Analyzing tables of statistical tests. Evolution. 1989;43:223–5.

41. Tamura K, Peterson D, Peterson N, Stecher G, Nei M, Kumar S. MEGA5: molecular evolutionary genetics analysis using maximum likelihood, evolutionary distance, and maximum parsimony methods. Mol Biol Evol. 2011;28:2731–9.

7

Pinpointing the vesper bat transposon revolution using the *Miniopterus natalensis* genome

Roy N. Platt II, Sarah F. Mangum and David A. Ray*

Abstract

Background: Around 40 million years ago DNA transposons began accumulating in an ancestor of bats in the family Vespertilionidae. Since that time, Class II transposons have been continuously reinvading and accumulating in vespertilionid genomes at a rate that is unprecedented in mammals. *Miniopterus* (Miniopteridae), a genus of long-fingered bats that was recently elevated from Vespertilionidae, is the sister taxon to the vespertilionids and is often used as an outgroup when studying transposable elements in vesper bats. Previous wet-lab techniques failed to identify *Helitrons*, TcMariners, or hAT transposons in *Miniopterus*. Limitations of those methods and ambiguous results regarding the distribution of piggyBac transposons left some questions as to the distribution of Class II elements in this group. The recent release of the *Miniopterus natalensis* genome allows for transposable element discovery with a higher degree of precision.

Results: Here we analyze the transposable element content of *M. natalensis* to pinpoint with greater accuracy the taxonomic distribution of Class II transposable elements in bats. These efforts demonstrate that, compared to the vespertilionids, Class II TEs are highly mutated and comprise only a small portion of the *M. natalensis* genome. Despite the limited Class II content, *M. natalensis* possesses a limited number of lineage-specific, low copy number piggyBacs and shares several TcMariner families with vespertilionid bats. Multiple efforts to identify *Helitrons*, one of the major TE components of vesper bat genomes, using de novo repeat identification and structural based searches failed.

Conclusions: These observations combined with previous results inform our understanding of the events leading to the unique Class II element acquisition that characterizes vespertilionids. While it appears that a small number of TcMariner and piggyBac elements were deposited in the ancestral *Miniopterus* + vespertilionid genome, these elements are not present in *M. natalensis* genome at high copy number. Instead, this work indicates that the vesper bats alone experienced the expansion of TEs ranging from *Helitrons* to piggyBacs to hATs.

Keywords: Miniopteridae, Vespertilionidae, TcMariner, Helitron, Transposable element

Background

Transposable elements (TEs) are genetic elements with the ability to mobilize throughout a host genome. Often TE copies are generated as a result of the mobilization process and TEs can end up occupying large portions of mammalian genomes. For example, between 45 and 70 % of the human genome is occupied by TEs [1, 2]. TEs are classified into two major classes based on their mobilization mechanism. Class I elements, also known as retrotransposons, mobilize as an RNA intermediate that is reverse transcribed back into the genome. These elements are referred to as "copy and paste" elements since they generate identical copies of themselves upon insertion. Retrotransposons are further classified into Long Terminal Repeats (LTRs), Long INterspersed Elements (LINEs), and Short INterspersed Elements (SINEs). Class II elements, also known as DNA transposons, mobilize via a transposase enzyme. During mobilization, the terminal inverted repeat-containing DNA transposons physically excise from the genome and re-integrate at another locus.

* Correspondence: david.4.ray@gmail.com
Department of Biological Sciences, Texas Tech University, Box 43131, Lubbock, TX 79409-3131, USA

However, in addition to these canonical "cut and paste" DNA transposons, *Helitrons* and Mavericks mobilize through other mechanisms that do not fully excise the template TE. As a result, these Class II elements are "copy and paste" transposons since they mobilize through a single DNA strand excised from the parent locus.

In general, retrotransposons are much more common in mammalian genomes than DNA transposons. For example, 43 % of the human genome is derived from retrotransposons vs. 3 % from DNA transposons [2]. In addition to being less frequent, transposons are often found in genomes as heavily mutated insertions; indicating long periods of inactivity. The single major exception to this general trend is the presence of recently inserted Class II elements in the genomes of vespertilionid bats [3, 4]. As much as 6 % of the *Myotis lucifugus* genome is derived from recently active *Helitrons* [5], ~3.5 % from cut and paste transposons [6], and half of all recent TE accumulation appears to come from Class II elements [7].

To understand the timing and evolutionary implications of this unique activity, we must first identify the taxonomic distribution and accumulation patterns of the elements involved. Previous work focusing on the initial horizontal transfer or reactivation of Class II elements in vespertilionids indicated that *Helitrons* are restricted to the vespertilionid lineage [8] and only a limited number of cut and paste transposon families are found beyond Vespertilionidae [6]. These results were based on comparisons of vespertilionids to several non-vesper bats including *Miniopterus*, a genus of long-fingered bats recently elevated to familial level from Vespertilionidae [9]. For example, using internal PCR primers, Ray et al. [6] tried to amplify piggyBac, hAT, and TcMariner elements in a panel of chiropteran including *Artibeus jamaicensis*, *Balionycteris* sp., *Corynorhinus rafinesquii*, *Eptesicus furinalis*, *Hipposideros cervinus*, *Kerivoula papillosa*, *Macroglossus sobrinus*, *Miniopterus* sp., *Myotis austroriparius*, *My. horsfieldii*, *Natalus stramineus*, *Nycticeius humeralis*, *Pteronotus parnellii*, *Rhinolophus borneoensis*, and *Thyroptera tricolor*. Results indicated that TcMariner elements were only present in vespertilionids (*C. rafinesquii*, *E. furinalis*, *K. papillosa*, *Myotis austroriparius*, *Myotis horsfieldii*, and *N. humeralis*). hATs and piggyBacs were only found in *Myotis* species, with the exception of one piggyBac (piggyBac2_ML) that was amplified in *Myotis* sps. and *Miniopterus* but was absent in other all other samples including the non-*Myotis* vespertilionids [6]. Probe-based hybridization failed to identify *Helitrons* in *Miniopterus* or any other non-vesper bats [8].

Modern genome assembly and sequencing techniques provide many advantages for TE discovery over wet-lab based techniques. Mispriming, in the case of PCR, or reduced hybridization efficiency, in the probe-based analyses, could easily allow elements to be missed in any or all of these genomes. In addition, these methods rely on a priori knowledge of TE content in order to build primers/probes for loci of interest. The recent release of the *Miniopterus natalensis* genome [10] allows these questions to be answered more precisely and with independent and unbiased data. Here, we characterize the repetitive portion of the *M. natalensis* genome with an emphasis on Class II elements in order to understand the acquisition of these Class II TEs in bats.

Methods

Repeats were identified in the *Miniopterus natalensis* genome using de novo methods and TEs were fully validated [11] as detailed below. Putative repeats were identified using RepeatModeler [12] and the current *M. natalensis* assembly (Genbank accession GCA_001595765.1). The RepeatModeler repeats were masked with RepeatMasker [13] using all known Chiropteran TEs (-species "Chiroptera") to remove repeats that have already been described in other bat species. Those repeats that were ≥80 % similar to known elements across more than 50 % of their length were excluded from downstream analyses. The remaining elements were considered possible *Miniopterus*-specific elements. To manually validate these repeats, they were used as BLASTn v2.2.27 [14] queries against the *M. natalensis* genome. BLASTn hits were restricted to those with E values greater than 1e-10. For each repeat, the forty loci most similar to the BLASTn query were extracted from the genome along with 500 or more bases of flanking sequence and aligned using MUSCLE v3.8.1551 [15]. Repeats with less than 10 BLASTn hits were culled from further analysis. For the remaining repeats, majority-rule consensus sequences were generated for each alignment using BioEdit v7.2.5 [16]. Elements that contained single copy DNA on both the 5' and 3' end were considered to be complete. If an alignment ended within a repetitive portion, the consensus sequence was generated across the entire repetitive portion of the alignment and this new consensus sequence was used as a query in subsequent BLASTn rounds. This process was iterated until all de novo repeats were fully represented.

Beyond RepeatModeler searches, attempts were made to identify low copy number and highly divergent *Helitrons* using HelitronScanner [17]. HelitronScanner searches the genome for 5' and 3' terminal sequences associated with *Helitrons*. Terminal sequences are then paired with their closest partner. Those falling within a set distance are considered putative *Helitrons*. Default parameters were used in HelitronScanner searches except for the scoring threshold, which was raised from a default of 5 to 10. As a control, a copy of the *M. natalensis* genome was shuffled using EMBOSS's shuffleseq (v6.6.0 [18]), and run in parallel using the same parameters. A series of BLAT [19], and BLAST searches were used to validate putative *Helitrons* that resulted from Helitron-Scanner queries.

All novel repeats were classified based on structural hallmarks (ex. poly-A tails, target site duplications, terminal inverted repeats, etc.) and homology to other TEs present in RepBase (accessed 1 April 2016 [20]). For larger elements, intact open reading frames (ORFs) were identified with ORF Finder [21]. Elements were classified using the 80-80-80 rule [22] and designated based on standard naming conventions implemented by RepBase [20]. For example, two SINEs in *M. natalensis* meet the 80-80-80 thresholds when compared to the canonical VES SINE, but each varies from one another by 5 % at the nucleotide level and contain diagnostic indels. In this case, both SINEs are recognized as members of the separate subfamilies of VES: VES-1_MNa and VES-2_MNa. After classification, the *M. natalensis* repeats were combined with all known mammal TEs from RepBase and used as a customized library to annotate the *M. natalensis* genome. For comparative purposes genomes from closely related bat species, were analyzed using identical RepeatMasker settings to provide a better estimate of the TE dynamics during the *Miniopterus* and Vespertilionidae divergence. These taxa include *Myotis lucifugus* (GCA_000147115.1), *Eptesicus fuscus* (GCA_000308155.1) and *Pteronotus parnellii* (GCA_000147115.1) and were chosen based on their phylogenetic relationships. Repeat accumulation profiles for all taxa were generated using the Kimura 2-parameter distance [23] between the RepeatMasker library and homologous loci in the genome. Highly mutable CpG sites [24] were excluded from distance calculations. Elements belonging to the same superfamily were binned based on their genetic distances. Distances were rounded down to the nearest full percentage. For comparison, average genetic distances between genomic TEs and the consensus library TE were calculated for all DNA transposons occupying more than 10 Kb of any bat genome.

To identify TEs specific to *M. natalensis*, repeats identified by RepeatModeler and successfully validated, were used as BLASTn queries against all other genomes in the NCBI Genomes (chromosomes) database. *M. natalensis* was excluded (NCBI Taxa ID 9432) from these searches. The most closely related species to *M. natalensis* in the NCBI Genome database are the vespertilionids, *Myotis lucifugus*, *Myotis brandtii*, *Myotis davidii*, and *Eptesicus fuscus*. *Pteronotus parnellii* (family Mormoopidae), serves as an outgroup to a monophyletic clade comprising Vespertilionidae + Miniopteridae [25]. Repeats were classified based on the species distribution of the 50 best BLASTn hits. If the best hits for a repeat belonged to a vespertilionid or *P. parnellii*, the *M. natalensis* repeat was assumed to have been active in the common ancestor of these taxa. If, however, the best hits were to species other than a vespertilionid or *P. parnellii*, then the TE has a distribution among species that does not follow the species tree. If no hits were found to other species, it was assumed that these elements are only found in *M. natalensis* and are lineage-specific. BLASTn hits were only considered if they had an E value greater than 1*e*-10 and were more than 80 % similar across 80 % of the length of the *M. natalensis* query.

Results

RepeatModeler analysis of the unmasked *Miniopterus natalensis* genome identified 396 putative repetitive sequences. After removing elements with homology to known chiropteran TEs, simple repeats, and low copy number elements 52 putative TEs remained. Of these, 13 were so heavily mutated in the *M. natalensis* genome that generating a consensus sequence was not feasible. The remaining 39 elements were fully validated and classified. In all: 10 LTRs, 2 SINEs, 2 LINEs, and 25 DNA transposons were identified. All LTR elements were solo LTRs of less 1,100 bp. These LTRs were classified as ERV1 (gammaretroviruses) or ERV3s (spumaviruses) based on the size of their target site duplicates. The two SINEs were variants of the VES family of SINEs common in many bats [26, 27]. The two LINEs belonged to the LINE-1 superfamily and were full length, with intact ORF2s, but contained premature stop codons in ORF1 of the consensus elements. Three non-autonomous piggyBac elements were recovered and verified via their TTAA target site duplications. Finally, 22 elements in the TcMariner superfamily were identified including three potentially autonomous elements. BLASTp results from ORFs in these transposons revealed similar domain organization in each. ORFs ranged in length from 493 to 594 amino acids and two of the three contained a helix-turn-helix, Tc5 transposase, and DDE-like integrase domain while the third lacked the initial helix-turn-helix domain. All TcMariner elements had terminal inverted repeats of 12-26 bps that began with CAG and TA target site duplications.

HelitronScanner was used to identify low copy number *Helitrons* that would have been culled based on the filtering criteria for the RepeatModeler data. As a negative control, searches for *Helitrons* were run in parallel on *M. natalensis*, and a shuffled version of the *M. natalensis* genome. HelitronScanner identified 10 elements ranging in size from 2,351 to 14,820 bps in the *M. natalensis* genome and none in the shuffled genome. Several steps were taken to confirm these as true *Helitrons*. First, these elements were used as BLASTn queries against the *M. natalensis* genome to determine copy number. Other than the original locus, no significant hits were found indicating these putative *Helitrons* were single copy. Next, we used BLAT to compare the putative *Helitrons* to the *Myotis lucifugus* genome. In nine of the 10 cases, full-length elements were found, but none overlapped with known *Myotis lucifugus Helitrons*, in the tenth case, no homologous sequence was found in *Myotis lucifugus*. Next, putative *Helitrons* were compared to all known TEs in RepBase. The putative *Helitrons* identified by HelitronScanner lacked homology to other known *Helitrons*. Finally, ORFs were identified with ORF Finder. The largest ORF from each putative *Helitron* was used as a BLASTp query. None of these searches identified domains associated with *Helitrons* (ex. Zinc-finger domains, replicase, helicase, etc. [28]) and a majority failed to recover significant hits to any known protein. Based on these results, the sequences recovered by HelitronScanner are likely artifacts of the search methodology and not true *Helitrons*, since these loci are single-copy, present in the *Myotis lucifugus* genome, lack homology to other known *Helitrons*, and lack ORFs expected in *Helitrons*.

To identify lineage-specific elements, the validated TEs were compared to all known genomes in the NCBI genomes database and classified as lineage-specific, ancestral, or disjunct based on the 50 best BLASTn hits. In all, six elements were specific to *M. natalensis*, five solo LTRs and one non-autonomous piggyBac. Seventeen of the validated elements were found in other vespertilionid bats, including eight transposons in the TcMariner superfamily. The best BLASTn hits for seven elements were to non-chiropteran taxa. Of these, six were cut and paste transposons (5 Tiggers and 1 piggyBac) and one was a LTR. All five Tiggers are elements previously identified in other non-chiropteran taxa and thus represent ancient transposons. One element, Tigger1_MNa shared similarity to more than twenty insertions in the brown kiwi (*Apteryx australis*) genome. All hits were ≥ 97 % similar across ≥ 92 % of the entire *M. natalensis* Tigger1_Mna element. Since our de novo analysis only masked chiropteran-specific elements, these elements, known from other non-chiropteran taxa, were not identified in the initial masking procedures. The closest BLASTn hit to the remaining nine elements was to *Pteropus alecto*, a pteropodid bat. The pteropodid bats are only distantly related to *Miniopterus* among bats and some elements likely represent subfamilies diverged from TEs in the ancestral bat genome. These elements were reclassified as "ancestral".

Individual TE insertions in the *M. natalensis* genome were annotated using the final validated TE library that was combined with all known mammalian repeats in RepBase. For comparison, *Myotis lucifugus*, *E. fuscus*, and *P. parnellii* were processed alongside *M. natalensis*. All four bat genomes contained similar quantities of TEs ranging from 24-27.5 % (Table 1). Class II content was more variable between species than any of the retrotransposon categories. Cut and paste transposons comprised only 1.52 % of

the *M. natalensis* genome and less than 0.01 % was derived from *Helitrons*. In general, DNA transposon content in *M. natalensis* was more similar to the outgroup, *P. parnellii*, than to the vespertilionids (Table 1). The repeat accumulation profile for *M. natalensis* (Fig. 1a) indicates that a significant majority of Class II elements are heavily mutated when compared to the presumed ancestral sequence, indicating long periods of inactivity within the genome. In fact, TEs in the *M. natalensis* genome appear to be accumulating less rapidly than in the past. *M. natalensis* and *P. parnellii* (Fig. 1b) both show declining accumulation of Class II elements and negligible *Helitron* content. Both vespertilionid bats show appreciable levels of *Helitron* content and recent accumulation of cut and paste elements (Fig. 1c and d).

In addition to accumulation profiles, average genetic distances between consensus elements and TE loci were calculated for all DNA transposons that occupied more than 10 Kb in any of the bat genomes examined (Additional file 1: Table S1). In all, 248 different DNA transposons met these criteria. A portion of this data is presented in Table 2. Genetic distances can be used as a relative metric for age and combined with presence or absence in other species to understand TE dynamics within this group. The most common superfamily of TEs in the genomes examined were hATs (133 of 248 elements). Generally, hATs fell into two categories; 1) they were present in some combination of vespertilionid genomes or, 2) they were found in *M. natalensis* and also identified in the vesper bats and the outgroup, *P. parnellii*. Based on genetic distance, nhAT-100_EF was the most recent hAT transposon in *M. natalensis* (Table 2). This transposons, nhAT-100_EF, was present in all four taxa examined and genetic distances fell within a limited range (18.69-19.75 %). *Helitrons* followed a similar pattern to hATs; the two *Helitrons* identified in *M. natalensis* were shared by all three other species and heavily mutated (Table 2). TcMariner transposons, in general, were shared among all analyzed taxa, with a limited number of exceptions. Two elements were not identified in *P. parnellii*. Four older elements with average genetic distances greater than 23 % were found only in *P. parnellii*. Interestingly this analysis, which relied on RepeatMasker searches, identified a single element restricted to *M. natalensis* (nTIGGER-7_MNa). The BLAST searches used to identify lineage-specific repeats (described above) identified a single homologous sequence in another bat, *Rhinolophus ferrumequinum* (99 % query coverage, 85 % identity, 2e-57 E value), but not to other vesper bats. It is possible that this element is specific to *M. natalensis* since it was only found at one locus in one other species. In either case, since *R. ferrumequinum* was not in our RepeatMasker searches, the distribution of this element among the taxa examined appears reasonable. Finally, most unclassified

Table 1 Transposable element content. The number of bases and percent of the genome derived from transposable elements was calculated in four species of bats. The percentage of the genome occupied by transposable elements was calculated based on the total genome size, excluding ambiguous regions or scaffold gaps ("N"s)

Classification	Miniopterus natalensis		Myotis lucifugus		E. fuscus		P. parnelli	
	Bases	Percentage	Bases	Percentage	Bases	Percentage	Bases	Percentage
Transposable elements	415,627,321	23.95 %	518,680,444	27.50 %	478,933,702	26.58 %	383,285,246	24.76 %
Class I Retrotransposons	388,593,157	22.39 %	424,243,455	22.50 %	383,040,593	21.26 %	346,459,100	22.37 %
Long Terminal Repeats	69,316,646	4.00 %	72,931,404	3.88 %	71,532,426	3.97 %	68,573,030	4.43 %
ERV	1,092,720	0.06 %	1,038,965	0.06 %	1,149,410	0.06 %	1,499,592	0.10 %
ERV1	23,526,841	1.36 %	28,053,951	1.49 %	26,324,280	1.46 %	19,669,702	1.27 %
ERV2	431,937	0.02 %	7,857,605	0.42 %	4,951,316	0.27 %	391,300	0.03 %
ERV3	41,929,575	2.42 %	34,495,447	1.83 %	37,663,204	2.09 %	45,513,437	2.94 %
Gypsy	357,161	0.02 %	220,101	0.01 %	272,666	0.02 %	600,987	0.04 %
LTR	1,978,412	0.12 %	1,265,335	0.07 %	1,171,550	0.07 %	898,012	0.05 %
Long INterspersed Elements	241,612,217	13.92 %	242,431,627	12.85 %	210,106,281	11.66 %	225,554,475	14.57 %
L1	240,801,801	13.88 %	241,785,916	12.82 %	209,396,239	11.63 %	224,541,665	14.51 %
L2	63,018	0.00 %	42,706	0.00 %	53,584	0.00 %	158,866	0.01 %
Penelope	3,390	0.00 %	2,619	0.00 %	1,994	0.00 %	3,371	0.00 %
R4	24,859	0.00 %	14,178	0.00 %	21,390	0.00 %	35,098	0.00 %
RTE	467,222	0.03 %	425,242	0.02 %	428,374	0.02 %	431,357	0.03 %
RTEX	246,560	0.01 %	156,754	0.01 %	198,740	0.01 %	376,073	0.02 %
Tx1	5,367	0.00 %	4,212	0.00 %	5,960	0.00 %	8,045	0.00 %
Short INterspersed Elements	77,664,294	4.47 %	108,880,424	5.77 %	101,401,886	5.63 %	52,331,595	3.37 %
Unclassified	141,822	0.01 %	113,971	0.01 %	119,602	0.01 %	197,376	0.01 %
tRNA	77,501,269	4.46 %	108,745,685	5.76 %	101,255,190	5.62 %	52,075,143	3.36 %
7SL	1,048	0.00 %	1,775	0.00 %	1,914	0.00 %	3,786	0.00 %
5S	20,155	0.00 %	18,993	0.00 %	25,180	0.00 %	55,290	0.00 %
Class II DNA transposons	26,535,664	1.53 %	91,629,080	4.85 %	92,568,583	5.14 %	36,073,984	2.34 %
Cut and Paste	26,433,314	1.52 %	47,434,627	2.51 %	35,693,046	1.98 %	35,940,177	2.33 %
Kolobok	10,135	0.00 %	8,065	0.00 %	10,145	0.00 %	16,113	0.00 %
MuDR	13,048	0.00 %	12,651	0.00 %	13,221	0.00 %	21,955	0.00 %
PiggyBac	366,671	0.02 %	261,766	0.01 %	941,162	0.05 %	117,137	0.01 %
TcMar-Mariner	7,537,182	0.43 %	7,941,486	0.42 %	10,197,575	0.57 %	11,885,374	0.77 %
hAT	18,506,278	1.07 %	39,210,659	2.08 %	24,530,943	1.36 %	23,899,598	1.55 %
Rolling circle	102,350	0.01 %	44,194,453	2.34 %	56,875,537	3.16 %	133,807	0.01 %
Helitrons	102,350	0.01 %	44,194,453	2.34 %	56,875,537	3.16 %	133,807	0.01 %
Unknown	498,500	0.03 %	2,807,909	0.15 %	3,324,526	0.18 %	752,162	0.05 %

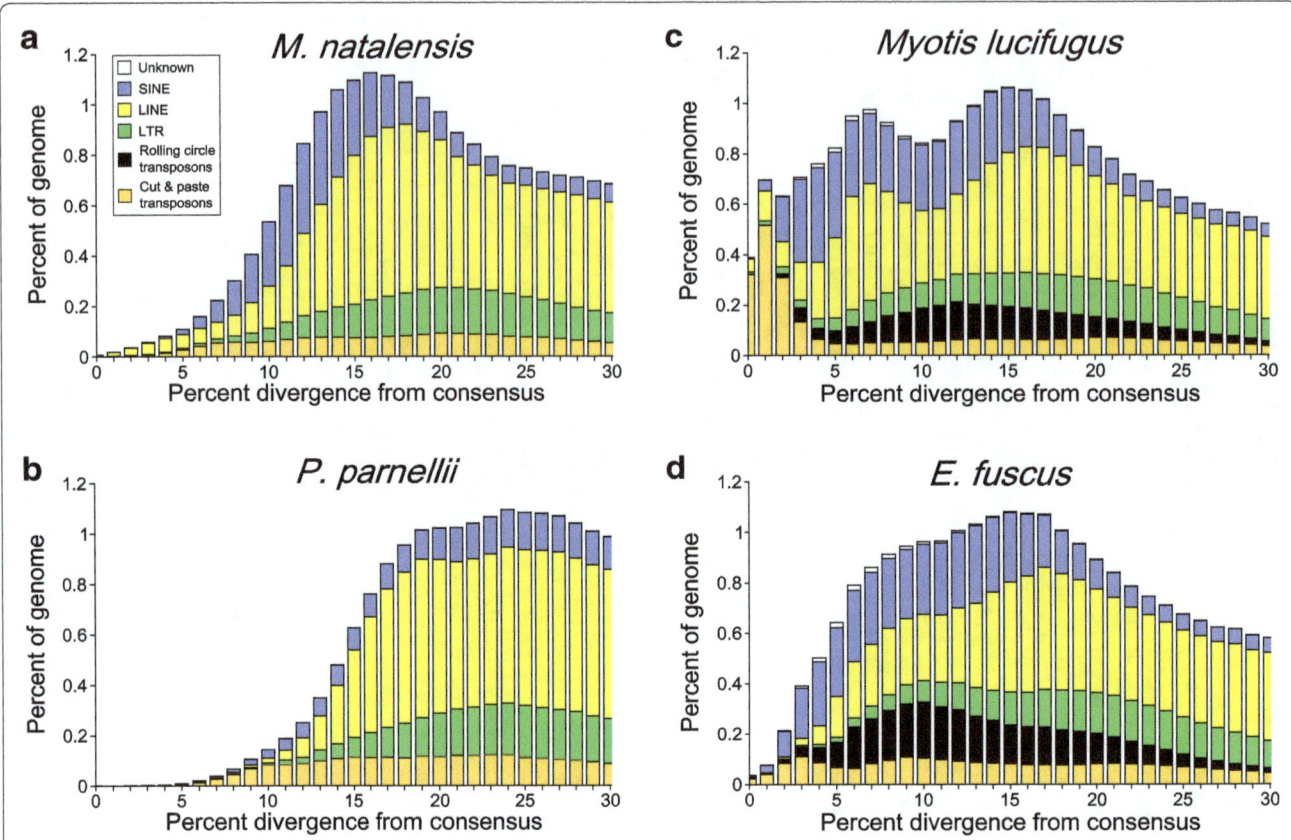

Fig. 1 Transposable element accumulation profiles in **a** *Miniopterus natalensis*, **b** *Pteronotus parnellii*, **c** *Myotis lucifugus*, and **d** *Eptesicus fuscus*. Kimura 2-parameter genetic distances were calculated between each repeat in the genome and the putative consensus for its subfamily. Distance values were binned based on transposable element type to visualize the accumulation of transposable elements over time. Due to their high mutation rate, CpG sites were excluded from genetic distance calculations

DNA transposons, Kolobok, and MuDRs were ancestral elements with high genetic diversity and present in all four taxa.

Discussion

Active DNA transposons are rare in mammals. To date, only the vespertilionid family of bats are known to have significant levels of active Class II elements. *Miniopterus* is the sole genus of the recently elevated family Miniopteridae, the sister family to Vespertilionidae [9]. Previous studies indicated that *Miniopterus* lacks the *Helitrons* found in vesper bats and may harbor limited piggyBac activity [6, 8]. Based on these results, it has been assumed that the horizontal transfer of DNA transposons occurred in an ancestral vespertilionid bat subsequent to the divergence of *Miniopterus*. Complete analysis of the *M. natalensis* genome generally supports previous conclusions with slight modifications, namely that limited Class II accumulation of TcMariner and piggyback elements indicate their presence in the *Miniopterus* + vespertilionid ancestor. It is possible that biases introduced with sequencing chemistries, genome assembly

methods, and bioinformatics analyses negatively influence the recognition of repetitive sequences. Highly repetitive sequences with low nucleotide diversity represent a significant problem for genome assembly methods. In addition, the culling of very low copy number elements (n = <10) from the initial de novo repeat identification with RepeatModeler could bias estimations slightly downward. While these influences are expected to be minimal, they cannot be accurately quantified and all results should be interpreted with these caveats in mind.

Species distribution of TEs identified in M. natalensis

De novo identification of TEs and manual curation identified several elements that are novel or exhibit interesting taxonomic distributions. Tigger1_MNa shared homology with twenty insertions in the brown kiwi genome and is closely associated with the consensus sequence for TIGGER1 originally identified in the human genome. These two consensus elements (Tigger1_MNa and TIGGER1) share almost 97.5 % similarity despite individual insertions being heavily mutated in the respective genomes [29] (Table 2). To demonstrate horizontal transfer between *M.*

Table 2 The average Kimura 2-parameter, genetic distance was calculated among all insertions for each element. Highly mutable CpG sites were excluded from distance calculations

Super Family	Element	Within group genetic distance (average)			
		Miniopterus natalensis	Myotis lucifugus	Eptesicus fuscus	Pteronotus parnellii
hAT	nhAT-100_EF	18.69	18.94	19.16	19.75
Helitron	Helitron1Nb_Mam	29.66	30.09	30.45	30.5
Helitron	Helitron3Na_Mam	32.74	32.87	33.57	33.93
PiggyBac	nPiggyBac-2_MNa	1.35	NA	NA	NA
PiggyBac	piggyBac2b_Mm	NA	1.91	NA	NA
PiggyBac	nPiggyBac-1_MNa	2.03	NA	NA	NA
PiggyBac	piggyBac1_Mm	NA	NA	2.18	NA
PiggyBac	npiggyBac-2_EF	NA	NA	2.86	NA
PiggyBac	npiggyBac-1_EF	NA	4.16	4.77	NA
PiggyBac	npiggy1_Mm	NA	NA	5.3	NA
PiggyBac	piggyBac_2a_Mm	NA	6.78	NA	NA
PiggyBac	nPiggyBac-3_MNa	7.58	NA	NA	NA
PiggyBac	piggyBac2_Mm	8.16	12.35	38.09	NA
TcMariner	nTIGGER-7_MNa	8.08	NA	NA	NA
TcMariner	nTIGGER-12_MNa	8.98	9.54	9.76	NA
TcMariner	nTIGGER-18_MNa	9.99	9.9	10.33	NA
TcMariner	TIGGER-1_Mna	13.63	14.41	14.6	15.66
TcMariner	TIGGER1	14	14.31	14.84	15.96

Distances were only calculated if the element occupied more than 10 kilobases in a genome. For species were elements were absent or occupied less than 10 kilobases of their genome, values are given as "NA"s. A limited number of transposons are shown here. A complete table displaying the average genetic distances of all elements is provided as Additional file 1: Table S1

natalensis and the brown kiwi, an element must have a disjunct phylogenetic distribution and high sequence similarity in multiple species beyond what is expected based on divergence times [30]. The BLASTn results for Tigger1_MNa seem to support a disjunct distribution, but its heavy mutation load may be within expectations based on a neutral mutation rate and the respective divergence times of these taxa [30]. Other factors giving the appearance of a disjunct species distribution, such as sequence contamination in the kiwi genome, cannot be conclusively ruled out.

BLAST searches identified several elements specific to the *M. natalensis* genome indicating their emergence sometime in the last 37.5 [31] to 43 my [9]. Five of these are LTRs but one non-autonomous piggyBac DNA transposon (npiggyBac-3_Mna) is specific to *M. natalensis* based on comparisons to all currently available genomes. npiggyBac-3_Mna was present in the *M. natalensis* genome at low frequency (577 copies). In addition to npiggyBac-3_Mna, previous work noted that a small region associated with *Myotis lucifugus* piggyBac2_ML (bp 1,536-2,340) was also present in *Miniopterus* sp. [6]. Analysis of the entire *M. natalensis* genome indicates that the piggyBac2_ML fragment amplified by Ray et al. [6] is present in the *M. natalensis* genome as part of the larger piggyBac2_Mm element. RepBase does not recognize piggyBac2_ML (accessed 1 April 2016). Instead, it contains piggyBac2_Mm, the *Microcebus murinus* counterpart to piggyBac2_ML that is presumed to have been horizontally transferred between *Microcebus murinus* and *Myotis lucifugus* [32]. To be consistent with RepBase naming conventions, we refer to piggyBac2_ML from Ray et al. [6] as piggyBac2_Mm. In all, RepeatMasker identified fewer than 80 piggyBac2_Mm loci occupying 58,499 bps in the *M. natalensis* genome.

These results suggest that the PCR-based analyses of Ray et al. [6] were accurate in their identification of piggyBac2_Mm distribution among chiropterans. In that work, however, piggyBac2_Mm was absent in non-*Myotis* vesper bats. RepeatMasker results identify piggyBac2_Mm in *E. fuscus*, but in a heavily mutated and truncated form (Table 2) implying that piggyBac2_Mm elements in *E. fuscus* are ancestral elements misidentified as piggyBac2_Mm. The presence of closely related piggyBacs in *Myotis lucifugus* and *M. natalensis* could be explained by two possible scenarios: horizontal transfer of piggyBac2_Mm between *M. natalensis* and a *Myotis* sp. or invasion of piggyBac2_ML into the *Miniopterus* + vespertilionid ancestral genome, and subsequent loss in the lineage leading to *Eptesicus*. The genus *Myotis* occupies a basal clade within Vespertilionidae [33] meaning that if piggyBac2_Mm was present as a single or few copies, a single loss could explain the presence of piggyBac2_Mm in *Myotis* and *M. natalensis*, but not *Eptesicus*. Further supporting this scenario, piggyBac2_Mm contains more genetic diversity (8.16-12.35 %; Table 2) than other piggyBac elements that are limited to single species. It is likely that piggyBac2_Mm is an older subfamily of elements and may even be one of the first transposons to invade the bat genomes. On the other hand, horizontal transfer of piggyBac2_Mm involving *Myotis lucifugus* and *Microcebus murinus* (the mouse lemur) has been reported previously [32]. The distribution of these three genera (*Microcebus*, *Miniopterus*, and *Myotis*) all include portions of Africa and/or Madagascar, which allows for the possibility of such transfers in ancestral species (assuming similar ancestral distributions). Based on the current data, piggyBac2_Mm likely represents an invasion in an ancestral bat genome followed by a loss in *E. fuscus* (Fig. 2). In either case, *M. natalensis*, *Myotis lucifugus*, and *E. fuscus* each have lineage-specific, highly similar piggyBac transposons indicating some level of transposition in these genomes (Table 2).

TE invasions in Vespertilionidae and Miniopteridae

Just under 30 % of the *M. natalensis* genome is derived from TEs. Though there is evidence of lineage-specific accumulation, the *M. natalensis* genome appears to have experienced minimal DNA transposon activity when compared to vespertilionids (Fig. 1). Around 26.5 Mb (1.5 %) of the genome is derived from Class II elements compared to ~5 % in the vespertilionid bats (Table 1). The bulk of these DNA transposon in *M. natalensis* are cut and

paste DNA transposons, specifically hATs which account for 70 % of all transposon content. Several observations indicate the hAT elements were deposited in a distantly related ancestor of these taxa. First, analysis of transposons in primate genomes identified significant transposon activity from TcMariners, piggyBacs, and hATs during the eutherian radiation 81-150 mya, hATs being the dominant transposon [34]. Second, the most abundant transposon in *M. natalensis*, hATs, were highly mutated and present in all the Vespertilionidae, *M. natalensis* and *P. parnellii*; indicating ancestral accumulation (Table 2). Third, the quantity of cut and paste transposons in *M. natalensis* is more similar to *P. parnellii* than its more closely related vesper relatives. Fourth, the methods used herein were capable of finding lineage-specific elements yet only one new piggyBac was identified (compared to five LTRs). Based on these results, it seems clear that the bulk of cut and paste DNA transposons were deposited prior to the Chiropteran divergence meaning that at least 70 % of all DNA transposon activity in *M. natalensis* is ancestral.

Helitrons are not as common as cut and paste transposons in the *M. natalensis* genome, occupying less than 100 Kb. Two *Helitrons* (Helitron1Nb_Mam and Helitron3Na_Mam; Table 2) appear to have been active prior to the emergence of Chiroptera based on their presence in the taxa examined. HelitronScanner, failed to identify Helitron1Nb_Mam and Helitron3Na_Mam, likely due to the high mutation load they carry (>30 % on average; Table 2). The failure to identify novel *Helitrons* through structural searches and the low copy numbers of ancestral *Helitrons* identified via homology makes it reasonable to conclude

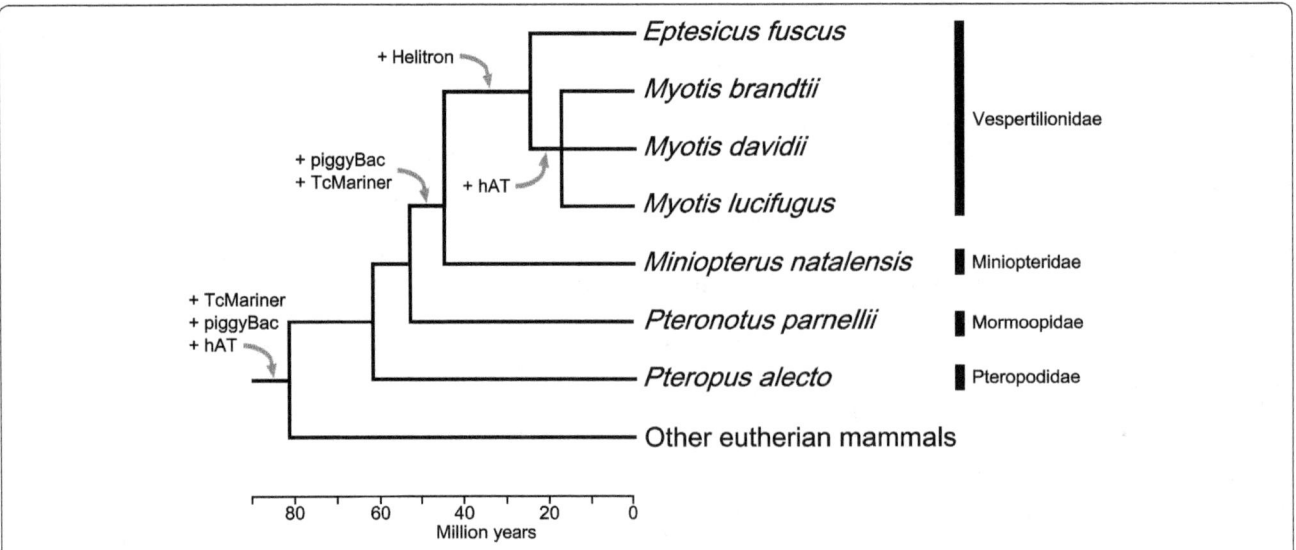

Fig. 2 A simplified tree depicting relationships among specimens examined. Time since divergence for each species is Time Tree of Life divergence estimate [39] from timetree.org. Relationships within *Myotis* are unresolved due to conflicting mitochondrial and nuclear phylogenies [40]. The gain of relevant, active transposons are plotted on respective nodes

that the *Helitrons* invasion into the vesper bats occurred subsequent to their divergence from *Miniopterus*.

The lack of significant cut and paste transposon accumulation and the absence of *Helitrons*, allows us to place more precise taxonomic and temporal limits on the DNA transposon invasion of an ancestral bat genome (Fig. 2). The presence of a limited number of TcMariner and piggyBac families present in *M. natalensis* and the vespertilionids seems to indicate that the acquisition of DNA transposons began just before the divergence of *Miniopterus* and the vespertilionids. *Helitrons* and hATs were introduced into an ancestral vespertilionid subsequent to the divergence of *Miniopterus*. Lineage-specific cut and paste DNA transposons reached much higher copy numbers in the vespertilionids genomes (Fig. 1c and d) than in the *M. natalensis* genome (Fig. 1a).

Conclusions

The results presented here confirm and expand upon previous findings regarding the distribution of DNA transposons in bats [3–6] and suggest several avenues of research. For example, if an ancestral *Miniopterus* + vespertilionid bat was exposed to DNA transposons, what factors were responsible for the differential accumulation in the daughter lineages? How have genomic defense mechanisms against TEs evolved in presence/absence of DNA transposons [35]? What vectors are responsible for transferring Class II elements to these bats [36]? Finally, what role do TEs play in the generation of taxonomic and genomic diversity? The rapid diversification of the vespertilionid bats is temporally associated with the acquisition of DNA transposons [7]. Individual TE insertions are generally neutral or deleterious, but instances of exaptation are known (reviewed in [37]). Beyond individual TE insertions, TE activity in general may be beneficial, allowing species to rapidly adapt to changing environments more quickly than relying on point mutations alone [38]. *Miniopterus* and the vespertilionids may represent extremes in the possible diversity of mammalian TE repertoires in sister taxa. By taking advantage of these contrasting compositions, it may be possible to answer specific questions regarding TEs and their role in genome evolution.

Additional files

Additional file 1: The average Kimura 2-parameter, genetic distance was calculated among all insertions for each element. Highly mutable CpG sites were excluded from distance calculations. Distances were only calculated if the element was occupied more than 10 kilobases in a species. For species where elements were absent or occupied less than 10 kilobases of their genome, values are given as "NA"s. Calculations for all meeting these critera are shown here. Table 2 in the manuscript is reduced fascimilie of this table containing only elements specifically discussed in the text. (XLSX 22 kb)

Additional file 2: Novel *Miniopterus natalensis* transpsoable elements were identified using de novo methods and manual curation. Sequences are in FastA format. (FAS 39 kb)

Abbreviations
TEs, LINE, SINE, LTR, mya, ORFs, Kb

Funding
This work was supported by the National Science Foundation (DEB-1355176) to DAR. Additional support was provided by College of Arts and Sciences at Texas Tech University.

Authors' contributions
DAR, SM, and RNP participated in all aspects of this manuscript. All authors read and approved the final manuscript.

Competing interests
The authors declare that they have no competing interests.

References
1. de Koning APJ, Gu W, Castoe TA, Batzer MA, Pollock DD. Repetitive elements may comprise over two-thirds of the human genome. PLoS Genet. 2011;7(12):e1002384. doi:10.1371/journal.pgen.1002384.
2. Lander, et al. Initial sequencing and analysis of the human genome. Nature. 2001;409(6822):860–921. http://www.nature.com/nature/journal/v409/n6822/full/409860a0.html.
3. Ray DA, Pagan HJT, Thompson ML, Stevens RD. Bats with hATs: Evidence for recent DNA transposon activity in genus *Myotis*. Mol Biol Evol. 2007;24(3): 632–9. doi:10.1093/molbev/msl192.
4. Pritham EJ, Feschotte C. Massive amplification of rolling-circle transposons in the lineage of the bat *Myotis* lucifugus. Proc Natl Acad Sci. 2007;104(6): 1895–900. doi:10.1073/pnas.0609601104.
5. Thomas J, Phillips CD, Baker RJ, Pritham EJ. Rolling-circle transposons catalyze genomic innovation in a mammalian lineage. Genome Biol Evol. 2014;6(10):2595–610. doi:10.1093/gbe/evu204.
6. Ray DA, Feschotte C, Pagan HJT, Smith JD, Pritham EJ, Arensburger P, et al. Multiple waves of recent DNA transposon activity in the bat, *Myotis lucifugus*. Genome Res. 2008;18(5):717–28. doi:10.1101/gr.071886.107.
7. Platt RN, Vandewege MW, Kern C, Schmidt CJ, Hoffmann FG, Ray DA. Large numbers of novel miRNAs originate from DNA transposons and are coincident with a large species radiation in bats. Mol Biol Evol. 2014;31(6): 1536–45. doi:10.1093/molbev/msu112.
8. Thomas J, Sorourian M, Ray D, Baker RJ, Pritham EJ. The limited distribution of Helitrons to vesper bats supports horizontal transfer. Gene. 2011;474(1–2): 52–8. http://dx.doi.org/10.1016/j.gene.2010.12.007.
9. Miller-Butterworth CM, Murphy WJ, O'Brien SJ, Jacobs DS, Springer MS, Teeling EC. A family matter: Conclusive resolution of the taxonomic position of the long-fingered bats, *Miniopterus*. Mol Biol Evol. 2007;24(7):1553–61. doi: 10.1093/molbev/msm076.
10. Eckalbar WL, Schlebusch SA, Mason MK, Gill Z, Parker AV, Booker BM et al. Transcriptomic and epigenomic characterization of the developing bat wing. Nat Genet. 2016;advance online publication. doi:10.1038/ng.3537.
11. Platt RN, Blanco-Berdugo L, Ray DA. Accurate transposable element annotation is vital when analyzing new genome assemblies. Genome Biol Evol. 2016;8(2):403–10. doi:10.1093/gbe/evw009.
12. Smit A, Hubley R. RepeatModeler Open-1.0. 2008-2015.
13. Smit A, Hubley R, P G. RepeatMasker Open-4.0. 2013-2015.
14. Altschul SF, Madden TL, Schäffer AA, Zhang J, Zhang Z, Miller W, et al. Gapped BLAST and PSI-BLAST: a new generation of protein database search programs. Nucleic Acids Res. 1997;25(17):3389–402. doi:10.1093/nar/25.17.3389.
15. Edgar RC. MUSCLE: multiple sequence alignment with high accuracy and high throughput. Nucleic Acids Res. 2004;32(5):1792–7. doi:10.1093/nar/gkh340.
16. Hall T. BioEdit version 7.0.0. 2004.
17. Xiong W, He L, Lai J, Dooner HK, Du C. HelitronScanner uncovers a large overlooked cache of Helitron transposons in many plant genomes. Proc Natl Acad Sci. 2014;111(28):10263–8. doi:10.1073/pnas.1410068111.

18. Rice P, Longden I, Bleasby A. EMBOSS: The European Molecular Biology Open Software Suite. Trends Genet. 2000;16(6):276–7.

19. Kent WJ. BLAT–the BLAST-like alignment tool. Genome research. 2002;12(4):656-64. doi:10.1101/gr.229202. article published online before march 2002.

20. Bao W, Kojima KK, Kohany O. Repbase Update, a database of repetitive elements in eukaryotic genomes. Mob DNA. 2015;6(1):1–6. doi:10.1186/s13100-015-0041-9.

21. Wheeler DL, Barrett T, Benson DA, Bryant SH, Canese K, Chetvernin V, et al. Database resources of the National Center for Biotechnology Information. Nucleic Acids Res. 2007;35 suppl 1:D5–12. doi:10.1093/nar/gkl1031.

22. Wicker T, Sabot F, Hua-Van A, Bennetzen JL, Capy P, Chalhoub B, et al. A unified classification system for eukaryotic transposable elements. Nat Rev Genet. 2007;8(12):973–82.

23. Kimura M. A simple method for estimating evolutionary rates of base substitutions through comparative studies of nucleotide sequences. J Mol Evol.16(2):111-20. doi:10.1007/bf01731581.

24. Xing J, Hedges DJ, Han K, Wang H, Cordaux R, Batzer MA. Alu Element Mutation Spectra: Molecular Clocks and the Effect of DNA Methylation. J Mol Biol. 2004;344(3):675–82. http://dx.doi.org/10.1016/j.jmb.2004.09.058.

25. Eick GN, Jacobs DS, Matthee CA. A nuclear DNA phylogenetic perspective on the evolution of echolocation and historical biogeography of extant bats (Chiroptera). Mol Biol Evol. 2005;22(9):1869–86. doi:10.1093/molbev/msi180.

26. Ray DA, Pagan HJ, Platt RN, Kroll AR, Schaack S, Stevens RD. Differential SINE evolution in vesper and non-vesper bats. Mob DNA. 2015;6(1):1–10. doi:10.1186/s13100-015-0038-4.

27. Platt RN, Zhang Y, Witherspoon DJ, Xing J, Suh A, Keith MS, et al. Targeted capture of phylogenetically informative Ves SINE Insertions in genus *Myotis*. Genome Biol Evol. 2015;7(6):1664–75. doi:10.1093/gbe/evv099.

28. Thomas J, Pritham EJ. Helitrons, the eukaryotic rolling-circle transposable elements. Microbiol Spectr. 2015;3(4). doi:10.1128/microbiolspec.MDNA3-0049-2014.

29. Smit AF, Riggs AD. Tiggers and DNA transposon fossils in the human genome. Proc Natl Acad Sci U S A. 1996;93(4):1443–8.

30. Schaack S, Gilbert C, Feschotte C. Promiscuous DNA: horizontal transfer of transposable elements and why it matters for eukaryotic evolution. Trends Ecol Evol. 2010;25(9):537–46. doi:10.1016/j.tree.2010.06.001.

31. Lack JB, Roehrs ZP, Stanley CE, Ruedi M, Van Den Bussche RA. Molecular phylogenetics of *Myotis* indicate familial-level divergence for the genus *Cistugo* (Chiroptera). J Mammal. 2010;91(4):976–92. doi:10.1644/09-mamm-a-192.1.

32. Pagan HJT, Smith JD, Hubley RM, Ray DA. PiggyBac-ing on a Primate genome: Novel elements, recent activity and horizontal transfer. Genome Biol Evol. 2010;2:293–303. doi:10.1093/gbe/evq021.

33. Roehrs ZP, Lack JB, Van Den Bussche RA. Tribal phylogenetic relationships within Vespertilioninae (Chiroptera: Vespertilionidae) based on mitochondrial and nuclear sequence data. J Mammal. 2010;91(5):1073–92. doi:10.1644/09-mamm-a-325.1.

34. Pace JK, Feschotte C. The evolutionary history of human DNA transposons: Evidence for intense activity in the primate lineage. Genome Res. 2007;17(4):422–32. doi:10.1101/gr.5826307.

35. Vandewege MW, Platt RN, Ray DA, Hoffmann FG. Transposable element targeting by piRNAs in Laurasiatherians with distinct transposable element histories. Genome Biol Evol. 2016. doi:10.1093/gbe/evw078.

36. Gilbert C, Chateigner A, Ernenwein L, Barbe V, Bézier A, Herniou EA et al. Population genomics supports baculoviruses as vectors of horizontal transfer of insect transposons. Nat Commun. 2014;5. doi:10.1038/ncomms4348.

37. Feschotte C, Pritham EJ. DNA Transposons and the Evolution of Eukaryotic Genomes. Annu Rev Genet. 2007;41:331–68. doi:10.1146/annurev.genet.40.110405.090448.

38. Oliver KR, Greene WK. Mobile DNA and the TE-Thrust hypothesis: supporting evidence from the primates. Mob DNA. 2011;2(1):1–17. doi:10.1186/1759-8753-2-8.

39. Hedges SB, Marin J, Suleski M, Paymer M, Kumar S. Tree of Life Reveals Clock-Like Speciation and Diversification. Mol Biol Evol. 2015;32(4):835–45. doi:10.1093/molbev/msv037.

40. Stadelmann B, Lin LK, Kunz TH, Ruedi M. Molecular phylogeny of New World Myotis (Chiroptera, Vespertilionidae) inferred from mitochondrial and nuclear DNA genes. Mol Phylogenet Evol. 2007;43(1):32–48. http://dx.doi.org/10.1016/j.ympev.2006.06.019.

8

Functional characterization of the active *Mutator*-like transposable element, *Muta1* from the mosquito *Aedes aegypti*

Kun Liu[1] and Susan R. Wessler[2*]

Abstract

Background: *Mutator*-like transposable elements (MULEs) are widespread with members in fungi, plants, and animals. Most of the research on the MULE superfamily has focused on plant MULEs where they were discovered and where some are extremely active and have significant impact on genome structure. The maize *MuDR* element has been widely used as a tool for both forward and reverse genetic studies because of its high transposition rate and preference for targeting genic regions. However, despite being widespread, only a few active MULEs have been identified, and only one, the rice *Os3378*, has demonstrated activity in a non-host organism.

Results: Here we report the identification of potentially active MULEs in the mosquito *Aedes aegypti*. We demonstrate that one of these, *Muta1*, is capable of excision and reinsertion in a yeast transposition assay. Element reinsertion generated either 8 bp or 9 bp target site duplications (TSDs) with no apparent sequence preference. Mutagenesis analysis of donor site TSDs in the yeast assay indicates that their presence is important for precise excision and enhanced transposition. Site directed mutagenesis of the putative DDE catalytic motif and other conserved residues in the transposase protein abolished transposition activity.

Conclusions: Collectively, our data indicates that the *Muta1* transposase of *Ae. aegypti* can efficiently catalyze both excision and reinsertion reactions in yeast. Mutagenesis analysis reveals that several conserved amino acids, including the DDE triad, play important roles in transposase function. In addition, donor site TSD also impacts the transposition of *Muta1*.

Keywords: Transposable elements, *Mutator*-like elements (MULE), *Aedes aegypti*, Yeast assay, Target site duplication (TSD), Transposase

Background

Transposable elements (TEs) are mobile fragments of DNA that can move from one locus to another in the host genome, often replicating in the process. TEs usually make up the largest fraction of eukaryotic genomes; accounting for almost half of the human genome, and more than 70% of the genomes of some grass species [1]. Based on their transposition intermediate, eukaryotic TEs are divided into two classes. Class 1 elements utilize an RNA intermediate in the transposition reaction while the intermediate for most class 2 elements is the element itself that is mobilized by a 'cut and paste' mechanism [2].

A TE family is composed of one or more transposase-encoding autonomous elements and up to several thousand nonautonomous elements that do not encode functional transposase. Family members share the same terminal inverted repeats (TIRs) and target site duplications (TSDs) of the same length allowing them to move in *trans* by utilizing the transposase encoded by the autonomous family member(s) [3].

Prior studies have classified class 2 TEs into upwards of 19 superfamilies on the basis of the relatedness of the element-encoded transposase [4]. The original *Mutator* element, now called *MuDR*, was first isolated from a maize strain as the agent responsible for its high forward mutation rate [5, 6]. Subsequently, members of this superfamily, called collectively *Mutator*-like transposable elements (MULEs), were found in other plants, and in

* Correspondence: susan.wessler@ucr.edu
[2]Department of Botany and Plant Sciences, University of California, Riverside, CA 92521, USA
Full list of author information is available at the end of the article

fungi, protozoans, and in a variety of animals, (from insects to fish to other metazoans) [3, 7–9]. Typical features of MULEs include long terminal inverted repeats (TIRs) (>100 bp) and an 8-10 bp TSD [10]. Nonautonomous family members often contain a variety of sequences between the TIRs, including fragments from host genes; such elements are called Pack-MULEs [11]. To date, only a few active MULEs have been identified, including *MuDR, Hop1*, Jittery, *TED*, and *Os3378* [6, 12–15]. Importantly, only the rice *Os3378* element has been shown to transpose in a heterologous host [15].

Most MULE transposases contain a N-terminal zinc finger DNA binding motif and a conserved C-terminal DDE domain, which has been shown to be the catalytic core for transposition reactions in other superfamilies but not to date for MULEs [16, 17]. Phylogenetic studies indicate that the DDE domain of MULE transposases is closely related to the IS256 family, which is present in diverse prokaryotes [18]. Each residue of the predicted DDE motif of IS256 has been shown to be necessary for transposition [19]. Most MULE transposases also harbor a CH motif and a W residue between the second D and E of the DDE domain, which are also conserved in the *hAT* superfamily. Mutagenesis analyses in *hAT* elements demonstrated the importance of these residues for transposition [20, 21]. Whether these residues are also important for the MULE transposase has not as yet been determined.

TE-based genetic tools have facilitated our deep understanding of the biology of both plants and animals, however, such tools are not currently available in the mosquito species, including *Aedes aegypti, Aedes albopictus, Culex quinquefasciatus* and *Anopheles gambiae*, which can spread dengue fever, yellow fever, chikungunya, zika, and many other diseases that are responsible for over one million deaths annually [22–25]. Four exogenous TEs that transpose at high frequencies in *Drosophila melanogaster* (*Hermes, Mos1, Minos, piggyBac*) were found to move rarely or not at all in the germ line of mosquitoes as very few germinal mutations were detected [26, 27]. To explain this result, it was hypothesized that mosquitoes have a strong genome defense system that could effectively recognize and silence foreign TEs [28]. Therefore endogenous active TEs could be effective mutagens from generation to generation as they might be able to evade genome surveillance. Availability of the genome sequence enabled the search for potentially active TEs in the mosquito genomes [29–32]. In this study we show that *Muta1*, identified by computer-assisted analysis of the *Ae. aegypti* genome, is an active TE capable of both insertion and excision in a yeast transposition assay. Site-directed mutagenesis analysis revealed that transposition activity in yeast was influenced by disruption of several conserved residues and by the presence of TSDs at the donor site. With

characteristics such as high transposition activity, precise excision, and no target sequence preference, *Muta1* could be crafted into an effective tool for forward mutation analyses in mosquitoes.

Results
Distribution of the MULE superfamily in mosquito genomes
To identify potentially active MULEs, we first assessed the abundance and distribution of MULEs in the genomes of *Ae. aegtpti, Ae albopictus, C quinquefasciatus*, and *An. gambiae* [29–32]. The conserved MULE DDE domain sequences were used as queries in TBlastN searches against the *Ae. aegypti* genome through the TARGeT program, which is designed for TE discovery [20, 33]. After removal of duplicate hits, sequences with significant similarity (e- value $<10^{-15}$) to the MULE DDE domain were identified and used to build a phylogenetic tree. As a result, no significant hits were identified in the *An. gambiae* genome, whereas 141, 105, and 10 putative MULE DDE domain sequences were identified from the *Ae. aegtpti, Ae albopictus*, and *C quinquefasciatus* genomes, respectively (Additional file 1: Figure S1).

Full-length elements were defined by comparison of sequences in the same branch of the phylogenetic tree where adjacent sequences share high similarity that extends beyond the DDE domain. Sequences near the boundary of similarity were examined manually for the long TIRs and 8-10 bp TSDs that are features of MULEs. The 10 hits in *C. quinquefasciatus* were likely remnants of ancient insertion events as they are highly divergent in sequence and lack identifiable TIR and TSD. The quality of the *Ae. albopictus* genome assembly prevented our search for full-length elements. In the *Ae. aegypti* genome, 31 full-length elements sorted into 14 families, each with > 95% sequence similarity (*Muta1-14*, Fig. 1).

Characterization of the MULE superfamily in *Ae. aegypti*
All of the 14 MULE families identified from *Ae. aegypti* have 8-9 bp TSD, 12 of 14 have TIRs >100 bp and two have TIRs <60 bp. Ten of the 12 long TIR families contain short subterminal tandem repeats (9-15 bp). All families except *Muta6* also contain derivative nonautonomous elements (Additional file 2: Table S1), which share high sequence similarity in their TIR and subterminal regions, but carry heterogeneous internal sequences including fragments of host genes.

The phylogenetic tree also reveals the evolutionary relationships of the 14 families. Among the 5 major groups resolved on the tree, group B includes half of the 14 families, while group D and E contains one family each (Fig. 1). Furthermore, long branches indicate extensive sequence differences between families. For example, although *Muta5* and *Muta8* belong to group A and locate to adjacent clades, no significant nucleotide similarity

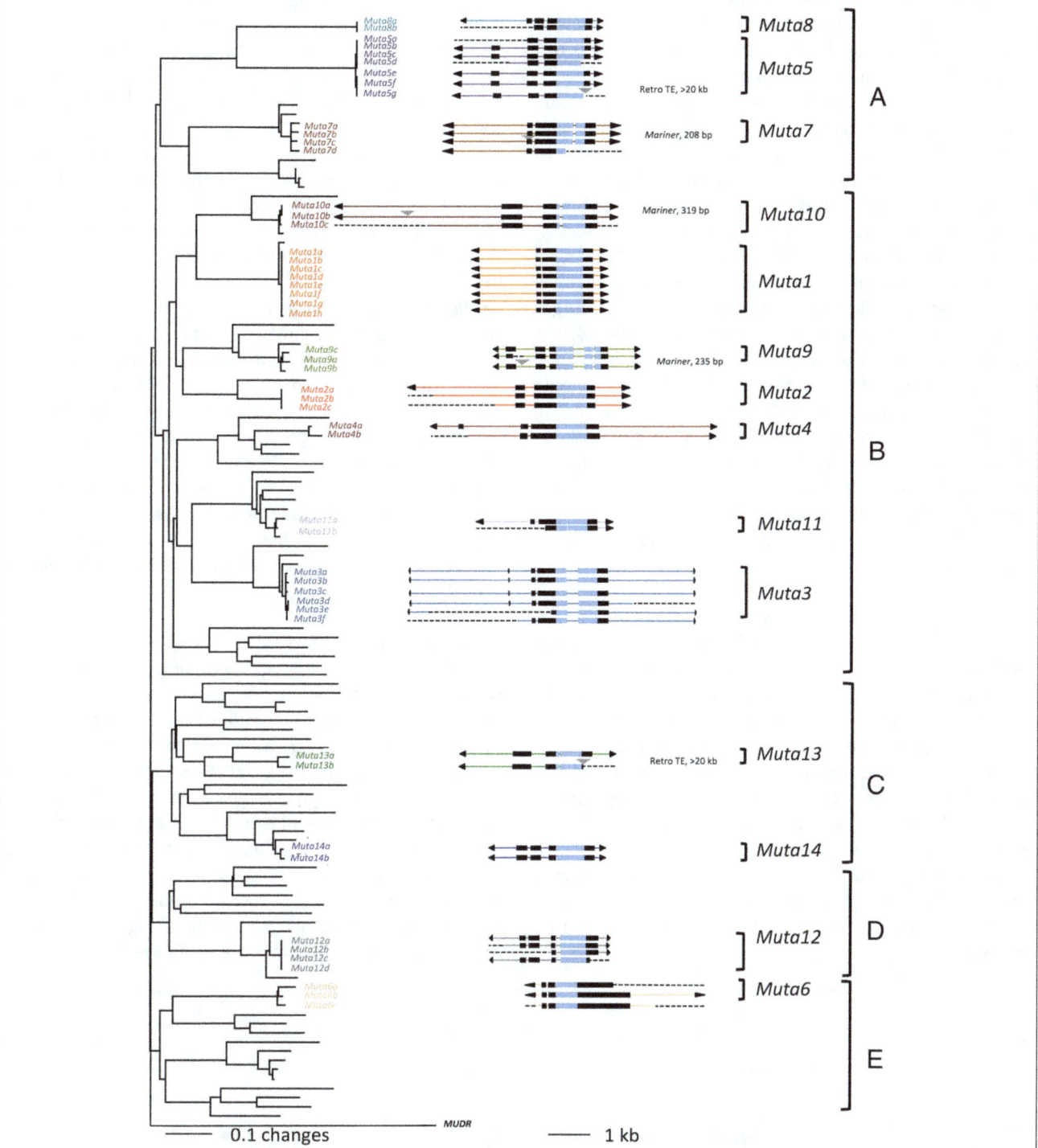

Fig. 1 Phylogenetic tree and structure of MULEs in *Ae. aegypti*. Neighbor-joining tree generated from a multiple alignment of 141 conceptually translated catalytic domains from transposase proteins with bootstrap values calculated from 1000 replicates. For element structures: TIRs are black triangles, exons are black boxes, DDE domains are blue boxes, introns are lines connecting boxes, colored lines indicate within family identity of noncoding regions, other TE insertions are gray triangles above elements, and dashed lines are missing sequences caused by gaps, deletions or large insertions. The maize *MuDR* transposase is used as an outgroup. The *Ae. aegypti Mutator* transposases are classified into 5 major lineages **a–e**

was detected between these elements. In another example, nucleotide similarity between *Muta3* and *Muta11* is restricted to the DDE region (71% identity, ~390 bp) and the TIRs (76% identity, ~ 110 bp).

After removal of other TE insertions and manual correction of frameshifts caused by small insertions and deletions, the 31 full-length elements were predicted to encode transposase proteins ranging in size from 416 to 554 residues (Additional file 2: Table S1). Comparison of conceptually translated transposases identified conserved regions, other than the DDE domain, that could have functional significance. The transposases of 13 families were predicted to harbor a FLYWCH type zinc finger DNA binding domain in the N-terminus, while *Muta6* was predicted to harbor a SWIM type zinc finger DNA binding domain in the C-terminus. Each family contains at least one putative full-length copy in which the coding region is not interrupted by frameshifts or stop codons. Of particular interest to this study, 3 families (*Muta1*, *Muta3* and *Muta5*) include at least 2 members that are identical or nearly identical (>99% identity). These features indicate recent and possible ongoing activity of multiple MULE families in *Ae. aegypti*.

Identification of the active *Muta1* family

Of the 14 families, *Muta1* appeared to be the best candidate for an active element. The family contains 7 identical copies and an 8th copy with only two point mutations in predicted noncoding DNA. *Muta1* is 3198 bp, flanked by TSDs of 8 bp or 9 bp, TIRs of 145 bp comprised of a 10 bp imperfect palindromic terminal motif with the 5th and 6th nucleotide unpaired, and 9 copies of a 12 bp subterminal tandem repeat separated by 3-4 bp spacers (Fig. 2, open and solid arrows, respectively). Because of the complexity of the TIR structure, *Muta1* can be classified as a type2 *Foldback* TE [34]. The *Muta1* transposase is predicted to be 504-amino acid transposase and encoded by two exons.

Over 300 putative nonautonomous elements derived from *Muta1* were detected in the *Ae. aegypti* genome (e- value <10^{-10}), with 171 flanked by perfect TSDs of 8 bp or 9 bp (Additional file 3: Figure S2A). Most of the derivative elements share the TIR sequence with *Muta1*, but the TIRs are often truncated, with variable copies of the subterminal repeats (Fig. 3). There are multiple copies of some derivative elements; for example, there are 4 copies of *Muta1NA1* (>98% identical). Most derivative elements contain variable internal sequences and share sequence similarity with only the TIR of *Muta1*. For about 20% of the derivative elements these variable internal regions can be aligned with sequences from host genes, much like previously described Pack-MULEs [11]. For example, the 1623 bp *Muta1NA3* contains a 276 bp fragment from a serine/threonine-protein kinase gene (97.5% identical, e-value < 1e-51) (Fig. 3). Among the 171 insertion sites, 110 (64.3%) are located in gene bodies or within 5kb upstream or downstream of genes. In a control dataset of 171 randomly selected genomic sites (see methods) only 49 (28.7%) were located in the same regions (Additional file 3: Figure S2B), suggesting that *Muta1* may have an insertion preference for genic regions as was previously reported for plant MULEs [6].

Muta1 can transpose in yeast

A yeast transposition assay was employed to determine whether *Muta1* transposase is able to catalyze the movement of natural and/or artificial nonautonomous elements. In prior studies members of five superfamilies (MULE, *Tc1/mariner*, *hAT*, *PIF/Harbinger* and *piggyBac*) were shown to transpose in yeast [15, 35–38]. Our yeast assay consisted of two plasmid constructs: an expression vector containing the *Muta1* transposase coding sequence downstream of the galactose inducible *GAL1* promoter, and a reporter vector containing a nonautonomous element inserted in, and blocking expression of, the *ADE2* gene (Additional file 4: Figure S3A). We first

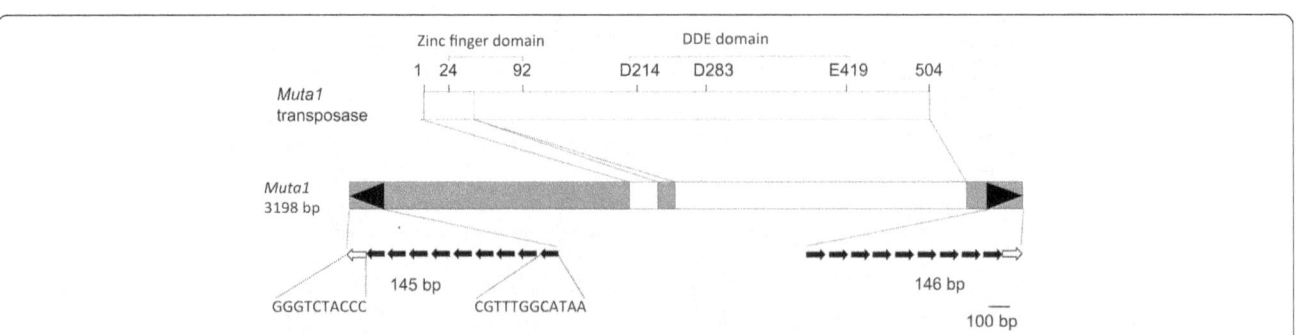

Fig. 2 Structural features of the *Muta1* element and its transposase. The eight virtually identical *Muta1* elements contain noncoding regions (shaded) and the coding region (white boxes) for the predicted 504-residue transposase with the predicted zinc and catalytic (DDE) domains discussed in the text. Structural features of *Muta1* include its distinctive long TIR (black arrowheads) whose substructure, expanded at the bottom, includes the 10 bp terminal palindromic motif (open arrow) and the 12 bp subterminal tandem repeats (black arrows) with linker DNA of 3-4bp represented by gaps between solid arrows

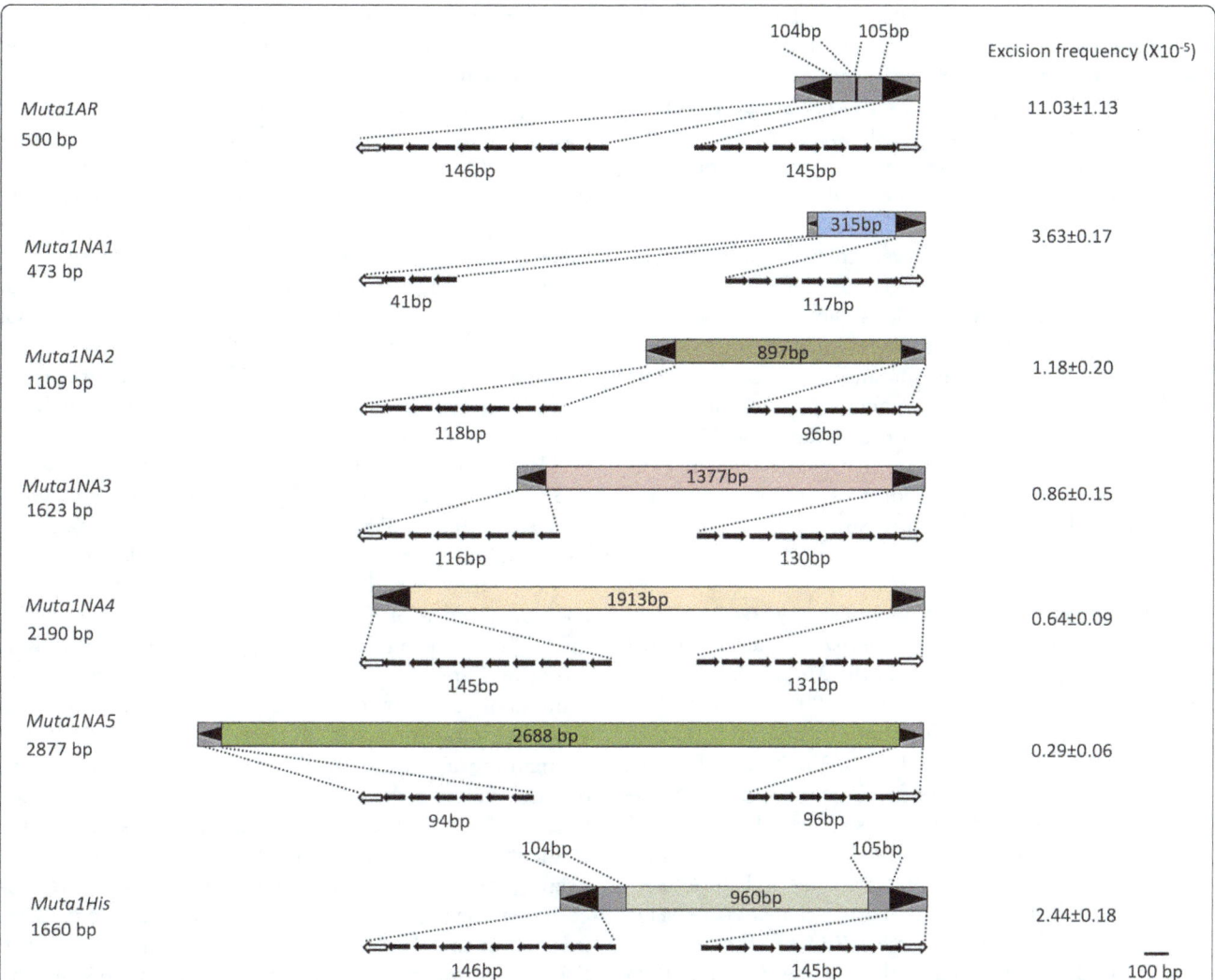

Fig. 3 Structural features of nonautonomous *Muta1* elements used in this study. *Muta1NA1* through *Muta1NA5* are natural elements; *Muta1AR* and *Muta1His* are artificial. Element lengths and internal sequences are indicated. Black arrowheads represent the TIRs, which include the terminal palindromic motif (open arrow) and subterminal tandem repeats (solid arrows). Gray shaded regions are sequences derived from *Muta1*; colored regions of each element indicate the diverse origin of internal sequences. Excision frequencies from the *ADE2* reporter in yeast assays are on the right

tested the artificial *Muta1AR* element which contains 250 bp from each end of *Muta1* (Fig. 3). Transposase mediated TE excision restored *ADE2* expression and resulted in reversion of adenine auxotrophy to prototrophy (Additional file 4: Figure S3B, Additional file 5: Figure S4A). Excision events were validated by PCR amplification of the *ADE2* empty site from revertants (Additional file 6: Figure S5A).

In subsequent experiments, the reporter plasmid was modified by inserting several natural nonautonomous elements (*Muta1NA1-5*, Fig. 3, Additional file 5: Figure S4B-F) into the *ADE2* coding region. Despite differences in TIR length and the length and sequence of internal regions, *Muta1* transposase was able to mobilize all of these elements, but none with a frequency as high as the artificial element *Muta1AR*.

Integration events catalyzed by *Muta1* transposase were assayed by first constructing the nonautonomous *Muta1HIS* element containing a yeast selectable marker *HIS3* gene flanked by 250 bp of the *Muta1* termini (Fig. 3). Integration of *Muta1HIS* into yeast chromosomes was assayed in plasmid-free cells following selection with 5-FOA, (see Methods, Additional file 4: Figure S3D). Comparison of the frequency of *ADE2* revertants retaining the *HIS3* marker and are 5-FOA resistant (2.71×10^{-6} His$^+$ 5-FOA^R cells/total cells) to the frequency of total *ADE2* revertants (2.44×10^{-5} Ade$^+$ cells/total cells) indicated that about 11% of the *Muta1HIS* elements excised from the donor plasmid had reintegrated in yeast chromosomes (Table 1). In another assay, *ADE2* colonies isolated directly from the *Muta1HIS* excision assay were tested to determine if they were also

His$^+$ and 5-FOA^R. Of 300 revertant colonies, 41 (14%) could proliferate on selective plates, in agreement with the results of the first approach (Table 1).

To determine the precise insertion sites, polymorphic bands on transposon display gels were recovered [39], sequenced and mapped to yeast chromosomes (Additional file 6: Figure S5B). Sixty of 62 sites had significant matches with yeast genomic sequence, while two matched plasmid sequences. Fourteen of the 60 genomic sites were in gene bodies while 27 were within 1kb of genes. Thus 41/60 insertions (68%) were in gene-rich regions [40]. Amplification and sequencing of each site revealed the presence of TSDs with 8 bp for 21 events and 9 bp for 39 events (Additional file 7: Table S2). Consensus sequences generated from the yeast insertion sites and the 171 sites for *Muta1* derivative elements in the *Ae. aegypti* genome indicated little or no sequence preference for insertion (Additional file 8: Figure S6).

The *Muta5* element does not transpose in yeast

The successful transposition of *Muta1* in yeast prompted us to perform a similar analysis of the *Muta5* family which contains 3 identical copies, a 4th with >99% sequence identity, and 3 copies with large deletions or insertions (Fig. 1). *Muta5* is 3496 bp, flanked by TIRs of 151 bp and TSDs of 8 bp or 9 bp and is predicted to contain 3 exons that encode a 554-amino acid transposase. The TIR of *Muta5* contains 9 copies of a 15 bp subterminal tandem repeat and an 8bp terminal motif (Additional file 9: Figure S7A). There are over 200 *Muta5* derivative elements (e- value <10^{-15}) in the *Ae. aegypti* genome, with 93 flanked by 8 bp or 9 bp TSD. Taken together the features of the *Muta5* family strongly suggested that it was an active element. However, in the yeast transposition assay the *Muta5* transposase was unable to catalyze transposition of *Muta5AR* (Additional file 9: Figure S7A), which contains 250 bp from the ends of *Muta5* (Additional file 9: Figure S7B).

Table 1 *Muta1* integration in yeast

Assay	Frequency
Muta1His integration[a]	
ade2$^+$ excisant	2.44 × 10^{-5} Ade$^+$ cells/total cells
His$^+$ 5-FOA^R excisant	2.71 × 10^{-6} His$^+$ 5FOA^R cells/total cells
	ratio of reintegration = 11%
Muta1His integration[b]	
ade2$^+$ excisant	300 colonies
His$^+$ 5-FOA^R excisant	41 colonies
	ratio of reintegration = 14%

[a]Measured by selection for presence of *Muta1His* and absence of plasmid following excision from *ADE2*
[b]Measured by analysis of independent Ade$^+$ revertants to maintain *Muta1His* but lacking plasmid

TSD at donor site affects *Muta1* transposition in yeast

Successful transposition of *Muta1* in yeast facilitated the analysis of the importance of its features by quantifying the impact of mutations on transposition quality and frequency. With regard to the role of the TSDs, although the nonautonomous constructs used in the assays described thus far lacked flanking TSDs, their reinsertion still generated TSDs of 8 bp or 9 bp. To examine the impact of TSD length or sequence on excision frequency of *Muta1AR*, three versions of 8 bp (TTCAATAG, CGATTCAA and GGTAACTC) or 9 bp (ATTCAATAG, TCGATTCAA and CGGTAACTC) TSDs were tested. Addition of 8 bp or 9 bp TSDs at the donor site increased *Muta1AR* excision frequency by ~7 fold and ~3 fold, respectively when compared to the controls lacking TSDs (Table 2). Similarly, introduction of TSDs flanking *Muta1-HIS* increased reintegration by about 40% and 90% for 8 bp TSDs or 9 bp TSDs respectively (Table 2). For both excision and integration, TSD sequence had little impact.

We next addressed the question of whether the presence or absence of TSDs at the donor site impacts the quality of excision by analyzing so-called transposon footprints. Specifically, class 2 TEs often leaves a footprint upon excision consisting of a few nucleotides or small rearrangements at the site of excision site [41]. Formation of footprints involves DNA repair of sequences flanking the excised element. To assess the impact of TSDs on the repair of excision sites, the donor element construct was modified so that all excision footprints (not only those that maintain the reading frame) could be analyzed. First, nonautonomous elements were inserted in the 5' UTR of *ADE2* (Additional file 4: Figure S3C). Second, because insertion of *Muta1AR* in the 5'UTR resulted in leaky *ADE2* expression, we substituted the longer *Muta1NA1* (Fig. 3), which, in the absence of transposase, blocked *ADE2* expression. When either 8 bp or 9 bp TSDs were added to *Muta1NA1*, about 90% of revertants were precise (Table 2), meaning that the element was removed as well as a single copy of the TSD. The quality of excision appeared to be independent of TSD sequence (Additional file 10: Figure S8B-G). In contrast, absence of donor site TSDs reduced perfect excision to only 10% of the *ADE2* revertant colonies sequenced (Table 2). Most excision sites reflected loss of a few nucleotides from either side of the flanking DNA. Occasionally, part of the TIR (up to 13 bp) was left after excision and repair (Additional file 10: Figure. S8A).

Mutagenesis of *Muta1* transposase

The catalytic domains of all characterized transposases of class 2 TEs contain a DDE/D amino acid triad [20] and mutagenesis studies confirmed its functional significance in the *piggyBac*, *Mariner* and *hAT* superfamilies [38, 42, 43]. Alignment with the transposase of other

Table 2 Impact of TSD on transposition

TSD length	TSD sequence	Excision frequency (X 10 $^{-5}$)	Reintegration frequency (%)[a]	precise excision /examined excision
0bp		11.03 ± 1.13	13.9 ± 1.8	4/38
	TTCAATAG	63.34 ± 9.12	20.88 ± 1.53	25/29
8bp	CGATTCAA	76.33 ± 11.53	18.22 ± 2.90	27/31
	GGTAACTC	68.53 ± 12.16	19.15 ± 2.39	24/27
	ATTCAATAG	31.52 ± 5.27	27.83 ± 3.01	27/29
9bp	TCGATTCAA	36.43 ± 6.12	24.35 ± 1.98	28/30
	CGGTAACTC	28.38 ± 3.11	29.01 ± 2.76	26/30

[a]Measured by analysis of independent Ade^{+} revertants to maintain *Muta1His* but lacking plasmid

active MULEs showed that the DDE triad in *Muta1* corresponds to D214, D283 and E419 (Fig. 4a). To determine if these conserved sites play key roles in transposition, site-directed mutagenesis was performed. Transposition activity was completely abolished when D214, D283 or E419 was mutated to alanine (Fig. 4b, Additional file 5: Figure S3I-K). In contrast, mutation of nonconserved sites, including E129, E188, E239, W313 and D473 to alanine had little impact. Although E373 is not a conserved site, mutation to alanine also completely abolished transposition activity.

The functional significance of two additional highly conserved residues in *Muta1*, H307 and W401, were

tested. In a prior study these residues were shown to be essential for transposition of the *hAT* superfamily member *Hermes* [21, 43]. Mutation of the corresponding *Hermes* H and W residues (H268 and W319) to alanine completely abolished transposition activity. Similarly for the *Muta1* transposase, mutation of H307 or W401 to alanine abolished activity (Fig. 4b). Analysis of *Hermes* transposase also showed that the W319 residue was likely necessary for the correct positioning of flanking DNA during the excision reaction, and that other aromatic residues can partially substitute for this function [21]. When W401 of *Muta1* was mutated to phenylalanine, transposition activity was reduced by 79% (Fig. 4b,

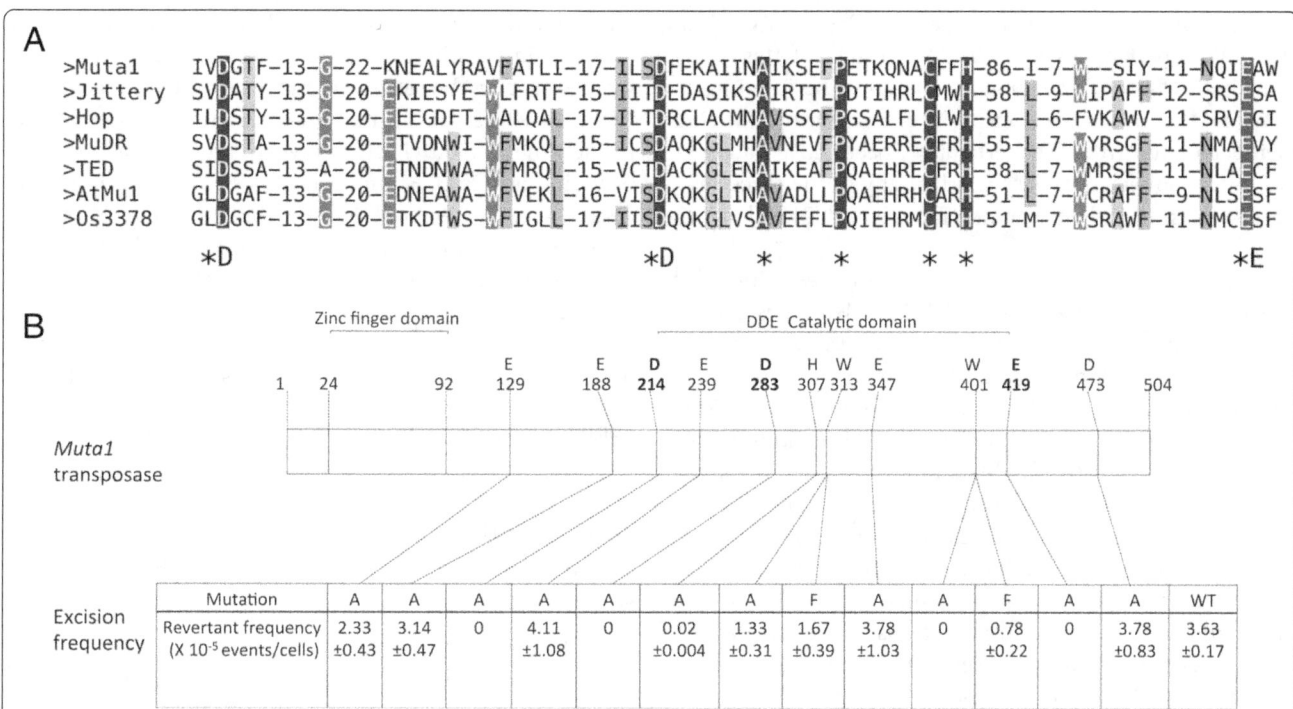

Fig. 4 MUSCLE alignment of the DDE domain in MULE transposases and the impact of mutations. **a** MUSCLE alignment of the DDE domain in *Muta1, MuDR, TED, atMu1, Os3378, Jittery,* and *Hop*. Shaded residues have related physical or chemical properties with darker shading denoting more conservation. Asterisks denote residues conserved in all sequences. **b** Schematic of the 504 amino acid Muta1 transposase with positions of the putative FLYWCH type zinc finger DNA binding domain (24–92 aa), and DDE triad (residues D214, D283, E419). Designated amino acids were mutated to alanine (*A*) or to both alanine and phenylalanine (*F*) resulting in the excision frequencies shown below. See text for details

Additional file 5: Figure S3L) and the frequency of precise excision dropped from 90% to approximately 48% (8 bp donor site TSD, 14/29 events) and 50% (9 bp donor site TSD, 15/30 events) (Additional file 10: Figure S8H&I), suggesting that the *Muta1* W401F mutation also led to inaccurate positioning. These results confirmed the importance of the putative DDE motif, the conserved H307, W401 and identified the nonconserved E373 as a potentially important residue for transposase function.

Discussion

The MULE superfamily is widespread in eukaryotic genomes and is closely related to prokaryotic IS*256* elements. However, although it is also found in the genomes of many insects, no active elements have been reported in insect species. In this study, we performed a thorough search for potentially active MULEs in the *Ae. aegypti* genome and demonstrate that *Muta1* encodes a transposase that catalyzes the excision and reinsertion of nonautonomous derivative elements in yeast. With the availability of this heterologous transposition assay, the function of the conserved MULE DDE domain and the role of TSD in transposition were tested.

The DDE/D domain is proposed to be the catalytic core involved in transposition of class 2 TEs and has been identified in all superfamilies [20]. Prior to this study, the functional significance of this domain had been experimentally validated only for members of the *piggyBac*, *Tc1/Mariner*, and *hAT* superfamilies [38, 42, 43]. Results of this study provide the first experimental evidence for the importance of the DDE motif in the transposition reaction of *Muta1*, a member of the MULE superfamily. Specifically, transposition was completely abolished when any of the three residues were mutated to alanine (Fig. 4b, Additional file 5: Figure S3I-K).

In addition to the DDE triad, other residues critical for transposition were identified. W401 of *Muta1* is a conserved residue that is also found in the *hAT* superfamily [20, 43]. Crystallographic analysis and in vitro biochemical assays showed that the corresponding W318 residue in *Hermes* functions in the positioning of flanking DNA, which ensures that the double strand break occurs at the correct position when an element excises from flanking DNA [21]. Other aromatic residues partially substitute for its function, however the mutant transposase generated additional species of intermediates in double strand break repair than the wild-type transposase, suggesting that inaccuracy in the position of the cleavage site may be the cause by these mutations [21]. For *Muta1*, the W401A mutation completely abolished transposition activity while the W401F mutation resulted in a 79% reduction of transposition frequency (Fig. 4b, Additional file 5: Figure S3L) and caused inaccurate excision as the frequency of precise excision dropped from 90% to ~50%

(Additional file 10: Figure S8H&I). Taken together these data suggest that this conserved tryptophan residue is likely playing a similar role in *hAT* and MULE transposases, which is to correctly position flanking DNA for the excision reaction. In addition to the W residue, a CxxH motif is also shared between MULE and *hAT* elements, the *Hermes* H268 was found to be located close to the DDE active center and involved in the interaction with TIRs [21]. Mutation of the corresponding H307 to alanine resulted in a 99.5% reduction of *Muta1* transposition activity (Fig. 4b), suggesting the importance of this residue for *Muta1* transposase function. Taken together, our study provides experimental evidence to support the close evolutional relationship reported previously between the MULE and *hAT* superfamilies [20].

Prior to this study, the only MULE shown to transpose in a heterologous host was the rice *Os3378* element [15]. Because both *Muta1* and *Os3378* have demonstrated activity in yeast assays that employed very similar experimental design and nonautonomous elements of similar size, comparison of assay results may be informative (Table 3). Excision frequencies of the 500 bp *Muta1AR*, as high as 6940 events per 10^7 cells (Fig. 2), is ~320 fold higher than *Os3378NA469* (469bp). For both elements, excision frequencies are increased by the presence of donor site TSD with *Muta1AR* enhanced by ~7 fold (8 bp) and 3 fold (9 bp), and *Os3378NA469* enhanced by ~17 fold with TSDs of 9 bp (Table 3). About 80% of *Os3378* integration sites in the yeast genome were located in gene bodies or within 1 kb of flanking regions of genes while *Muta1* had a slightly lower ratio of 68%. In summary *Muta1* shows very similar transposition behavior as *Os3378*, and the higher activity of *Muta1* makes it a better tool for the future study of MULE transposition, for example, the biochemical process of excision and integration, and how Pack-MULEs capture host gene fragments [11].

The presence of donor site TSDs impacts the quality of *Muta1*-mediated excision events as various footprints

Table 3 Comparison of *Muta1* and *Os3378* transposition in yeast

	Muta1	OS3378
Excision frequency no TSD (X 10^{-7})[a]	1103	1.2
Excision frequency 8bp TSD (X 10^{-7})[a]	6940	-
Excision frequency 9bp TSD (X 10^{-7})[a]	3211	20.2
Reintegration frequency No TSD[b]	13.90%	59.26%
Reintegration frequency 8bp TSD[b]	19.41%	-
Reintegration frequency 9bp TSD[b]	27.06%	39.28%
Percentage of reinsertion in gene rich regions	80%	68%

[a]Based on the excision of *Muta1AR* or *Os3378NA469* from the coding region of *ADE2*
[b]Based on the reintegration of *Muta1His* or *Os3378NA469*
- Corresponding experiment was not reported

were generated without donor TSD (Additional file 10: Figure S8A). The predominance of small deletions (1-4 bp) suggests that the *Muta1* transposase cuts outside the TIR. In contrast, with the presence of either 8 bp or 9 bp TSD at the donor site, most excision events were precise and the actual TSD sequence did not seem to matter (Additional file 10: Figure S8 B-G). Similar behavior was also observed for IS*256*, the prokaryotic TE family related to MULEs, and for the one other MULE tested, Os3378. Reduction of TSD from 8bp to 6bp eliminated precise excision of IS*256* and reduced the *Os3378* precise excision frequency from 97.44% to 82.05% [15, 19]. For IS*256* it was hypothesized that precise excision is achieved through a transposase-independent replication slippage mechanism that requires a short stretch of homologous DNA with a minimum length of 8 bp [44]. In our assay, the absence of donor TSDs resulted in a 90% reduction in precise excision (Additional file 10: Figure S8A-G), which suggests a role for TSDs in promoting precise excision.

Thirty-one full-length MULEs that group into 14 families were identified in the *Ae. aegypti* genome. Several families have identical or nearly identical full-length copies including *Muta1* (7 identical), *Muta5* (3 identical), and *Muta3* (2 with only 2 noncoding SNPs). Although the existence of identical genomic copies is a feature of active TEs, *Muta5* was unable to catalyze the movement of nonautonomous derivative elements in yeast (Additional file 9: Figure S7). One explanation for our success with *Muta1* but failure with *Muta5* is that the latter has 3 predicted exons while the former has 2. More predicted exons would increase the chances of incorrectly assembling the actual/functional *Muta5* transposase.

Accumulation of seven identical copies of *Muta1* in *Ae. aegypti* suggests that this element may still be active or that it has some success evading the genome surveillance system shown previously to effectively silence exogenous TEs [26–28]. In this regard, it may be possible to engineer *Muta1* to make it an effective endogenous mutagen. Like the *MuDR* system in maize, where the genome has numerous copies of nonautonomous *Mu* elements, there are over 300 *Muta1* derivative elements in the *Ae. aegypti* genome (Additional file 3: Figure S2A). Although mobility in the yeast assay does not guarantee mobility in the host, if *Muta1* was able to mobilize even a subset of these elements as it does in yeast, it could be an effective tool for high frequency insertional mutagenesis, especially when coupled with its preference for genic insertions and a lack of target sequence preference (Additional file 3: Figure S2B, Additional file 8: Figure S6).

Conclusions

This is the first report of the transposition of a non-plant MULE, *Muta1*, in a heterologous system and provides the first experimental evidence for the functional significance of the DDE domain in the transposition reaction in the MULE superfamily. High frequency transposition in a yeast assay facilitated the determination of *Muta1* transposition features including precise excision, genic targeting with no sequence preference and the impact of TIR and TSD for insertion and excision. Taken together, *Muta1* may be a valuable tool for forward genetics in mosquitoes.

Methods
Identification of MULEs in *Ae. aegypti*

The mosquito genomes used in this study: AaegL3 build for *Ae. aegypti* (https://www.vectorbase.org/organisms/aedes-aegypti/liverpool/aaegl3); AaloF1 build for *Ae. albopictus* (https://www.vectorbase.org/organisms/aedes-albopictus/foshan/aalof1); CpipJ2 build for *C. quinquefasciatus* (https://www.vectorbase.org/organisms/culex-quinquefasciatus/johannesburg/cpipj2) and AgamP4 build for *An. gambiae* (https://www.vectorbase.org/organisms/anopheles-gambiae/pest/agamp4). The conserved MULE DDE domain from all eukaryotes [20] was used as query to search the mosquito genomes by TBLASTN, as implemented in the TARGeT pipeline [33] with an E-value cutoff of 0.001. Flanking DNA sequences with 10 kb upstream and downstream of the matched region were retrieved. The ends of a putative element were determined by aligning two closely related elements with their 20 kb flanking sequences, TIR boundaries and TSDs were manually identified. Coding capacity of each element was predicted by the GENSCAN program (http://genes.mit.edu/GENSCANinfo.html).

To identify *Muta1* derivative nonautonomous elements, 50 bp from each end of *Muta1* was used in a BLASTN search with TARGeT [33] using default parameters. One hundred bp of flanking DNA sequences were retrieved for manual verification of the TIR and TSD of each derivative element. Fifty bp flanking each element were used for BLASTN searches against the *Ae. aegypti* genome (AaegL3 build) to determine the genomic location and compared to the genome annotation (release AaegL3.3, https://www.vectorbase.org/organisms/aedes-aegypti/liverpool/aaegl3) to determine the adjacent genes. For the control data set, 171 genome coordinates across the 4,757 scaffolds were randomly generated, and compared to the genome annotation (release AaegL3.3) to determine the surrounding sequences and genes. The random insertion sites generation used 1,000 replicates to estimate the expected number of insertions (and standard deviations) in each category.

Yeast construct construction

Genomic DNA of individual *Ae. aegypti* mosquito (Liverpool strain, obtained from Dr. Atkinson, UC Riverside) was extracted using the DNeasy Blood &

Tissue Kit (Qiagen). The two exons of *Muta1* predicted by GENSCAN program were cloned from genomic DNA and fused through overlap PCR (all primer sequences are available in Additional file 11: Table S3). The complete transposase coding sequence was then cloned into the Gateway cloning vector pENTR and transferred to destination vector pGAL415-ccdb [45] with LR Clonase (Invitrogen) to generate the pGAL415-ccdbMuta1 plasmid (Additional file 3: Figure S2A).

Two hundred fifty bp from each end of *Muta1* were fused by overlap PCR to generate *Muta1AR*. The *HIS3* fragment containing the yeast *HIS3* coding sequence, *HIS3* 5' and 3' UTR, and *HIS3* promoter was cloned from vector pGAL415-ccdb [45]. The *HIS3* fragment was then fused with 250 bp from each end of *Muta1* through overlap PCR to generate *Muta1HIS*. *Muta1NA1-5* elements were cloned directly from genomic DNA. All nonautonomous elements were inserted in the *Hpa*I site of *ADE2* for the exon excision assay or the *Xho*I site for 5' UTR excision assay through homologous recombination in yeast as previously described [37]. Donor site TSDs were introduced by adding corresponding TSD sequences in primers (Additional file 11: Table S3).

For *Muta5* assay, plasmid pGAL415-ccdbMuta5 was constructed in the same way as pGAL415-ccdbMuta1, and *Muta5NA* was constructed by overlap PCR (Additional file 11: Table S3).

Transposition assay

The yeast transposition assay using *Saccharomyces cerevisiae* strain DG2523 and the pWL89a vector was described previously [35, 36]. Transformation was performed using the Frozen-EZ Yeast Transformation kit (Zymo research). For excision assays, transformants were grown in 5 ml liquid media of CSM -leu-ura with 2% dextrose. After growth to saturation (36 h), cells were washed twice with 5 ml water, resuspended in 0.5 ml water and plated onto CSM -his-leu-ade with 2% galactose. Colonies were counted after incubation at 30°C for 15 days and viable counts were made by plating 100 μl of a 1×10^5 and 1×10^6 dilution on YPD plates.

For the reintegration assay, cells were grown to saturation in 5ml liquid CSM -leu-ura with 2% dextrose, cells were washed twice with 5 ml sterile water, resuspended in 0.5 ml water and plated onto CSM -leu-ura-ade with 2% galactose plate and CSM -his-leu + 5-FOA with 2% galactose plates. Colonies were counted after incubation at 30°C for 15 days, and viable counts were made by plating 100 μl of a 1×10^5 and 1×10^6 dilution on YPD plates. In another approach, individual Ade+ *Muta1HIS* excision revertant colonies isolated directly from plates of CSM -his-leu-ade with 2% galactose were streaked on CSM -his + 5-FOA plates to calculate the reintegration frequency.

Excision and reinsertion analysis

For footprint analysis, colony PCR was performed on *ADE2* revertant colonies using primers (Table S3) flanking the insertion sites. PCR products were gel extracted (Zymoclean Gel DNA Recovery Kit) and sequenced. For reinsertion analysis, transposon display was conducted [39]. Genomic DNA was extracted from revertant colonies using the Yeastar genomic DNA kit (Zymo research); DNA samples were digested by *Bfa*I followed by adapter ligation. Pre-amplification and selective amplification were used to amplify the sequences between *Muta1* TIR and the *Bfa*I adapter sequence. Amplicons consisting of flanking sequences of the reinsertion sites and part of *Muta1* TIR were resolved on a 4% agarose gel, and polymorphic fragments were recovered and sequenced. Flanking sequences were mapped to the yeast genome (S288C, http://yeastgenome.org/) and the reinsertion sites were determined with regard to the closest genomic features. The insertion site analysis figure was made using the program Pictogram http://genes.mit.edu/pictogram.html.

Mutagenesis of *Muta1* transposase

Site-directed mutagenesis was used to generate mutant versions of *Muta1* transposase. One pair of primers (Additional file 11: Table S3) was used for each mutation site, and pGAL415-ccdbMuta1 plasmid was used as template. PCR products were digested with *Dpn1* to remove template, and the resulting plasmid was sequenced to confirm that mutations occurred as expected.

Additional files

Additional file 1: Figure S1. Phylogeny and MULE copy number of mosquito and fruit fly species. Phylogenetic relationship and approximate divergence time of *Ae.aegypti, Ae. albopictus, C. quinquefasciatu,s* and *An. gambiae.* Genome size and copy number of MULE DDE domain in each species is shown. (TIFF 292 kb)

Additional file 2: Table S1. Summary of 14 MULE families in *Ae. aegypti.* (DOCX 81 kb)

Additional file 3: Figure S2. Features of *Muta1* derivative elements in *Ae. aegypti.* (A) 171 *Muta1* derivative elements divide into 5 groups based on size. Within each group, the number of elements with 8 bp TSD and 9 bp TSD are shown in grey and black, respectively. (B) Distribution of *Muta1* derivative element insertion sites in the *Ae. aegypti* genome. Mean ± s.d., n = 1,000 (for control). Number of insertion sites in the *Ae. aegypti* genome and control data set is shown in black and grey, respectively. (TIF 4425 kb)

Additional file 4: Figure S3. Yeast transposition assay constructs. (A) Structures of pMuta1_PAG415 and pWL89Ae. AmpR, ampicillin resistance gene; ori, *E. coli* replication origin; Pgal1, GAL1 promoter; CYC1 ter, terminator; CEN, centromere sequences of yeast chromosomes; ARS, autonomous replication site. Dashed lines indicate the position of nonautonomous element insertions, in the 5'UTR and coding region respectively. Black arrows indicate the positions of primers used for PCR analysis in Figure S3A. (B) Excision from coding region of *ADE2*. (C) Excision from 5' UTR of *ADE2*. (D) Reintegration. In the parental strain, pWL89A carries *Muta1HIS* in the coding region of *ADE2*. Reintegration is assayed by selecting cells that retain the *HIS* marker in Muta1HIS when the parental plasmid is excluded by 5-FOA treatment, which is toxic to Ura[+] cells. (TIF 602 kb)

Additional file 5: Figure S4. *ADE2* revertant colonies from the yeast transposition assay. (A-G) Excision activity of nonautonomous elements from the *ADE2* coding region. (H) Negative control, *Muta1AR* excision from the *ADE2* coding region is tested on plates without galactose. (I-L) *Muta1AR* excision from the *ADE2* coding region is tested with mutant transposases. (TIF 3512 kb)

Additional file 6: Figure S5. Analysis of excision and reinsertion events. (A) PCR analysis of the *Muta1NA1* excision sites from ADE2 revertants using flanking primers. Expected band size is 820bp (control) or 350bp, with or without *Muta1NA1*, respectively. (B) Transposon display analysis of *Muta1HIS* reinsertion in the yeast genome. DNA bands are amplicons consisting of flanking sequences of the reinsertion sites and part of the TIR. PWL89A-Muta1HIS vector is used as control. Arrows indicate the polymorphic bands that represent the insertion of *Muta1HIS* in different genomic locations. (TIF 413 kb)

Additional file 7: Table S2. Target Site Duplications (TSDs) and locations of *Muta1* transpositions in yeast. (DOCX 96 kb)

Additional file 8: Figure S6. Seqlogo of insertion sites of *Ae. aegypti Muta1* derivative elements and reintegration sites in yeast. Both 8bp TSDs (A) and 9bp TSDs (B) and their 7bp flanking sequences are analyzed, insertion preference is shown as a pictogram (height of letter indicates percentage of each nucleotide at that position) and the frequencies of preferred nucleotides, if any, are shown. (TIF 1361 kb)

Additional file 9: Figure S7 Structural features of *Muta5* and excision assay results. (A) Structural features of *Muta5*. White boxes are coding regions, shaded boxes are noncoding regions, and triangles are the TIR. Within the TIR, black arrows represent 9 copies of the 15bp subterminal tandem repeat, open arrow represents the 8bp terminal motif. The putative 554-residue transposase is predicted to harbor a zinc finger domain and the catalytic (DDE) domain. The artificial *Muta5AR* element contains 250 bp from each end of *Muta5*. (B) Excision frequencies of *Muta5AR* and *Muta5AR* from the *ADE2* reporter in the yeast assay. *Muta1* serves as the positive control. (TIF 373 kb)

Additional file 10: Figure S8. Footprints from *Muta1NA1* excision events. Arrows indicate the *Muta1NA1* insertion, length and sequence of the TSD in each assay (shown above the arrows). TSD or sequences derived from TSDs are in bold, sequences derived from *Muta1NA1* are underlined, the number of recovered events is on the right. (A) Footprints of *Muta1NA1* excision from the *ADE2* 5′ UTR without donor site TSD. (B-D) Footprints of *Muta1NA1* excision from the *ADE2* 5′ UTR with different 8 bp TSD sequences. (B) TTCAATAG; (C) CGATTCAA; (D) GGTAACTC. (E-G) Footprints of *Muta1NA1* excision from the *ADE2* 5′ UTR with different 9 bp TSD sequences. (E) TCGATTCAA, (F) CGGTAACTC, (G) ATTCAATAG. (H-I) Footprints of *Muta1NA1* excision from the *ADE2* 5′UTR with the transposase W401F mutation. (H) the 8bp TSD TTCAATAG was used. (I) the 9bp TSD ATTGAATAG was used. (TIF 665 kb)

Additional file 11: Table S3. Primers used in this study. (DOCX 124 kb)

Abbreviations
MULE: Mutator-like transposable elements; PCR: Polymerase chain reaction; TE: Transposable element; TIR: Terminal inverted repeat; TSD: Target site duplication

Acknowledgements
We thank Jim Burnette, Lu Lu, Brad Cavinder, Peter Atkinson and Jason Stajich for their technical assistance and advice.

Funding
This research funded by a grant from the W.M. Keck Foundation to SRW.

Authors' contributions
KL performed the sequence analysis, designed and carried out the yeast vector construction and transposition assays. SRW conceived the study, participated in its design and coordination and helped to draft the manuscript. Both authors read and approved the final manuscript.

Competing interests
The authors declare that they have no competing interests.

Author details
Graduate Program in Botany and Plant Sciences, University of California, Riverside, CA 92521, USA. [2]Department of Botany and Plant Sciences, University of California, Riverside, CA 92521, USA.

References
1. Finnegan DJ. Eukaryotic transposable elements and genome evolution. Trends Genet. 1989;5:103–7.
2. Wicker T, Sabo F, Hua-Van A, Bennetzen JL, Capy P, Chalhoub B, Flavell A, Leroy P, Morgante M, Panaid O, Paux E, SanMiguel P, Schulman AH. A unified classification system for eukaryotic transposable elements. Nat Rev Genet. 2007;8:973–82.
3. Feschotte C, Pritham EJ. DNA transposons and the evolution of eukaryotic genomes. Annu Rev Genet. 2007;41:331–68.
4. Bao W, Jurka MG, Kapitonov VV, Jurka J. New Superfamilies of Eukaryotic DNA Transposons and Their Internal Divisions. Mol Biol Evol. 2009;26(5):983–93.
5. Robertson DS. Characterization of a mutator system in maize. Mutat Res. 1978;51:21–8.
6. Lisch D. Regulation of the *Mutator* System of Transposons in Maize. Methods Mol Biol. 2013;1057:123–42.
7. Neuveglise C, Chalvet F, Wincker P, Gailardin C, Casaregola S. *Mutator*-like element in the yeast *Yarrowia lipolytica* displays multiple alternative splicings. Eukaryot Cell. 2005;4:615–24.
8. Pritham EJ, Feschotte C, Wessler SR. Un-expected diversity and differential success of DNA transposons in four species of entamoeba protozoans. Mol Biol Evoanl. 2005;22:1751–63.
9. Lopes FR, Silva JC, Benchimol M, Costa GG, Pereira GA, Carareto CM. The protist *Trichomonas vaginalis* harbors multiple lineages of transcriptionally active *Mutator*-like elements. BMC Genomics. 2009;10:330.
10. Lisch D. Mutator transposons. Trends Plant Sci. 2002;7:498–504.
11. Jiang N, Bao Z, Zhang X, Eddy SR, Wessler SR. Pack-MULE transposable elements mediate gene evolution in plants. Nature. 2004;431:569–73.
12. Chalvet F, Grimaldi C, Kaper F, Langin T, Daboussi MJ. *Hop*, an active *Mutator*-like element in the genome of the fungus *Fusarium oxysporum*. Mol Biol Evol. 2003;20:1362–75.
13. Xu Z, Yan X, Maurais S, Fu H, O'Brien DG, Mottinger J, Dooner HK. *Jittery*, a *Mutator* distant relative with a paradoxical mobile behavior: Excision without reinsertion. Plant Cell. 2004;16:1105–14.
14. Li Y, Harris L, Dooner HK. *TED*, an autonomous and rare maize transposon of the *Mutator* superfamily with a high gametophytic excision frequency. Plant Cell. 2013;25:3251–65.
15. Zhao D, Ferguson A, Jiang N. Transposition of a rice *Mutator*-like element in the yeast *Saccharomyces cerevisiae*. Plant Cell. 2015;27:132–48.
16. Babu MM, Iyer LM, Balaji S, Aravind L. The natural history of the WRKY-GCM1 zinc fingers and the relationship between transcription factors and transposons. Nucleic Acids Res. 2006;34(22):6505–20.
17. Nesmelova IV, Hackett PB. DDE transposases: Structural similarity and diversity. Adv Drug Deliv Rev. 2010;62(12):1187–95.
18. Eisen JA, Benito MI, Walbot V. Sequence similarity of putative transposases links the maize *Mutator* autonomous element and a group of bacterial insertion sequences. Nucleic Acids Res. 1994;22:2634–6.
19. Loessner I, Dietrich K, Dittrich D, Hacker J, Ziebuhr W. Transposase-dependent formation of circular IS*256* derivatives in *Staphylococcus epidermidis* and *Staphylococcus aureus*. J Bacteriol. 2002;184:4709–14.
20. Yuan YW, Wessler SR. The catalytic domain of all eukaryotic cut-and-paste transposase superfamilies. Proc Natl Acad Sci U S A. 2011;108:7884–9.
21. Hickman AB, Ewis HE, Li X, Knapp JA, Laver T, Doss AL, Tolun G, Steven AC, Grishaev A, Bax A, Atkinson PW, Craig NL, Dyda F. Structural basis of hAT transposon end recognition by Hermes, an octameric DNA transposase

from Musca domestica. Cell. 2014;158:353–67.

22. Calisher CH. Persistent Emergence of Dengue. Emerg Infect Dis. 2005;11:738–9.

23. Marchette NJ, Garcia R, Rudnick A. Isolation of Zika virus from Aedes aegypti mosquitoes in Malaysia. Am J Trop Med Hyg. 1969;18:411–5.

24. Aviles G, Sabattini MS, Mitchell CJ. Peroral susceptibility of *Aedes albifasciatus* and *Culex pipiens* complex mosquitoes (*Diptera: Culicidae*) from Argentina to western equine encephalitis virus. Rev Saude Publica. 1990; 24(4):265–9.

25. Lindsay SW, Wilkins HA, Zieler HA, Daly RJ, Petrarca V, Byass P. Ability of Anopheles gambiae mosquitoes to transmit malaria during the dry and wet seasons in an area of irrigated rice cultivation in The Gambia. J Trop Med Hyg. 1991;94(5):313–24.

26. Fraser MJ. Insect Transgenesis: Current Applications and Future Prospects. Annu Rev Entomol. 2012;57:267–89.

27. O'Brochta DA, Sethuraman N, Wilson R, Hice RH, Pinkerton AC, Levesque CS, Bideshi DK, Jasinskiene N, Coates CJ, James AA, Lehane MJ, Atkinson PW. Gene vector and transposable element behavior in mosquitoes. J Exp Biol. 2003;206:3823–34.

28. Scali C, Nolan T, Sharakhov I, Sharakhova M, Crisanti A, Catteruccia F. Post-integration behavior of a *Minos* transposon in the malaria mosquito *Anopheles stephensi*. Mol Genet Genomics. 2007;278:575–84.

29. Nene V, et al. Genome sequence of *Aedes aegypti*, a major arbovirus vector. Science. 2007;316:1718–23.

30. Chen XG, et al. Genome sequence of the Asian Tiger mosquito, *Aedes albopictus*, reveals insights into its biology, genetics, and evolution. Proc Natl Acad Sci U S A. 2015;112:5907–15.

31. Arensburger P, Megy K, Waterhouse RM, Abrudan J, Amedeo P, Antelo B, Bartholomay L, Bidwell S, Caler E, Camara F. Sequencing of Culex quinquefasciatus establishes a platform for mosquito comparative genomics. Science. 2010;330:86–8.

32. Holt RA, et al. The genome sequence of the malaria mosquito Anopheles gambiae. Science. 2002;298:129–49.

33. Han YJ, Burnette JM, Wessler SR. TARGeT: A web-based pipeline for retrieving and characterizing gene and transposable element families from genomic sequences. Nucleic Acids Res. 2009;37, e78.

34. Bingham PE, Zachar Z. Retrotransposons and the FB transposon from *Drosophila melanogaster*. Mobile DNA Washington D.C: American Society for Microbiology Press. 1989;485–502.

35. Yang G, Weil CF, Wessler SR. A rice *Tc1/mariner*-like element transposes in yeast. Plant Cell. 2006;18:2469–78.

36. Weil CF, Kunze R. Transposition of maize *Ac/Ds* transposable elements in the yeast *Saccharomyces cerevisiae*. Nat Genet. 2000;26:187–90.

37. Hancock CN, Zhang F, Wessler SR. Transposition of the *Tourist*-MITE *mPing* in yeast: an assay that retains key features of catalysis by the class 2 *PIF/Harbinger* superfamily. Mob DNA. 2010;1:5.

38. Mitra R, Fain-Thornton J, Craig NL. piggyBac can bypass DNA synthesis during cut and paste transposition. Embo J. 2008;27:1097–109.

39. Biedler J, Qi Y, Holligan D, Della Torre A, Wessler SR, Tu Z. Transposable element (TE) display and rapid detection of TE insertion polymorphism in the Anopheles gambiae species complex. Insect Mol Biol. 2003;12(3):211–6.

40. Lynch M, Sung W, Morris K, Coffey N, Landry CR, Dopman EB, Dickinson WJ, Okamoto K, Kulkarni S, Hartl DL, Thomas WK. A genome-wide view of the spectrum of spontaneous mutations in yeast. Proc Natl Acad Sci U S A. 2008;8105(27):9272–7.

41. Sutton WD, Gerlach WL, Schwartz D, Peacock WJ. Molecular analysis of *Ds* controlling element mutations at the *Adhl* locus of maize. Science. 1983;223:1265–8.

42. Brillet B, Bigot Y, Auge-Gouillou C. Assembly of the *Tc1* and mariner transposition initiation complexes depends on the origins of their transposase DNA binding domains. Genetica. 2007;130(2):105–20.

43. Zhou L, Mitra R, Atkinson PW, Hickman AB, Dyda F, Craig NL. Transposition of *hAT* elements links transposable elements and V(D)J recombination. Nature. 2004;432:995–1001.

44. Hennig S, Ziebuhr W. A transposase-independent mechanism gives rise to precise excision of IS*256* from insertion sites in *Staphylococcus epidermidis*. J Bacteriol. 2008;190:1488–90.

45. Alberti S, Gitler AD, Lindquist S. A suite of gateway (R) cloning vectors for high-throughput genetic analysis in *Saccharomyces cerevisiae*. Yeast. 2007;24:913–9.

9

Globular domain structure and function of restriction-like-endonuclease LINEs: similarities to eukaryotic splicing factor Prp8

M. Murshida Mahbub[1], Saiful M. Chowdhury[2*] and Shawn M. Christensen[1*]

Abstract

Background: R2 elements are a clade of early branching Long Interspersed Elements (LINEs). LINEs are retrotransposable elements whose replication can have profound effects on the genomes in which they reside. No crystal or EM structures exist for the reverse transcriptase (RT) and linker regions of LINEs.

Results: Using limited proteolysis as a probe for globular domain structure, we show that the protein encoded by the *Bombyx mori* R2 element has two major globular domains: (1) a small globular domain consisting of the N-terminal zinc finger and Myb motifs, and (2) a large globular domain consisting of the RT, linker, and type II restriction-like endonuclease (RLE). Further digestion of the large globular domain occurred within the RT. Mapping these RT cleavages onto an updated model of the R2Bm RT indicated that the thumb of the RT was largely protected from proteolytic cleavage. The crystal structure of the large globular domain of Prp8, a eukaryotic splicing factor, was a major template used in building the R2Bm RT model, particularly the thumb region. The large fragment of Prp8 consists not only of a RT similar to R2Bm, but also an RLE and a linker connecting the two regions. The linker sequences adjacent to the RLE in LINEs and Prp8 share a set of two important α-helices and a (presumptive) knuckle/ββα structural motif that are closely associated with the thumb. The RLEs of LINEs and Prp8 share a unique catalytic core residue spacing as well as other key residues.

Conclusions: The protein encoded by RLE LINEs consists of two major globular domains. The larger of the two globular domain contains the RT, linker, and RLE and is similar to the large fragment of the spliceosomal protein Prp8. The similarities are suggestive of possible common ancestry.

Keywords: Transposable element (TE), Line, Non-LTR retrotransposon, Target primed reverse transcription (TPRT), Reverse transcriptase, RNA splicing

Background

Long INterspersed Elements (LINEs), also called non-LTR retrotransposons, are a major class of retrotransposable elements. LINEs package their transcribed RNA into ribonucleoprotein particles (RNP) using element encoded proteins translated from the mRNA being packaged. LINEs insert their genetic material back into the host genome at a new location by target primed reverse

transcription (TPRT) [1–5]. TPRT is initiated by cleavage of one of the target chromosomal strands by an element encoded DNA endonuclease. The free 3′-OH DNA end generated by the DNA endonuclease is used to prime reverse transcription of the element RNA, thus inserting a new DNA copy of element into the host genome.

All LINEs are believed to require the same basic activities to integrate: RNA binding activity, DNA binding activity, DNA endonuclease activity, reverse transcriptase (RT) activity, and completion of integration by second strand synthesis. There are two major groups of LINEs. The two groups share a common RT and a IAP/gag-like

---
* Correspondence: schowd@uta.edu; shawnc@uta.edu
[2]Department of Chemistry and Biochemistry, University of Texas at Arlington, 700 Planetarium Place, Room 130, Arlington, TX 76010, USA
[1]Department of Biology, University of Texas at Arlington, 501 S. Nedderman Drive, Room 337, Arlington, TX 76010, USA

CCHC zinc-knuckle. The two groups differ in their open reading frame (ORF) structures, RNA binding domains, DNA binding domains, and DNA endonuclease domains used to form the element RNP and to integrate into the host DNA.

The earlier branching group has a single ORF. The ORF encodes a multifunctional protein with N-terminal zinc finger and Myb motifs, an RT, a gag-knuckle like motif, and a type II restriction-like endonuclease (RLE) with a restriction endonuclease like fold (REL) (reviewed in [6, 7]). This group of LINEs is generally site-specific during integration. The insect R2 element is a well-studied example of this early branching LINE group.

The later branching group has two open reading frames. The second open reading is similar to that of the earlier branching group. It encodes an apurinic-apyrimidinic family endonuclease (APE), a RT, and the gag knuckle-like motif (reviewed in [8–12]). The mammalian L1 element is a well-studied example of this later branching LINE group.

While crystal structures exist for the APE endonuclease and for the protein product of the first ORF of APE LINEs, no crystal or cryo-EM structures exist for the RLE LINEs, nor for the regions common between the two groups of LINEs [13–18]. Our previous paper reported a protein threading model for the restriction-like endonuclease of R2 elements [19]. This paper reports the globular domain structure of R2Bm as probed by limited proteolysis. An updated model of the R2 RT is also presented along with an analysis of the linker region between the RT and the endonuclease. The R2 proteolytic data, in conjunction with sequence-structure alignments of the RT, linker, and RLE, indicate that RLE LINEs share a number of commonalities with the large fragment of Prp8, a highly conserved eukaryotic splicing factor that has a RT domain and an RLE domain, beyond those already discovered and discussed [20–22].

Results

Mapping and sequencing LysC protease resistant fragments of R2Bm protein

In order to probe the globular domain structure of R2Bm, R2Bm protein was subjected to limited proteolysis by one of several proteases. LysC, which cleaves on the C-terminal side of lysine residues, was one of these proteases. There are 42 lysine residues in the expressed and purified R2Bm protein. Aliquots from the digestion reaction were pulled at different time points and the reactions terminated. The digestion profile of R2 protein cleaved by LysC at the different time points were analyzed by SDS-PAGE (Fig. 1a). At least nine major bands (LA-LI) were observed. Some of these bands appeared early in the time course (e.g., LA, LC, LF, and LG), while other bands appeared at later time points (e.g., LE, LH, and LI). Collectively, these bands represent protease resistant

R2Bm fragments. The protease resistant fragments were excised from the gel, acetylated, and then digested to completion with trypsin. The peptides resulting from the trypsin digest were sequenced by nano-LC-ESI mass spectrometry. The original N-terminal end(s) of the protease resistant fragment (i.e., those ends resulting from LysC cleavage) were identifiable as they had been acetylated. The N-terminal ends resulting from LysC cleavage in bands LA-LI are reported in Fig. 1b. The y and b ion series that allowed the N-terminal peptide identification are given. The MS/MS spectrum in support of the peptide identification are provided in the Additional file 1: S1A. The internal peptides resulting from further trypsin cleavage of the LysC resistant bands were similarly sequenced by MS/MS (Additional file 1: S1B).

The approximate C-terminal end of LysC protease resistant fragments LA-LI were determined by sequencing of the internal peptides and by the apparent molecular weight of the original protease resistant bands on SDS PAGE gels given the experimentally determined N-terminal end. The peptide sequencing data derived from bands LA-LI have been mapped back onto the linear domain structure of R2Bm and are summarized in Fig. 1c. A more detailed amino acid break down of the different subdomains of the R2Bm ORF can be found in Additional file 2: S6 and in Fig. 5. Please note that the ORF and numbering is for the R2 protein generated from our R2 protein expression construct (ΔNR2Bm) which is slightly amino-terminally truncated compared to the genbank entry for R2Bm.

Full-length R2Bm (118 kD) was quickly processed by LysC to form a large ~89 kD LA band and shorter ~29 kD LF and ~22 kD LG bands. The LF band was found to have a fragment with alternative N-terminal ends that mapped near the beginning of the R2Bm ORF, at amino acid residues four and seven—a serine (S4) and a glutamic acid (E7), respectively. Internal peptides of the LF fragment included the ZF and Myb domains and ended within −1, a conserved basic region involved in RNA binding [23]. The fragment from band LG was similar to the LF fragment, except that fragment LG was ~60 amino acids shorter. Fragment LG had an N-terminal end that mapped to amino acid R64 of the R2Bm ORF, removing most of the ZF from the fragment. The C-terminal end of the LG fragment appeared to be similar to LF.

The polypeptide that constituted the large ~89 kD LA band had two alternative N-terminal ends, R242 and C256. The LA fragment spanned from −1 to the end of the ORF. The LA fragment contained the entire RT, the endonuclease, and the linker region connecting the RT and endonuclease domains.

Another large prominent band appeared along with band LA at the earlier time points: band LC. The fragment from band LC consisted of part of the RT, starting within RT6, at amino acids S595 and H609, and ending at the end of the

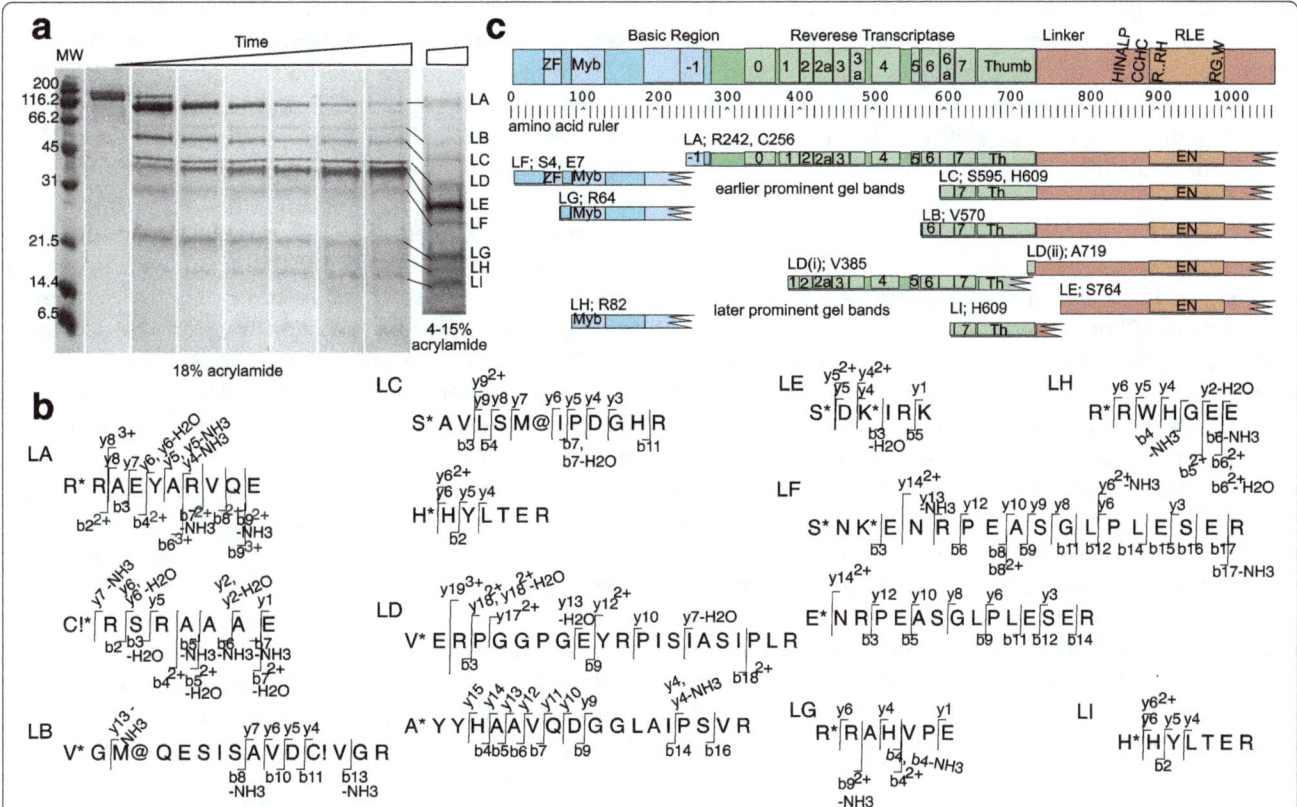

Fig. 1 Mapping and sequencing LysC protease resistant fragments of R2Bm protein. **a** R2Bm protein was digested with LysC protease and analyzed by SDS-PAGE. Major observed bands were designated LA-LI. The triangle represents a time course of LysC digestion. The molecular weight (MW) marker values are given. **b** Identification of proteolytic fragments of R2Bm protein. Bands from panel A were cut out, further processed, and analyzed by nano-LC-ESI-MS/MS sequencing. The N-terminal of the band producing fragment was identified by acetylation. Internal peptides were sequenced as well. The y and b ions that identified the N-terminal end are indicated. Symbols: * = acetylation; @ = oxidation;! = carbamidomethylation. The spectrum is given in the supplemental data. **c** Map of the band purified R2Bm fragments. A detailed diagram of the R2Bm open reading frame (ORF) is given along with an amino acid ruler. The boundaries of the R2Bm proteolytic resistant fragments LA-LI are mapped below, along with the amino acid and primary sequence position of the first amino acid of the fragment. The C-terminal ends were not exactly pinpointed but were roughly determined using the apparent MW from the SDS-PAGE gel and by the coverage of internal peptides sequenced by nano-LC-ESI-MS/MS. The major earliest and latest appearing gel bands are roughly grouped together in the map. Abbreviations: zinc finger (ZF) and restriction-like endonuclease (RLE)

R2Bm ORF. Like the LA fragment, the LC fragment contained the endonuclease domain in addition to the RT.

Band LB was present at low amounts across the time series in Fig. 1a. At different protease ratios, however, band LB was only prominent at earlier time points (data not shown). Fragment LB had about 30 more amino acids of the RT than did fragment LC. Fragment LB is likely processed into fragment LC.

At later time points, fragments LA, LB, and LC were further processed. Band LD consisted of two non-overlapping fragments, LD(i) and LD(ii), of about the same size. In the 18% gel the LD fragments ran as a single band, while on the gradient gel a doublet was observed (Fig. 1a). The first LD fragment, LD(i), consisted of the bulk of the RT, from V385, which was located in RT1, through most of the thumb. The second LD fragment, LD(ii) started near the end of the thumb at amino acid A719 and continued through the end of the ORF.

Fragment LD(ii) contained the endonuclease and the linker region that connects the endonuclease to the RT.

Fragment LC gets cleaved at K763 to generate bands LE and LI. Band LI consisted of the N-terminal portion of fragment LC with an N-terminal end of H609. The fragment in band LE had an N-terminal end of S764 and contained the linker and RLE. Band LE was a major late appearing band that accumulated over time. Fragments LC and LD(ii) are likely both processed into fragment LE.

Band LH consisted of a fragment with an N-terminal end located at the beginning of the Myb domain at amino acid R82. The polypeptide appeared to be derived from fragments LF and/or LG but was further truncated at the N-terminal end.

As fragments from the RT and the ZF/Myb regions of the ORF were processed into smaller polypeptides, those polypeptides became difficult to resolve and visualize on SDS-PAGE, especially on preparative gels. Depending upon the gel percentage and band location, an excised gel slice can still contain signal from bands just above or below that area. In the later time points, the background between bands increases due to non-banding polypeptides. We did not trust our ability to identify bands and N-terminal ends below about 18 kD.

Mapping and sequencing of GluC protease resistant fragments of R2Bm protein

The second protease used to probe globular domain structure of R2Bm was GluC. GluC cleaves on the C-terminal side of glutamic acid residues, and to a lesser extent (100-fold) on aspartic acid residues. There are 69 glutamic acid residues and 47 aspartic acid residues in the R2Bm protein. Aliquots were pulled from the digestion reaction at different time points and terminated. The digestion profile of R2Bm protein cleaved by GluC at the different time points was analyzed by SDS-PAGE (Fig. 2a). The protease resistant bands visualized on the SDS-PAGE were labeled GA-GK. The A-K designators, however, do not necessarily equate to an equivalent LysC resistant R2 fragment, as the designators are by order of apparent-molecular-weight and not by R2 ORF region. The protease resistant fragments were excised from the gel, processed, and sequenced by nano-LC-ESI mass spectrometry. The y and b ion series that allowed the N-terminal peptide identification are given for each band (Fig. 2b and Additional file 3: S2). A map of the internal peptides found in each band are reported in Additional file 4: S3.

GluC, like LysC, quickly cleaved the R2Bm protein into a large fragment of about 87 kD (band GA) and a small fragment of about 30 kD (band GH). The large fragment, GA, consisted of the RT, the linker, and the RLE (Fig. 2c). The small fragment consisted of the N-terminal region of the R2Bm protein. The protein fragments isolated from bands GB-GF were, to a first approximation, further truncations of the GA fragment, where the truncating cleavages were located within the RT. The most prominent of these fragments and bands were GC, GE, and GF. Bands GE and GF appeared late in the time course. The fragments isolated from bands GJ and GK were, to a first approximation, further truncations of band GH. As band GH disappeared, band GJ became more prominent. As band GJ disappeared, band GK appeared. The two bands marked GG were prominent on the 18% acrylamide gel because of a band compression artifact. The GG area contained faint bands and a diffuse smear on the gradient gel. The lower of the two bands appeared to be GluC, while the upper band could not be ascertained. Band GI also could not be ascertained.

There were a number of alternative N-terminal ends found for fragment GA: L252, M279, T281, and A300

(Fig. 2b, c). Fragments GH and GJ also were found to have several alternative N-terminal ends: N8, A12, and R21. In order to aid in interpreting the N-terminal ends of the protease resistant R2Bm fragments (especially early and late cleavage determinations) and to attain a more comprehensive accounting of cleavages that did not give rise to readily observable bands, an experiment was performed where GluC cleavages were detected at a given time point without separating individual proteolytic fragments. Instead of fractionating the fragments, the terminated protease reaction was run into the SDS-PAGE gel for only a few millimeters. A fairly large section of gel near the wells was then excised and processed for cleavage detection (Fig. 2d). This technique of running the reaction minimally into the gel is a near equivalent to direct detection in solution (i.e., no gel fractionation). For technical reasons (see materials and methods), however, it was necessary to have the proteolysis reaction processed through a gel slice. Each column of boxes below the ORF map is a potential GluC cleavage site (D/E), or rather the amino acids immediately following a GluC cleavage site that would become acetylated if the preceding D or E residue were cleaved. Each progressive row is a (longer) time point with identified cleavages reported as a heat map of peptide spectral match (psm) values for each site for each time point. In the heatmap data, there appeared to have been several pre-existing R2Bm N-terminal ends present in the R2Bm protein preparation as N-terminal signals at positions P36, P185, and S786 were detected in the zero time point on the heatmap. No major bands on the SDS-page gels, however, were attributable to these fragments.

Comparing the heatmap results (Fig. 2d) with the data derived from the SDS-PAGE bands (Fig. 2b) provided an extra window into the relative cleavability and timing of several important cleavage sites. It appeared that the major early cleavage events were near the start of the RT. Cleavage at E278 was the most robust cleavage event and gave rise to fragments GA and GH. The cleavage event in domain −1 at position E251 was also a major cleavage event. Cleavage at E251 peaked midway through the digestion reaction as band GJ become prominent. Cleavage at E251 occurred in the full protein as well as in a C-terminal truncation of fragment GH. The T281 and A300 N-terminal ends of band GA appeared to be the result of later cleavage events (at E280 and E299, respectively) that further truncated the original GA fragment.

Another major cleavage event in Fig. 2d was an early event located at E507. Amino acid E507 is within RT4, and cleavage at this location resulted fragment GC. Two other prominent cleavage locations in Fig. 2d, E614 and E649, were later cleavage events and gave rise to fragments GE and GF, respectively.

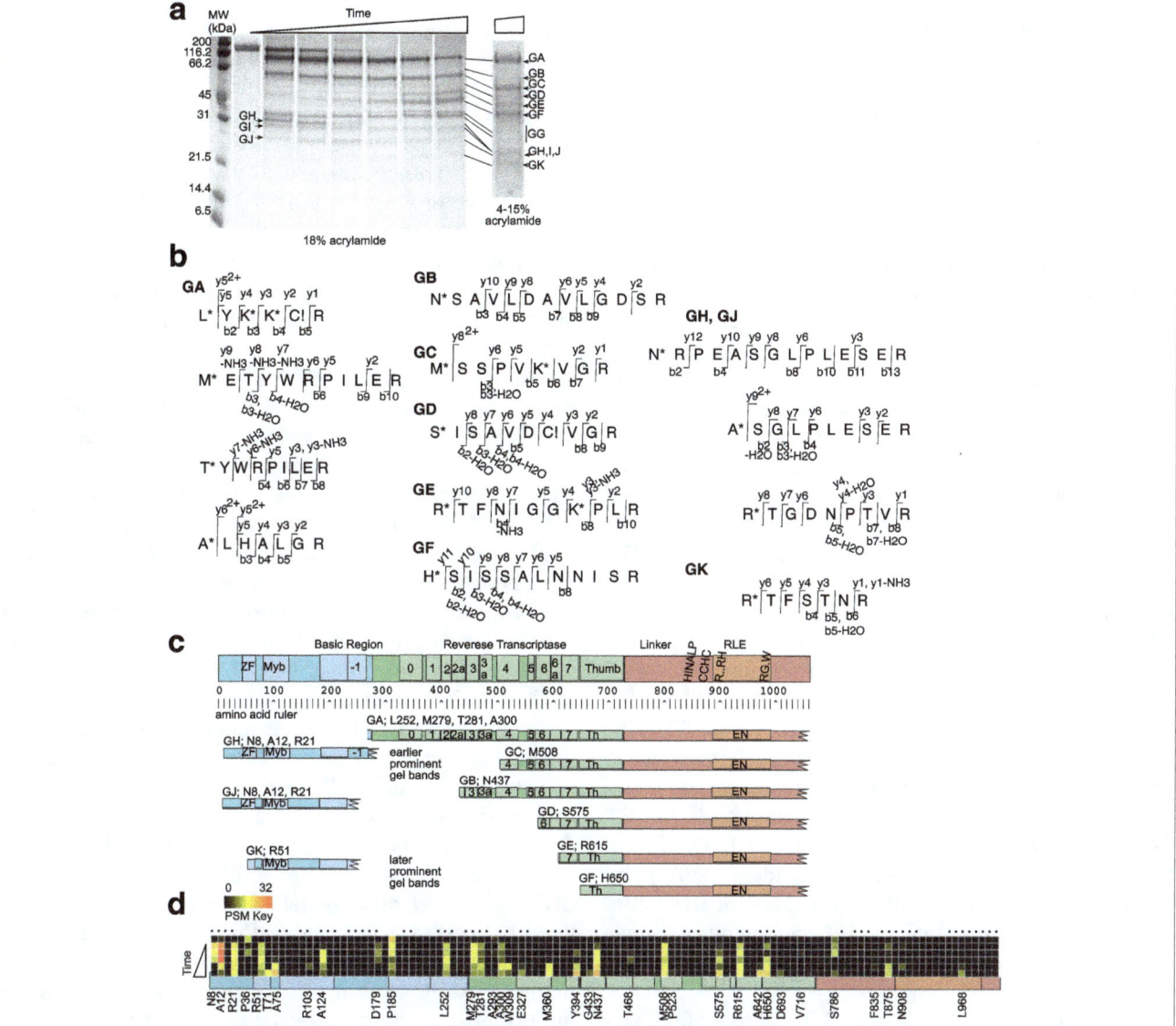

Fig. 2 Mapping and sequencing GluC protease resistant fragments of R2Bm protein. Symbols and abbreviations are as in Fig. 1. **a** R2Bm protein was digested with GluC protease and analyzed by SDS-PAGE. Major observed bands were designated GA-GK. **b** Identification of proteolytic fragments of R2Bm protein. Bands from panel A were cut out, further processed, and analyzed by nano-LC-ESI-MS/MS sequencing. The N-terminal of the band producing fragment was identified by acetylation. **c** Map of the band purified R2Bm fragments. The major GluC generated R2Bm fragments detected in panel A are mapped below the ORF diagram and rulers. **d** Heatmap of GluC cleavages found in non-fractionated digestion reactions of R2Bm protein across time. Each column of boxes represents a GluC cleavage site. GluC cleaves after an E residue, indicated by a dot above the column, or a D residue (no dot). The positions of the amino terminal ends generated by observable GluC cleavages are given below the boxes. The number of peptide spectral matches (# PSM) are color coded as shown in the key. The R2Bm ORF is diagramed below the heatmap. Each row is a different time point, with the top row being the zero time point (no GluC) and the bottom row being eight hours. The triangle represents increasing time of GluC digestion

Amino acid E614 is located in RT6, and E649 is located at the beginning of the thumb of the RT. The N-terminal ends of fragments GC and GF were confirmed by Edman degradation (data not shown). Other cleavages were observed within the RT in Fig. 2d, not all of which gave rise to major stable fragments visible on the SDS-PAGE gel. Interior RT fragments (i.e., those not associated with the linker and RLE) were either heterogeneous in nature or unstable such that bands were not observed on an SDS-PAGE gel.

The N-terminal ends of the GH and GJ fragments, like GA, were ragged. The GH and GJ fragments had N-terminal ends of N8, A12, and R21. While all three positions were robust in Fig. 2d, cleavage at E11 to generate the A12 end was the most prominent. It should be noted, however, that the original N-terminal end of the R2 protein was not tracked as the combination of proteases used in generating the peptides for MS/MS sequencing generated peptides too small to be readily detected. There was likely a time dependent shortening of the N-terminal ends in the E7-E50 region of the GH to GJ to GK progression that we were unable to fully quantify. A list of all cleavage sites and early/late data is given in Additional file 5: S7.

Protein threading model of the R2Bm RT and mapping of the protease cleavages onto the model

It has been nearly 20 years since a model of the R2 RT has been generated using homology modeling and protein threading [24]. The updated RT model shown in Fig. 3 (and Additional file 6: S4) was constructed using the Phyre 2.0 protein modeling server [25]. The model spanned amino acid residues Y246-P754 of the expressed ΔNR2Bm protein and spanned from the end of −1 through the thumb of the RT [19]. The initial residues, Y246-E263, and the final residues, R736-P754, were modeled ab initio by the modeling program. Residues V264-V735, however, were modeled with high homology confidence using four known protein structures as templates: 5hhl (chain A), 5g2X (chain C), 4i43 (chain B) and 1khv (chain A) (Fig. 3a) [21, 22, 26, 27]. The first two templates are group II intron RTs: the cryo-EM structure of lactococcal group II intron LtrA protein and the crystal structure of the *Eubacterium rectale* group II intron RT. The third template is the RT found in the eukaryotic splicing factor Prp8. The fourth template is the caliciviral RNA dependent RNA polymerase. Only the high confidence regions of the R2Bm RT were kept in the final model; the ab initio regions were deleted from the 3D depictions presented in Figs. 3b-e.

The region between −1 and RT0 (V264-P322) was modeled solely from the RNA dependent RNA polymerase (RdRP) but was of high confidence. The region from RT0 through RT2a (I323-R449) was built using the two group II intron RTs and the RNA dependent RNA polymerase. The RT3-RT6 area (K450-L602) was modeled using the group II intron structures, RNA dependent RNA polymerase, and Prp8. The area between RT6 and RT7 was modeled only from Prp8. RT7 was modeled by the group II intron structures as well as Prp8. The thumb was modeled using only the Prp8 crystal structure as a template.

A ribbon diagram of the R2Bm RT model is presented in Fig. 3b. The R2Bm RT assumed the canonical hand-like configuration, with fingers, palm, and thumb regions, and was overall similar to RdRP [28–31]. A word of caution is warranted, homology models are not crystal or cryo-EM structures. The models are comparatively quite crude and resemble their individual templates.

That said, homology modeling tools have improved greatly and model with high confidence can be quite informative when one lacks a high resolution structure. The thumb region (1305-1375) was very long and prominent in R2Bm. The −1, index finger (276-288), and middle finger formed one of two bulbous regions as in RdRP. The pinky finger (RT0) formed the second bulbous region. Just behind the index and middle finger was the ring finger (RT1 β-strands). The RT2 α-helix was positioned behind RT0. The region spanning from −1 to RT0 (yellow in the ribbon diagram) includes the index finger (276-288) and the palm-traversing α-helix (298-314).

The index finger and RT0 are connected by the palm traversing α-helix, a feature shared between RdRP, Prp8, and, apparently, LINE polymerases. Telomerases have the index finger α-helix but lack RT0 (the pinky finger) as well as the palm-traversing α-helix (structural overlays of R2Bm RT with PDB ID 3du5 data not shown) [32]. In group II intron RTs, the index finger and palm traversing helix are not present (PDB ID 5hhl and 5g2x) [22, 26]. Group II intron RTs do, however, have an RT0 and an extension to the RT0 termed NTD, both positioned on the pinky finger side [22, 26].

The index finger region is important for the polymerization functions. A monoclonal antibody directed against the vicinity of the index finger of the hepatitis C virus RdRP was found to inhibit both primer-dependent and de novo RNA synthesis [33].

The pinky finger region is also important for polymerization. The RT0 of R2Bm and group II intron RTs share a set of antiparallel α-helices connected by a loop [22, 26, 34, 35]. In RdRP the RT0 homologue is the "G-loop," or "motif G." The G-loop functions in template-RNA binding and translocation [28, 36]. A monoclonal antibody directed against the G-loop was found to be inhibitory to primer-dependent RNA synthesis but not de novo RNA synthesis [33]. The RT0 domain of RLE LINEs contains a PGPD motif in the loop. The PGPD motif, when mutated in R2Bm, abolished template jumping activity of the RT and reduced, to some extent, overall polymerization activity [23]. Template jumping activity is also observed in RdRP, Mauriceville retroplasmid, and group II intron RTs [37–39]. Mutation of the PGPD motif in R2Bm also reduced the binding to the 5′ and 3′ PBM RNAs [23]. The group II intron protein's RT0 and its extension (the NTD) are involved in binding DIVa of the group II intron RNA [22, 40, 41]. The interaction between RT0 and DIVa is required for positioning the intronic-RNA-template for reverse transcription (TPRT), but it is not strictly essential for splicing [40].

RLE LINEs, telomerase, and group II introns possess RNA binding domains upstream (N-terminal) of the

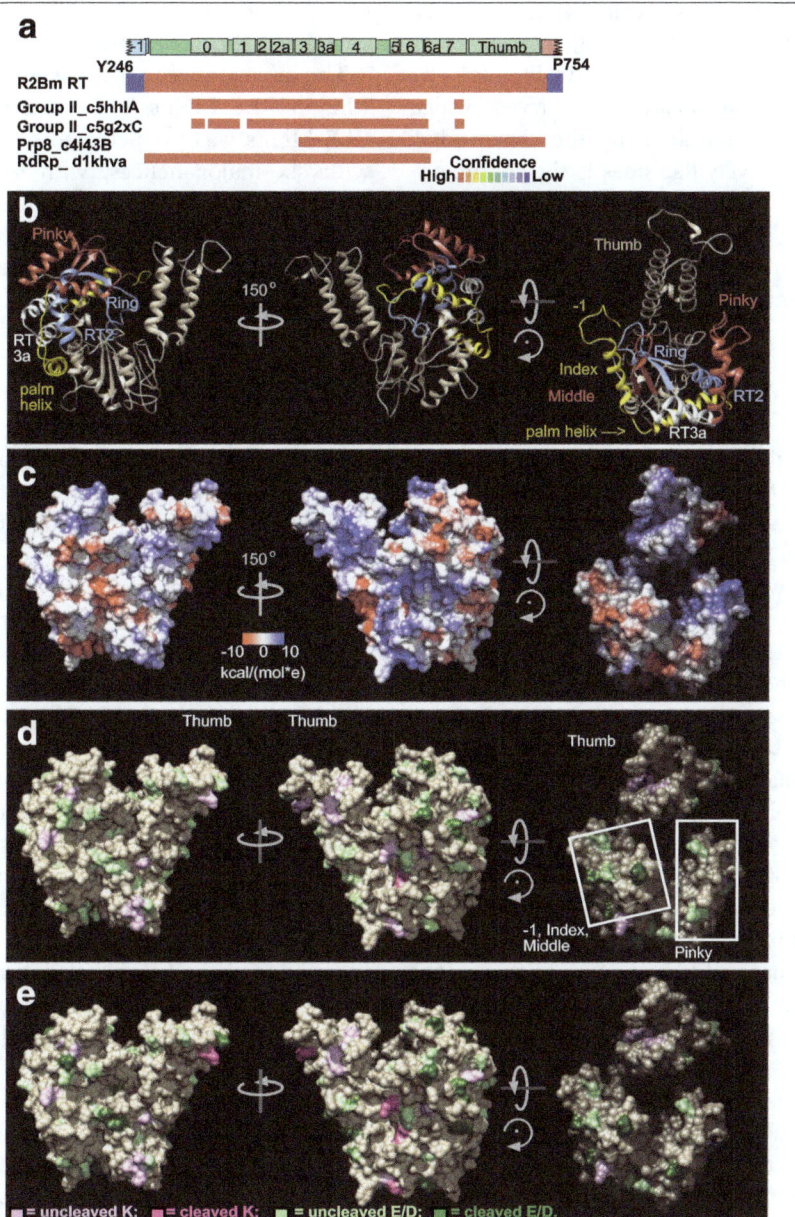

Fig. 3 Modeling of the R2Bm RT domain and mapping of the proteolytic cleavages onto the RT model. **a** R2Bm RT model construction and confidence report from Phyre2. 5hhl: crystal structure of the RT domain of the group II intron encoded protein from *Eubacterium rectale*. 5g2xC: the maturase protein in the cryo-EM structures of a spliced *Lactococcus lactis* group II A intron RNP. 4i43B: the splicing factor Prp8 protein large domain crystal structure. 1khv: the crystal structure of rabbit hemorrhagic disease virus RNA-dependent RNA polymerase. **b** Ribbon model representation of R2Bm RT with several key regions highlighted. The pinky finger (RT0) is colored red, as is the middle finger of RT4. The region spanning from a portion of the −1 to RT0 is in yellow. This region includes a remnant of the −1 loop, the index finger α-helix, and the α-helix that traverses palm. The ring finger (RT1) is in blue, as is the RT2 α-helix. **c** Coulombic surface rendering of the R2Bm model. **d** Early proteolytic cleavage sites mapped onto the R2Bm RT model. Dark green coloring marks glutamic acid and aspartic acid residues that were cleaved. Pale green marks glutamic acid and aspartic acid residues that were not cleaved. Pink coloring marks lysine residues that were cleaved. Pale purple marks lysine residues that were not cleaved: See key in panel E. **e** Early plus later proteolytic cleavage sites mapped onto the R2Bm RT model. Markings are as in panel D

reverse transcriptase. In the case of the group II intron protein, the N-terminal domain is an extension of RT0 and resides on that side of the RT (the pinky finger side). The extended RT0 and IFD bind to DIVa of the intron RNA [22, 40]. In R2, the RNA binding domain −1 is on the opposite side of the fingers from RT0. The remnants of domain −1 is on the index finger side. Mutations in −1 abolished 5′ and 3′ PBM RNA binding [23]. Telomerases also contain an RNA binding region upstream of the RT that is involved in binding RNA [32, 42, 43].

A coulombic surface map is presented in Fig. 3c. The R2Bm RT adopts an overall shape of a curved wedge with the backside of the thumb being the sharp edge. One of the two comparatively flat sides is the thumb-to-RT0 face. This face has a small central acidic patch surrounded by mostly hydrophobic residues in the model. The other fairly flat side is the thumb to index finger side and is predominantly basic. The third side is rounded. It spans from the index finger to RT0 and has a central vertical streak of acidic residues running through a central streak of (mostly) hydrophobic residues. The streaks are centered below the ring finger. The hydrophobic regions, and perhaps the acidic patches/streaks within them, are potential areas of further protein-protein interactions.

The R2Bm RT model was used for mapping the earlier cleavages (Fig. 3d) as well as all of the cleavages (early plus later, Fig. 3e) for both LysC and GluC proteases. LysC cleaves on the C-terminal side of K residues. There are 18 K residues in the R2Bm RT model, six of which are cleaved to some degree. Cleavage in the ab initio regions are included in the cleavage count, although the ab initio sequences have been deleted from the 3D models in the figure. GluC cleaves on the C-terminal of E residues and much less often on the C-terminal side of D residues. There are 30 E residues in the R2Bm RT, 14 of which are cleaved to some degree. There are 26 D residues in the RT, six of which are weakly cleaved. Most of the early cleavages mapped to the −1 ab initio regions (not shown), the index finger, and the tip of the middle finger. There was also a cleavage on the basic face between the thumb and −1. Some of the next cleavages were also on the basic face as well as on the RT0 protrusion and on the knife edge (the backside) of the thumb. Most of the prominent thumb was protected from cleavage. Later cleavages were found on the secondary structures just behind where the first cleavages were (i.e., the regions behind the index finger α-helix) and on the flat hydrophobic thumb-to-RT0 face inside the acidic patch.

The large fragment of the eukaryotic splicing factor Prp8 and restriction-like endonuclease bearing LINEs share a common set of sequence motifs and structure

RTs share a common set of sequence domains, numbered 1-7, and a thumb region [34, 35, 44–47]. The thumb usually contains a three-helix bundle. In addition to the thumb and RT1 through RT7, the RT of LINEs contains insertions: 0, 2a, 3a, and 6a. Several of these insertions are present in other eukaryotic RTs (Additional file 7: S5A-D and [34, 35, 44–46]). The RT domain of Prp8 is very similar to that of LINEs, having 0, 2a, 3a, 4a, and 6a insertions. The telomerase RT

encodes 2a, and 3a. The RT of group II intron proteins encodes 0, 2a, 3a, 4a, and 7a.

The area between the reverse transcriptase and the RLE in RLE LINEs is the linker region. The linker in RLE LINEs was predicted to be predominantly α-helical with six major helices, with some groups having 2-3 additional helices (Additional file 7: S5). A weak scoring helix also was often observed in the highly-conserved (presumptive) gag-knuckle (see below). The region downstream of the RT in APE LINEs were more diverse, with 5 -14 predicted helical regions. The crystal structure and EM-structures of Prp8 have about 13 helices. β-strands were less prevalent in the linker of RLE LINEs than in APE LINEs (about 0-2 vs 4-6). Among the RLE LINEs, only Utopia may contain comparatively high number of linker β-strands. Several clades of APE LINEs encode an RNaseH domain downstream of the RT (reviewed in [9]).

A multiple alignment and Ali2D secondary structure prediction is presented for the most conserved portion of the linker for RLE LINEs, APE LINEs, and Prp8 (Fig. 4a). Near the end of the linker region of RLE LINEs is the highly conserved IAP/gag-like CCHC zinc-knuckle motif with a spacing of $CX_{2-3}CX_{7-8}HX_4C$ (Fig. 4a). In R2Bm the IAP/gag-like CCHC zinc-knuckle is located at amino acids 863-883. The spacing of the cysteines and histidine in the motif is similar to that of IAP domains, although a bit smaller, or gag-knuckles, although a bit larger [48]. IAP domains form a ββα structure around zinc ion [48]. A gag-knuckle is β-strand followed by a knuckle (a sharp turn) with a less structured finish (e.g., coil with bends) [48]. The zinc ion is coordinated by the C and H residues of the motif [48]. The β-strands and α-helix are generally short. The canonical structure for an IAP domain is indicated above the R2Bm sequence listed in Fig. 4a as is the predicted (Ali2D) secondary structure for the linker region of RLE LINEs. For many RLE LINEs a short α-helix was predicted near the H residue (Additional file 7: S5). A β-strand was occasionally predicted near the first C residue using JPRED for RLE LINEs (data not shown). In many of the RLE LINE clades (e.g., R2, Dong, NeSL, and Utopia) there was a conserved R residue (R867) between the first two C residues.

APE LINEs, although lacking a downstream RLE, have a linker region that also ends with the IAP/gag-like CCHC zinc-knuckle. For APE LINEs, the area near the H was often predicted to be a β-strand.

It is not clear if Prp8 had a IAP/gag-like CCHC zinc-knuckle at one time or not. No CCHC motif exist in Prp8. Where the knuckle would be is not reliably aligned by sequence alignment programs to the LINE IAP/gag-like CCHC zinc-knuckle, or rather that there are several ways to align the region. In Fig. 4a, the $CX_{2-3}C$ (866-810

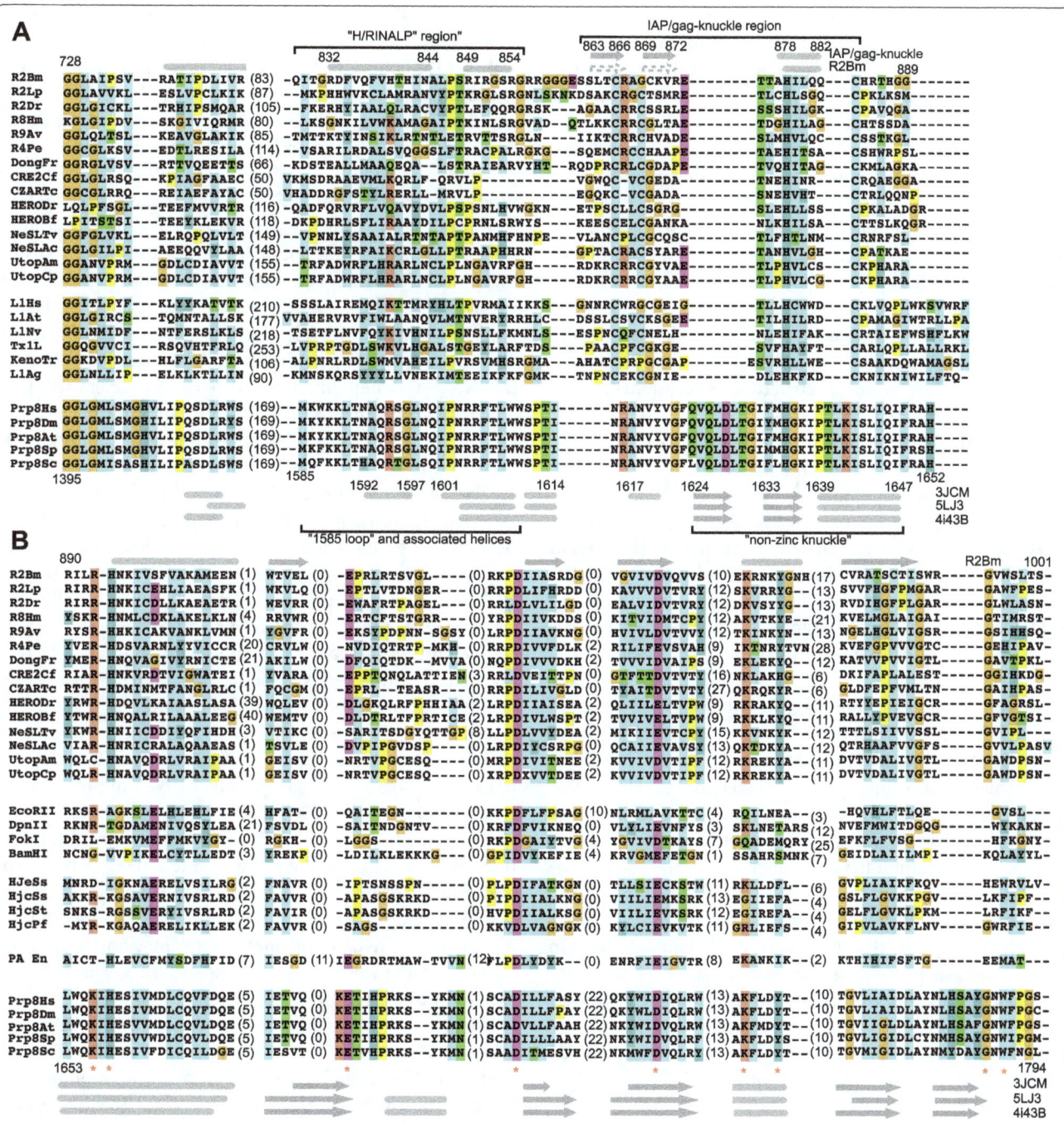

Fig. 4 Structure based alignments of linker and RLE regions. **a** Multiple sequence alignment and Ali2D prediction of the most conserved region of the linker. The Ali2D predicted secondary for R2Bm protein are marked above the R2Bm sequence. The canonical ββα structure of a gag-knuckle/IAP is also presented. The secondary structure of Prp8 is given below the Prp8 sequences. The Prp8 secondary structures are from several reported crystal and cryo-EM structures: 3JCM, 5LJ3, 4i43B. Rounded bars are α-helices and arrows are β-strands. **b** Multiple sequence alignment RLE from LINEs, type II restriction enzymes, archeal Holliday junction resolvases, influenza PA endonuclease, and Prp8. Abbreviations: R2Bm = *Bombyx mori* (M16558.1); R2Lp = *Limulus polyphemus* (AF015814.1); R2Dr = *Danio rario* (34392533); R8Hm = *Hydra magnipapillata*; R9Av = *Adenata vaga* (ACV95454.1); R4Pe = *Parascaris equorum* (AAB02297.1); DongFr = *Fugu rubripes*; CRE2Cf = *Crithidia fasciculata*; ZARTc = *Trypanosoma cruzi*; HERODr = *Danio rario*; HEROBf = *Branchiostoma floridae*; NeSLTv = *Trichomonas vaginalis*; NeSLAc = *Acanthamoeba castellanii*; UtopAm = *Alligator mississippiensis* Utopia; UtopCp = *Crocodylus porosus* Utopia; Prp8Hs = *Homo sapiens* (NP_006436.3); Prp8Dm = *Drosophila melangaster* (NP_610735.1); Prp8At = *Arabidopsis thaliana* (Q9T0I6); Prp8Sc = *Saccharomyces cerevisiae* (P33334); Prp8Sp = *Schizosaccharomyces pombe* (O14187); PA En = Influenza virus PA endonclease (3HW4). R8Hm, DongFr, CRE2Cf, CZARTc, HERODr, HEROBf, NeSLTv, NeSLAc, UtopAm, and UtopCp sequences were collected from Repbase [69]. Holiday junction resolvases Ssol Hje (1ob8) and Ssol Hjc (1hh1) are from *Sulfolobus solfataricus*. Holiday junction resolvase Stok Hjc (2eo0) is from *Sulfolobus tokodaii* str. 7 Pfur Hjc (1gef) is from *Pyrococcus furiosus*

of R2Bm) of the zinc knuckle motif has been aligned with NRAN(1615-1620) in Prp8. In this configuration the R(1616) of Prp8 aligns to a conserved R found at the start of the zinc knuckle (R867 in R2Bm) in a number of the RLE LINEs. Residue 1620 (Y) in Prp8 then aligns with the conserved Y in Utopia (position 870 in R2Bm). In Prp8 there are two conserved H residues which could potentially line up with the H of the LINE CCHC motif: H1635 and H1652. In Fig. 4a we aligned the H1635 to the H of the LINE CCHC motif as H1652 was too close to the endonuclease. An alternative and perhaps better method of aligning Prp8 and LINEs in this area is by structure. Prp8 contains a non-zinc knuckle with a $\beta\beta\alpha$ structure that is positioned at 1624-1647. The Prp8 non-zinc knuckle might be a structural equivalent to the LINE IAP/gag-like CCHC zinc-knuckle.

In addition to the knuckle, there are two helices upstream of the knuckle in both Prp8 and LINEs that align well in sequence alignments. The two predicted α-helices upstream of knuckle in LINEs tended to be separated by LP. In R2 elements the sequence at the end of the first helix was highly conserved, being KXRI-NALP(840-847) or similar. In R2Bm the sequence was HTHINALP (see also reference [24]). The two helices prior to the knuckle appear to be present in both RLE and APE LINEs. In the APE LINE L1Hs, the sequence upstream of the gag knuckle that PROMALS3D aligned to the R2 KXRINALP was TMRYHLTP of HMKKCSSSLIAREMQIKTTMRYHLTP. The HMKKC is not shown in the alignment. Conversion of HMKK to AAAA and SSS to AAA reduced retrotransposition activity [5]. In the alignment SSS is below position 828 of R2Bm. A recombinant C-terminal 180 amino acid containing peptide (from SSS to the end of the ORF) bound RNA nonspecifically, but a mutation of the CCHC motif within the peptide did not affect RNA binding [49]. In the full-length protein, however, mutations of the conserved cysteines of the CCHC motif affected RNP formation and knocked out retrotransposition activity in cell culture assays [5, 50]. In Prp8, the non-zinc knuckle is predicted to make contact with mRNA in the U4/U6.U5 tri-snRNP complex [51].

The helices upstream of Prp8's knuckle include an important loop (1585 loop, sometimes called the α-finger) that is important for binding RNA [51, 52]. The R2RLE LINE KXRINALP(840-847) helix equivalent in Prp8 was located at Prp8 residues 1592-1602 of the 1585 loop. The loop and helix region was found to be dynamic in Prp8 [21, 51–53]. In the U4/U6.U5 tri snRNP (cryo-EM structure 3JCM) the area forms a loop (QFKK, 1586-1589) plus an α-helix (HAQRTG, 1592-1597) [51]. The loop residues contact RNA and the Dib1 protein and were involved in branch point selection [51, 52]. After branching (cryo-EM structure 5LJ3), the area is not helical [54]. In the crystal structure (c4i43B), which lacks RNA, this area is unresolved and thus is likely unstructured [21].

RLE LINEs, like R2Bm, encode a restriction-like DNA endonuclease downstream of the IAP/gag-like CCHC zinc-knuckle (Fig. 4b). The DNA endonuclease found in RLE LINEs was found to have a fairly canonical $\alpha\beta\beta\beta\alpha\beta$ restriction endonuclease-like fold, although it had a unique variant of the PD-(D/E)XK catalytic core [19]. The catalytic K, which is usually near the D/E residue in the third β-sheet, was found to be located much farther away in LINE RLE. The catalytic K in the LINE RLE is the first K in the KX_2KY motif. The second K is less conserved across R2 elements and across RLE LINEs. The motif is located in the second α-helix [19]. The Y of the KX_2KY, when mutated, also reduces cleavage [19]. The catalytic K in Prp8 is located in an identical position as the RLE of LINEs. The Y residue is also present in Prp8 and is identically positioned relative to the catalytic K. The second K of the LINE KX_2KY is not present. The similarities between the Prp8 RLE and the LINE RLE go beyond the endonuclease fold and the positioning of the catalytic residues. At the far end of the endonuclease fold, just beyond the fourth β-strand, is a mutually conserved GXW motif. At the other end of the RLE fold—at the beginning of the first α-helix—is a conserved H residue and a conserved K residue. In R2Bm the equivalent is RH. Mutating the RH residues in R2Bm severely reduces DNA binding and DNA cleavage [19]. At the end of the first β-strand of both Prp8 and LINE RLEs is a conserved D/E that also appears to be unique to these two groups. Except for a 22 amino acid insertion between β-sheets 2 and 3 of Prp8, both Prp8 and LINE endonucleases are about the same size. The Prp8 endonuclease appears to have the amino acid residues needed for the cleavage activity, but the residues do not appear to be involved in metal coordination in the crystal structure; rather, these residues stabilize the polypeptide loop blocking the active site [21].

Comparative bar diagrams comparing R2Bm, Prp8, and LtrA ORF structure are presented in Fig. 5b. The RT, Linker, and RLE are highlighted in green, maroon, and orange, respectively. The areas of the ORF that have been aligned, homology modeled, or deemed structurally equivalent are also indicated. The secondary structure present in the linker of R2Bm and Prp8 is depicted. The areas of the linker that align in sequence (HINALP/1585-loop regions) or by structure (knuckle) are indicated in maroon. Ribbon diagrams are also given for RT-RLE for R2Bm (Fig. 5c) and Prp8 (Fig. 5d) using the same color scheme as the bar diagrams. A structural overlay (ribbon diagrams) of R2Bm RT and RLE onto Prp8 RT-RLE is presented in Fig. 5e. A second overlay between R2Bm and Prp8 is presented in Fig. 5f in which the R2Bm RT is a surface model and is colored an in Fig. 3e. Also colored in Fig. 5f is the Prp8 non-zinc

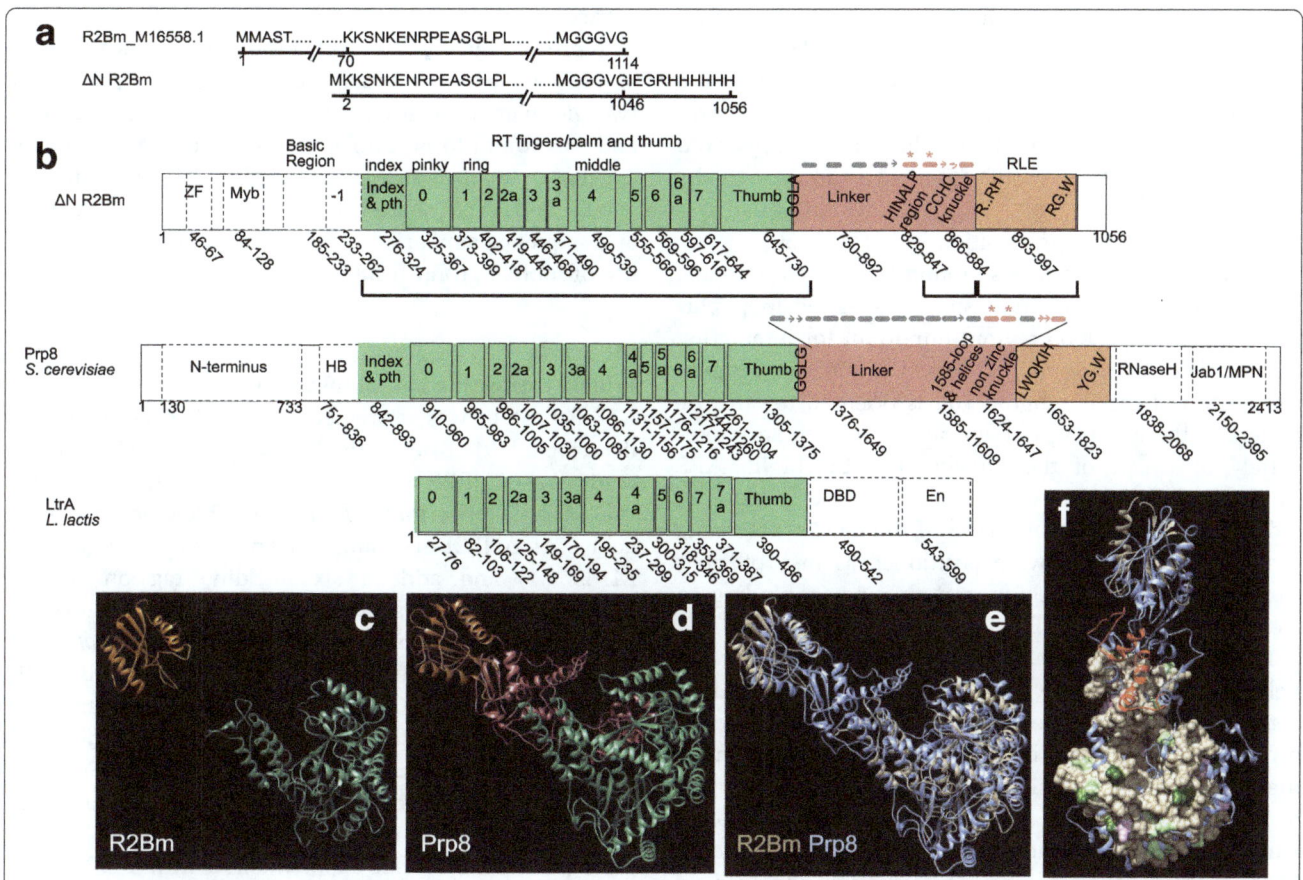

Fig. 5 A Comparison of domain architechture of R2Bm, Prp8, and LtrA group II intron proteins. **a** The amino acid equivalences between ΔNR2Bm and that of the full length R2Bm ORF (genbank entry M16558.1) is such that the 2nd amino acid of the ΔNR2Bm ORF is 70th amino acid of full length R2Bm. **b** The ORF structure of ΔNR2Bm, *S. cerevisiae* Prp8, and *L. lactis* LtrA are presented as a colored bar diagrams. The RT is green, the linker is a maroon, and the RLE is orange. The RT sequence-motif blocks are indicated with approximate primary amino acid numbers. The locations of the fingers are also shown. The bar diagrams are not drawn to scale. The domains and numbering for Prp8 and LtrA group II intron are largely from [22, 52, 70, 71]. Brackets indicate regions of R2Bm that align well with corresponding region of Prp8. The RT and the RLE align well in sequence and by structure modeling with Prp8. In the linker, the sequence of (and around) the colored α-helices (rounded bars) with an asterisk align when well when anchored by the reverse transcriptase and the RLE. The colored β-strands and the α-helix lacking an asterisk do not align well but (may) form a knuckle that is structurally similar. The secondary structure is predicted in R2Bm, but known in Prp8 (3JCM). **c** Ribbon diagrams of the R2Bm RT and RLE models. The ribbons have been colored as in the the cooresponding bar diagram. **d** Ribbon diagrams of the Prp8 cryo-EM structure (3JCM, RT to RLE) is shown in matching color bar diagram. **e** Ribbon overlay of R2Bm (tan) with Prp8 large fragment (blue). **f** The R2Bm reverse transcriptase model as colored in Fig. 3e, a ribbon model of the R2Bm RLE and the large fragment of Prp8 (blue with red 1585 loop and knuckle). Abbreviations: palm traversing helix (pth); DNA binding domain (DBD); Endonuclease (EN). All other abbreviations and symbols are as in previous figures

knuckle and the 1585 loop region (both red). The Prp8 knuckle and 1585 loop regions are positioned near the top of the thumb. The knuckle also is positioned near the RLE. It sits between the thumb and RLE. The 1585 loop is oriented toward the fingers of the RT. The R2Bm RT equivalents (HINALP region and knuckle) may also be closely associated with the thumb. Having the linker associated with the thumb would explain, in part, why the thumb is protected from cleavage.

Discussion

The R2Bm protein was found to be comprised of two major globular domains: the ZF/Myb/−1 N-terminal domain and the RT/linker/RLE superdomain. The index

finger of the RT and the −1 region were the most accessible areas for protease cleavages to occur, indicating that these regions might represent flexible conformational-switch areas that may help coordinate the nucleic acid binding and cleavage activities of the two globular domains. The ZF/Myb/−1 region is present among all of the early branching LINE elements with a variable number of ZFs and Myb motifs [55]. The primary variability in the RT/linker/RLE superdomain was in the linker, with Cre elements often having a deletion relative to R2 and Utopia having an insertion. The linker was predicted to be largely α-helical across all RLE LINEs and, we hypothesize, closely associated with both the RT thumb and the RLE, similar to Prp8. The linker of R2 contained several highly conserved sequence motifs and secondary structures, most notably the presumptive-ββα-forming IAP/gag-like CCHC zinc-knuckle. Just upstream of the knuckle in R2 elements are several well aligned and predicted α-helices separated by the highly conserved KXRINALP(840-847) motif. The two appear to be present in both RLE and APE LINEs. Mutations in this region affected retrotransposition in L1Hs [5, 50].

The linker of Prp8 was also found to have an non-zinc knuckle structure and an upstream dynamic loop plus helix region (1585 loop region) important for interacting with nucleic acids [51–53]. In Prp8 the 1585 loop region sits on top of the RT thumb and is oriented toward the fingers of the RT. It would appear tha the region immediately upstream of the knuckle in RLE LINE, APE LINEs, and Prp8 might be structurally conserved and, to a degree, functionally conserved. If the helices preceding the knuckle in LINEs were positioned as in Prp8, it is easy to envision the region participating in binding to element RNA or target DNA.

The gag knuckle-like motif and associated upstream helices might promote switching between polymerase active and endonuclease active conformations of the R2 protein in response to binding insertion reaction intermediates. Our proteolysis study was done in the absence of nucleic acids. In the absence of RNA, the R2 protein would be expected to adopt a conformation that would have characteristics of the conformation involved in second-strand cleavage. It is possible that in the presence of RNA or DNA our results would differ from those presented as the nucleic acid might block some sites from being cleaved while presenting other newly accessible sites due to protein conformational changes induced by nucleic acid binding.

The RT of Prp8 has been noted to share similarities to RdRP and to the RTs encoded by mobile group II introns and LINEs [26, 34]. However, because Ppr8 and the group II intron protein both function as splicing maturases, the similarity to group II introns has been stressed. The RLE LINE RT, however, appears to be more similar to Prp8 given the presence of an index finger and a palm traversing helix although our phylogenetic studies (not reported) using the RT were inconclusive. In addition to the RT both Prp8 and RLE LINEs have a RLE. While it has been noted in the literature that the large fragment of Prp8 contains an RLE, the connection to LINEs had not been presented beyond noting that LINEs also contain an RT and an RLE. Using sequence and structure comparisons, we may have been able to infer insights that are not yet forthcoming in phylogenetic trees. In this paper, we have shown that the LINE RT, linker, and RLE share more points of commonality to the large fragment of Prp8 than does the group II intron maturase.

Conclusions

The protein encoded by RLE LINEs was shown to consist of two major globular domains. The larger of the two globular domain contained the RT, linker, and RLE and was found to be similar to the large fragment of the spliceosomal protein Prp8. The RLE, RT, and linker of LINEs and Prp8 shared a greater degree of structural and sequence similarity to each other than to the maturase of mobile group II introns.

Methods

Protein expression and purification

R2Bm protein was expressed and purified as previously described [19]. Briefly, the R2Bm protein used in this study was ΔNR2Bm. The ΔNR2Bm construct removes the variable N-terminal found in R2 elements (amino acid 2 of ΔNR2Bm = amino acid 70 of genbank entry M16558.1) and adds a six histidine tag on the C-terminal end of the protein [19]. The ΔNR2Bm expression construct was put into BL21 *Escherichia coli* cells. Five hundred milliliter cultures were grown in LB broth, expressed with IPTG, lysed, and the soluble material purified over a Talon affinity column (Clontech #635501). The R2Bm protein was eluted off the column in 50 mM HEPES pH 7.5, 100 mM NaCl, 50% glycerol, 0.1% triton X-100, 150 mM imidazole. Proteins were stored in elution buffer supplemented with 1 mM DTT (final concentration) at −20 °C. R2 protein was quantified by SYPRO Orange (Sigma #S5692) staining of samples run on SDS-PAGE relative to a BSA standard (Biorad #500-0202). All quantitations were done using Fiji software analysis of digital photographs [56].

Limited proteolysis of R2Bm protein and processing of the polypeptides

Limited digestion of purified R2 protein was carried out in the absence of nucleic acids using a trace amount of GluC (NEB, #P8100S) or LysC (Promega, #V1671) protease. Digestion was stopped using SDS loading buffer (to a final concentration of 50 mM Tris-Cl, pH 8.8 final; 4% SDS; 10 mM DTT) and heated. Proteolytic fragments were carbamidomethylated (55 mM final, Alfa aesar, #A14715) in the loading buffer in the dark at room temperature for 30 min prior to loading onto a precast (Biorad criterion polyacrylamide gel, 18% and 4-15%) SDS-PAGE gel [57]. The resolved protein fragments were stained with colloidal coomassie blue (Invitrogen, #LC6025). Prominent bands from across the proteolytic time course were excised from the gel, cut into 1 mm pieces, and destained using 25 mM NH$_4$HCO$_3$/50% Acetonitrile (ACN). Gel pieces were shrunk with 100% ACN (VWR, #BDH6002-4) and dried by Speed Vac (Eppendorf) [58–60].

The primary amines, including the amino-terminal end of the proteolytic fragments, were acetylated in the gel slice using 15% acetic anhydride (Sigma, #320102) for five hours at room temperature within the individual excised gel fragments [61, 62]. Acetylation was stopped by adding 1 M NH_4HCO_3 (Sigma, #40867) solution [61, 62]. After 20 min, the gel pieces were shrunk by 100% ACN.

The dried gel pieces were swelled in 25 mM NH_4HCO_3 containing trypsin or GluC for 1.5 h at 4 °C and any unabsorbed NH_4HCO_3 solution was then discarded [58–60]. The gel pieces were covered with 25 mM NH_4HCO_3 and the in-gel digestion was carried out overnight at 37 °C. Peptides from in-gel digestion reaction were collected in the supernatant. Additional extractions with 0.1% formic acid (FA) (Sigma, #399388) and 50% ACN/0.1% FA were also collected and added to the supernatant [58–60]. The supernatant was dried in a Speed Vac and purified over C18 zip tip using standard procedures [58–60].

To catch any major cleavage sites that did not result in isolatable SDS-page bands, limited proteolysis reactions were run on an SDS-PAGE gel for a very short time so as to not resolve bands, rather keeping them clustered near the well. The top portions of these lanes were excised and processed as above so as to remove triton and otherwise prepare the polypeptides for mass spectrometry, thus avoiding the precipitation of the larger R2Bm protein fragments that occurs if the polypeptide processing (for mass spectrometry) was done in solution instead of in-gel. This abbreviated in-gel procedure is roughly equivalent to a direct "in-solution" detection of cleavage sites.

Mass spectrometry and Edman degradation
The eluted peptides were resuspended in 0.1% FA for sequencing by nanoLC-ESI-MS/MS using a Thermo Scientific LTQ Velos Pro ion trap mass spectrometer. R2 peptides were identified using Thermo Proteome Discoverer software (version 2.0); a database of R2Bm protein fragments was created, and a peptide was assigned as either N-terminal end or internal peptide based on the position of acetyl groups in the peptide sequence [61]. The internal peptides generated after trypsin (second) digestion will lack an acetyl group at the N-terminal end, as acetylation is performed prior to the second protease digestion step.

Amino-terminal sequencing of the separated proteolytic fragments was used to map the protease cleavage sites back onto the primary sequence of R2 and thus delimit globular domain boundaries. The internal peptides were also identified from the MS/MS spectrum. The internal peptide coverage and sequence were used to help verify the peptide location within the R2 ORF and to act as a rough estimation of the C-terminal boundary of the fragment, along with SDS-PAGE

estimation of the fragment's molecular weight. The Glu-C cleavage heatmap was generated using Gitools [63].

For detection by Edman degradation, an SDS-PAGE gel was electrophoresed onto a PVDF membrane. Excised bands on the PVDF membrane were sent to UT Southwestern proteomics core for Edman sequencing.

3D modeling and multiple sequence alignments
The Phyre 2.0 protein fold recognition server was used to model the RT domain of R2Bm protein [25]. The intensive mode with default parameters of Phyre 2.0 were used. Different lengths of R2Bm sequence upstream and downstream of RT domain were submitted for modeling to find the sequence window that modeled the best. Model visualization was aided by UCSF Chimera package [64].

The PROMALS3D server was used for structure based alignment with minor manual adjustments [65–67]. Seventy five LINE sequences were aligned first in PROMALS3D server that included 31 RLE LINE and 44 APE LINEs. Using this LINE alignment as constraint, an extended alignment was built with three nMat proteins, five group II introns, seven RVT genes and 10 Prp8 proteins. The secondary structure was plotted on the multiple sequence alignment using the Ali2D program of the MPI bioinformatic tool kit [67, 68].

Additional files

Additional file 1: Lys C mapping data. (ZIP 983 kb)

Additional file 2: R2Bm sequence and domain boundaries. (DOCX 28 kb)

Additional file 3: Glu C mapping data (amino terminal identification). (PDF 287 kb)

Additional file 4: Glu C mapping data (internal peptides). (PDF 541 kb)

Additional file 5: List of Glu C and Lys C sites with early vs. later designations. (XLSX 482 kb)

Additional file 6: Pdb file of R2Bm RT model. (PDB 307 kb)

Additional file 7: Alignment files. (ZIP 5764 kb)

Abbreviations
APE: Apurinic-apyrimidinic family endonuclease; LINE: Long interspersed element (also called non-LTR retrotransposon); ORF: Open reading frame; RLE: Restriction-like endonuclease; RNP: Ribonucleoprotein particles; RT: Reverse transcriptase; TPRT: Target primed reverse transcriptase

Acknowledgements
We would like to thank members of the Chowdhury lab, especially Abu Hena Mostafa Kamal, as well as the staff at the Shimadzu Center for Advanced Analytical Chemistry for collective discussions and help with mass spectrometry. We would like to thank members of the Christensen lab for critical reading of the manuscript. Finally, we would like to thank Micki Christensen for copy editing.

Funding
This work was supported, in part, by the National Science Foundation [0950983] and by a University of Texas Arlington Research Enhancement grant, both awarded to Shawn Christensen. Additional support was from a Phi Sigma grant awarded to Murshida Mahbub. The Thermo Scientific LTQ Velos Pro ion trap mass spectrometer was purchased maintained with startup funds awarded to Saiful M. Chowdhury.

Authors' contributions

MM performed all of the experiments, including mass spectrometry and bioinformatics. She helped with designing the experiments and with interpreting the results. M also helped to write the paper. SMC was the mass spectrometry expert. He provided experimental design help with how to prepare the protein for mass spectrometry and how to determine the N-terminal ends. He provided expert help in interpreting the mass spectra. S provided the mass spectrometer and the funds related to running the mass spectrometer. SMC was the transposable element expert. He broadly designed all aspects of the project, including experimental design, bioinformatics, and result interpretation. He wrote the paper and provided the funding for MM. All authors read and approved the final manuscript.

Competing interests

The authors declare that they have no competing interests.

References

1. Luan DD, Korman MH, Jakubczak JL, Eickbush TH. Reverse transcription of R2Bm RNA is primed by a nick at the chromosomal target site: a mechanism for non-LTR retrotransposition. Cell. 1993;72:595–605.

2. Christensen SM, Ye J, Eickbush TH. RNA from the 5′ end of the R2 retrotransposon controls R2 protein binding to and cleavage of its DNA target site. Proc Natl Acad Sci U S A. 2006;103:17602–7.

3. Christensen SM, Eickbush TH. R2 target-primed reverse transcription: ordered cleavage and polymerization steps by protein subunits asymmetrically bound to the target DNA. Mol Cell Biol. 2005;25:6617–28.

4. Feng Q, Moran JV, Kazazian HHJ, Boeke JD. Human L1 retrotransposon encodes a conserved endonuclease required for retrotransposition. Cell. 1996;87:905–16.

5. Moran JV, Holmes SE, Naas TP, DeBerardinis RJ, Boeke JD, Kazazian HHJ. High frequency retrotransposition in cultured mammalian cells. Cell. 1996;87:917–27.

6. Eickbush TH, Eickbush DG. Integration, regulation, and long-term stability of R2 Retrotransposons. Microbiol Spectr. 2015;3:MDNA3-0011. doi: 10.1128/microbiolspec.MDNA3-0011-2014.

7. Eickbush TH. R2 and related site-specific non-long terminal repeat Retrotransposons. In: Craig NL, Craigie R, Gellert M, Lambowitz AM, editors. Mobile DNA II. Washington, DC: ASM Press; 2002. p. 813–35.

8. Moran JV, Gilbert N. Mammalian LINE-1 Retrotransposons and related elements. In: Craig NL, Craigie R, Gellert M, Lambowitz AM, editors. Mobile DNA II. Washington, DC: ASM Press; 2002. p. 836–69.

9. Zingler N, Weichenrieder O, Schumann GG. APE-type non-LTR retrotransposons: determinants involved in target site recognition. Cytogenet Genome Res. 2005;110:250–68.

10. Richardson SR, Doucet AJ, Kopera HC, Moldovan JB, Garcia-Perez JL, Moran JV. The influence of LINE-1 and SINE Retrotransposons on mammalian genomes. Microbiol Spectr. 2015;3:MDNA3-0061. doi: 10.1128/microbiolspec.MDNA3-0061-2014.

11. Fujiwara H. Site-specific non-LTR retrotransposons. Microbiol Spectr. 2015;3:MDNA3-0001. doi: 10.1128/microbiolspec.MDNA3-0001-2014.

12. Babushok DV, Kazazian HH. Progress in understanding the biology of the human mutagen LINE-1. Hum Mutat. 2007;28:527–39. doi: 10.1002/humu.20486.

13. Weichenrieder O, Repanas K, Perrakis A. Crystal structure of the targeting endonuclease of the human LINE-1 retrotransposon. Structure. 2004;12:975–86. doi: 10.1016/j.str.2004.04.011.

14. Repanas K, Zingler N, Layer LE, Schumann GG, Perrakis A, Weichenrieder O. Determinants for DNA target structure selectivity of the human LINE-1 retrotransposon endonuclease. Nucleic Acids Res. 2007;35:4914–26. doi: 10.1093/nar/gkm516.

15. Khazina E, Weichenrieder O. Non-LTR retrotransposons encode noncanonical RRM domains in their first open reading frame. Proc Natl Acad Sci U S A. 2009;106:731–6.

16. Schneider AM, Schmidt S, Jonas S, Vollmer B, Khazina E, Weichenrieder O. Structure and properties of the esterase from non-LTR retrotransposons suggest a role for lipids in retrotransposition. Nucleic Acids Res. 2013;41:10563–72. doi: 10.1093/nar/gkt786.

17. Januszyk K, Li PW, Villareal V, Branciforte D, Wu H, Xie Y, Feigon J, Loo JA, Martin SL, Clubb RT. Identification and solution structure of a highly conserved C-terminal domain within ORF1p required for retrotransposition of long interspersed nuclear element-1. J Biol Chem. 2007;282:24893–904.

18. Khazina E, Truffault V, Buttner R, Schmidt S, Coles M, Weichenrieder O. Trimeric structure and flexibility of the L1ORF1 protein in human L1 retrotransposition. Nat Struct Mol Biol. 2011;18:1006–14.

19. Govindaraju A, Cortez JD, Reveal B, Christensen SM. Endonuclease domain of non-LTR retrotransposons: loss-of-function mutants and modeling of the R2Bm endonuclease. Nucleic Acids Res. 2016;44:3276–87. doi: 10.1093/nar/gkw134.

20. Dlakić M, Mushegian A. Prp8, the pivotal protein of the spliceosomal catalytic center, evolved from a retroelement-encoded reverse transcriptase. RNA. 2011;17:799–808. doi: 10.1261/rna.2396011.

21. Galej WP, Oubridge C, Newman AJ, Nagai K. Crystal structure of Prp8 reveals active site cavity of the spliceosome. Nature. 2013;493:638–43. doi: 10.1038/nature11843.

22. Qu G, Kaushal PS, Wang J, Shigematsu H, Piazza CL, Agrawal RK, Belfort M, Wang HW. Structure of a group II intron in complex with its reverse transcriptase. Nat Struct Mol Biol. 2016;23:549–57. doi: 10.1038/nsmb.3220.

23. Jamburuthugoda VK, Eickbush TH. Identification of RNA binding motifs in the R2 retrotransposon-encoded reverse transcriptase. Nucleic Acids Res. 2014;42:8405–15. doi: 10.1093/nar/gku514.

24. Burke WD, Malik HS, Jones JP, Eickbush TH. The domain structure and retrotransposition mechanism of R2 elements are conserved throughout arthropods. Mol Biol Evol. 1999;16:502–11.

25. Kelley LA, Mezulis S, Yates CM, Wass MN, Sternberg MJ. The Phyre2 web portal for protein modeling, prediction and analysis. Nat Protoc. 2015;10:845–58. doi: 10.1038/nprot.2015.053.

26. Zhao C, Pyle AM. Crystal structures of a group II intron maturase reveal a missing link in spliceosome evolution. Nat Struct Mol Biol. 2016;23:558–65. doi: 10.1038/nsmb.3224.

27. Ng KK, Cherney MM, Vazquez AL, Machin A, Alonso JM, Parra F, James MN. Crystal structures of active and inactive conformations of a caliciviral RNA-dependent RNA polymerase. J Biol Chem. 2002;277:1381–7. doi: 10.1074/jbc.M109261200.

28. Wu J, Liu W, Gong P. A structural overview of RNA-dependent RNA polymerases from the Flaviviridae family. Int J Mol Sci. 2015;16:12943–57. doi: 10.3390/ijms160612943.

29. Lu G, Gong P. A structural view of the RNA-dependent RNA polymerases from the Flavivirus genus. Virus Res. 2017;234:34–43. doi: 10.1016/j.virusres.2017.01.020.

30. Lu G, Gong P. Crystal structure of the full-length Japanese encephalitis virus NS5 reveals a conserved methyltransferase-polymerase interface. PLoS Pathog. 2013;9:e1003549. doi: 10.1371/journal.ppat.1003549.

31. Thompson AA, Peersen OB. Structural basis for proteolysis-dependent activation of the poliovirus RNA-dependent RNA polymerase. EMBO J. 2004;23:3462–71. doi: 10.1038/sj.emboj.7600357.

32. Gillis AJ, Schuller AP, Skordalakes E. Structure of the Tribolium Castaneum telomerase catalytic subunit TERT. Nature. 2008;455:633–7. doi: 10.1038/nature07283.

33. Nikonov A, Juronen E, Ustav M. Functional characterization of fingers subdomain-specific monoclonal antibodies inhibiting the hepatitis C virus RNA-dependent RNA polymerase. J Biol Chem. 2008;283:24089–102. doi: 10.1074/jbc.M803422200.

34. Lambowitz AM, Belfort M. Mobile bacterial group II Introns at the crux of eukaryotic evolution. Microbiol Spectr. 2015;3:MDNA3-0050. doi: 10.1128/microbiolspec.MDNA3-0050-2014.

35. Zimmerly S, Wu L. An unexplored diversity of reverse Transcriptases in bacteria. Microbiol Spectr. 2015;3:MDNA3-0058. doi: 10.1128/microbiolspec.MDNA3-0058-2014.

36. Shu B, Gong P. Structural basis of viral RNA-dependent RNA polymerase catalysis and translocation. Proc Natl Acad Sci U S A. 2016;113:E4005–14. doi: 10.1073/pnas.1602591113.

37. Arnold JJ, Cameron CE. Poliovirus RNA-dependent RNA polymerase (3Dpol) is sufficient for template switching in vitro. J Biol Chem. 1999;274:2706–16.

38. Chen B, Lambowitz AM. De novo and DNA primer-mediated initiation of cDNA synthesis by the mauriceville retroplasmid reverse transcriptase involve recognition of a 3′ CCA sequence. J Mol Biol. 1997;271:311–32.

39. Mohr S, Ghanem E, Smith W, et al. Thermostable group II intron reverse transcriptase fusion proteins and their use in cDNA synthesis and next-generation RNA sequencing. RNA. 2013;19:958–70. doi: 10.1261/rna.039743.113.

40. SQ G, Cui X, Mou S, Mohr S, Yao J, Lambowitz AM. Genetic identification of potential RNA-binding regions in a group II intron-encoded reverse transcriptase. RNA. 2010;16:732–47.

41. Watanabe K, Lambowitz AM. High-affinity binding site for a group II intron-

encoded reverse transcriptase/maturase within a stem-loop structure in the intron RNA. RNA. 2004;10:1433–43.

42. Huang J, Brown AF, Wu J, Xue J, Bley CJ, Rand DP, Wu L, Zhang R, Chen JJ, Lei M. Structural basis for protein-RNA recognition in telomerase. Nat Struct Mol Biol. 2014;21:507–12. doi: 10.1038/nsmb.2819.

43. Mitchell M, Gillis A, Futahashi M, Fujiwara H, Skordalakes E. Structural basis for telomerase catalytic subunit TERT binding to RNA template and telomeric DNA. Nat Struct Mol Biol. 2010;17:513–8. doi: 10.1038/nsmb.1777.

44. Wyatt HDM, West SC, Beattie TL. InTERTpreting telomerase structure and function. Nucleic Acids Res. 2010;38:5609–22. doi: 10.1093/nar/gkq370.

45. Gladyshev EA, Arkhipova IR. A widespread class of reverse transcriptase-related cellular genes. Proc Natl Acad Sci U S A. 2011;108:20311–6.

46. Xiong Y, Eickbush TH. Origin and evolution of retroelements based upon their reverse transcriptase sequences. EMBO J. 1990;9:3353–62.

47. Arkhipova IR, Batzer MA, Brosius J, Feschotte C, Moran JV, Schmitz J, Jurka J. Genomic impact of eukaryotic transposable elements. Mob DNA. 2012;3:19.

48. Krishna SS, Majumdar I, Grishin NV. Structural classification of zinc fingers: survey and summary. Nucleic Acids Res. 2003;31:532–50.

49. Piskareva O, Ernst C, Higgins N, Schmatchenko V. The carboxy-terminal segment of the human LINE-1 ORF2 protein is involved in RNA binding. FEBS Open Bio. 2013;3:433–7. doi: 10.1016/j.fob.2013.09.005.

50. Doucet AJ, Hulme AE, Sahinovic E, et al. Characterization of LINE-1 ribonucleoprotein particles. PLoS Genet. 2010;6:e1001150. doi: 10.1371/journal.pgen.1001150.

51. Wan R, Yan C, Bai R, Wang L, Huang M, Wong CC, Shi Y. The 3.8 Å structure of the U4/U6.U5 tri-snRNP: insights into spliceosome assembly and catalysis. Science. 2016;351:466–75. doi: 10.1126/science.aad6466.

52. Nguyen TH, Galej WP, Bai XC, Oubridge C, Newman AJ, Scheres SH, Nagai K. Cryo-EM structure of the yeast U4/U6.U5 tri-snRNP at 3.7 Å resolution. Nature. 2016;530:298–302. doi: 10.1038/nature16940.

53. Bertram K, Agafonov DE, Liu WT, Dybkov O, Will CL, Hartmuth K, Urlaub H, Kastner B, Stark H, Lührmann R. Cryo-EM structure of a human spliceosome activated for step 2 of splicing. Nature. 2017;542:318–23. doi: 10.1038/nature21079.

54. Galej WP, Wilkinson ME, Fica SM, Oubridge C, Newman AJ, Nagai K. Cryo-EM structure of the spliceosome immediately after branching. Nature. 2016;537: 197–201. https://doi.org/10.1038/nature19316.

55. Shivram H, Cawley D, Christensen SM. Targeting novel sites: the N-terminal DNA binding domain of non-LTR retrotransposons is an adaptable module that is implicated in changing site specificities. Mob Genet Elements. 2011;1:169–78.

56. Schindelin J, Arganda-Carreras I, Frise E, et al. Fiji: an open-source platform for biological-image analysis. Nat Methods. 2012;9:676–82. doi: 10.1038/nmeth.2019.

57. Lane LC. A simple method for stabilizing protein-sulfhydryl groups during SDS-gel electrophoresis. Anal Biochem. 1978;86:655–64.

58. Havlis J, Thomas H, Sebela M, Shevchenko A. Fast-response proteomics by accelerated in-gel digestion of proteins. Anal Chem. 2003;75:1300–6.

59. Shevchenko A, Wilm M, Vorm O, Mann M. Mass spectrometric sequencing of proteins silver-stained polyacrylamide gels. Anal Chem. 1996;68:850–8.

60. Shevchenko A, Tomas H, Havlis J, Olsen JV, Mann M. In-gel digestion for mass spectrometric characterization of proteins and proteomes. Nat Protoc. 2006;1:2856–60. doi: 10.1038/nprot.2006.468.

61. Chowdhury SM, Munske GR, Yang J, Zhukova D, Nguyen H, Bruce JE. Solid-phase N-terminal peptide enrichment study by optimizing trypsin proteolysis on homoarginine-modified proteins by mass spectrometry. Rapid Commun Mass Spectrom. 2014;28:635–44. doi: 10.1002/rcm.6820.

62. Celic I, Masumoto H, Griffith WP, Meluh P, Cotter RJ, Boeke JD, Verreault A. The sirtuins hst3 and Hst4p preserve genome integrity by controlling histone h3 lysine 56 deacetylation. Curr Biol. 2006;16:1280–9.

63. Perez-Llamas C, Lopez-Bigas N. Gitools: analysis and visualisation of genomic data using interactive heat-maps. PLoS One. 2011;6:e19541. doi: 10.1371/journal.pone.0019541.

64. Pettersen EF, Goddard TD, Huang CC, Couch GS, Greenblatt DM, Meng EC, Ferrin TE. UCSF chimera–a visualization system for exploratory research and analysis. J Comput Chem. 2004;25:1605–12. doi: 10.1002/jcc.20084.

65. Pei J, Grishin NV. PROMALS3D: multiple protein sequence alignment enhanced with evolutionary and three-dimensional structural information. Methods Mol Biol. 2014;1079:263–71. doi: 10.1007/978-1-62703-646-7_17.

66. Larkin MA, Blackshields G, Brown NP, et al. Clustal W and Clustal X version 2. 0. Bioinformatics. 2007;23:2947–8. doi: 10.1093/bioinformatics/btm404.

67. Larsson A. AliView: a fast and lightweight alignment viewer and editor for large datasets. Bioinformatics. 2014;30:3276–8. doi: 10.1093/bioinformatics/btu531.

68. Alva V, Nam SZ, Söding J, Lupas AN. The MPI bioinformatics toolkit as an integrative platform for advanced protein sequence and structure analysis. Nucleic Acids Res. 2016;44:W410–5. doi: 10.1093/nar/gkw348.

69. Bao W, Kojima KK, Kohany O. Repbase update, a database of repetitive elements in eukaryotic genomes. Mob DNA. 2015;6:11. doi: 10.1186/s13100-015-0041-9.

70. Galej WP, Nguyen THD, Newman AJ, Nagai K. Structural studies of the spliceosome: zooming into the heart of the machine. Curr Opin Struct Biol. 2014;25:57–66. doi: 10.1016/j.sbi.2013.12.002.

71. Blocker FJ, Mohr G, Conlan LH, Qi L, Belfort M, Lambowitz AM. Domain structure and three-dimensional model of a group II intron-encoded reverse transcriptase. RNA. 2005;11:14–28. doi: 10.1261/rna.7181105.

10

Chicken (*Gallus gallus*) endogenous retrovirus generates genomic variations in the chicken genome

Jinmin Lee[1], Seyoung Mun[1], Dong Hee Kim[2], Chun-Sung Cho[3], Dong-Yep Oh[4] and Kyudong Han[1*]

Abstract

Background: Transposable elements (TEs) comprise ~10% of the chicken (*Gallus gallus*) genome. The content of TEs is much lower than that of mammalian genomes, where TEs comprise around half of the genome. Endogenous retroviruses are responsible for ~1.3% of the chicken genome. Among them is *Gallus gallus* endogenous retrovirus 10 (GGERV10), one of the youngest endogenous retrovirus families, which emerged in the chicken genome around 3 million years ago.

Results: We identified a total of 593 GGERV10 elements in the chicken reference genome using UCSC genome database and RepeatMasker. While most of the elements were truncated, 49 GGERV10 elements were full-length retaining 5′ and 3′ LTRs. We examined in detail their structural features, chromosomal distribution, genomic environment, and phylogenetic relationships. We compared LTR sequence among five different GGERV10 subfamilies and found sequence variations among the LTRs. Using a traditional PCR assay, we examined a polymorphism rate of the 49 full-length GGERV10 elements in three different chicken populations of the Korean domestic chicken, Leghorn, and *Araucana*. The result found a breed-specific GGERV10B insertion locus in the Korean domestic chicken, which could be used as a Korean domestic chicken-specific marker.

Conclusions: GGERV10 family is the youngest ERV family and thus might have contributed to recent genomic variations in different chicken populations. The result of this study showed that one of GGERV10 elements integrated into the chicken genome after the divergence of Korean domestic chicken from other closely related chicken populations, suggesting that GGERV10 could be served as a molecular marker for chicken breed identification.

Keywords: Retrotransposon, Full-length GGERV10, Genomic variation, Molecular marker, Incomplete lineage sorting

Background

Transposable elements (TEs) are frequently referred to as "junk DNA" in the host genome and compose a major portion of most vertebrate genomes [1]. They are classified as DNA transposons and retrotransposons according to their mobilization methods. DNA transposons integrate into the host genome through a "cut and paste" mechanism but retrotransposons propagate using a "copy and paste" mechanism [2]. TEs have played a role in generating genomic variation, genetic novelty and contributed to speciation and evolutionary transitions in the vertebrate lineage [3]. Several different vertebrate genomes have been sequenced and published [3]. One of them is chicken (*Gallus gallus*) and its size is ~1.2 billion base pairs, which is approximately one third of the size of the most of mammalian including human genome [4, 5]. Unlike most mammalian genomes, TE content is remarkably low in the chicken genome [4–6]. There are various different TE groups in the chicken genome, which include chicken repeat 1 (CR1), long interspersed element 2 (LINE2), endogenous retrovirus (ERV), long terminal repeat (LTR) element, and DNA transposon [4]. Among them, ERVs comprise approximately 1.3% of the chicken genome. This element was originated from exogenous retroviral infection through germ-line cells [4, 7, 8]. ERVs is known to be transmitted vertically in the host genome and propagated through reinfection and retrotransposition

* Correspondence: kyudong.han@gmail.com
[1]Department of Nanobiomedical Science & BK21 PLUS NBM Global Research Center for Regenerative Medicine, Dankook University, Cheonan 330-714, Republic of Korea
Full list of author information is available at the end of the article

events [9]. Avian ERVs are classified into three major exogenous retroviral classes (class I to III), according to *pol* amino acid sequences [10], and consist of four internal coding regions: group-specific antigen (*gag*), protease gene (*pro*), RNA-dependent DNA polymerase gene (*pol*), and envelope gene (*env*), which are flanked by LTRs [11–13]. However, most ERVs are lack of the envelope protein domain due to accumulated mutations (insertion, deletion, and substitution) in the elements and/or negative selection in the host genome [14, 15]. Recently, it was suggested that a retrovirus without *env* gene could be complemented through co-infection with a retrovirus which has a functional *env* [16].

Huda et al. constructed a GGERV phylogenetic tree of fourteen distinct GGERV families based on reverse transcriptase (RT) sequences. GGERV10 element, the youngest ERV family, was integrated into the chicken genome about 0–3 million years ago [8]. Full-length GGERV elements include intact *gag* and *pol* genes, which are necessary for the propagation of the elements. The result of the study showed that GGERV10 family was recently integrated into the chicken genome and proposed that the element could be retrotranspositionally active in the chicken genome.

The LTR sequences of ERV element contain an internal promoter and regulatory sequences (e.g., transcription factor binding site). Therefore, ERVs could alter the expression of host genes by introducing alternative splicing or regulating gene expression in a tissue-specific manner [17]. In fact, it was reported that ERV associated-gene regulation changed the phenotype of its host; *Araucana* lays a blue egg. ERV, locating on the 5' flanking region of *SLCO1B3* gene in the chicken genome, controls the egg color [18].

In this study, we identified 49 full-length GGERV10 elements in the chicken reference genome (galGal4, Nov. 2011) using a combined method of computational data mining, manual inspection, and experimental validation. Through polymorphism test of the elements, we found that one of them is a Korean breed-specific ERV. This element could be used as a molecular marker for Korean domestic chicken. In sum, we suggest that GGERV10 elements have contributed to the genomic variation of different chicken breeds and could be used as a molecular markers for chicken breed identification.

Results and discussion
Identification of GGERV10 insertions
To investigate genomic variation caused by the insertion of GGERV10 family, we computationally extracted 593 putative GGERV10 elements from the chicken (*Gallus gallus*) reference genome, based on RepeatMasker annotation (http://www.repeatmasker.org/cgi-bin/WEBRepeatMasker). Then, we manually inspected them and

divided them into three groups: full-length GGERV10 elements, solo-LTRs, and truncated GGERV10 elements. 49, 483, and 61 elements were grouped into full-length GGERV10 elements, solo-LTRs, and truncated GGERV10 elements, respectively. However, the truncated 61 copies were excluded from our data because either or both LTR sequence(s) were missed in them (Table 1). We further examined full-length GGERV10 elements or solo-LTRs, which were probably derived from homologous recombination between LTRs. The remaining 532 GGERV10elements were grouped into five subfamilies, based on their LTR sequence. The LTR sequence variations were annotated by Repbase (http://www.girinst.org/repbase/index.html): GGERV10A, GGERV10B, GGERV10C1, GGERV10C2, and GGERV10D [19]. As shown in Table 2, GGERV10C2 is most abundant while GGERV10B is least abundant in the chicken genome. We examined the chromosomal distribution of GGERV10 and the result showed a high density of the GGERV10 elements on chromosomes 1, 2, and Z. In addition, we calculated the number of GGERV10 insertions per Mbp for each chromosome, and chromosome Z showed the highest insertion/Mbp, shown in Additional file 1: Table S1.

To examine whether the GGERV10 elements have target site preference for their integration, we investigated target site duplications (TSDs) of each of the 532 GGERV10 element including full-length GGERV10 elements and solo-LTRs. TSDs are a hallmark of retrotransposition events. As shown in Additional file 2: Table S2 and Additional file 3: Table S3, there were no target site preferences for GGERV10 insertion.

Diagnostic sequence characteristics between GGERV10 LTRs
To understand the characteristic of full-length GGERV10 elements, we examined the average length of each LTR sequence. Among the GGERV10 subfamilies, GGERV10B showed the longest LTR sequence with an average of 382 bp. In contrast, the LTR sequence of GGERV10A family was shortest and the averaged size was 295 bp (Table 2). We investigated sequence variations in GGERV10 subfamily by comparing LTR sequences of full-length GGERV10 elements. LTR sequences with a deletion more than 50 bp were excluded for this analysis due to a technical difficulty to align them with other LTR elements. Additional file 4: Figure S1 shows the multiple sequence alignment of LTR sequences (Additional file 5). Interestingly,

Table 1 Summary of GGERV10 elements

Classification	Number of loci
Computationally extracted GGERV10 loci	593
Full-length GGERV10 elements	49
Solo-LTR GGERV10 elements	483
Truncated GGERV10 elements	61

Table 2 Characterization of GGERV10 subfamilies

Classification		Copy number	Number of full-length	Number of solo-LTRs	Average length of each LTR subfamily
GGERV10 subfamilies	GGERV10A	27	7	20	295
	GGERV10B	25	13	12	382
	GGERV10C1	117	6	111	329
	GGERV10C2	251	10	241	336
	GGERV10D	112	13	99	332

the full-length GGERV10 elements were divided into two distinct groups, depending on diagnostic sequence characteristics. The first group contained GGERV10A and GGERV10B which shared the 'E' region. However, they were distinguished from each other based on 'A' and 'B' regions. In addition, there was 24-nt duplication (5'-GCGTAGCGAGGGAAACGAGGTGTG-3') in the GGERV10A subfamily.

GGERV10C1, GGERV10C2, and GGERV10D subfamilies were grouped by sharing the 'F' region. We further examined the sequence structure of the second group. The result showed that 'H' region was shared between GGERV10C1 and GGERV10C2 subfamilies while the 'C' region was shared between GGERV10C1 and GGERV10D subfamilies. However, 'D' and 'G' regions were unique in GGERV10C2 and GGERV10D subfamilies, respectively. Interestingly, we found a unique sequence feature on GGERV10_76 and GGERV10_205 elements. For example, the 5' LTR sequence of GGERV10B_76 was matched with the GGERV10D LTR consensus sequence whereas its 3' LTR sequence was matched with the GGERV10B LTR consensus sequence. The 5' LTR sequence of GGERV10C2_205 was matched with the GGERV10C2 LTR consensus sequence whereas its 3' LTR sequence was matched with the GGERV10C1 LTR consensus sequence. Although GGERV10B_76 and GGERV10C2_205 LTR consist of a chimeric structure, we could not find the evidence of a chimeric structure in their body sequence regions (*gag-pro-pol-env*). The GGERV10 elements with a chimeric sequence could be generated by template switching between homologous LTR sequences.

A previous study reported that GGERV10 LTR elements carried fixed dinucleotide terminal inverted repeats, 'TG' and 'CA', in the 5' and 3' end of their LTR sequences [8]. In this study, we identified GGERV10 LTR-specific terminal inverted repeats, 'TGTTG' and 'CAACA' at its 5' and 3' end, respectively, as shown in Additional file 4: Figure S1.

Genetic distance between GGERV10 elements

The time of a proviral integration can be estimated based on LTR divergence and intactness of proviral open reading frames (ORFs) [17]. The comparison of LTR sequences is the standard method to estimate the age of full-length ERV insertion [20]. It is well known that the nucleotide difference between the 5' and 3' LTR sequences

of a single GGERV10 element resulted from point mutations after insertion [21]. Therefore, the nucleotide difference between the 5' and 3' LTR sequences could be used to estimate the ERV insertion time [22]. To estimate the age of the GGERV10 subfamilies, we performed the NET-WORK analysis [23], based on the evolutionary divergence between all LTR sequences of each subfamily (Additional file 6: Table S4). Using a nucleotide mutation rate of 0.19% per million year (myr) [24], the age of each GGERV10 subfamily was calculated and the result showed that GGERV10B is the youngest GGERV10 subfamily; its estimated age was 3.70 myr.

We also tried to reconstruct the phylogenetic relationships between the full-length GGERV10 LTRs, using a neighbor-joining phylogeny. As we expected, the 5' and 3' LTR sequences of each GGERV10 element were highly similar to each other. In addition, our phylogenetic analysis based on 5' and 3' LTR sequences of GGERV10 elements grouped them into five different subfamilies, which is consistent with Repbase data [25] (Fig. 1).

Genomic environment of full-length GGERV10 integration regions

To determine the genomic environment of full-length GGERV10 integration regions, we analyzed the GC content and gene density of genomic regions flanking them (Additional file 2: Table S2). We calculated the GC content in 20-kb windows centered on each GGERV10 locus. The GC content of the flanking regions was, on average, 40.91%, which is lower than the average GC content of the chicken reference genome, 42.92% [26]. It indicates that full-length GGERV10 elements exist in AT-rich regions. We also analyzed the gene density in the 2 Mb of flanking genomic sequences centered on each full-length GGERV10 element. The average gene density of the flanking regions was about 3.83 genes per Mb, which was much lower than that of the chicken genome (an average of 20.41 genes per Mb). The 93.8% (46/49) of full-length GGERV10 elements locate in the intergenic region but only three elements reside in the intronic region. Based on the results, we state that full-length GGERV10 elements preferentially locate in the genomic regions with a high AT content but a low gene density.

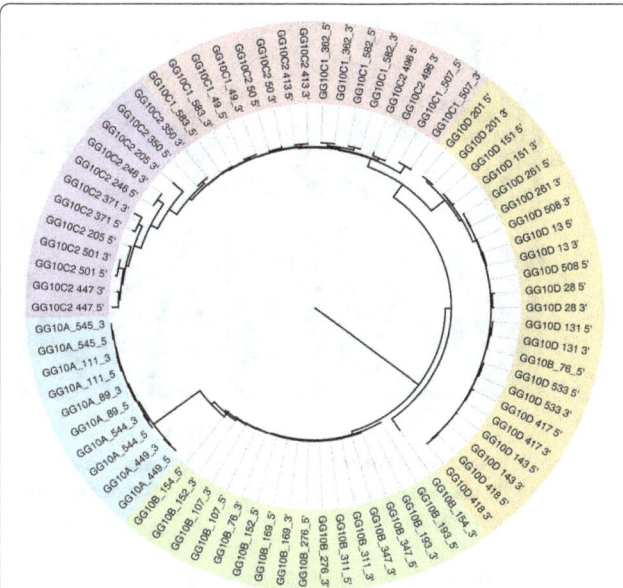

Fig. 1 Phylogenetic relationship between the GGERV10 elements. Based on the LTR sequence, neighbor-joining phylogenetic tree of full-length GGERV10 elements was constructed. Evolutionary distances were constructed using the Kimura 2-parameter method [38]. The result of bootstrap calculations (bootstrap value >70%) based on 1,000 replications is shown. The black bar indicates 0.005 nucleotide substitutions per nucleotide position

Genomic structure of GGERV10 elements

Structurally or functionally intact ERVs contain *gag*, *pro/pol*, and *env* genes but most of the ERVs have not preserved the internal sequences. Over time, integrated ERV copies accumulate nucleotide substitutions or frameshift mutations [27]. In addition, homologous recombination occurs between the two LTRs of each element, leading to a solo-LTR [28].

Using RetroTector10 program [29], we evaluated the genomic structure and function of full-length GGERV10 elements. The program is able to identify open reading frames (ORFs) in chicken ERV elements. The result showed that none of the full-length GGERV10 elements have retained intact *gag*, *pro/pol*, and *env* genes. Most of the full-length GGERV10 elements were deficient in *pro/pol* and *env* genes. The 31 out of the 49 (63.2%) full-length GGERV10 elements retained the primer-binding site (pbs) and *gag* gene. However, 15 (30.6%) full-length GGERV10 elements contained mutations in the *gag* gene, which were frameshift mutations caused either by insertion or deletion, and the remaining three full-length GGERV10 elements had deficient pbs (Additional file 7: Table S5). Interestingly, all GGERV10B elements contained a polypurine tract in the internal *env* gene, which is served as a primer for the synthesis of the second (plus) DNA strand following reverse transcription [30]. In addition, six out of seven GGERV10A elements had an aspartyl protease (PR) in the internal *pro* gene, which

is required for the processing of the Gag precursor, and had a reverse transcriptase in the internal *pol* gene, which is required for reverse transcription of RNA into DNA [31]. Furthermore, we investigated the LTR sequences of full-length GGERV10 elements using TRANSFAC® to identify putative transcription factor binding sites within the LTR sequences. As shown in Additional file 8: Figure S2, the LTR sequences contain 28 different transcription factor binding sites (Additional file 9). The result showed that all of the full-length GGERV10 elements are retrotranspositionally incapable in the chicken genome. However, they might be able to regulate gene expression of the neighboring genes by offering transcription factor binding sites.

Polymorphism of full-length GGERV10 elements

To check for presence/absence polymorphisms of the 49 full-length GGERV10 elements in the 9 chicken genomic DNA samples (3 for the Korean domestic chicken, 3 for Leghorn, and 3 for *Araucana*), we conducted polymerase chain reaction (PCR) amplification of each full-length GGERV10 locus by using the locus-specific designed primers (Additional file 10: Table S6). The result showed that there are three possible states at a GGERV10 locus: absence of the GGERV10 element, presence of the GGERV10 element, and presence of the solo-LTR generated by the homologous recombination between 5′ and 3′ LTRs. 18.4% of full-length GGERV10 elements were polymorphic in the three different chicken breeds of the Korean domestic chicken, Leghorn, and *Araucana*. The polymorphism level was 28.6% (2/7), 46.1% (6/13), and 7.7% (1/13) for GGERV10A, GGERV10B, and GGERV10D, respectively. In contrast, GGERV10C1 and C2 subfamilies showed no polymorphism in the chicken breeds.

Molecular markers for identification of chicken breeds

One of *Araucana*-specific GGERV10A insertions locates in the 5′ flanking region of *SLCO1B3* gene and is responsible for the blue eggshell color in *Araucana*. It suggests that GGERV10 elements could be served as a genetic marker [32]. It suggests the possibility that any of the full-length GGERV10 elements could be breed-specific locus. As our polymorphism test showed that three of the 49 full-length GGERV10 elements, GGERV10B_107, GGERV10B_193, and GGERV10B_311, are polymorphic in the chicken breeds, we further examined them using PCR with 80 chicken-DNA samples from three different chicken breeds (40 Korean domestic chicken, 20 Leghorn, and 20 *Araucana*). Through the PCR assay, we found that GGERV10B_107 and GGERV10B_193 elements are insertionally polymorphic in the 80 chicken-DNA samples (data not shown) while GGERV10B_311 locus had one more state, a deletion event at the pre-insertion site of the element. As shown in Fig. 2, GGERV10B_311 element is Korean domestic chicken

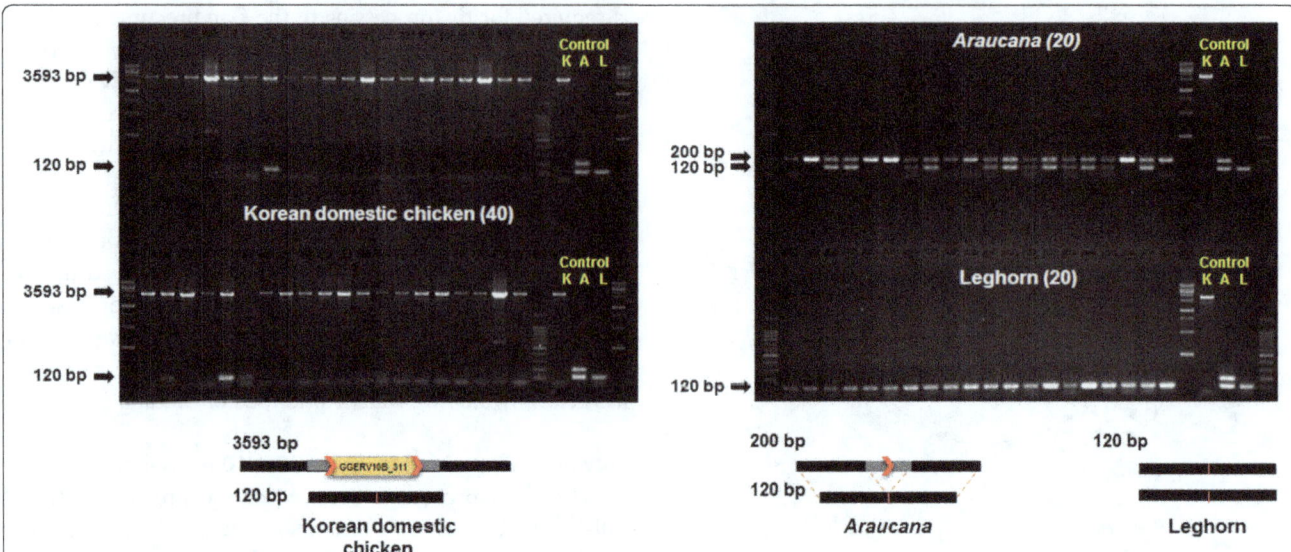

Fig. 2 Polymorphic pattern of the GGERV10B_311 locus in three chicken breeds. PCR amplification was conducted with 80 chicken DNA samples from three different chicken breeds (40 Korean domestic chicken, 20 Leghorn, and 20 *Araucana*). GGERV10B_311 (3,593 bp) insertion was present only in Korean domestic chicken (*left*) and small deletion allele (120 bp) was also detected. Two amplicon of *Araucana* indicates the absence of the GGERV10B_311 element and small deletion. Additionally, leghorn has only small deletion (*right*). Korean domestic chicken (K), *Araucana* (A), and leghorn (L)

breed-specific (Additional file 11: Table S7). In the *Araucana* samples, a polymorphic pattern was observed at the pre-insertion site of GGERV10B_311element; one of the two different PCR products was the expected size for the case where GGERV10B_311 element is absent but the other one was smaller than the expected size. The Leghorn breed produced only one type of the PCR products which were smaller than the expected size for the case without GGERV10B_311 insertion. To verify the unexpected PCR results at the GGERV10B_311 locus, we sequenced the PCR products and performed sequence alignment of the region (Additional file 12). The result found that 80 bp deletion event occurred in the pre-insertion site of GGERV10B_311 element and the GGERV10B_311 element is Korean domestic chicken-specific.

Incomplete lineage sorting events were previously reported to explain genetic polymorphism created by retrotransposons and retrotransposon-mediated deletions between closely related species [33–36]. In this study, a discordant PCR amplification pattern was shown at GGERV10B_311 locus, and incomplete lineage sorting between the three chicken breeds well explains the unexpected PCR result (Fig. 3). As shown in Additional file 13: Figure S3, a 80-bp deletion seemed to occur before the divergence of the Korean domestic chicken, Leghorn, and *Araucana* breeds. After the divergence of *Araucana* and the common ancestor of the Korean domestic chicken and Leghorn, the 80-bp deletion was still polymorphic in all of the three breeds. Then, the

Korean domestic chicken was diverged from Leghorn, and the 80-bp small deletion was finally fixed in the Leghorn. Later, the GGERV10B insertion occurred only in the Korean domestic chicken breed. However, we cannot rule out that Leghorn species is artificially selected in farm due to modern commercial strain. Therefore, the evolution scenario could be modified or strongly supported if more chicken breeds are used in the further experiment.

Conclusions

In this study, we characterized GGERV10 family, one of the youngest GGERV families in the chicken genome. The chicken reference genome contains a total of 593 GGERV10 elements but among them, only 49 elements are full-length. GGERV10 elements are retrotranspositionally inactive in the chicken genome because they are lack of intact genes necessary for the retrotransposition. However, they have a potential to regulate the expression of the neighboring genes as they retain 23 transcription factor binding sites. To identify breed-specific GGERV10 locus, the 49 full-length GGERV10 loci were subjected to a traditional PCR using 80 genomic DNAs isolated from the Korean domestic chicken, Leghorn, and *Araucana* as PCR template. Through the assay, GGERV10B insertion was identified to be Korean domestic breed-specific. This locus could be used to distinguish the Korean domestic chicken from other breeds of Leghorn and *Araucana*. This study supports that TEs including ERVs could be used as a molecular

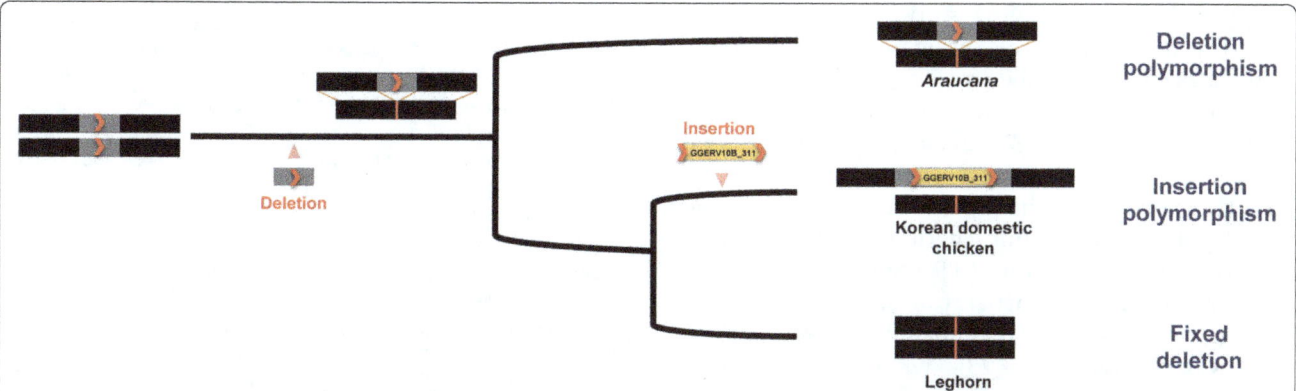

Fig. 3 Schematic of incomplete lineage sorting in the GGERV10B_311 locus. The small deletion occurred before the divergence of the *Araucana* and other breeds and was still polymorphic at the time of speciation. Subsequently, the *Araucana* had maintained deletion polymorphism. After the divergence of the Korean domestic chicken and Leghorn, Korean domestic chicken-specific GGERV10B_311 element insertion event occurred and maintained insertional polymorphism. However, the deletion allele was fixed in the Leghorn species. The *Gray* box, *red* arrow, and *red* line indicate small deletion region, TSD, and deletion point, respectively

marker for species identification due to their virtually homoplasy-free phylogenetic character [37].

Methods
Computational analysis for GGERV10 loci of chicken (*Gallus gallus*)
To identify GGERV10 elements in the chicken genome, we extracted 593 GGERV10 loci from the Chicken reference genome (ICGSC Gallus_gallus-4.0/galGal4; Nov. 2011 assembly) by using UCSC Table Browser utility (http://genome.ucsc.edu/) and then, we identified full-length GGERV10 loci by RepeatMasker (http://www.repeatmasker.org/cgi-bin/WEBRepeatMasker). Finally, a total of 49 full-length GGERV10 loci were analyzed about their genomic features. First, we extracted each 10 kb sequences on 5′ and 3′ flanking region of full-length GGERV10 loci using the Chicken BLAT search Tool (https://genome.ucsc.edu/cgi-bin/hgBlat). Using these sequences, we calculated GC contents based on EM-BOSS GeeCee server (http://emboss.bioinformatics.nl/cgi-bin/emboss/geecee). We also examined the gene density in the flanking sequences of the GGERV10 candidates. Each 2 Mb sequence of both flanking region of each GGERV10 locus was extracted and the number of genes were counted in these sequences using the National Center for Biotechnology Information Map Viewer utility (http://www.ncbi.nlm.nih.gov/mapview/map_search.cgi?taxid=9031&build=102.0).

PCR amplification and sequence analysis
To confirm insertion of GGERV10 identified through computational analysis, we performed PCR in chicken genomic DNA panel. Chicken genomic DNA panel was composed of 9 chicken genomic DNA samples (3 Korean domestic chicken, 3 leghorn, and 3 blue-egg shell chicken). The panel was provided from National Institute of Animal Science (Korea). Oligonucleotide Primer set for PCR amplification of each identified GGERV10 locus was designed through Primer3 (http://bioinfo.ut.ee/primer3-0.4.0/primer3/) and Oligocalc (http://www.basic.northwestern.edu/biotools/oligocalc.html) programs. Primer information is summarized in Additional file 10: Table S6. PCR amplification was performed in 20 μL reaction volume using 10-20 ng template DNA, 200 nM of each oligonucleotide primer, and 10 μL of master mixture of 2X *EF Taq* Pre mix4 (SolGent, Seoul, Republic of Korea) containing DNA polymerase, PCR buffer, dNTP, tracking dye, and 5X Band Doctor™. PCR amplification was carried out by following process: an initial denaturation step of 5 min at 95 °C, followed by 35 cycles of 1 min at 95 °C, 40 sec at the optimal annealing temperature and optimal time depending on PCR product size for extension at 72 °C, followed by a final extension step of 10 min at 72 °C. Bio-rad™ iCycler thermocycler (Biorad, Munich, Germany) was used for PCR amplification. Amplified PCR products were loaded on a 1.5% agarose gel for electrophoresis, stained by EcoDye Nucleic acid staining solution (BIOFACT, Daejeon, Korea), and visualized with UV fluorescence. Four out of 49 GGERV10 candidates contains poly (N) stretches in the chicken sequence. So, these loci were sequenced and determined by using the BigDye Terminator v3.1 Sequencing Kit (Applied Biosystems, FosterCity, CA, USA) through ABI 3500 Genetic analyzer (Applied Biosystems).

Phylogenetic analysis
To perform phylogenetic analysis, GGERV10 subfamily consensus sequences were generated using the module MegAlign available in the DNA Star program (DNA STAR Inc.,Wisconsin). And aligned GGERV10 elements with this consensus sequence using the software BioEdit version 7.0.5.3 (Hall, 1999). Molecular Evolutionary

Genetics Analysis (MEGA) software 6 was used to construct phylogenetic tree using the neighbor-Joining method. Each node of the tree was estimated based on 1000 bootstrap. The bootstrap analysis was performed according to the Kimura-2-parpameter distance (Kimura, 1980).

Furthermore, to estimate evolutional age of each GGERV10 subfamily, full-length GGERV10 subfamilies were aligned based on LTR sequence except a few GGERV10 copies had partial truncated LTR. The putative age of each GGERV10 subfamilies were calculated with NETWORK 4.611 [23]. We used a nucleotide mutation rate of 0.2 ~ 0.26% per site per myr, assuming that ERVs accumulate mutations at the neutral evolution rate after their insertion.

Transcription factor binding site search in GGERV10 LTR

To analyze putative transcription binding sites in consensus sequences of GGERV10 subfamily, we used TRANSFAC® Professional 7.4.1 (http://genexplain.com/transfac/) with threshold 0.95.

RetroTector analysis

RetroTector10 program (http://retrotector.neuro.uu.se/pub/queue.php?show=submit), a platform-independent java program package, was used to investigate genomic structure of full-length GGERV10 candidates in the chicken genome. It includes three basic modules: (i) Prediction of LTR candidates, (ii) Prediction of chains of conserved retroviral motifs fulfilling distance constraints and (iii) Attempted reconstruction of the original retroviral protein sequences, combining alignment, codon statistics, and properties of protein ends.

Additional files

Additional file 1: Table S1. Chromosome density of GGERV10 elements in the chicken genome. (XLSX 13 kb)

Additional file 2: Table S2. Summary of full-length GGERV10 elements. (XLSX 14 kb)

Additional file 3: Table S3. Summary of solo-LTR copies. (XLSX 41 kb)

Additional file 4: Figure S1. Alignment of LTR sequences between the full-length GGERV10 elements. Using the BioEdit program, 5' and 3' LTR sequences from 38 full-length GGERV10 elements were aligned. Shared sequences among five subfamilies were indicated by colored boxes (A, B, C, D, E, F, G, and H). Blue boxes of both ends indicate GGERV10 family-specific terminal inverted repeats of LTR region. Orange boxes in the 'A' region indicate 24-nt duplication. (TIF 11339 kb)

Additional file 5: LTR sequences of the full-length GGERV10 elements. (FAS 34 kb)

Additional file 6: Table S4. Age estimation of full-length GGERV10 elements. (XLSX 10 kb)

Additional file 7: Table S5. Investigation of full-length GGERV10 elements. (XLSX 18 kb)

Additional file 8: Figure S2. Investigation of putative transcription factor binding sites within the LTR sequence. Colored boxes indicate putative transcription factor binding sites in the LTR consensus sequence from GGERV10 subfamilies. Five GGERV10 subfamilies have shared or specific transcription factor binding sites. (TIF 10395 kb)

Additional file 9: LTR consensus sequences of each full-length GGERV10 subfamilies. (FAS 1 kb)

Additional file 10: Table S6. Primer information for the PCR amplification of full-length GGERV10 elements. (XLSX 16 kb)

Additional file 11: Table S7. Condition for the PCR amplification of GGERV10B_311 element. (XLSX 11 kb)

Additional file 12: Sequence alignment of GGERV10B_311 locus in three chicken breeds. (FAS 10 kb)

Additional file 13: Figure S3. Sequence comparison of GGERV10B_311 locus in three chicken breeds. Sequence alignment of GGERV10B_311 locus in three chicken breeds shows complex genomic feature. Purple boxes indicate primer sequences for GGERV10B_311 locus. Each colored box indicates three breeds: Korean domestic chicken (yellow), *Araucana* (green), and leghorn (blue). Blue box presents target site duplication (TSD) sequence. (TIF 8189 kb)

Abbreviations

CR1: Chicken repeat1; *env*: Envelope gene; ERV: Endogenous retrovirus; *gag*: Group-specific antigen; GGERV: *Gallus gallus* endogenous retrovirus; LINE: Long Interspersed element; LTR: Long terminal repeat; MIR: Mammalian interspersed repeat; ORFs: Open reading frames; pbs: Primer-binding site; PCR: Polymerase chain reaction; *pol*: RNA-dependent DNA polymerase gene; *pro*: Protease gene; RT: Reverse transcriptase; TE: Transposable element; TSDs: Target site duplications

Acknowledgments
We thank Dr. J. Lee for his useful comments during preparation of the manuscript.

Funding
This research was supported by the Bio & Medical Technology Development Program of the National Research Foundation (NRF) funded by the Ministry of Science, ICT & Future Planning (2016M3A9B6026776).

Authors' contributions
JL, SM, DK, CC, DO, and KH conceived and designed the experiments. KH, JL, and SM performed the experiments. JL and SM performed the computational analysis. JL, SM, DK, CC, DO, and KH analyzed the data. DK, DO, and KH contributed reagents/materials/analysis tools. JL, SM, DO, and KH wrote the paper. All authors read and approved the final manuscript.

Competing interests
The authors declare that they have no competing interests.

Author details
¹Department of Nanobiomedical Science & BK21 PLUS NBM Global Research Center for Regenerative Medicine, Dankook University, Cheonan 330-714, Republic of Korea. ²Department of Anesthesiology and Pain Management, College of Medicine, Dankook University, Cheonan 330-714, Republic of Korea. ³Department of Neurosurgery, College of Medicine, Dankook University, Cheonan 330-714, Republic of Korea. ⁴Gyeongsangbuk-Do Livestock Research Institute, Yeongju 750-871, Republic of Korea.

References
1. Mandal PK, Kazazian Jr HH. SnapShot: vertebrate transposons. Cell. 2008; 135(1):192–192.e1.
2. Cordaux R, Batzer MA. The impact of retrotransposons on human genome evolution. Nat Rev Genet. 2009;10(10):691–703.
3. Bohne A, Brunet F, Galiana-Arnoux D, Schultheis C, Volff JN. Transposable elements as drivers of genomic and biological diversity in vertebrates. Chromosome Res. 2008;16(1):203–15.

4.   International Chicken Genome Sequencing C. Sequence and comparative analysis of the chicken genome provide unique perspectives on vertebrate evolution. Nature. 2004;432(7018):695–716.

5.   Wong GK, Liu B, Wang J, Zhang Y, Yang X, Zhang Z, Meng Q, Zhou J, Li D, Zhang J, et al. A genetic variation map for chicken with 2.8 million single-nucleotide polymorphisms. Nature. 2004;432(7018):717–22.

6.   Hughes AL, Piontkivska H. DNA repeat arrays in chicken and human genomes and the adaptive evolution of avian genome size. BMC Evol Biol. 2005;5:12.

7.   Mason AS, Fulton JE, Hocking PM, Burt DW. A new look at the LTR retrotransposon content of the chicken genome. BMC Genomics. 2016;17:688.

8.   Huda A, Polavarapu N, Jordan IK, McDonald JF. Endogenous retroviruses of the chicken genome. Biol Direct. 2008;3:9.

9.   Weiss RA. The discovery of endogenous retroviruses. Retrovirology. 2006;3:67.

10. Jern P, Sperber GO, Blomberg J. Use of endogenous retroviral sequences (ERVs) and structural markers for retroviral phylogenetic inference and taxonomy. Retrovirology. 2005;2:50.

11. Spencer TE, Palmarini M. Endogenous retroviruses of sheep: a model system for understanding physiological adaptation to an evolving ruminant genome. J Reprod Dev. 2012;58(1):33–7.

12. Griffiths DJ. Endogenous retroviruses in the human genome sequence. Genome Biol. 2001;2(6):REVIEWS1017.

13. Khodosevich K, Lebedev Y, Sverdlov E. Endogenous retroviruses and human evolution. Comp Funct Genomics. 2002;3(6):494–8.

14. Mun S, Lee J, Kim YJ, Kim HS, Han K. Chimpanzee-specific endogenous retrovirus generates genomic variations in the chimpanzee genome. PLoS One. 2014;9(7):e101195.

15. Feschotte C, Gilbert C. Endogenous viruses: insights into viral evolution and impact on host biology. Nat Rev Genet. 2012;13(4):283–96.

16. Magiorkinis G, Gifford RJ, Katzourakis A, De Ranter J, Belshaw R. Env-less endogenous retroviruses are genomic superspreaders. Proc Natl Acad Sci U S A. 2012;109(19):7385–90.

17. Bolisetty M, Blomberg J, Benachenhou F, Sperber G, Beemon K. Unexpected diversity and expression of avian endogenous retroviruses. MBio. 2012;3(5): e00344-00312.

18. Wang Z, Qu L, Yao J, Yang X, Li G, Zhang Y, Li J, Wang X, Bai J, Xu G, et al. An EAV-HP insertion in 5′ Flanking region of SLCO1B3 causes blue eggshell in the chicken. PLoS Genet. 2013;9(1):e1003183.

19. Bao W, Kojima KK, Kohany O. Repbase Update, a database of repetitive elements in eukaryotic genomes. Mob DNA. 2015;6:11.

20. Jha AR, Nixon DF, Rosenberg MG, Martin JN, Deeks SG, Hudson RR, Garrison KE, Pillai SK. Human endogenous retrovirus K106 (HERV-K106) was infectious after the emergence of anatomically modern humans. PLoS One. 2011;6(5): e20234.

21. Dangel AW, Baker BJ, Mendoza AR, Yu CY. Complement component C4 gene intron 9 as a phylogenetic marker for primates: long terminal repeats of the endogenous retrovirus ERV-K(C4) are a molecular clock of evolution. Immunogenetics. 1995;42(1):41–52.

22. Han GZ. Extensive retroviral diversity in shark. Retrovirology. 2015;12:34.

23. Bandelt HJ, Forster P, Rohl A. Median-joining networks for inferring intraspecific phylogenies. Mol Biol Evol. 1999;16(1):37–48.

24. Zhang G, Li C, Li Q, Li B, Larkin DM, Lee C, Storz JF, Antunes A, Greenwold MJ, Meredith RW, et al. Comparative genomics reveals insights into avian genome evolution and adaptation. Science. 2014;346(6215):1311–20.

25. Kohany O, Gentles AJ, Hankus L, Jurka J. Annotation, submission and screening of repetitive elements in Repbase: RepbaseSubmitter and Censor. BMC Bioinformatics. 2006;7:474.

26. Rao YS, Chai XW, Wang ZF, Nie QH, Zhang XQ. Impact of GC content on gene expression pattern in chicken. Genet Sel Evol. 2013;45:9.

27. Lander ES, Linton LM, Birren B, Nusbaum C, Zody MC, Baldwin J, Devon K, Dewar K, Doyle M, FitzHugh W, et al. Initial sequencing and analysis of the human genome. Nature. 2001;409(6822):860–921.

28. Stoye JP. Endogenous retroviruses: still active after all these years? Curr Biol. 2001;11(22):R914–6.

29. Sperber G, Lovgren A, Eriksson NE, Benachenhou F, Blomberg J. RetroTector online, a rational tool for analysis of retroviral elements in small and medium size vertebrate genomic sequences. BMC Bioinformatics. 2009;10 Suppl 6:S4.

30. Steinbiss S, Willhoeft U, Gremme G, Kurtz S. Fine-grained annotation and classification of de novo predicted LTR retrotransposons. Nucleic Acids Res. 2009;37(21):7002–13.

31. Carre-Eusebe D, Coudouel N, Magre S. OVEX1, a novel chicken endogenous retrovirus with sex-specific and left-right asymmetrical expression in gonads. Retrovirology. 2009;6:59.

32. Oh D, Son B, Mun S, Oh MH, Oh S, Ha J, Yi J, Lee S, Han K. Whole genome re-sequencing of three domesticated chicken breeds. Zool Sci. 2016;33(1):73–7.

33. Han K, Lee J, Meyer TJ, Wang J, Sen SK, Srikanta D, Liang P, Batzer MA. Alu recombination-mediated structural deletions in the chimpanzee genome. PLoS Genet. 2007;3(10):1939–49.

34. Callinan PA, Wang J, Herke SW, Garber RK, Liang P, Batzer MA. Alu retrotransposition-mediated deletion. J Mol Biol. 2005;348(4):791–800.

35. Han K, Lee J, Meyer TJ, Remedios P, Goodwin L, Batzer MA. L1 recombination-associated deletions generate human genomic variation. Proc Natl Acad Sci U S A. 2008;105(49):19366–71.

36. Xue B, He L. An expanding universe of the non-coding genome in cancer biology. Carcinogenesis. 2014;35(6):1209–16.

37. Suh A, Paus M, Kiefmann M, Churakov G, Franke FA, Brosius J, Kriegs JO, Schmitz J. Mesozoic retroposons reveal parrots as the closest living relatives of passerine birds. Nat Commun. 2011;2:443.

38. Tamura K, Peterson D, Peterson N, Stecher G, Nei M, Kumar S. MEGA5: molecular evolutionary genetics analysis using maximum likelihood, evolutionary distance, and maximum parsimony methods. Mol Biol Evol. 2011;28(10):2731–9.

Evolutionary history of LTR-retrotransposons among 20 Drosophila species

Nicolas Bargues and Emmanuelle Lerat[*]

Abstract

Background: The presence of transposable elements (TEs) in genomes is known to explain in part the variations of genome sizes among eukaryotes. Even among closely related species, the variation of TE amount may be striking, as for example between the two sibling species, *Drosophila melanogaster* and *D. simulans*. However, not much is known concerning the TE content and dynamics among other Drosophila species. The sequencing of several Drosophila genomes, covering the two subgenus *Sophophora* and *Drosophila,* revealed a large variation of the repeat content among these species but no much information is known concerning their precise TE content. The identification of some consensus sequences of TEs from the various sequenced *Drosophila* species allowed to get an idea concerning their variety in term of diversity of superfamilies but the used classification remains very elusive and ambiguous.

Results: We choose to focus on LTR-retrotransposons because they represent the most widely represented class of TEs in the *Drosophila* genomes. In this work, we describe for the first time the phylogenetic relationship of each LTR-retrotransposon family described in 20 Drosophila species, compute their proportion in their respective genomes and identify several new cases of horizontal transfers.

Conclusion: All these results allow us to have a clearer view on the evolutionary history of LTR retrotransposons among *Drosophila* that seems to be mainly driven by vertical transmissions although the implications of horizontal transfers, losses and intra-specific diversification are clearly also at play.

Keywords: LTR-retrotransposons, Drosophila, Horizontal transfer, Transposable element dynamics

Background

It is now clearly established that the presence of transposable elements (TEs), which can make up a large and variable proportion of eukaryotic genomes, explains in part the variations of genome sizes [1–3]. Even among closely related species, the variation of TE amount may be striking, as it is the case for the two sibling species, *Drosophila melanogaster* and *D. simulans*. Indeed, it has been shown since a long time that *D. melanogaster* harbors around three times more TEs than *D. simulans* although both species share a lot of similarities like a similar worldwide geographical distribution or the fact that they are almost phenotypically identical [4, 5]. In a previous study, we have analyzed 12 LTR retrotransposons and three non-LTR retrotransposons described in detail in *D. melanogaster* and known to present variations in copy number among natural populations of *D. simulans* [6]. We determined their copy numbers and structures in the related species of the *melanogaster* subgroup *D. simulans*, *D. sechellia*, and *D. yakuba*. Our results showed that *D. melanogaster* appears like a special case among these other drosophila species with a lot of full-length and potentially active copies whereas more ancient and degraded sequences were present in the three other species. This was pointing out the fact that relying only on one genome from one given species is not enough to fully understand the dynamics of TEs in related species.

Not much is known concerning the TE content and dynamics among Drosophila species expected in *D.*

* Correspondence: emmanuelle.lerat@univ-lyon1.fr
CNRS, UMR 5558, Laboratoire Biométrie et Biologie Evolutive, Université de Lyon, Université Claude Bernard Lyon 1, F-69622 Villeurbanne, France

melanogaster. The sequencing of 11 other Drosophila genomes, covering the two subgenus *Sophophora* and *Drosophila*, revealed a large variation of the euchromatic repeat content among these species, going from ~2.7% in *D. simulans* and *D. grimshawi* to ~25% in *D. ananassae* [7]. Some years latter, the sequencing of eight additional species from the same subgenus from the consortium modENCODE (https://www.hgsc.bcm.edu/arthropods/drosophila-modencode-project) was performed but no much information is known concerning the TE content in these last species. The only studies that made the effort to decipher TE dynamics in Drosophila species other than the model species *D. melanogaster* were either centered on particular species like for example on *D. buzzatii* and *D. mojavensis* [8] or on a particular type of TEs like the exploration of the dynamics of mariner DNA transposons [9], Roo and RooA LTR retrotransposons [10] or of DINE-1 elements [11]. The presence of consensus sequences in the Repbase database [12] of TEs from the various sequenced *Drosophila* species is certainly helping to get an idea concerning their variety in term of diversity of families inside these species. However, the classification remains particularly elusive and ambiguous. For example, from the name and annotation, it is not possible to tell the difference between the LTR-retrotransposons BEL1 and BEL-1 from *D. virilis* nor it is possible to consider that the element Gypsy-1 present in *D. rhopaloa* is homologous to the element Gypsy-1 in *D. ficusphila*.

A clearer view on TE evolutionary relationship among *Drosophila* species is thus needed to understand how TEs can be maintained in genomes and what mechanisms make them diversify inside a genome. This is what we have intended to perform in this work and to do so, we choose to focus on LTR-retrotransposons because they are known to usually represent the most widely represented class of TEs in the *Drosophila* genomes for which we have the sequences [7, 13] and because they can be subject to numerous horizontal transfers in these species [6, 14–18]. Moreover, our previous work has confirmed the existence of sequence variants, especially in *D. simulans*, that could have emerged from recombination between closely related families, giving a lead toward a mechanism of formation of new families that remains to be explored [6]. However, to be able to determine such events, it is indispensable to have a clear idea about the evolutionary links among the various families present in the *Drosophila* genomes.

LTR-retrotransposons are one of the main subclasses among the elements transposing by a "copy-and-paste" mechanism via an RNA-intermediate [19]. They possess Long Terminal Repeat (LTR) sequences at their extremities and usually present two open reading frames encoding for the proteins necessary for their transposition, especially the *gag* and *pol* genes. According to the protein domain order found in the *pol* gene, three superfamilies have been described: *Ty1/Copia*, *Ty3/Gypsy*, and *BEL/Pao* [20]. In this work, we have defined new reference sequences corresponding to consensus of families never described until now in the species from the *melanogaster* subgroup using a *de novo* approach and we used, in addition, described reference elements to 1) determine accurately their phylogenetic positions inside each superfamily; 2) detect horizontal transfers events, especially among the *melanogaster* subgroup and identify potential losses and intra-specific diversification events; 3) compute their proportion in their respective genomes. Our results allowed us to determine for each family from the 20 Drosophila species to which exact group inside the superfamilies they belong. Although vertical transmission, along with losses and intra-specific diversity, seem to be the most common scenario to explain the phylogenetic pictures we observed, we also identified some new cases of HTs especially among certain species from the *melanogaster* subgroup and detected some new groups of TEs that are absent from *D. melanogaster*, *D. simulans*, *D. sechellia*, and *D. erecta*. All these results allow us to have a clearer view on the evolutionary history of LTR retrotransposons among these 20 *Drosophila* species.

Results and discussion
Identification of new reference elements in the species from the *melanogaster* subgroup

For the four drosophila species from the *melanogaster* subgroup, we obtained 1501 candidates for *D. yakuba*, 603 for *D. simulans*, 1681 for *D. sechellia*, and 766 for *D. erecta* when using the LTRharvest program on their genome assemblies. We then retained the sequences corresponding to real LTR-retrotransposons, the remaining being false positives. The proportion of these false positives was quite high (between 77 and 85%) compared to what was expected by the use of LTRharvest on the *D. melanogaster* genome [21, 22]. This could be due to the fact that in *D. melanogaster*, the TEs correspond to mainly full-length elements whereas in the other species full-length elements are more rare, as it has been observed when analyzing 12 LTR-retrotransposons [6].

We thus retained 217 sequences in *D. yakuba*, 103 in *D. simulans*, 325 in *D. sechellia*, and 178 in *D. erecta*. For each species, we clustered the sequences in group of families, according to the 80-80-80 rule, for which we constructed a consensus. We thus obtained 54 different consensus (or references) for *D. yakuba*, 46 for *D. simulans*, 59 for *D. sechellia*, and 22 for *D. erecta*. To determine if some of them were already described families we compared them to the consensus present in the Repbase database as well as sequences present in Flybase and

Genbank, or described only in the literature [18]. We considered a consensus to be already described if we were able to find a match with more than 98% nucleotidic identity with a described element in the same species. In the case of *D. yakuba*, 25 consensus appeared to be new families, 43 in *D. simulans*, 57 in *D. sechellia*, and 21 in *D. erecta*. These results indicate that a lot of the identified references correspond to new elements from known families in *D. melanogaster* but never described for in these four species to date. This implies that a still large amount of unknown references need to be discovered and that the databases, and especially Repbase, are not exhaustive for these particular species. Interestingly, a large proportion of them (17 in *D. yakuba*, 25 in *D. simulans*, 32 in *D. sechellia*, and four in *D. erecta*) displayed a very high percentage identity (over 95% on average) with elements from *D. melanogaster*, giving some hints toward potential HTs that need to be explored, since the global % identity is about 93.6% between *D. melanogaster* and both *D. simulans* or *D. sechellia*, and about 68% between *D. melanogaster* and both *D. yakuba* and *D. erecta*.

In total, we thus have identified 141 new reference elements (or families) from these four *Drosophila* species over a total of 206 elements (see Additional file 1 for their fasta sequences), with 76 of them corresponding to elements almost identical to references from *D. melanogaster* (i. e. with a mean % identity over 98% when comparing the entire nucleotidic sequences). Additionally, we found two new elements from the *Ty1/Copia* superfamily in *D. melanogaster* corresponding to COPIA2bis and new_Xanthias, although this genome is particularly well studied and annotated.

Phylogenetic analyses of the three main superfamilies of LTR retrotransposons reveal a dynamics mainly constituted by vertical transmissions, several cases of horizontal transfers but also internal species diversification

We have built reference sequences, representative of a given family, corresponding to consensus obtained from the alignment of copies detected using a *de novo* approach in the four species of the *melanogaster* subgroup. However, this approach did not allow us to retrieve the sequences of all the known LTR retrotransposon families in some species, probably due to the lack of full-length copies for these missing elements. Indeed, a drawback of a *de novo* approach like *LTRharvest* is that it is only able to identify full-length or nearly full-length elements, with two conserved LTRs at each extremity. We thus added the missing families whose sequences were present in Repbase, corresponding to 35 sequences in *D. yakuba*, 15 sequences in *D. simulans*, 10 sequences in *D. sechellia*, and one sequence in *D. erecta* (see Additional

file 2: Table S1 and Table 1 for the number of reference sequences in each superfamily for each species).

For each of the three main superfamilies, *Ty1/Copia*, *BEL/Pao* and *Ty3/Gypsy*, we then reconstructed the phylogeny of the various families based on the pol protein, which contains the most conserved enzymatic domains in the LTR-retrotransposons [19, 20].

The phylogenetic analysis of Ty1/Copia families reveals a history made of a majority of vertical transmissions, with many losses

Families from the *Ty1/Copia* are not the most diversified in the *drosophila* species (Table 1). However, *D. ananassae* and *D. willistoni* present the highest number of different references with respectively eight and nine families, whereas no reference has been identified for *D. erecta* and *D. rhopoloa*. The phylogenetic tree based on the pol protein is represented on Fig. 1. Three major clades, highlighted in pink, green, and yellow, well supported by the bootstrap values, can be separated in several subclades. Some of them correspond to the classical known groups, which are COPIA (in light orange), 1731 (in yellow), and XANTHIAS (in blue green) [20]. We also were able to identify some new subclades that we named COPIABIS in orange, due to its proximity to COPIA, COPIA2 in light blue, COPIA2BIS in light green, and NEW XANTHIAS in dark green. The

Table 1 Number of reference sequences from each superfamily

Species	Ty1/Copia	BEL/Pao	Ty3/Gypsy	Total
D. ananassae	8	22	44	74
D. biarmipes	1	4	14	19
D. bipectinata	3	12	26	41
D. elegans	2	18	44	64
D. erecta	0	2	20	22
D. eugracilis	3	5	14	22
D. ficusphila	1	5	13	19
D. grimshawi	4	3	8	15
D. kikkawai	1	1	5	7
D. melanogaster	6	7	48	61
D. mojavensis	4	9	10	23
D. persimilis	3	14	16	33
D. pseudoobscura	2	7	26	35
D. rhopaloa	0	7	17	24
D. sechellia	6	9	51	66
D. simulans	6	8	44	58
D. takahashi	2	22	20	44
D. virilis	1	6	18	25
D. willistoni	9	21	61	91
D. yakuba	4	12	44	60

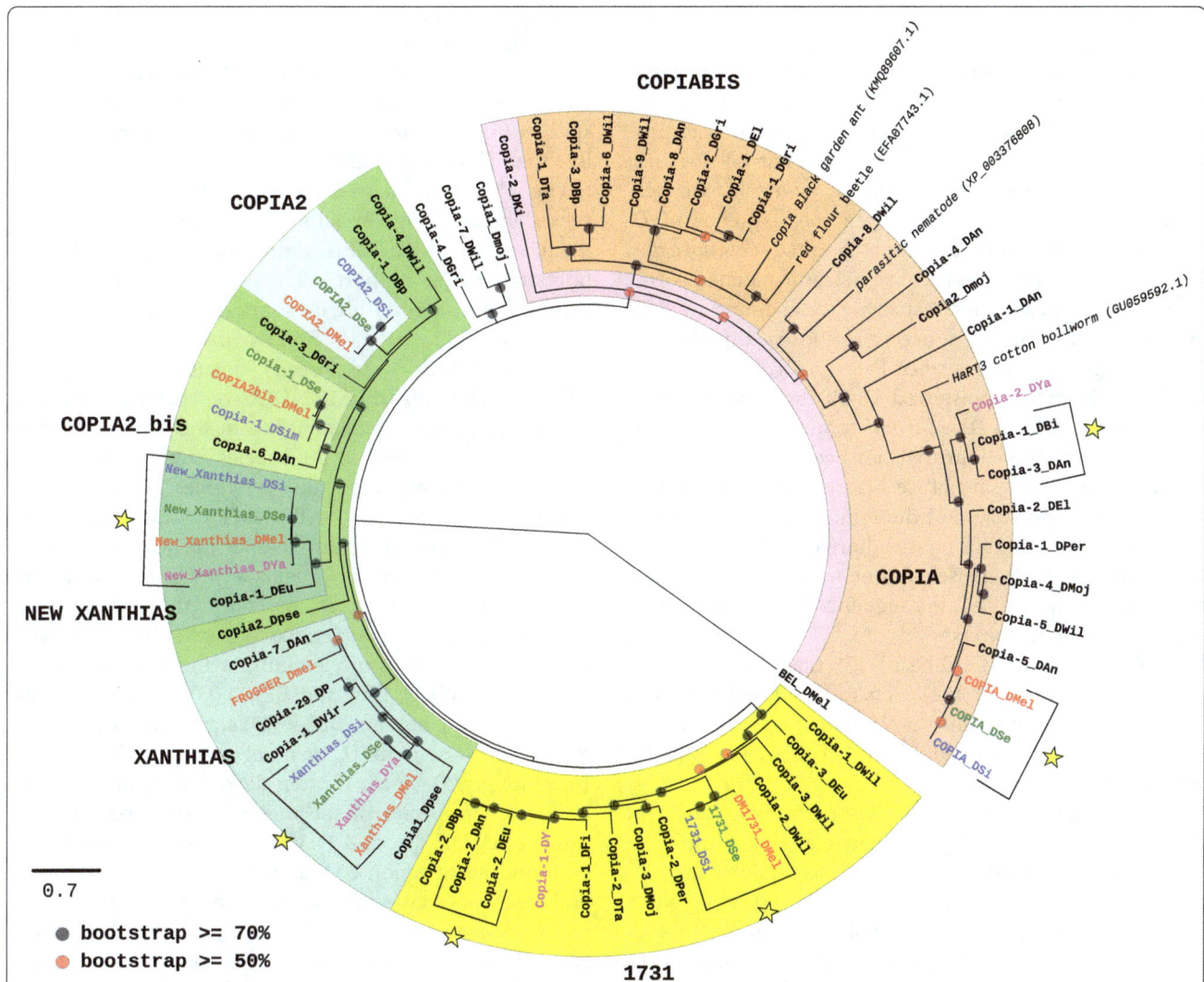

Fig. 1 Maximum likelihood phylogenetic tree based on the polyprotein amino acid sequences of *Ty1/Copia* elements. Only bootstrap values greater than 50% (*red dots*) or greater than 70% (*black dot*) are indicated. The tree has been rooted by the BEL element from *D. melanogaster*. The names of the species are abbreviated as follows: DAn, *D. ananassae*; DBi, *D. biarmipes*; DBp, *D. bipectina*; DEl, *D. elegans*; DEu, *D. eugracilis*; DFi, *D. ficusphila*; DGri, *D. grimshawi*; DKi, *D. kikkawai*; DMel, *D. melanogaster* (*in red*); DMoj/Dmoj, *D. mojavensis*; DPer/DP, *D. persimilis*; Dpse, *D. pseudoobscura*; DSe, *D. sechellia* (*in green*); DSi, *D. simulans* (*in blue*); DTa, *D. takahashi*; DVir, *D. virilis*; DWil, *D. willistoni*; Dya/DY, *D. yakuba* (*in pink*). Four sequences from other organisms are included. *Yellow stars* represent cases of confirmed horizontal transfers (see details in Additional file 3: Figure S1a)

COPIABIS subclade has the particularity to present no sequence from the *melanogaster* subgroup species. The species harboring families from this subclade are restricted to *D. ananassae*, *D. elegans*, *D. bipectinata*, *D. takahashi*, *D. willistoni*, and *D. grimshawi*. This patchy distribution among species from both Drosophila and Sophophora subgenus could indicate that elements from this subclade have been lost in the other species.

An interesting point concerning the families present in the species of the *melanogaster* subgroup is that their nucleotidic sequences are particularly similar, especially among the three species *D. melanogaster*, *D. simulans*, and *D. sechellia*, and sometimes *D. yakuba*. In order to determine if this could be due to HT events, we tested

the hypothesis for the elements COPIA2, COPIA2bis, new_Xanthias, Xanthias, 1731 and COPIA. In the cases of new_Xanthias, Xanthias, 1731 and COPIA, the VHICA method [23] allowed us to confirm HT events between *D. melanogaster* and *D sechellia* (COPIA, 1731, Xanthias and new_Xanthias) and between *D. yakuba* and *D. simulans* (Xanthias and new_Xanthias) (Additional file 3: Figure S1a). Excepted new_Xanthias and Xanthias, these HTs were already documented in previous studies for the same species [14, 17, 23]. Two new cases of HTs were also detected implying species outside the *melanogaster* subgroup for the elements Copia-3_DAn from *D. ananassae* and Copia-1_DBi from *D. biarmipes* on one hand, and the elements Copia-2_DAn from *D. ananassae*, and Copia-2_Deu from *D. eugracilis* on the other hand (Additional file 3: Figure S1a). We however did not observe any HT implying one of the numerous families from *D. willistoni* and the Copia element of *D. melanogaster*, contrary to what was proposed [24]. To make sure that it could not be due to a missing reference in Repbase for *D. willistoni*, we performed a blastn search in the genome sequence of *D. willistoni* using the Copia element from *D. melanogaster*. However, we did not find any significant matches corresponding to a nearly identical sequence as found by PCR approaches [24]. Such a situation is not that unusual. Indeed, the genome sequence of *D. melanogaster* is empty of the horizontally transferred DNA transposon P that was introduced from *D. willistoni* some decades ago [25] simply because the strain that has been sequenced is an old lab strain taken in nature before the HT happened [26]. It is thus possible that the sequenced genome of *D. willistoni* is not harboring the horizontally transferred Copia sequence otherwise present in several other natural populations. Indeed, the sequenced strain of *D. willistoni*, Gd-H4-1, corresponds to a population from Guadeloupe Island (Caribbean) [7] that has not been tested in the work of [24]. A similar observation has been made in South American populations of *D. willistoni* in which no evidence of the HT of Copia was detected [27].

Globally, the pattern of species presence/absence in the phylogenetic tree displayed on Fig. 1 is compatible with a large majority of vertical transmissions for elements from the *Ty1/Copia* superfamily. Indeed, some TEs may have been lost in several of the analyzed species. For example, the lack of *Ty1/Copia* sequences in *D. erecta* was confirmed by the blastn searches using reference sequences from the other species from the *melanogaster* subgroup on its genome sequence (Additional file 4: Figure S2a). Some families are also absent from *D. yakuba* (1731, COPIA2bis, and Frogger, see Additional file 4: Figure S2a). The Copia-2_DYa from *D. yakuba* displayed hits of degraded fragments present in *D. melanogaster*, *D. simulans*, and *D. sechellia*, whereas the Copia-1_DY element is present in these species. Thus, it is likely that the first element was lost in the other species of the *melanogaster* subgroup, *D. yakuba* excepted.

The phylogenetic analysis of BEL/Pao reveals several cases of intra-species diversification and a majority of vertical transmissions among Drosophila families

This group has been shown to be reduced to metazoan species contrary to other LTR-retrotransposon groups, which suggests that it could have arisen early in the metazoan evolution [20, 28]. The families from this group are more numerous than those from the *Ty1/Copia* superfamily (Table 1). All species harbor several

families from this superfamily, the species with the most numerous number of families being *D. ananassae*, *D. takahashii*, and *D. willistoni*, with respectively 22, 22 and 21 families, which is not particularly the case for the species from the *melanogaster* subgroup, which contain less than a dozen of families.

The phylogenetic tree based on the pol protein of these families (Fig. 2) allowed to distinguish the two main known clades BEL and PAO [20, 28] with high bootstrap value supports. The BEL clade can also be subdivided in several highly supported subclades among which DIVER2, DIVER, BATUMI/MAX, ROO/ROOA and BEL, which were already documented [28], and BELMONDO and BELMONDO2 representing two new subclades. Reference elements from these two new subclades are not present in the species from the *melanogaster* subgroup excepted three families present in *D. yakuba* that belong to the BELMONDO2 subclade. We checked by blastn searches whether the absence of homologous elements in the other species of the *melanogaster* subgroup was real or only the reflexion of unidentified complete reference sequences. We were able to detect traces of elements in the other species in the case of BEL-3_DYa and BEL-4_DYa (Additional file 4: Figure S2b) but for BEL-5_DYa, no homologous sequence is present in *D. melanogaster* and *D. sechellia*. Interestingly, *D. erecta* is often devoid of families present in the other species of the *melanogaster* subgroup (Batumi, BEL, DIVER, and DIVER2) or only remnants can be found in its genome (Max and Ninja). There are several cases of what could be considered as recent emergences of new families inside a species (Fig. 2). A recent emergence corresponds to a clade of several different families inside a given species. This indicates, like for paralogous genes inside host gene families, that diversification events appeared after speciation events, inside the considered species. All these events are restricted to three species corresponding to those with the highest number of families: *D. ananassae* (BEL-21_DAn and BEL-22_DAn; BEL-19_DAn and BEL-2_DAn; BEL-10_DAn, BEL-11_DAn and BEL-12_DAn; BEL-6_DAn and BEL-18_DAn), *D. willistoni* (BEL-11_DWil and BEL-19_DWil) and *D. takahashii* (BEL-4_DTa and BEL-19_DTa). These elements could correspond to sequence variants, as it was observed for several LTR-retrotransposons in *D. simulans* and *D. sechellia* [6].

Some cases of HTs have been verified and mainly concern species from the *melanogaster* subgroup. Indeed, we were able to validate recent HTs between *D. yakuba* and *D. simulans* (BEL, DIVER, DIVER2, and Max), between *D. melanogaster* and *D. sechellia* (BEL and DIVER), more ancient HTs between an ancestor of *D. yakuba* and an ancestor of *D. sechellia*/*D. simulans* (Ninja), and between ancestor of *D. yakuba* and an

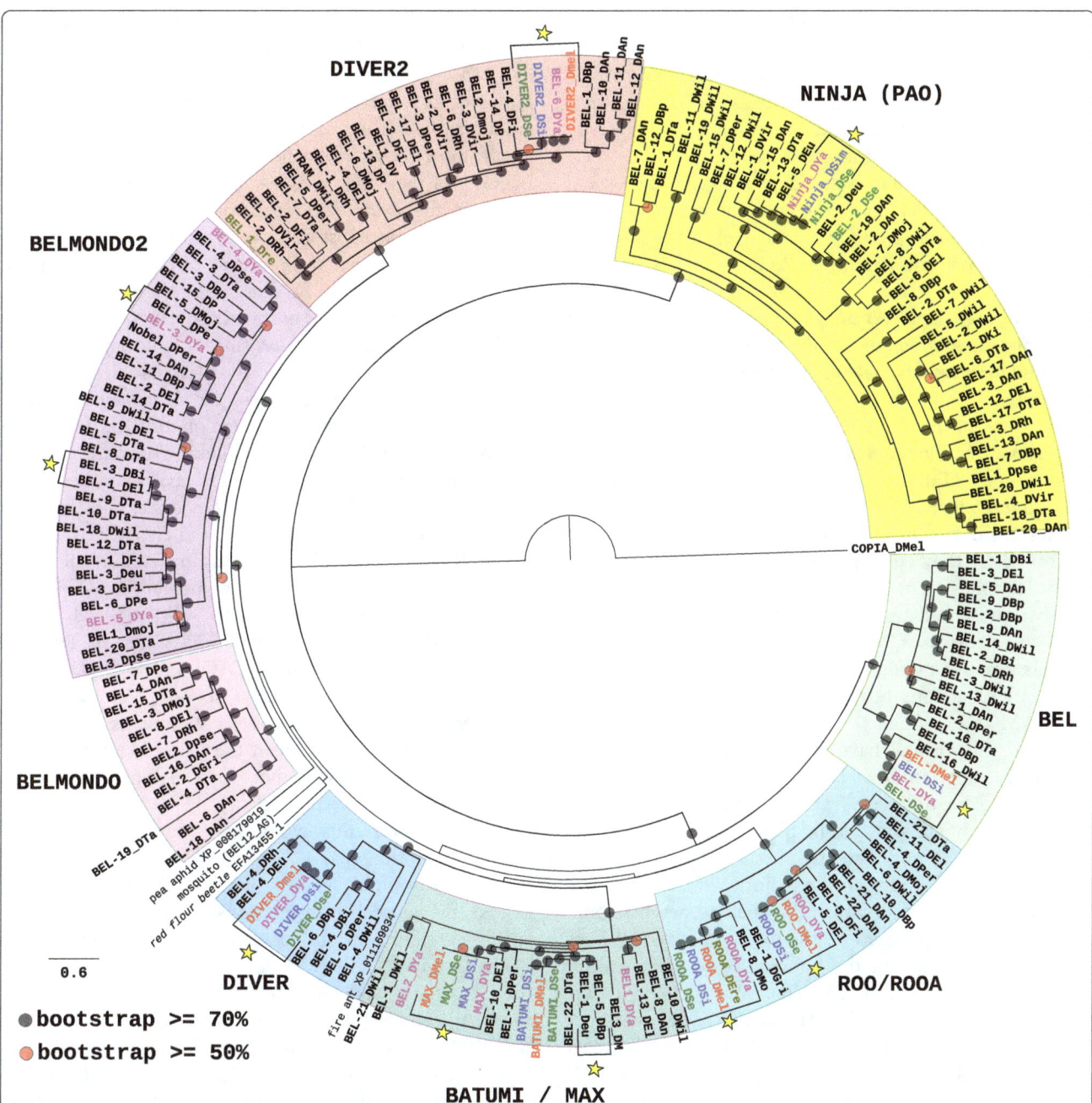

Fig. 2 Maximum likelihood phylogenetic tree based on the polyprotein amino acid sequences of *BEL/Pao* elements. Only bootstrap values greater than 50% (*red dots*) and greater than 70% (*black dot*) are indicated. The tree has been rooted by the COPIA element from *D. melanogaster*. The names of the species are abbreviated as follows: DAn, *D. ananassae*; DBi, *D. biarmipes*; DBp, *D. bipectina*; DEl, *D. elegans*; DEre, *D. erecta* (*in yellow*); DEu/Deu, *D. eugracilis*; DFi, *D. ficusphila*; DGri, *D. grimshawi*; DKi, *D. kikkawai*; DMel, *D. melanogaster* (*in red*); DMir, *D. miranda*; DMoj/Dmoj/Dmo/DM, *D. mojavensis*; DPer/Dpe/DP, *D. persimilis*; DPse/Dpse, *D. pseudoobscura*; DRh, *D. rhopaloa*; DSe, *D. sechellia* (*in green*); DSi, *D. simulans* (*in blue*); DTa, *D. takahashi*; DVir/DV, *D. virilis*; DWil, *D. willistoni*; DYa, *D. yakuba* (*in pink*). Four sequences from other insects are included. *Yellow stars* represent cases of confirmed horizontal transfers (see details in Additional file 3: Figure S1b)

ancestor of *D. sechellia/D. simulans/D. melanogaster* (ROO and ROOA) (Additional file 3: Figure S1b). Some of these events have been previously documented concerning ROO, BEL, Max, DIVER, and DIVER2 [10, 14, 16, 17]. However, the HTs of Ninja between the ancestors of *D. yakuba* and *D. sechellia/D. simulans* and of

ROOA between the ancestors of *D. yakuba* and the other four species of the *melanogaster* subgroup were not documented before. We also detected three new cases of HTs implicating *D. yakuba* and *D. persimilis* (BEL-8_DPer, and BEL-3_DYa elements), *D. biarmipes* and *D. elegans* (BEL-3_DBi and BEL-1_DEl elements), and *D. eugracilis* and *D. bipectinata* (BEL-1_DEu and BEL-5_DBp elements) (Additional file 3: Figure S1b; Fig. 2). Then, although we can confirm HT events for some of the elements from the *BEL/Pao* group, their number is not very important compared to the diversity of sequences present in these species.

All these results indicate that duplications and losses of elements but not HT may have been the main drivers of the evolution of this group of TEs to explain the observed phylogenetic patterns among the 20 Drosophila species.

The phylogenetic analyses of Ty3/Gypsy families underline a highly diversified group of families in which many HT are identified for elements the most closely related to retroviruses

Elements from the *Ty3/Gypsy* superfamily correspond to the most diversified and numerous TEs in the *Drosophila* genomes (Table 1). Their diversity is particularly striking in *D. willistoni*, *D. ananassae*, *D. elegans* and the species from the *melanogaster* subgroup (excepted *D. erecta*) where they correspond to between 41 and 61 different families. The *Ty3/Gypsy* elements, which are widely represented among the eukaryotes, are closely related to retroviruses, some of them being even considered as real retroviruses like the Gypsy element, or at least as endogenous retroviruses like the elements Tirant, ZAM and Idefix in *D. melanogaster* [29–32].

Several big groups have been identified among the *Ty3/Gypsy* superfamily of Drosophila [20]. They have in common to be different from the *Ty3/Gypsy* chromoviruses, which are present in plants, fungi and vertebrates. Since it is not currently possible to determine by the name of the families in the majority of the Drosophila species the group to which they belong, we first built a phylogenetic tree based on the pol proteins of all families to have an idea of the boundaries of these groups (Additional file 5: Figure S3). This allowed us to define three groups (Group 1 "OSVALDO/ULYSSES", Group 2 "MICROPIA/SACCO", and Group 3 "errantiviridae/412") for which we built three separated phylogenetic trees in order to have more resolved nodes and high statistical bootstrap values.

The phylogenetic tree based on the pol protein of families from the Group 1 "OSVALDO/ULYSSES" presents five main subgroups supported by strong bootstrap values (Fig. 3). Three of them correspond to already known clades (OSVALDO, ULYSSES, and ISIS), whereas the two last correspond to new clades (OSIRIS and ISIS-like). Originally, the elements Osvaldo [33] and Isis [34] were first described in *D. buzzati*, a species closely related to *D. mojavensis* from the repleta group, and Ulysses was first described in *D. virilis* from the same species complex [35]. This may explain why the species from the *melanogaster* subgroup are not well represented in Group 1 since only one family exists

for *D. melanogaster* (GYPSY12), two for *D. sechellia* (GYPSY12_Dse and GYPSY6_Dse), two for *D. simulans* (GYPSY12_Dsi and Gypsy-13_Dsim), and none for *D. erecta*. However, *D. yakuba* possess 14 families, that are distributed among each of the clades, indicating that they do not originate from recent diversification inside this species. The majority of these elements are absent from the other species from the *melanogaster* subgroup, which could indicate that they have been lost in these last species (Additional file 4: Figure S2c). Indeed, the presence of these families in *D. yakuba* do not seem to be due to HT events, at least not with the species analyzed in this work. Then, these families could be present in *D. yakuba* since a long time. All other Drosophila species harbor more or less families from this group (from two in *D. biarmipes* and *D. fichusphila*, to 17 in *D. ananassae*) with the exception of *D. kikkawai* in which no family has been identified. The only two HT events that were confirmed in this analysis concern the GYPSY12 elements between *D. melanogaster* and *D. sechellia*, and the elements Gypsy-1_Deu and Gypsy-22_DAn between *D. eugracilis* and *D. ananassae* (Additional file 3: Figure S1c). Several cases of recent emergences of new families inside a species can be pointed out (Fig. 3). They concern *D. willistoni* (Gypsy-1_DWil and Gypsy-61_DWil in the OSALDO clade, Gypsy-52_DWil and Gypsy-34_DWil in the ISIS clade), *D. persimilis* (Gypsy-7_DPer and Gypsy-11_DPer in the OSIRIS clade), *D. ananassae* (Gypsy-4_DAn and Gypsy-17_DAn in the OSVALDO clade) and *D. mojavensis* (Gypsy1_Dmoj, Gypsy4_Dmoj and Gypsy6_Dmoj in the OSVALDO clade).

To summarize, the evolutionary history of the elements from the group "OSVALDO/ULYSSES" among the 20 Drosophila species seems to be mainly represented by vertical transmissions, with cases of intraspecific duplications and losses but almost no HT.

The Group 2, "MICROPIA/SACCO", can be separated into two main clades in the phylogenetic tree based on the pol proteins (Fig. 4), one grouping elements of the MICROPIA/MDG3 type and the other corresponding to a new clade that we named SACCO. Two known subclades (BLASTOPIA and BICA), well supported by strong bootstrap values, are present near the two subclades MDG3 and MICROPIA. All Drosophila species harbor elements from this Group, *D. grimshawi* excepted. However, the number of families greatly varies from one species to another, going from two in *D. erecta*, to 35 in *D. willistoni*. In the species from the *melanogaster* subgroup, they are moderately numerous in *D. sechellia* and *D. yakuba* (with respectively 15 and 19 families) but less abundant in *D. melanogaster* and *D. simulans* (eight families in each). Interestingly, the families present in *D. yakuba* do not often have homologs in the other species from the *melanogaster* subgroup (Additional file 4: Figure S2c). Several families from *D.*

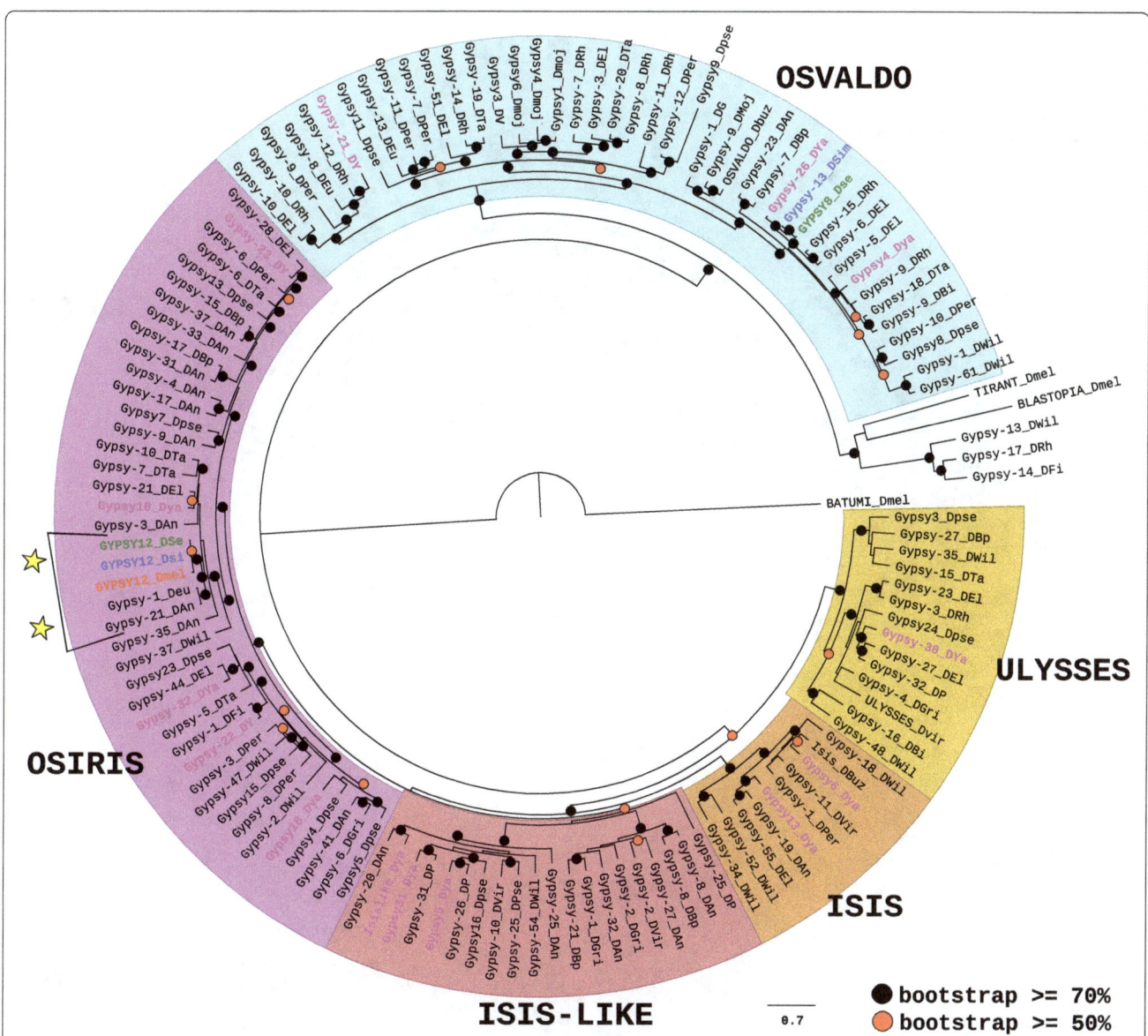

Fig. 3 Maximum likelihood phylogenetic tree based on the polyprotein amino acid sequences of *Ty3/Gypsy* elements from the group "OSVALDO/ULYSSES". Only bootstrap values greater than 50% (*red dots*) and greater than 70% (*black dot*) are indicated. The tree has been rooted by the Batumi element from *D. melanogaster* and we also added elements from the two other groups of *Ty3/Gypsy* (Tirant and BLASTOPIA from *D. melanogaster*). The names of the species are abbreviated as follows: DAn, *D. ananassae*; DBi, *D. biarmipes*; DBp, *D. bipectina*; Dbuz, *D. buzzatti*; DEl, *D. elegans*; DEu/Deu, *D. eugracilis*; DFi, *D. ficusphila*; DGri/DG, *D. grimshawi*; DKi, *D. kikkawai*; DMel, *D. melanogaster* (*in red*); DMoj/Dmoj, *D. mojavensis*; DPer/DP, *D. persimilis*; DPse/Dpse, *D. pseudoobscura*; DRh, *D. rhopaloa*; DSe, *D. sechellia* (*in green*); DSi, *D. simulans* (*in blue*); DTa, *D. takahashi*; DVir/DV, *D. virilis*; DWil, *D. willistoni*; DYa/DY, *D. yakuba* (*in pink*). *Yellow stars* represent cases of confirmed horizontal transfers (see details in Additional file 3: Figure S1c)

melanogaster (seven families), *D. simulans* (four families), *D. sechellia* (seven families) and *D. yakuba* (two families) seem to be implicated in cases of HT events among the *melanogaster* subgroup (See yellow stars in the Fig. 4 and results from VHICA analyses displayed in Additional file 3: Figure S1c). Except the HT event concerning the blastopia element between *D. melanogaster* and *D. sechellia*, these HT events were already documented before [16, 17, 23, 36]. Among all the families, only one case of intra-

specific diversification can be observed for the elements Gypsy-59_DWil and Gypsy-16_DWil.

Globally, elements from this group have mainly a history of vertical transmissions with few cases of HT identified that occurred between *D. melanogaster* and *D. sechellia*, and *D. yakuba* and *D. simulans*.

The Group 3 "errantiviridae/412" is the largest of all three groups by the number of families present in the analyzed species. Only one species, *D. persimilis*, is

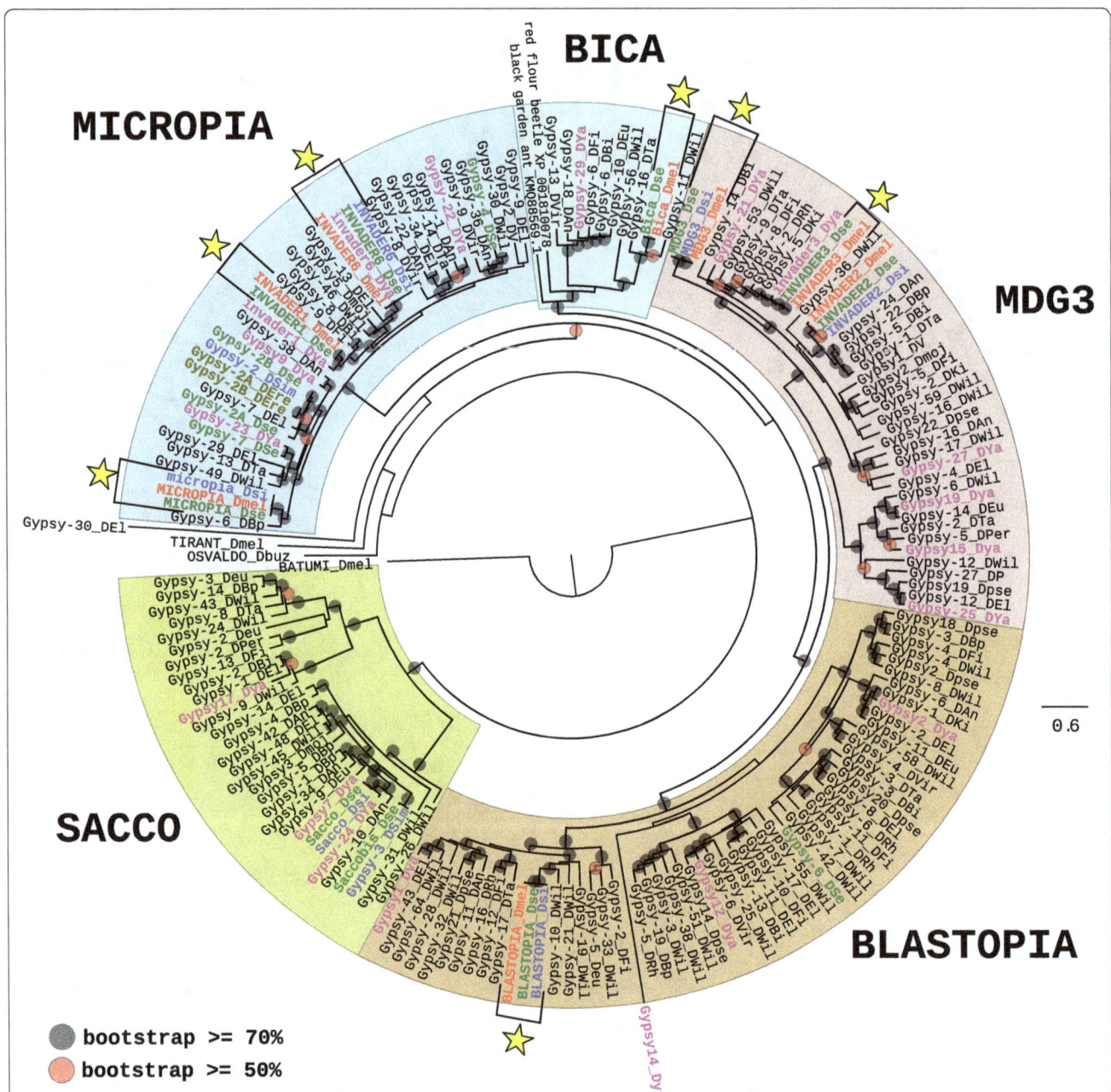

Fig. 4 Maximum likelihood phylogenetic tree based on the polyprotein amino acid sequences of *Ty3/Gypsy* elements from the group "MICROPIA/SACCO". Only bootstrap values greater than 50% (*red dots*) and greater than 70% (*black dot*) are indicated. The tree has been rooted by the Batumi element from *D. melanogaster* and we also added elements from the two other groups of *Ty3/Gypsy* (Tirant from *D. melanogaster* and Osvaldo from *D. buzzati*). The names of the species are abbreviated as follows: DAn, *D. ananassae*; DBi, *D. biarmipes*; DBp, *D. bipectina*; Dbuz, *D. buzzatti*; DEl, *D. elegans*; DEre, *D. erecta*; DEu/Deu, *D. eugracilis*; DFi, *D. ficusphila*; DMel, *D. melanogaster* (*in red*); Dmoj, *D. mojavensis*; DPer/DP, *D. persimilis*; Dpse, *D. pseudoobscura*; DRh, *D. rhopaloa*; DSe, *s. sechellia* (*in green*); DSi, *D. simulans* (*in blue*); DTa, *D. takahashi*; DVir/DV, *D. virilis*; DWil, *D. willistoni*; DYa/Dya, *D. yakuba* (*in pink*). Two sequences from other insects are included. *Yellow stars* represent cases of confirmed horizontal transfers (see details in Additional file 3: Figure S1d)

devoid of elements from this type. The number of families is however quite variable, going from only one in *D. ficusphila* or two in *D. kikkawai* and *D. rhopoloa*, to 32 in *D. melanogaster* and *D. simulans* or 34 in *D. sechellia*. Some of the families present in the three last species are not always present in *D. yakuba* and *D. erecta* like for example ACCORD, Pifo, and QUASIMODO. The elements Gypsy-8_Dsim and Gypsy-5_DSe seem also to be usually absent from the other species of the *melanogaster* subgroup, excepted *D. simulans* (for Gypsy-5_DSe) and *D. sechellia* (for Gypsy-8_Dsim) (Additional file 4: Figure S2c). In these cases, it is likely that these elements have been lost in the species where they cannot be found. The phylogenetic tree based on the pol proteins displayed four known clades: CHIMPO, 412/MDG1, 17.6, and GYPSY (Fig. 5). Excepted in the CHIMPO clade, the species from the *melanogaster* subgroup possess several families inside each clade. A large number of families from these species are involved in HT events (see yellow stars in the Fig. 5 and Additional file 3: Figure S1d, e, and f).

Several of them correspond to already described events in other works [6, 14, 16–18, 36, 37]. However, for some of the previously described elements involved in HTs among *D. melanogaster*, *D. simulans* and *D. yakuba* (Chimpo, Tabor and Chouto) we found that *D. sechellia* but also *D. ananassae* in the case of the Chouto element, may also be involved in HTs (Fig. 5, Additional file 3: Figure S1d, e, and f). We also detected new cases of HT events implicating species of the *melanogaster* subgroup like between *D. melanogaster* and *D. sechellia* (QUASIMODO2), *D.*

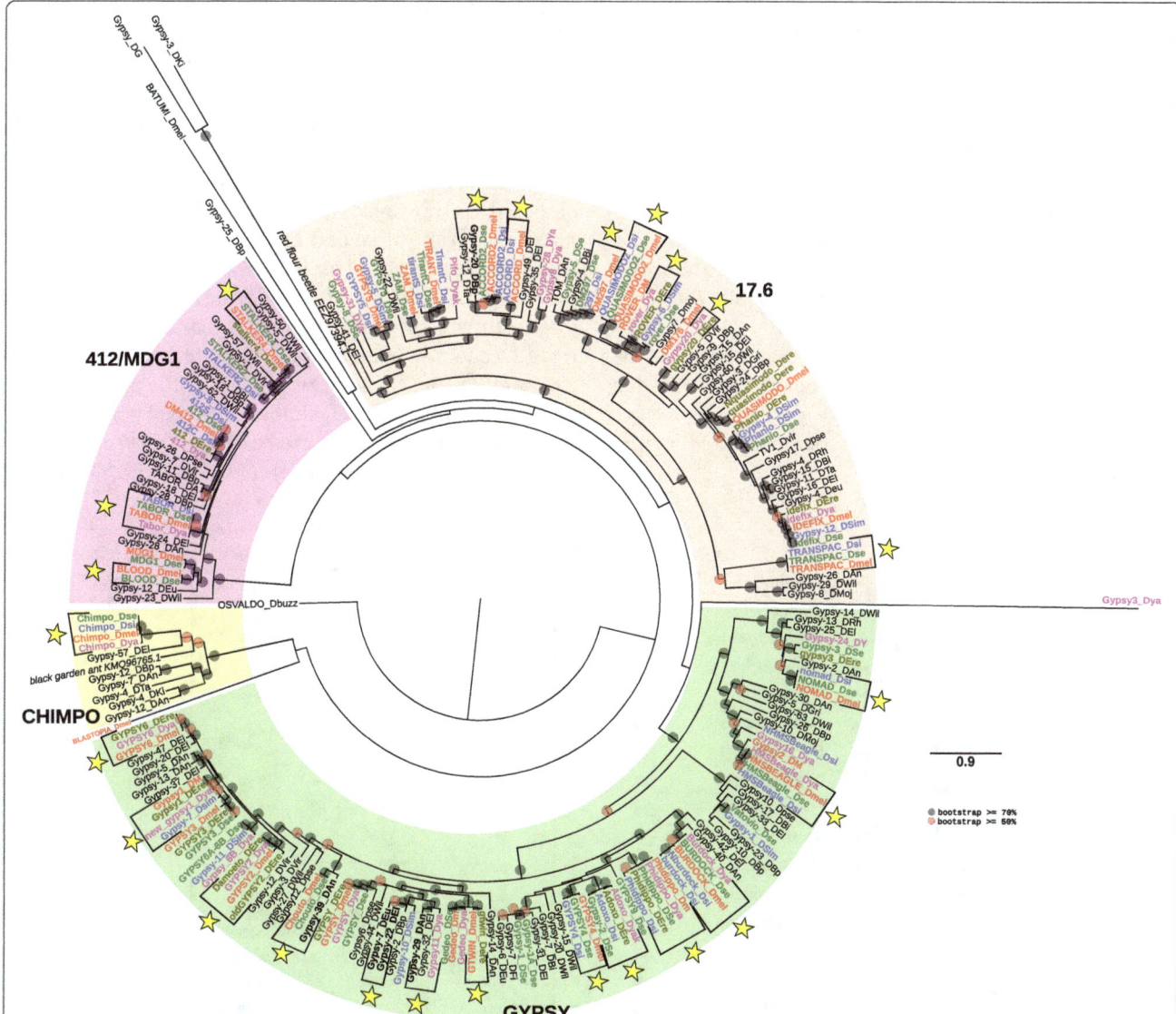

Fig. 5 Maximum likelihood phylogenetic tree based on the polyprotein amino acid sequences of *Ty3/Gypsy* elements from the group "errantiviridae/412". Only bootstrap values greater than 50% (*red dots*) and greater than 70% (*black dot*) are indicated. The tree has been rooted by the Batumi element from *D. melanogaster* and we also added elements from the two other groups of *Ty3/Gypsy* (Blastopia from *D. melanogaster* and Osvaldo from *D. buzzati*). The names of the species are abbreviated as follows: DAn, *D. ananassae*; DBi, *D. biarmipes*; DBp, *D. bipectina*; Dbuz, *D. buzzatti*; DEl, *D. elegans*; DEre, *D. erecta*; DEu/Deu, *D. eugracilis*; DFi, *D. ficusphila*; DGri/DG, *D. grimshawi*; DKi, *D. kikkawai*; Dmel/DM, *D. melanogaster* (*in red*); DMoj/Dmoj, *D. mojavensis*; Dpse, *D. pseudoobscura*; DRh, *D. rhopaloa*; Dse, *D. sechellia* (*in green*); Dsi, *D. simulans* (*in blue*); DTa, *D. takahashi*; DVir, *D. virilis*; DWil, *D. willistoni*; DY/Dya, *D. yakuba* (*in pink*). Two sequences from other insects are included. *Yellow stars* represent cases of confirmed horizontal transfers (see details in Additional file 3: Figure S1d and e)

melanogaster and *D. erecta* (gtwin), *D. melanogaster* and *D. yakuba* (Damoeto/GYPSY2), *D. yakuba* and *D. erecta* (gypsy20_Dya/gypsy20_DEre, rover and adoxo), *D. yakuba* and *D. ananassae* (Gypsy11_Dya/Gypsy-29_DAn). We also detected a case of HT between *D. elegans* and *D. eugracilis* (Gypsy-22_DEl and Gypsy-7_DEu), and between the ancestor of *D. bipectinata* and the ancestor of *D. melanogaster/D. simulans/D. sechellia* (ACCORD2/Gypsy-20_DBp). Based on the phylogenetic tree displayed in Fig. 5, we can observe four cases of intra-specific diversifications that happened in *D. willistoni* (Gypsy-5_DWil/Gypsy-50_DWil in the 412/MDG1 clade), and inside the GYPSY clade in *D. elegans* (Gypsy-47_DEl/Gypsy-20_DEl), in *D. ananassae* (Gypsy-5_DAn/Gypsy-13_DAn), and *D. bipectinata* (Gypsy-23_DBp/Gypsy-10_DBp).

In summary for the families of the group 3 "errantiviridae/412", the evolutionary history of these elements seem to have implied a substantial amount of HTs at least among the species from the *melanogaster* subgroup but also a majority of vertical transmissions, losses and few intra-specific events of diversification in the other species. The elements from this group being the most similar to retroviruses compared to other LTR-retrotransposons, it is thus possible than they may be more prone for HT than other type of elements due to their capacity to form virus-like particles or by being in some cases infectious, as demonstrated in the *melanogaster* subgroup [38].

Proportion of LTR retrotransposons is highly variable among the species but is not directly associated with genome size

We determined the proportion of each superfamily of LTR-retrotransposons in the assemblies of each species. The results are presented in the Fig. 6. We can see that for all species, elements from the *Ty1/Copia* superfamily are the less abundant, followed by the elements from the *BEL/Pao* superfamily, the elements from the *Ty3/Gypsy* superfamily being the most abundant. *D. ananassae* and *D. persimilis* are the species presenting the highest content of *BEL/Pao* elements with respectively 6.64 and 4.22%. Concerning the *Ty3/Gypsy* type elements, they are particularly abundant in *D. sechellia* (9.36%), *D. grimshawi* (9.77%) and *D. ananassae* (12.44%). In the case of the last species, the global abundance of LTR-retrotransposons (19.5% in total) is in agreement with the global estimate of repeats found in this assembly [7] or based on raw reads [39] although for this last estimate, the proportion of LTR-retrotransposons is lower that what we found. The values are more surprising in the case of *D. sechellia* and *D. grimshawi* for which the total estimate of repeat content originally described in the 12 genomes manuscript was quite low (respectively 3.67 and 2.84% [7]). Interestingly, in both cases, the proportion we observed is almost entirely attributable to only one family: Tabor in the case of *D. sechellia* (representing 3.15% of the genome) and Gypsy-5_DGri in the

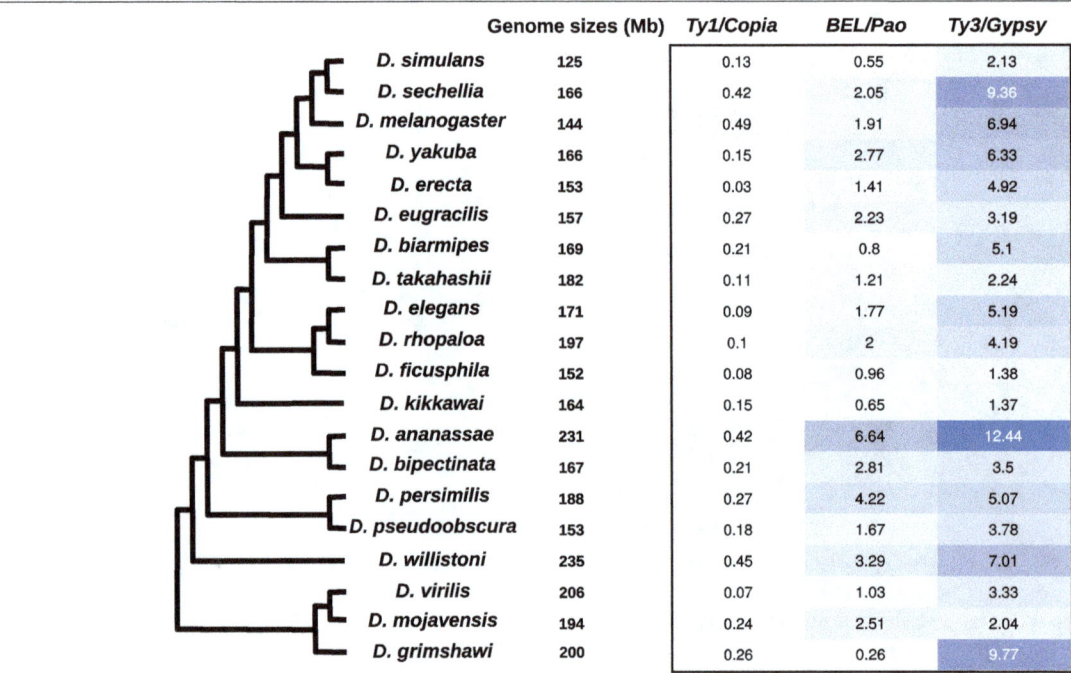

	Genome sizes (Mb)	Ty1/Copia	BEL/Pao	Ty3/Gypsy
D. simulans	125	0.13	0.55	2.13
D. sechellia	166	0.42	2.05	9.36
D. melanogaster	144	0.49	1.91	6.94
D. yakuba	166	0.15	2.77	6.33
D. erecta	153	0.03	1.41	4.92
D. eugracilis	157	0.27	2.23	3.19
D. biarmipes	169	0.21	0.8	5.1
D. takahashii	182	0.11	1.21	2.24
D. elegans	171	0.09	1.77	5.19
D. rhopaloa	197	0.1	2	4.19
D. ficusphila	152	0.08	0.96	1.38
D. kikkawai	164	0.15	0.65	1.37
D. ananassae	231	0.42	6.64	12.44
D. bipectinata	167	0.21	2.81	3.5
D. persimilis	188	0.27	4.22	5.07
D. pseudoobscura	153	0.18	1.67	3.78
D. willistoni	235	0.45	3.29	7.01
D. virilis	206	0.07	1.03	3.33
D. mojavensis	194	0.24	2.51	2.04
D. grimshawi	200	0.26	0.26	9.77

Fig. 6 Proportion (in %) in the genomes of the 20 Drosophila species of each superfamily of LTR retrotransposons. The intensity of the *blue colors* is proportional to the TE proportion. The species are presented according to the phylogenetic tree topology as proposed by Seetharam & Stuart 2013, and we have indicated the genome sizes of each sequenced genome

case of *D. grimshawi* (representing 8.62% of the genome). Since both families have been recently described (it is a new reference sequence described in this work for Tabor in *D. sechellia* and the one of *D. grimshawi* has been described in Repbase in 2011 after the publication of the genome sequence), it is possible that they were not detected by the initial TE annotation performed on the first assemblies in 2007.

For all the other species, we did not observe particular individual families with a high proportion. The global proportion of LTR-retrotransposons is rather the result of the cumulative sum of numerous different families. Indeed, we observed a significant positive correlation between the genome size (indicated in Fig. 6) and the number of references (i.e. the number of families as indicated in the last column of Table 1) present in a given species (Spearman correlation test $r = 0.56$, p-value = 0.01026). This is in contradiction with a previous work for which no correlation was observed between the genome size and the TE diversity among various eukaryotes [40]. However, in this last study, very distant organisms were considered going from fungi to animals, whereas in our case, all the considered species diverged at most 40 Myr ago. Interestingly, we did not observe a significant correlation between the genome size (Fig. 6) and the proportion of LTR-retrotransposons (Spearman correlation test $r = 0.41$, p-value = 0.07115) whereas it has been shown that the proportion of repeats is correlated with Drosophila genome size, in link with phylogenetic inertia [39]. It is possible that the lack of correlation comes from the fact that we are considering only LTR-retrotransposons. The non-LTR retrotransposons and DNA transposons can indeed represent significant proportions in some of the Drosophila species [7, 39].

Conclusion

In this work, we have for the first time been able to replace all families of LTR-retrotransposons from 20 species of Drosophila in a phylogenetic framework, allowing to clearly determine to which group inside each superfamily they belong. This will allow more detailed analyses concerning the specific evolution of particular families in different species. Indeed, it will now be possible to look more closely at specific families displaying sequence variants in some species to try understand how they were formed. For that, further analyses need to be performed like phylogenetic analyses based on other parts of the elements than the pol protein. This should help us determine if recombination between TE families in some species, like *D. willistoni* or *D. ananassae*, may explain why the number of families is so high in their genomes for example.

We also confirmed that HT events may occur for LTR-retrotransposons, mainly among some species from the *melanogaster* subgroup, but that they do not represent the most usual way in the evolutionary dynamics of LTR-retrotransposons since vertical transmissions, but also losses and intra-specific diversification play a large role.

Methods
Genomic data
The fasta genome sequences from the 20 Drosophila species were retrieved from the flybase website (ftp://ftp.flybase.net/genomes/) for *D. ananassae* (v1.04), *D. erecta* (v1.04), *D. grimshawi* (v1.3), *D. melanogaster* (v6.05), *D. mojavensis* (v1.04), *D. persimilis* (v1.3), *D. pseudoobscura* (v3.2), *D. sechellia* (v1.3), *D. simulans* (v2.01), *D. virilis* (v1.2), *D. willistoni* (v1.3), and *D. yakuba* (v1.3), and from the NCBI website (http://www.ncbi.nlm.nih.gov) for *D. biarmipes* (v2.0), *D. bipectinata* (v2.0), *D. elegans* (v2.0), *D. eugracilis* (v2.0), *D. ficusphila* (v2.0), *D. kikkawai* (v2.0), *D. rhopoloa* (v2.0), and *D. takahashii* (v2.0). These genomes have been obtained using different sequencing technologies and have various levels of qualities concerning the sequencing coverage and the assembly effort [7].

Identification of reference elements
Since *D. melanogaster* is a well annotated genome, we directly used the consensus sequences of LTR retrotransposons that are present in Repbase for this species [12]. To determine the reference elements of the other species from the *melanogaster* subgroup (*D. simulans, D. sechellia, D. yakuba*, and *D. erecta*), we first used the program LTRharvest [21] using the parameters settled for *D. melanogaster* since the program gave very good results for this species [22]. This program allows to identify potential complete LTR retrotransposons based on their structure. For each species, the candidates were then clustered using Uclust [41] with parameter -id 0.9. The sequences of each cluster were aligned using MUSCLE v3.8.31 [42] and the alignments were visualized with Seaview version 4.4.2 [43] to built a consensus for each cluster. Each consensus was manually corrected in regard to the other sequences to obtain a potentially "active" element with coding capacity. We also used each reference element to perform blastn [44] searches against the *D. simulans, D. sechellia, D. yakuba*, and *D. erecta* genomes to retrieve incomplete sequences of LTR retrotransposons not found by LTRharvest and to built consensus sequences using the alignments of the copies with low divergence compared to the reference sequence. We compared each reconstructed consensus with the sequences present in Repbase, Flybase and Genbank to identify already known elements and thus discriminating new characterized elements. We used the NCBI ORFfinder software (https://www.ncbi.nlm.nih.gov/orffinder/) to identify and retrieve the pol proteins. For

the other 15 *Drosophila* species, we retrieved the consensus sequences corresponding to the internal part of the elements from Repbase and used ORFfinder to identify and retrieve the pol proteins. For 15 of them, either no coding capacity was detected or the corresponding gene was corresponding to gag or env, and they were thus not included in the phylogenetic reconstructions. In total for the 20 Drosophila species, we obtained 563 sequences from the *Ty3/Gypsy* superfamily, 195 from the *BEL/Pao* superfamily, and 67 from the *Ty1/Copia* superfamily.

We used the BLASTN facility of flybase (http://flybase.org/blast/) to determine the absence of elements in species for which no reference elements were identified but which were present in closely related species of the *melanogaster* subgroup. We considered an element as "present" if a reference was described in the species (either in this work or from previous works) or if we detected few very long hits with high sequence identity (>90%), as "absent" if the blast searches lead to either no match or not significant ones (small fragments of less than 100 bp), with "traces" if the blast searches lead to numerous significant fragmented matches with low sequence identity (<90%).

Alignment and phylogenetic tree reconstruction

For each superfamily, *Ty3/Gypsy*, *BEL/Pao*, and *Ty1/Copia*, the protein sequences corresponding to pol of each reference element were aligned using MAFFT version 7 [45]. We added some sequences from a few other organisms available in Genbank (see figure legends). The non-informative sites in each alignment were removed using trimAL version 1.3 [46]. We determined the amino acid evolution model to be used in the phylogenetic reconstructions using ProtTest version 3 [47]. This analysis allowed us to reveal the same evolutionary model for protein evolution LG + I + G + F to best explain our data for each superfamily. Tree reconstructions were performed by maximum-likelihood method as implemented in PHYML 3.0 [48] with 100 bootstrap replicates using the LG + I + G + F evolutionary model. They were represented and edited using the FigTree software version 1.4.1 (Rambaut 2006–2013 http://tree.bio.ed.ac.uk/software/figtree/).

Confirmation of LTR retrotransposon horizontal transfer (HT)

Phylogenetic incongruences of TEs clustered with homologs from distant drosophila species or very short branches grouping different species that could indicate HT events, were analyzed by using the VHICA method [23]. Briefly, this method is based on the differences between the evolution rate at synonymous positions between TEs and a set of vertically transferred reference genes but also taking into account the codon usage bias.

For each compared pair of species, the correlation between the codon usage and the synonymous substitution rate is considered among reference genes assumed to be vertically transmitted. TEs with a significant deviation from host gene values are interpreted as potential horizontally transfered. To use VHICA, we performed the alignment of 30 orthologous genes among the 20 drosophila species using MACSE [49]. The list of the 30 genes correspond to a randomly selected subsample of the genes used by Wallau et al. [23] (Additional file 6: Table S2). The MACSE program was used to align the coding parts of the consensus TEs for which we had suspicion of HTs.

Proportion of LTR retrotransposons in the 20 Drosophila genomes

The RepeatMasker program (Smit et al. 1996–2010 http://www.repeatmasker.org) was used on the complete genome sequences of the 20 drosophila using a custom library corresponding to all identified reference elements. The .out output file was then parsed using one_code_to_find_them_all [50] to determine the proportion of each superfamily.

Additional files

> **Additional file 1:** Fasta sequences of the newly described reference TEs. (FASTA 959 kb)
>
> **Additional file 2: Table S1.** Repbase references (with internal part) for the all drosophila species with their associate superfamily and clade as found in the phylogenetic analyses (Figs. 1, 2, 3, 4 and 5). (PDF 311 kb)
>
> **Additional file 3: Figure S1.** Graphical matrix view generated by the VHICA method for HT cases in the A) *Ty1/Copia* superfamily, B) *BEL/Pao* and C,D,E) *Ty3/Gypsy* superfamilies. (PDF 779 kb)
>
> **Additional file 4: Figure S2.** Pattern of presence (black), absence (white) or traces (gray) of a given TE in the species of the *melanogaster* subgroup for A) Ty1/Copia, B) BEL/Pao, and C) Ty3/Gypsy superfamilies. (PDF 50 kb)
>
> **Additional file 5: Figure S3.** Maximum likelihood treee based on the pot proteins of all Ty3/Gypsy elements. (PDF 35 kb)
>
> **Additional file 6: Table S2.** Single copy orthologous genes from the Drosophila genomes used in the dS estimate in the VHICA method. (PDF 231 kb)

Abbreviations
HT: Horizontal transfer; LTR: Long terminal repeat; TE: Transposable element

Acknowledgements
This work was performed using the computing facilities of the CC LBBE/PRABI.

Funding
This work was supported by Agence Nationale de la Recherche [Exhyb ANR-14-CE19- 0016-01], the Fondation pour la Recherche Médicale [DEP20131128536] and the CNRS.

Authors' contributions
EL conceived the project; NB and EL performed the analyses and interpreted the results; EL wrote the manuscript. Both authors read and approved the final manuscript.

Competing interests
The authors declare that they have no competing interests.

References

1. Kidwell MG. Transposable elements and the evolution of genome size in eukaryotes. Genetica. 2002;115:49–63.
2. Tenaillon MI, Hollister JD, Gaut BS. A triptych of the evolution of plant transposable elements. Trends Plant Sci. 2010;15:471–8.
3. Elliott TA, Gregory TR. What's in a genome? the C-value enigma and the evolution of eukaryotic genome content. Philos Trans R Soc Lond B Biol Sci. 2015;370:20140331.
4. Dowsett AP, Young MW. Differing levels of dispersed repetitive DNA among closely related species of Drosophila. Proc Natl Acad Sci U S A. 1982;79: 4570–4.
5. Boulesteix M, Weiss M, Biémont C. Differences in genome size between closely related species: the *Drosophila melanogaster* species subgroup. Mol Biol Evol. 2006;23:162–7.
6. Lerat E, Burlet N, Biémont C, Vieira C. Comparative analysis of transposable elements in the melanogaster subgroup sequenced genomes. Gene. 2011; 473:100–9.
7. Clark AG, Eisen MB, Smith DR, Bergman CM, Oliver B, Markow TA, et al. Evolution of genes and genomes on the Drosophila phylogeny. Nature. 2007;450:203–18.
8. Rius N, Guillén Y, Delprat A, Kapusta A, Feschotte C, Ruiz A. Exploration of the *Drosophila buzzatii* transposable element content suggests underestimation of repeats in Drosophila genomes. BMC Genomics. 2016;17:344.
9. Wallau GL, Capy P, Loreto E, Hua-Van A. Genomic landscape and evolutionary dynamics of mariner transposable elements within the Drosophila genus. BMC Genomics. 2014;15:727.
10. de la Chaux N, Wagner A. Evolutionary dynamics of the LTR retrotransposons roo and rooA inferred from twelve complete Drosophila genomes. BMC Evol Biol. 2009;9:205.
11. Yang H-P, Barbash DA. Abundant and species-specific DINE-1 transposable elements in 12 Drosophila genomes. Genome Biol. 2008;9:R39.
12. Bao W, Kojima KK, Kohany O. Repbase update, a database of repetitive elements in eukaryotic genomes. Mob DNA. 2015;6:11.
13. Bergman CM, Quesneville H, Anxolabéhère D, Ashburner M. Recurrent insertion and duplication generate networks of transposable element sequences in the *Drosophila melanogaster* genome. Genome Biol. 2006;7:R112.
14. Sánchez-Gracia A, Maside X, Charlesworth B. High rate of horizontal transfer of transposable elements in Drosophila. Trends Genet. 2005;21:200–3.
15. Loreto ELS, Carareto CM a, Capy P. Revisiting horizontal transfer of transposable elements in Drosophila. Heredity (Edinb). 2008;100:545–54.
16. Bartolomé C, Bello X, Maside X. Widespread evidence for horizontal transfer of transposable elements across Drosophila genomes. Genome Biol. 2009; 10:R22.
17. Modolo L, Picard F, Lerat E. A new genome-wide method to track horizontally transferred sequences: application to Drosophila. Genome Biol Evol. 2014;6:416–32.
18. Zanni V, Eymery A, Coiffet M, Zytnicki M, Luyten I, Quesneville H, et al. Distribution, evolution, and diversity of retrotransposons at the flamenco locus reflect the regulatory properties of piRNA clusters. Proc Natl Acad Sci. 2013;110:19842–7.
19. Havecker ER, Gao X, Voytas DF. The diversity of LTR retrotransposons. Genome Biol. 2004;5:225.
20. Llorens C, Muñoz-Pomer A, Bernad L, Botella H, Moya A. Network dynamics of eukaryotic LTR retroelements beyond phylogenetic trees. Biol Direct. 2009;4:41.
21. Ellinghaus D, Kurtz S, Willhoeft U. LTRharvest, an efficient and flexible software for de novo detection of LTR retrotransposons. BMC Bioinformatics. 2008;9:18.
22. Lerat E. Identifying repeats and transposable elements in sequenced

genomes: how to find your way through the dense forest of programs. Heredity (Edinb). 2010;104:520–33.
23. Wallau GL, Capy P, Loreto E, Le Rouzic A, Hua-Van A. VHICA, a New Method to Discriminate between Vertical and Horizontal Transposon Transfer: application to the Mariner Family within Drosophila. Mol Biol Evol. 2016;33: 1094–109.
24. Jordan IK, Matyunina LV, McDonald JF. Evidence for the recent horizontal transfer of long terminal repeat retrotransposon. Proc Natl Acad Sci U S A. 1999;96:12621–5.
25. Daniels SB, Peterson KR, Strausbaugh LD, Kidwell MG, Chovnick A. Evidence for horizontal transmission of the P transposable element between Drosophila species. Genetics. 1990;124:339–55.
26. Kaminker JS, Bergman CM, Kronmiller B, Carlson J, Svirskas R, Patel S, et al. The transposable elements of the *Drosophila melanogaster* euchromatin: a genomics perspective. Genome Biol. 2002;3: RESEARCH0084.
27. Rubin PM, Loreto EL, Carareto CM, Valente VL. The copia retrotransposon and horizontal transfer in *Drosophila willistoni*. Genet Res (Camb). 2011;93: 175–80.
28. de la Chaux N, Wagner A. BEL/Pao retrotransposons in metazoan genomes. BMC Evol Biol. 2011;11:154.
29. Kim A, Terzian C, Santamaria P, Pélisson A, Purd'homme N, Bucheton A. Retroviruses in invertebrates: the gypsy retrotransposon is apparently an infectious retrovirus of *Drosophila melanogaster*. Proc Natl Acad Sci U S A. 1994;91:1285–9.
30. Song SU, Gerasimova T, Kurkulos M, Boeke JD, Corces VG. An env-like protein encoded by a Drosophila retroelement: evidence that gypsy is an infectious retrovirus. Genes Dev. 1994;8:2046–57.
31. Leblanc P, Desset S, Giorgi F, Taddei AR, Fausto AM, Mazzini M, et al. Life cycle of an endogenous retrovirus, ZAM, in *Drosophila melanogaster*. J Virol. 2000;74:10658–69.
32. Akkouche A, Rebollo R, Burlet N, Esnault C, Martinez S, Viginier B, et al. Tirant, a newly discovered active endogenous retrovirus in Drosophila simulans. J Virol. 2012;86:3675–81.
33. Pantazidis A, Labrador M, Fontdevila A. The retrotransposon Osvaldo from *Drosophila buzzatii* displays all structural features of a functional retrovirus. Mol Biol Evol. 1999;16:909–21.
34. García Guerreiro MP, Fontdevila A. Molecular characterization and genomic distribution of Isis: a new retrotransposon of *Drosophila buzzatii*. Mol Genet Genomics. 2007;277:83–95.
35. Evgen'ev MB, Corces VG, Lankenau DH. Ulysses transposable element of Drosophila shows high structural similarities to functional domains of retroviruses. J Mol Biol. 1992;225:917–24.
36. de Setta N, Van Sluys M-A, Capy P, Carareto CMA. Multiple invasions of Gypsy and Micropia retroelements in genus Zaprionus and melanogaster subgroup of the genus Drosophila. BMC Evol Biol. 2009;9:279.
37. Ludwig A, Valente VLDS, Loreto ELS. Multiple invasions of Errantivirus in the genus Drosophila. Insect Mol Biol. 2008;17:113–24.
38. Schaack S, Gilbert C, Feschotte C. Promiscuous DNA: horizontal transfer of transposable elements and why it matters for eukaryotic evolution. Trends Ecol Evol. 2010;25:537–46.
39. Sessegolo C, Burlet N, Haudry A. Strong phylogenetic inertia on genome size and transposable element content among 26 species of flies. Biol Lett. 2016;12:20160407.
40. Elliott TA, Gregory TR. Do larger genomes contain more diverse transposable elements? BMC Evol Biol. 2015;15:69.
41. Edgar RC. Search and clustering orders of magnitude faster than BLAST. Bioinformatics. 2010;26:2460–1.
42. Edgar RC. MUSCLE: a multiple sequence alignment method with reduced time and space complexity. BMC Bioinformatics. 2004;5:113.
43. Gouy M, Guindon S, Gascuel O. SeaView version 4: a multiplatform graphical user interface for sequence alignment and phylogenetic tree building. Mol Biol Evol. 2010;27:221–4.
44. Altschul SF, Madden TL, Schäffer AA, Zhang J, Zhang Z, Miller W, et al. Gapped BLAST and PSI-BLAST: a new generation of protein database search programs. Nucleic Acids Res. 1997;25:3389–402.
45. Katoh K, Standley DM. MAFFT multiple sequence alignment software version 7: improvements in performance and usability. Mol Biol Evol. 2013; 30:772–80.
46. Capella-Gutiérrez S, Silla-Martínez JM. trimAl: a tool for automated

alignment trimming in large-scale phylogenetic analyses. Bioinformatics. 2009;25:1972–3.

47.   Darriba D, Taboada GL, Doallo R, Posada D. ProtTest 3: fast selection of best-fit models of protein evolution. Bioinformatics. 2011;27:1164–5.

48.   Guindon S, Dufayard J-F, Lefort V, Anisimova M, Hordijk W, Gascuel O. New algorithms and methods to estimate maximum-likelihood phylogenies: assessing the performance of PhyML 3.0. Syst Biol. 2010;59:307–21.

49.   Ranwez V, Harispe S, Delsuc F, Douzery EJP. MACSE: Multiple Alignment of Coding SEquences accounting for frameshifts and stop codons. Plos One. 2011;6:e22594.

50.   Bailly-Bechet M, Haudry A, Lerat E. "One code to find them all": a perl tool to conveniently parse RepeatMasker output files. Mob DNA. 2014;5:13.

12

Evolutionary history of the *mariner* element *galluhop* in avian genomes

Natasha Avila Bertocchi[1,2*], Fabiano Pimentel Torres[1,2], Analía del Valle Garnero[1,2], Ricardo José Gunski[1,2] and Gabriel Luz Wallau[3]

Abstract

Background: Transposable elements (TEs) are highly abundant genomic parasites in eukaryote genomes. Although several genomes have been screened for TEs, so far very limited information is available regarding avian TEs and their evolutionary histories. Taking advantage of the rich genomic data available for birds, we characterized the evolutionary history of the *galluhop* element, originally described in *Gallus gallus*, through the use of several bioinformatic analyses.

Results: *galluhop* homologous sequences were found in 6 of 72 genomes analyzed: 5 species of Galliformes (*Gallus gallus, Meleagris gallopavo, Coturnix japonica, Colinus virginianus, Lyrurus tetrix*) and one Buceritiformes (*Buceros rhinoceros*). The copy number ranged from 5 to 10,158, in the genomes of *C. japonica* and *G. gallus* respectively. All 6 species possessed short elements, suggesting the presence of Miniature Inverted repeats Transposable Elements (MITEs), which underwent an ancient massive amplification in the *G. gallus* and *M. gallopavo* genomes. Only 4 species showed potential MITE full-length partners, although no potential coding copies were detected. Phylogenetic analysis of reconstructed coding sequences showed that *galluhop* homolog sequences form a new *mariner* subfamily, which we termed *Gallus*. Inter-species and intragenomic *galluhop* distance analyses indicated a high identity between the consensus of *B. rhinoceros* and the other 5 related species, and different emergence ages of the element between the Galliformes species and *B. rhinocerus*, suggesting that horizontal transfer took place from Galliformes to a Buceritiformes ancestor, probably through an intermediate species.

Conclusions: Overall, our results showed that *mariner* elements have amplified to high copy numbers in some avian species, and that this transposition burst probably occurred in the common ancestor of *G. gallus* and *M. gallopavo*. In addition, although no coding sequences could be found currently, they probably existed, allowing an ancient massive MITE amplification in these 2 species. The other 4 species also have MITEs, suggesting that this new *mariner* family is prone to give rise to such non-autonomous derivatives. Last, our results suggest that a horizontal transfer event of a *galluhop* element occurred between Galliformes and Buceritiformes.

Keywords: Galluhop, Mariner, Avian genome, Horizontal transfer, MITEs, Genomic parasites

Background

Transposable elements (TEs) are widely distributed and abundant component of many eukaryotic genomes. TEs can be classified in two main classes, based on their transposition mechanism: Class I (moves through an RNA intermediate) and Class II (through a DNA intermediate) [1–3]. Successful proliferation of TEs in genomes is linked to their replicative and mobile capacity within the host genome and also between genomes [4, 5]. On the other hand, most of the time this mobility is neutral or deleterious to the host organism. New TE insertions in gene-coding regions or in upstream/downstream positions can have a drastic impact on flanking genes [6]. These highly similar and repetitive sequences throughout the genome also provide a substrate for ectopic recombination events that can lead to chromosomal inversions and deletions [7, 8]. However, an increasing body of evidence is showing that insertions of

* Correspondence: bertoccinatasha@gmail.com
[1]Programa de Pós-graduação em Ciências Biológicas, Universidade Federal do Pampa (Unipampa), São Gabriel, Rio Grande do sul 97300-000, Brazil
[2]Laboratório de Diversidade Genética Animal, Universidade Federal do Pampa (Unipampa), São Gabriel, Rio Grande do sul 97300-000, Brazil
Full list of author information is available at the end of the article

new TEs introduce variability and can sometime be adaptive for the host genome [9, 10].

TEs are an integral part of host genomes and hence are vertically transmitted to descendants through the male and female germ line DNA, and from ancestral to extant species in the course of evolution [11]. However, compelling evidence in a wide variety of taxa has increasingly revealed that Horizontal Transfer (HT), the exchange of genetic material between isolated sexual species, is an effective way in which TEs invade new genomes and colonize other species [11, 12]. Currently, around 2853 Horizontal Transposon Transfer (HTT) events have been reported [13]. The *mariner* family of Class II DNA transposons has the highest number of HTT cases reported (52) [13, 14]. Such events have been characterized in a wide variety of taxa, including insects and mammals [14–16]. In birds, considering all TE families of Class I and II, only seven HTT events have been reported so far: two retrotransposons (AviRTE), which took place between several bird species ancestors and human pathogenic nematodes [17].

Non-autonomous elements can emerge at any step of the TE "life cycle" through deletion or internal region degeneration, yet retain their transposition capacity in the presence of autonomous or coding copies. Internally deleted non-autonomous elements originating from Class II transposons are known as Miniature Inverted-repeat Transposable Elements (MITEs) [2]. These elements possess deletions or a degenerated coding region, but preserved Terminal Inverted Repeats (TIRs) which can be recognized by functional transposases [18, 19]. MITEs have been associated with several Class II superfamilies such as hAT, *P* and Tc1/mariner [20–22]. Usually MITEs reach higher copy numbers than their autonomous counterparts, a form of parasitism that may lead to the extinction of the entire TE family in the long term [23].

Although TEs are currently recognized as major players in genome evolution, in some taxa such as birds, knowledge of TEs is limited [24, 25]. One of the reasons for this gap has been the scarcity of available genome sequences, but since 2014, more than 70 draft whole genome sequences have become available [26]. Among the few studies focusing on TEs in bird genomes, a reduction in repetitive DNA was detected in sauropsids, perhaps due to the purifying selection pressure acting on metabolism optimization [25, 27, 28]. In particular, Class II TEs, which are abundant in other eukaryotic species, appear to show limited diversity in the few avian genomes studied so far: the chicken *Gallus gallus* and the wild turkey *Meleagris gallopavo* [24, 29].

Elements from the *Tc1-mariner* superfamily generally are 1.3 kb long, and contain TIRs of approximately 28 bp and a unique ORF (Open Reading Frame) which codes for a transposase [30, 31]. Because of the great diversity of the *mariner* family, these elements were classified in subfamilies based on phylogenetic analyses. The classification proposed by Rouault et al [32] includes 12 subfamilies (*mauritiana, cecropia, rosa, mellifera, lineata, capitata, irritans, briggsae, elegans, Atlantis* and *CRI*). Among the Class II TEs found in avian genomes, a *mariner*-like element termed *galluhop* was previously characterized [29, 33], but up to now no other study has focused on understanding its evolution in other avian species.

Here, we aimed to characterize the evolutionary history of *galluhop* homolog sequences found in available avian genomes. Our results showed that *galluhop*-like sequences compose a new *mariner* subfamily, which was exchanged between two bird taxa through horizontal transfer, probably mediated by an intermediate species.

Methods
Bioinformatic workflow
Genome search for galluhop homologs
The nucleotide sequence from the *galluhop* consensus described by Wicker et al. [33] was obtained from the Repbase database [33–35]. 72 avian genomes were available as of May 2016 (Additional file 1: Table S1). BLASTn searches were performed using the *galluhop* consensus sequence from Repbase, using default parameters. Only blast results with an E-value lower than e^{-10} were analyzed further. *In house* python scripts were used to retrieve all sequences and 200 base pairs of flanking sequences from each copy.

Sequence alignments of all copies plus flanking sequences from each species were performed with MAFFT v.7 [36] (Additional file 2: Figure S1).

Functional characterization
The resulting alignments were manually inspected and corrected in order to precisely identify TIRs and target site duplications. TIRs conservation was determined visually, using Weblogo [37]. After identification and definition of element copy boundaries, all copies were characterized by the presence of ORFs, using the OrfFinder script implemented in UGENE [38] and the script implemented in Emboss gertof (http://emboss.sourceforge.net/apps/cvs/emboss/apps/getorf.html) with the following parameters: -minsize 900 -find 1 -methionine Y. Copies were classified as i) possessing a predicted coding protein = > than 300 aa and conserved TIRs as potential autonomous copies; ii) possessing a potential coding protein <= than 300 aa and conserved TIRs as potential non-autonomous copies; iii) copies with a missing TIR but with ORFs = > than 300 aa as potential coding copies; and iv) elements with a missing TIR and ORFs <= 300 aa as partial elements (Additional file 2: Figure S1).

Nucleotide distance and phylogenetic analysis

In order to estimate the interspecies distance of TEs, we reconstructed the majority consensus ancestor element with all copies found per genome, using UGENE [38]. The Kimura 2 parameter (K2P) distance between all copies and their corresponding consensus sequence was estimated with the distmat script from the Emboss package (http://emboss.sourceforge.net/apps/release/6.6/emboss/apps/distmat.html) and histogram distribution plotted with ggplot2 [39] in the R environment [40]. Dating between *galluhop* consensus elements and copies within each genomes was performed according to the eq. T = k/2r [41]. T represents the divergence time between TEs, k is the divergence value between the TE consensus and copies, and r is the mean evolutionary rate for bird genomes [41]. We used species-specific evolutionary rates when available, or the closest relative rates: *Gallus gallus* 1.9×10^{-3}, *Meleagris gallopavo* 2.0×10^{-3}, *Buceros rhinoceros* 2.3×10^{-3}, *Lyrurus tetrix* 1.9×10^{-3}, and *C. virginianus* 1.9×10^{-3} [42]).

We also obtained the coding regions of 50 single-copy orthologous genes between the *B. rhinoceros* and *L. tetrix* genomes, and estimated the K2P distance in order to compare with the TE K2P distance. The OrthoDB database [43] was used to search single-copy orthologous genes found in all 52 available avian genomes analyzed in this database version. Due to the lack of data for *L. tetrix* in the database, we used the mRNA accession number of *B. rhinoceros* as the blastn query against the *L. tetrix* genome in order to obtain the gene sequence used for the latter.

Alignments of reconstructed *galluhop* coding region (almost complete ORF and partial for those composed only for MITEs) from all 6 species that possessed *galluhop* homolog sequences were performed, using a previously published transposase alignment covering most of the *mariner* subfamilies [41], using MAFFT v.7 [36].

Phylogenetic reconstruction was performed by maximum likelihood, using PHYML [44], and branch support was evaluated by SH-like support [45].

Results and discussion

galluhop homologs in bird genomes

Six of the 72 avian genomes analyzed harbored *galluhop*-like sequences (Table 1). Five of these are from species of the order Galliformes that diverged from each other at least 46 Mya (CI: 37 – 55 Mya – [46]): *Colinus virginianus*, *Coturnix japonica*, *Lyrurus tetrix*, *Gallus gallus* and *Meleagris gallopavo*. We also identified *galluhop*-related sequences in *Buceros rhinoceros* of the order Bucerotiformes, which diverged from Galliformes 98 Mya (CI: 92.1–104.0 Mya – [46]). Consensus of *gallohop per* species can be found in Additional file 3. Table 1 shows the Kimura 2 parameter distance between the consensus element from each species.

These elements reached a high copy number in both the *G. gallus* and *M. gallopavo* genomes, 10,158 and 8317 respectively. The remaining 4 species showed lower copy numbers, from 5 to 96 copies (Table 2). No potential autonomous or coding copies were found (Table 2). Four of 5 Galliformes species (*G. gallus*, *M. gallopavo*, *C. viginianus* and *L. tetrix*) showed elements with a similar size to the reference *galluhop* element deposited in Repbase (around 1300 bp; Table 1), although most of them showed two 12-bp insertions that prevented any transposase from being fully encoded (Fig. 1). The remaining species of Galliformes, *C. japonica*, showed only 5 short elements of 550 bp and conserved TIRs resembling MITEs in the assembly version analyzed. Last, *B. rhinocerus* showed 14 copies of 575 bp but with conserved TIRs and subterminal regions of the elements (Fig. 1 and Additional file 4 Figure S2). Most of the *galluhop*-like sequences found showed both imperfect TIRs (Additional file 4: Figure S2) and target site duplication (TSD) TA characteristic of *mariner* elements (Additional file 5: Figure S3).

Phylogenetic analysis using all *galluhop*-like consensus sequences and several sequences from the *mariner* subfamilies indicates that *galluhop*-like elements compose a new *mariner* subfamily, which we termed *Gallus* (Fig. 2). TEs from the *Gallus* family emerged in the ancestor of the order Galliformes (around 55–65 Mya) [46], increasing its copy number, particularly in the *G. gallus* and *M. gallopavo* genomes. Only non-autonomous copies of the *Gallus* subfamily were found possessing several mutations, multiple stop codons and changes in the element reading frame (Fig. 1). We also found a large number of short non-autonomous

Table 1 Kimura 2 parameter distance between each *galluhop* consensus sequence

M. gallopavo	L. tetrix	G. gallus	C. japonica	C. virginianus	B. rhinocerus	Species
						B. rhinocerus
					0.0921	C. virginianus
			0.1654	0.1843	C. japonica
			0.1691	0.0519	0.0755	G. gallus
	0.0416	0.141	0.0519	0.0571	L. tetrix
0.0382	0.0225	0.166	0.0535	0.0797	M. gallopavo

Table 2 Avian genomes with *galluhop* and characteristics of copies

Partial elements (160–1200 bp)[b]	Non-autonomous elements (~500–600 bp)	Full-length elements (~1200–1300 bp)	ORFs[a]	No. of copies	Assembly size (Mb)	Species	Order
29	9927	202	N	10,158	1046.93	*G. gallus*	Galliformes
0	8187	130	N	8317	1061.82	*M. gallopavo*	Galliformes
28	61	7	N	96	1171.86	*C. virginianus*	Galliformes
76	19	1	N	96	657.025[c]	*L. tetrix*	Galliformes
0	4	0	N	4	531.96[c]	*C. japônica*	Galliformes
0	14	0	N	14	1065.78	*B. rhinocerus*	Buceritiformes

[a]No ORFs were found in the analyzed elements
[b]Partial elements are copies with a missing TIR and ORFs <= 300 aa
[c]*L. tetrix* and *C. japonica* genomes have a smaller assembly size than most avian genomes, since they are only partially assembled. A new assembly version of the *C. japonica* genome is available, with a higher assembly size of 927.657 Mb – GCA_000511605.2, but it was not used in our study since it was released after we conducted all analyses in the previous assembly version

elements (around 500–600 bp) with preserved 5′ and 3′ regions of the element, including TIRs (Fig. 1 and Additional file 4: Figure S2), but with a large deletion compared with the full-length consensus element (Fig. 1). These shorter elements showed all the characteristics of MITEs [19] and amplified successfully in *G. gallus* and *M. gallopavo*, composing the large majority of *galluhop* copies found in these genomes (97.7% in *G. gallus* and 98.4% in *M. gallopavo* genomes). *C. virginianus* and *L. tetrix* also showed amplification of MITEs on a smaller scale, and *C. japonica* and *B. rhinocerus* possessed only MITEs elements and no trace of their possible autonomous counterparts (Fig. 1). Taken together, these findings suggest that MITEs originated independently in this new *mariner* subfamily, which probably

affected the fate of these elements leading to the extinction of the TE family in all avian genomes studied. This view is in agreement with the hypothesis that the emergence of superparasites such as MITEs can lead TE families/subfamilies to decay and disappear over time [19, 47].

galluhop intra- and interspecies evolution

The intragenomic divergence between each *galluhop* copy and its corresponding ancestor consensus sequence was calculated in order to infer the time frame of TE arrival and their amplification dynamics in each genome, except for *C. japonica*, due to the low copy number in this genome (Table 2). Making use of species-specific or the closest-relative evolution rates, we could estimate this dynamic in

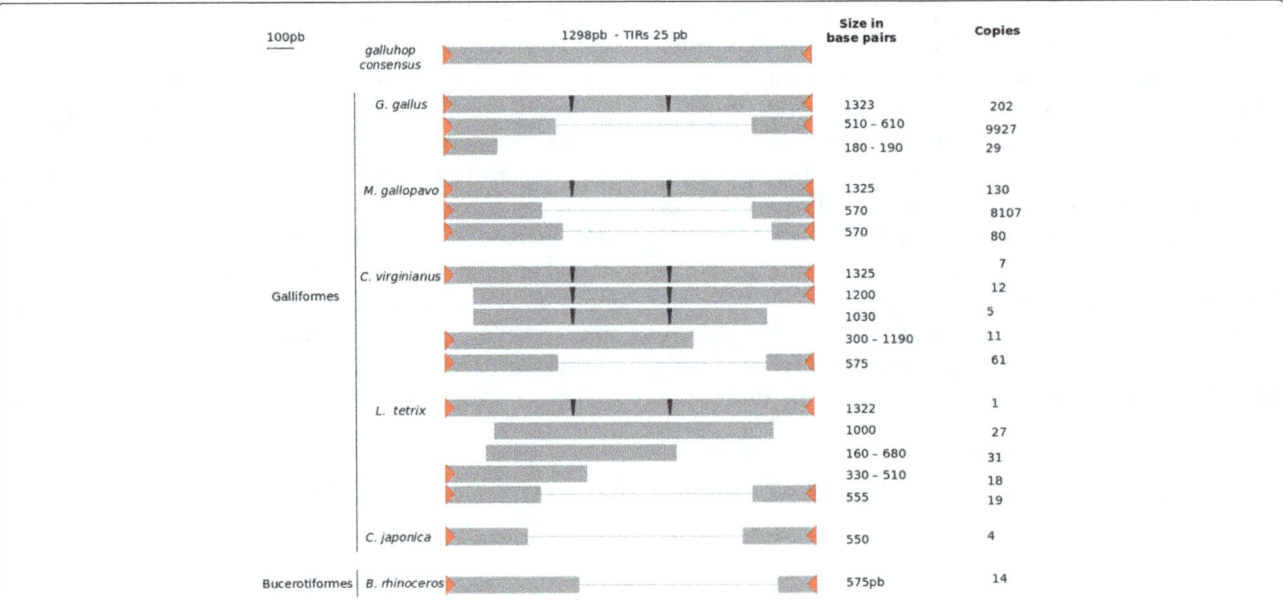

Fig. 1 Schematic representation of the reconstructed *galluhop* copies compared to the *galluhop* consensus. Regions of terminal inverted repeats shown in *red*, transposase coding region in *light gray*, and insertion region in *dark gray*. Order Galliformes: four genomes (*G. gallus*, *M. gallopavo*, *C. virginianus* and *L. tetrix*) showed potential complete partners although there are no potential coding copies, and *C. japonica* showed short elements. Order Bucerotiformes: *B. rhinoceros* showed only short elements

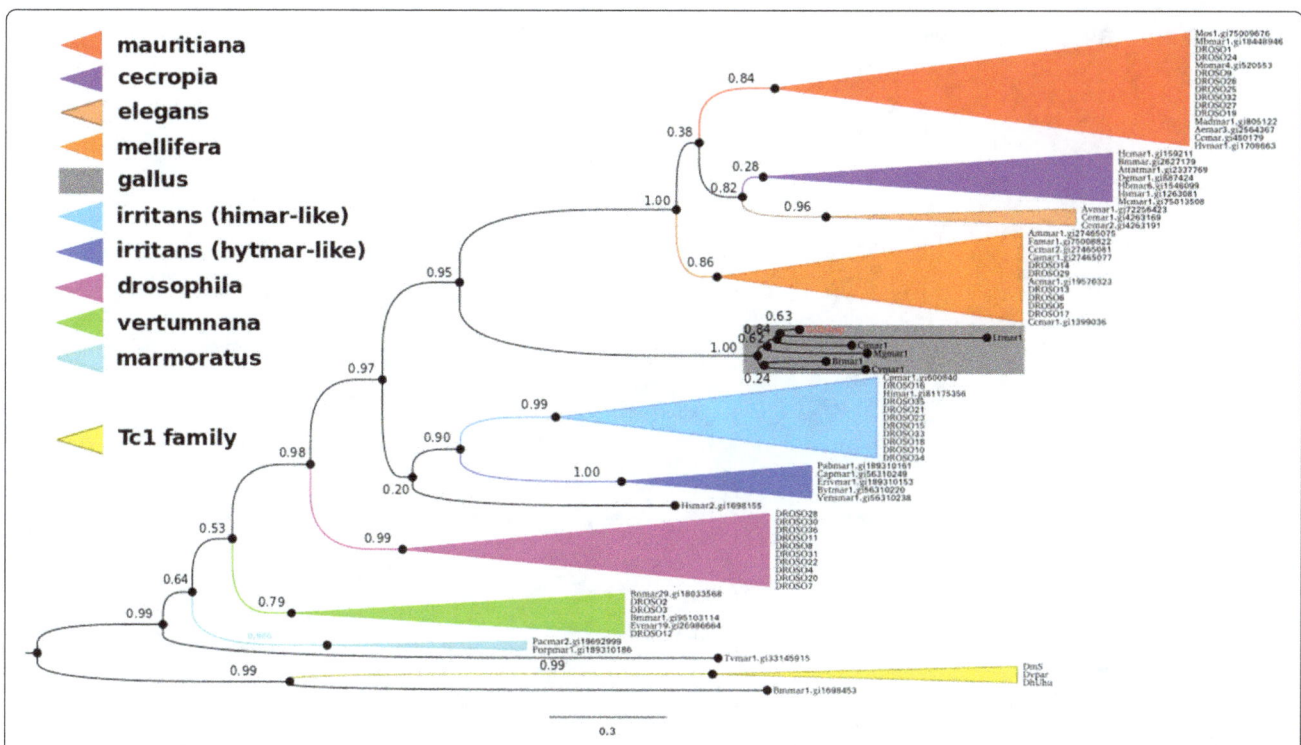

Fig. 2 Phylogeny of *mariner*-like transposases. Phylogeny of *mariner*-like transposases, by maximum likelihood using PHYML (Guindon and Gascuel 2003). *Clade colors* denote the different subfamilies of the *mariner* family, indicated to the *left* of the tree. In *gray*: the new subfamily, *Gallus*

millions of years ago (MYA). As seen in Fig. 3a and b, the species of Galliformes showed a wide distribution of element ages, with a single peak ocurring in *G. gallus* and *M. gallopavo* between 100 and 25 MYA (Fig. 3a), and two peaks in *C. virginianus* and *L. tetrix* at around 87.5 and 37.5 MYA and 37.5 and 18.75 MYA (Fig. 3b), suggesting that these elements are ancient parasites of galliformes genomes and increased in copy number through single or double amplification waves. However, the only buceritiformes species bearing *galluhop* elements, *B. rhinocerus*, showed a much younger element distribution ranging from 31.5 and 18.75 MYA, suggesting a single, more recent, amplification wave (Fig. 3b).

These differing amplification age distributions could be interpreted as due to the differing evolutionary rates between the species analyzed, and not due to different emergence and amplification dates of the TEs. The *B. rhinocerus* genome has the highest evolutionary rates of the species analyzed here, suggesting that if this bias is real, we would expect to observe lower than expected element ages in this species biasing our analysis. In order to evaluate if lower evolutionary rates could significantly change the estimates for *B. rhinocerus* elements, we used the evolutionary rate for water birds (1.6×10^{-3}) [42], which is one of the lowest estimates for birds, to estimate the *B. rhinocerus galluhop* invasion. Even so,

we obtained ages for *B. rhinocerus* elements between 31.2 and 43.7 MYA, which is still much younger than all estimates for the origin of *galluhop* in galliform genomes, suporting the hypothesis that *galluhop* emerged in *B. rhinocerus* more recently than in Galliformes.

Younger element ages in *B rhinocerus*, a species from the Neoaves, order Buceritiformes, combined with a patchy distribution of *galluhop* in the avian tree, found in only 5 additional galliform species (*C. virginianus*, *C. japonica*, *L. tetrix*, *G. gallus* and *M. gallopavo*) which diverged from *B. rhinocerus* around 85–98 MYA [46], suggests that probably a horizontal transfer event took place directly between the common ancestor of these taxa or through an intermediate species.

In order to gain additional insights about possible donor and receptor species, we first evaluated the evolutionary distance of species-specific TEs consensus sequences. Among all galliform consensus sequences, the distance at the nucleotide level varied from 0.0382 to 0.1654 (Table 1). The *B. rhinocerus* consensus showed a K2P distance of 0.0571 to 0.1843, being the lowest distance comparison with the *L. tetrix* consensus (Table 1). Second, we evaluated the evolutionary distance of the TEs consensus of *B. rhinocerus-L. tetrix* (K2P = 0.0571) with 50 single-

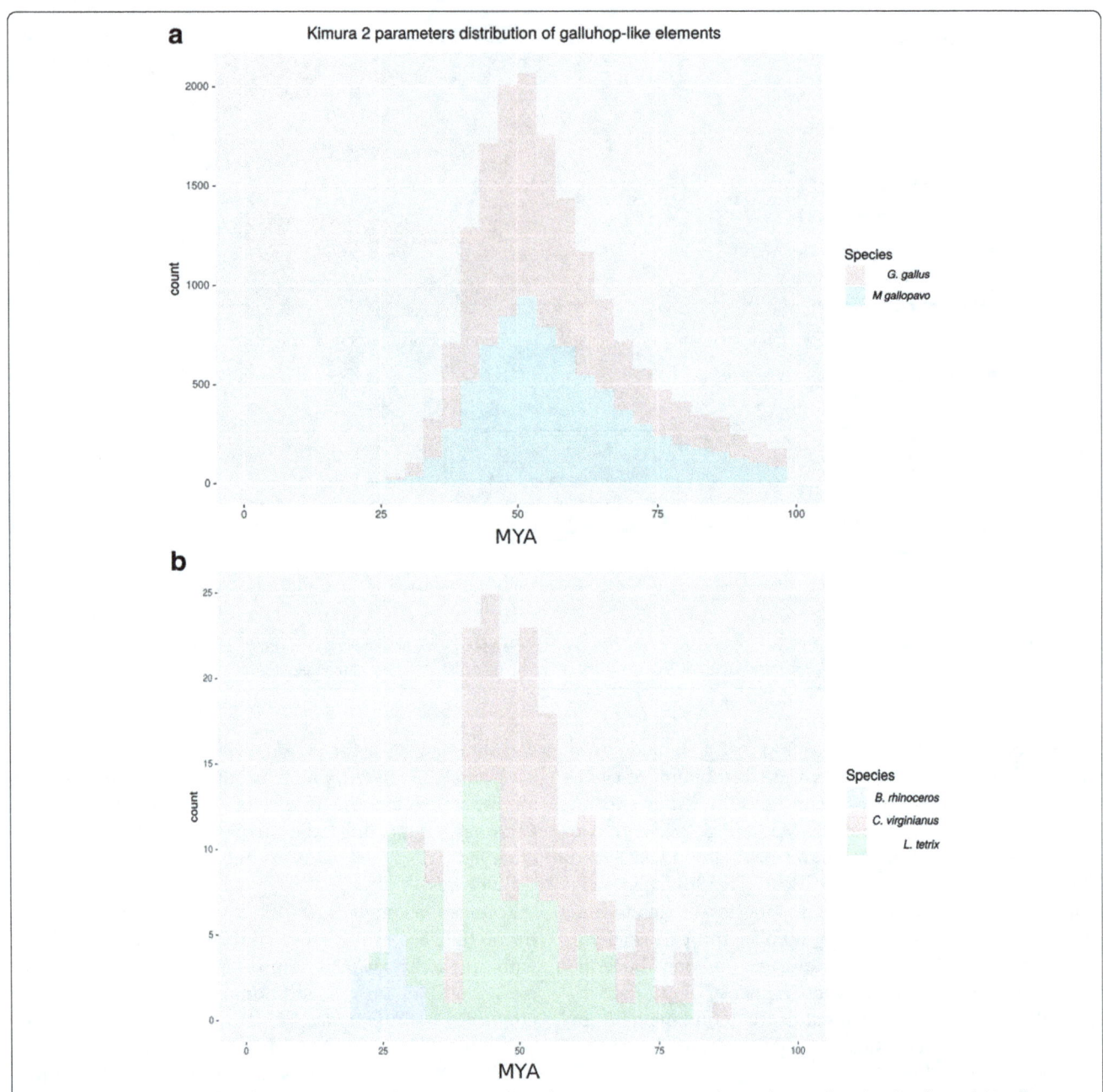

Fig. 3 Amplification dynamics of elements within each genome in million of years. **a** Intragenomic dating of copies found in *G. gallus* and *M. gallopavo*. **b** Intragenomic dating of copies found in *B. rhinocerus*, *C. virginianus* and *L. tetrix*

copy host genes of each species. This reasoning is based on the following principle: a similar or higher TE-host gene distance is expected if TEs were evolving by vertical transfer, since they would have had the same time to accumulate mutations as host genes. On the other hand, a shorter TE distance compared with the host-gene distance is expected if a horizontal transfer took place. Figure 4 depicts an expected host-gene normally distributed K2P distance, with a tail for more divergent host genes. Most of the genes have an average K2P distance, and few genes have extreme values of low and high K2P, which can be explained by the negative and positive selection acting on them. The TE K2P distance (red arrow) is shorter than 92% of all host genes analyzed (46 genes) and falls in the extreme lower range of K2P values of host genes. Although one can think of this as an indication that the TE is evolving vertically since it has a similar distance as some host genes, TEs evolve neutrally, so we would expect to see vertically inherited

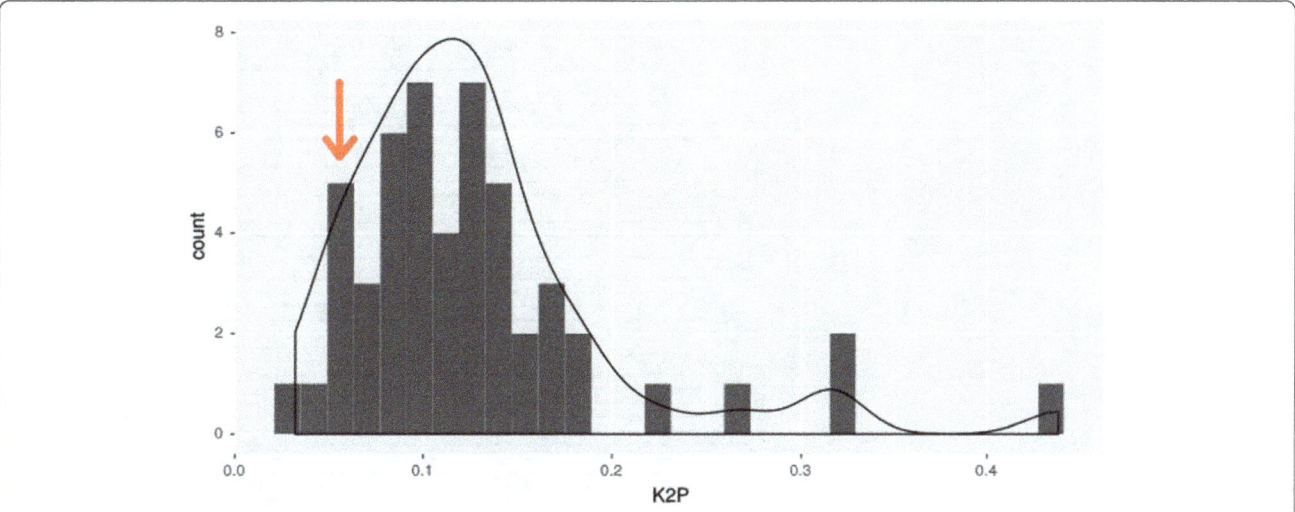

Fig. 4 Density plot of Kimura 2 parameter distance between *B. rhinocerus* and *L. tetrix*. K2P distance of 50 single-copy orthologous genes (*gray shading*) and consensus TEs (*red arrow*)

TEs within the average host gene distance or at the opposite extreme of the distribution. Therefore, those results are in agreement with the HT hypothesis between *L. tetrix* and *B. rhinocerus* ancestor.

One of the supporting lines of evidence which can shed light on time, direction and the presence of a possible intermediate species of an HTT event is the distribution of current and ancestors of the species involved and the element invasion dates. If host species have an overlapping distribution range and the estimates of element invasion are similar, then it is reasonable to suggest that HT occurred directly between them. Contrariwise, a non-overlapping range suggests that the HTT event occurred between the ancestors and different elements invasion ages through an intermediate species. *L. tetrix* and *B. rhinoceros* currently have distinct distribution ranges; the former is restricted to northern Eurasia, from the Swiss-Italian-French Alps to Scandinavia, Estonia and Russia; while the second occurs in Southeast Asia, including Borneo, Singapore, Malaysia and Thailand [48]. Fossils of other species of the genus *Lyrurus* and order Buceriti-formes were found in Bulgaria and dated to the Miocene epoch (20.44 to 7.24 MYA) [49–51], although recent genome-wide paleogeographic inferences are few and limited, so that the ancestral distribution ranges of these two species cannot yet be defined with certainty [52]. Based on *galluhop*-like sequence ages, we observed that this element invaded *B. rhinocerus* ~ 31 MYA in the early Oligocene epoch of the Cenozoic era, while it arose in the *L. tetrix* genome around 75-82 MYA. Taken together, our data support an ancient HTT event between the ancestor of Galliformes and Buceritiformes or through an intermediate species; the latter is the most probable hypothesis, since different element ages were found (Fig. 5).

Conclusions

The evolution of transposable elements usually shows complex patterns, such as patchy distributions within taxa, associated with a high similarity of TEs in host species that diverged long ago. The presence of such patterns can be explained by an exchange of TEs by these species or independent acquisitions from a third source, which characterizes a phenomenon known as HTT. HTT events have been reported throughout the eukaryote tree of life in recent years, and several of these events were reported for vertebrate species [13]. For instance, the SPIN transposon was found in more than 17 distantly related tetrapod species, including mammals as well as an African frog and a lizard, showing high similarity and patchy distribution [53, 54]. Despite these recent findings in vertebrates, only seven HTT events have been documented thus far, involving an avian clade and parasitic nematodes [17].

Here we evaluated the evolutionary history of the *mariner* element *galluhop* in Avian genomes. Our results shed new light on the phylogeny of the *mariner* family, describing a new subfamily termed *Gallus,* and highlights the successful amplification of MITEs of this subfamily in some avian genomes. We also report the first documented HTT event involving bird species. The analyses of the TE distribution, interspecies similarity and intragenomic element ages support the existence of the first HTT event between avian genomes.

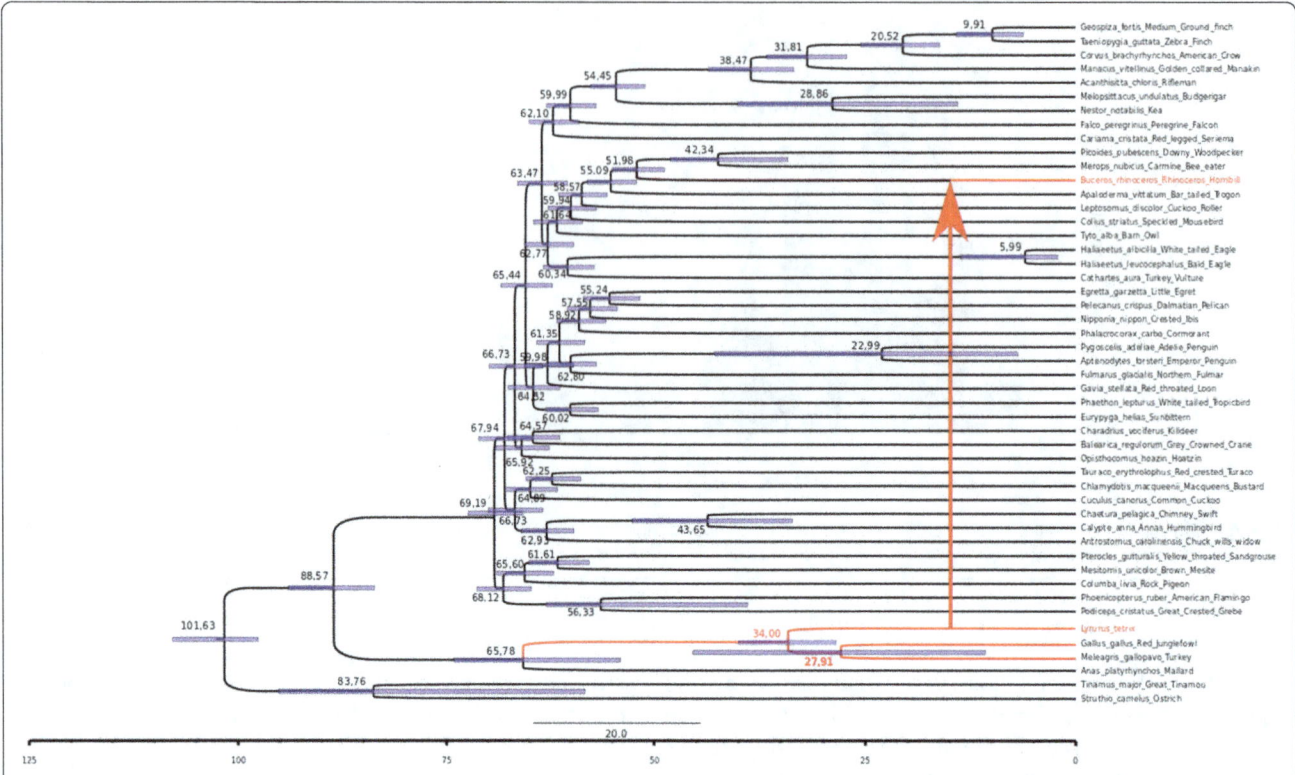

Fig. 5 *Horizontal* Transfer hypothesis of *galluhop* elements. Chronogram TENT avian tree from Jarvis et al. [42] with the addition of probable *L. tetrix* positioning and split data following TimeTree data [44] *Red* branches denote the evolutionary hypothesis of the *Gallus* subfamily vertical evolution in the *Galliformes*, and *horizontal* transfer from *L. tetrix* to *B. rhinoceros* ancestors. *X bar* below the tree denotes the time in millions of years. Number close to nodes are the mean estimate of ancestors and blue bars are 95% credible interval as estimated by Jarvis et al [42]

Additional files

Additional file 1: Table S1. GenBank access numbers for bird genomes. (XLSX 14 kb)

Additional file 2: Figure S1. Experimental design procedure showing steps of the analysis. Galluhop homologous sequences were found in 6 of 72 genomes analyzed. We analyzed the functional and structural characteristics and phylogenetic reconstruction of the putative transposases. (TIFF 4512 kb)

Additional file 3: Fasta consensus sequences. (FASTA 6 kb)

Additional file 4: Figure S2. Graphical representation of the conservation of terminal inverted repeats (TIRs). The TIRs 5′ and 3′ galluhop element in the six genomes generated with WebLogo [35]. Order Galliformes: *G. gallus* (A – A′),*M. gallopavo* (B – B′), *C. virginianus* (C – C′), *L. tetrix* (D – D′) and *C. japonica* (E – E′). Order Bucerotiformes: *B. rhinoceros* (F – F′). (TIFF 7747 kb)

Additional file 5: Figure S3. Graphical representation of the conservation of target site duplications (TSDs). The TSDs 5′ and 3′ galluhop element in the six genomes generated with WebLogo [35]. Order Galliformes: *G. gallus* (A – A′),*M. gallopavo* (B – B′), *C. virginianus* (C – C′), *L. tetrix* (D – D′) and *C. japonica* (E – E′). Order Bucerotiformes: *B. rhinoceros* (F– F′). (TIFF 4900 kb)

Abbreviations
HTT: Horizontal transposon transfer; K2P: Kimura 2 parameters; MITE: Miniature inverted repeats transposable element; ORF: Open reading frame; TE: Transposable element; TIR: Terminal inverted repeat; TSD: Target site duplication

Acknowledgements
The authors are grateful to Dr. Elgion Loreto for his inspiring comments on the manuscript, to Thays de Oliveira for assistance with the figures, and to the anonymous reviewers for their valuable comments to improve the manuscript.

Funding
Not applicable.

Authors' contributions
NB, GW, FT, AG and RG conceived and designed the experiments. NB and GW performed the computational analysis and wrote the paper. All authors read and approved the final manuscript.

Competing interests
The authors declare that they have no competing interests.

Author details
[1]Programa de Pós-graduação em Ciências Biológicas, Universidade Federal do Pampa (Unipampa), São Gabriel, Rio Grande do sul 97300-000, Brazil. [2]Laboratório de Diversidade Genética Animal, Universidade Federal do Pampa (Unipampa), São Gabriel, Rio Grande do sul 97300-000, Brazil. [3]Departamento de Entomologia, Instituto Aggeu Magalhães – FIOCRUZ-CPqAM, Recife, Pernambuco, Brazil.

References

1. Finnegan DJ. Eukaryotic transposable elements and genome evolution. Trends Genet. 1989. [cited 1989 Apr 1];5:103–7. Available from: http://www.ncbi.nlm.nih.gov/pubmed/2543105.

2. Wicker T, Sabot F, Hua-Van A, Bennetzen JL, Capy P, Chalhoub B, et al. A unified classification system for eukaryotic transposable elements. Nat Rev Genet. [Internet]. Nature Publishing Group; 2006 [cited 2006];8:973–82. Available from: http://www.nature.com/nrg/journal/vaop/ncurrent/full/nrg2165.html.

3. Kapitonov VV, Jurka J. A universal classification of eukaryotic transposable elements implemented in Repbase. Nat Rev Genet. 2008 [cited 2008 May 1]; 9:411–2; author reply 414. Available from: http://www.ncbi.nlm.nih.gov/pubmed/18421312.

4. Kidwell MGMG, Lisch DRDR. Perspective: transposable elements, parasitic DNA, and genome evolution. Evol Int J Org Evol. 2000 [cited 2000];55:1–24. Available from: http://www.ncbi.nlm.nih.gov/pubmed/11263730.

5. Silva JC, Loreto EL, Clark JB. Factors that affect the horizontal transfer of transposable elements. Curr Issues Mol Biol. 2004 [cited 2004 Jan 1];6:57–71. Available from: http://www.ncbi.nlm.nih.gov/pubmed/14632259.

6. Feschotte C. Transposable elements and the evolution of regulatory networks. Nat Rev Genet. 2008 [cited 2008 May 1];9:397–405. Available from: http://www.ncbi.nlm.nih.gov/pubmed/18368054.

7. Cáceres M, Puig M, Ruiz A. Molecular characterization of two natural hotspots in the Drosophila buzzatii genome induced by transposon insertions. Genome Res. 2001 [cited 2001 Aug 1];11:1353–64. Available from: http://www.ncbi.nlm.nih.gov/pubmed/11483576.

8. McVean G. What drives recombination hotspots to repeat DNA in humans? Philos Trans R Soc London Ser B, Biol Sci. 2010 [cited 2010 Apr 27];365: 1213–8. Available from: http://www.ncbi.nlm.nih.gov/pubmed/20308096.

9. Volff J-N. Turning junk into gold: domestication of transposable elements and the creation of new genes in eukaryotes. 2005 [cited 2005];28:913–22. Available from: http://dx.doi.org/10.1002/bies.20452.

10. Casola C, Hucks D, Feschotte C. Convergent domestication of pogo-like transposases into centromere-binding proteins in fission yeast and mammals. Mol Biol Evol. 2008 [cited 2008 Jan 16];25:29–41. Available from: http://www.ncbi.nlm.nih.gov/pubmed/17940212.

11. Schaack S, Gilbert C, Feschotte C. Promiscuous DNA: horizontal transfer of transposable elements and why it matters for eukaryotic evolution. Trends Ecol Evol. 2010 [cited 2010 Sep 28];25:537–46. Available from: http://www.ncbi.nlm.nih.gov/pubmed/20591532.

12. Wallau GL, Ortiz MF, Loreto ELS. Horizontal transposon transfer in eukarya: detection, bias, and perspectives. Genome Biol. Evol. [Internet]. Oxford University Press; 2011 [cited 2011];4:801–11. Available from: http://gbe.oxfordjournals.org/content/4/8/801.short.

13. Dotto BR, Carvalho EL, Silva AF, Duarte Silva LF, Pinto PM, Ortiz MF, et al. HTT-DB: horizontally transferred transposable elements database. Bioinforma. 2015 [cited 2015 Sep 1];31:2915–7. Available from: http://www.ncbi.nlm.nih.gov/pubmed/25940562.

14. Peccoud J, Loiseau V, Cordaux R, Gilbert C. Massive horizontal transfer of transposable elements in insects. Proc Natl Acad Sci U S A. 2017 [cited 2017 May 2];114:4721–6. Available from: http://www.ncbi.nlm.nih.gov/pubmed/28416702.

15. Wallau GL, Hua-Van A, Capy P, Loreto ELS. The evolutionary history of mariner-like elements in Neotropical drosophilids. Genetica [Internet]. Springer; 2010 [cited 2010];139:327–38. Available from: http://link.springer.com/article/10.1007/s10709-011-9552-6.

16. Oliveira SG, Bao W, Martins C, Jurka J. Horizontal transfers of Mariner transposons between mammals and insects. Mob DNA. 2012 [cited 2012 Sep 26];3:14. Available from: http://www.ncbi.nlm.nih.gov/pubmed/23013939.

17. Suh A, Witt CC, Menger J, Sadanandan KR, Podsiadlowski L, Gerth M, et al. Ancient horizontal transfers of retrotransposons between birds and ancestors of human pathogenic nematodes. Nat Commun. 2016 [cited 2016 Apr 21];7:11396. Available from: http://www.ncbi.nlm.nih.gov/pubmed/27097561.

18. González J, Petrov D. Genetics. MITEs–the ultimate parasites. Science. 2009 [cited 2009 Sep 11];325:1352–3. Available from: http://www.ncbi.nlm.nih.gov/pubmed/19745141.

19. Fattash I, Rooke R, Wong A, Hui C, Luu T, Bhardwaj P, et al. Miniature inverted-repeat transposable elements: discovery, distribution, and activity. Genome. 2013 [cited 2013 Sep 8];56:475–86. Available from: http://www.ncbi.nlm.nih.gov/pubmed/24168668.

20. Osborne PW, Luke GN, Holland PWH, Ferrier, DEK. Identification and characterisation of five novel miniature inverted-repeat transposable elements (MITEs) in amphioxus (Branchiostoma floridae). Int J Biol Sci. 2006 [cited 2006 Apr 10];2:54–60. Available from: http://www.ncbi.nlm.nih.gov/pubmed/16733534.

21. Yang G, Nagel DH, Feschotte C, Hancock CN, Wessler SR. Tuned for transposition: molecular determinants underlying the hyperactivity of a Stowaway MITE. Science. 2009 [cited 2009 Sep 11];325:1391–4. Available from: http://www.ncbi.nlm.nih.gov/pubmed/19745152.

22. Deprá M, Ludwig A, Valente VL, Loreto EL. Mar, a MITE family of hAT transposons in Drosophila. Mob DNA. 2012 [cited 2012 Aug 31];3:13. Available from: http://www.ncbi.nlm.nih.gov/pubmed/22935191.

23. Le Rouzic A, Capy P. Population genetics models of competition between transposable element subfamilies. Genetics. 2006 [cited 2006 Oct 3];174: 785–93. Available from: http://www.ncbi.nlm.nih.gov/pubmed/16888345.

24. Kordis D. Transposable elements in reptilian and avian (sauropsida) genomes. Cytogenet Genome Res. 2009 [cited 2009 Mar 6];127:94–111. Available from: http://www.ncbi.nlm.nih.gov/pubmed/20215725.

25. Kapusta A, Suh A. Evolution of bird genomes—a transposon's-eye view. Ann New York Acad Sci. 2017 [cited 2017 Feb 20];1389:164–85. Available from: http://www.ncbi.nlm.nih.gov/pubmed/27997700.

26. O'Brien SJ with Koepfli K-P, Paten B, Genome 10K Community of Scientists. The Genome 10K Project: a way forward. Annu Rev Anim Biosci. 2014 [cited 2014]; 3:57–111. Available from: http://www.ncbi.nlm.nih.gov/pubmed/25689317.

27. Organ CL, Shedlock AM, Meade A, Pagel M, Edwards SV. Origin of avian genome size and structure in non-avian dinosaurs. Nature. 2007 [cited 2007 Mar 8];446:180–4. Available from: http://www.ncbi.nlm.nih.gov/pubmed/17344851.

28. Zhang Q, Edwards SV. The evolution of intron size in amniotes: a role for powered flight? Genome Biol Evol. 2011 [cited 2011];4:1033–43. Available from: http://www.ncbi.nlm.nih.gov/pubmed/22930760.

29. International Chicken Genome Sequencing Consortium. Sequence and comparative analysis of the chicken genome provide unique perspectives on vertebrate evolution. Nature. 2004 [cited 2004 Dec 9];432:695–716. Available from: http://www.ncbi.nlm.nih.gov/pubmed/15592404.

30. Plasterk RH, Izsvák Z, Ivics Z. Resident aliens: the Tc1/mariner superfamily of transposable elements. Trends Genet. 1999 [cited 1999 Aug 1];15:326–32. Available from: http://www.ncbi.nlm.nih.gov/pubmed/10431195.

31. Shao H, Tu Z. Expanding the diversity of the IS630-Tc1-mariner superfamily: discovery of a unique DD37E transposon and reclassification of the DD37D and DD39D transposons. Genetics. 2001 [cited 2001 Nov 1];159:1103–15. Available from: http://www.ncbi.nlm.nih.gov/pubmed/11729156.

32. Rouault J-D, Casse N, Chénais B, Hua-Van A, Filée J, Capy P. Automatic classification within families of transposable elements: Application to the mariner Family. Gene. 2009 [cited 2009 Dec 15];448:227–32. Available from: http://www.ncbi.nlm.nih.gov/pubmed/19716406.

33. Wicker T, Robertson JS, Schulze SR, Feltus FA, Magrini V, Morrison JA, et al. The repetitive landscape of the chicken genome. Genome Res. 2005 [cited 2005 Jan 15];15:126–36. Available from: http://www.ncbi.nlm.nih.gov/pubmed/15256510.

34. Jurka J, Kapitonov V, Pavlicek A, Klonowski P, Kohany O, Walichiewicz J. Repbase Update, a database of eukaryotic repetitive elements. Cytogenet Genome Res. 2004 [cited 2004];110:462–7. Available from: http://dx.doi.org/10.1159/000084979.

35. Bao W, Kojima KK, Kohany O. Repbase Update, a database of repetitive elements in eukaryotic genomes. Mob DNA. 2015 [cited 2015 Jun 2];6:11. Available from: http://www.ncbi.nlm.nih.gov/pubmed/26045719.

36. Katoh K, Standley DM. MAFFT multiple sequence alignment software version 7: improvements in performance and usability. Mol Biol Evol. 2013 [cited 2013 Apr 16];30:772–80. Available from: http://www.ncbi.nlm.nih.gov/pubmed/23329690.

37. Crooks GE, Hon G, Chandonia JM, Brenner SE. WebLogo: a sequence logo generator. 2003 [cited 2003];14:1188–90. Available from: http://www.ncbi.nlm.nih.gov/entrez/query.fcgi?cmd=Retrieve&db=PubMed&dopt=Citation&list_uids=15173120.

38. Okonechnikov K, Golosova O, Fursov M. Unipro UGENE: a unified bioinformatics toolkit. Bioinformatics. 2012 [cited 2012 Apr 15];28. Available from: https://academic.oup.com/bioinformatics/article-lookup/doi/10.1093/bioinformatics/bts091.

39. Wickham H. ggplot2: elegant graphics for data analysis. Springer New York: Springer-Verlag New York Inc.; 2008 [cited 2008]. Available from: http://www.springer.com/us/book/9780387981413.

40. R Core Team R version 3.0. R: A language and environment for statistical computing. Found. Stat. Comput. Vienna, Austria. Vienna, Austria: R Foundation for Statistical Computing; 2012 [cited 2012]; Available from: https://www.r-project.org/.

41. Dan Graur, Wen-Hsiung Li. Fundamentals of Molecular Evolution. 2nd edn. Sunderland: Sinauer Associates Inc.; 2000. pp. 481. paperback. ISBN 0878932666. https://global.oup.com/academic/product/fundamentals-of-molecular-evolution-9780878932665?cc=br&lang=en&

42. Jarvis ED, Gilbert MTP, Wang J with Zhang G, Li C, Li Q, Li B, Larkin DM, Lee C, et al., Avian Genome Consortium. Comparative genomics reveals insights into avian genome evolution and adaptation. Science. 2014 [cited 2014 Dec 12];346:1311–20. Available from: http://www.ncbi.nlm.nih.gov/pubmed/25504712.

43. Zdobnov EM, Tegenfeldt F, Kuznetsov D, Waterhouse RM, Simão FA, Ioannidis P, et al. OrthoDB v9.1: cataloging evolutionary and functional annotations for animal, fungal, plant, archaeal, bacterial and viral orthologs. Nucleic Acids Res. [Internet]. 2017 [cited 2017 Jan 4];45. Available from: https://academic.oup.com/nar/article-lookup/doi/10.1093/nar/gkw1119.

44. Guindon S, Gascuel O. A Simple, Fast, and Accurate Algorithm to Estimate Large Phylogenies by Maximum Likelihood. Syst Biol. 2002 [cited 2002];52: 696–704. Available from: http://dx.doi.org/10.1080/10635150390235520.

45. Anisimova M, Gascuel O. Approximate likelihood-ratio test for branches: A fast, accurate, and powerful alternative. Syst Biol. 2006 [cited 2006 Aug 1];55: 539–52. Available from: http://www.ncbi.nlm.nih.gov/pubmed/16785212.

46. Hedges SB, Marin J, Suleski M, Paymer M, Kumar S. Tree of life reveals clock-like speciation and diversification. Mol Biol Evol. 2015 [cited 2015 Apr 3];32: 835–45. Available from: http://www.ncbi.nlm.nih.gov/pubmed/25739733.

47. Naito K, Cho E, Yang G, Campbell MA, Yano K, Okumoto Y, et al. Dramatic amplification of a rice transposable element during recent domestication. Proc. Natl. Acad. Sci. United States Am. 2006 [cited 2006 Nov 21];103:17620–5. Available from: http://www.ncbi.nlm.nih.gov/pubmed/17101970.

48. Rasmussen PC. Threatened Birds of Asia: The BirdLife International Red Data Book. Nigel J Collar Auk [Internet]. 2004 [cited 2004 Apr 1];121. Available from: http://www.jstor.org/stable/info/10.2307/4090426.

49. Boev Z, Kovachev D. Euroceros bulgaricus gen. nov., sp. nov. from Hadzhidimovo (SW Bulgaria) (Late Miocene) – the first European record of Hornbills (Aves: Coraciiformes). Geobios [Internet]. 2007 [cited 2007 Jan 1]; 40. Available from: http://linkinghub.elsevier.com/retrieve/pii/S0016699506001069.

50. Jarvis ED, Mirarab S, Aberer AJ, Li B, Houde P, Li C, et al. Whole-genome analyses resolve early branches in the tree of life of modern birds. Science. 2014 [cited 2014 Dec 12];346:1320–31. Available from: http://www.ncbi.nlm.nih.gov/pubmed/25504713.

51. Dyke GJ. Jiri Mlikovsky: Cenozoic birds of the world, part 1: Europe. J. Vertebr. Paleontol. [Internet]. Taylor & Francis Group; 2003 [cited 2003 Apr 11];23:258–258. Available from: http://www.tandfonline.com/doi/abs/10.1671/0272-4634(2003)23[258:CBOTWP]2.0.CO;2.

52. Claramunt S, Cracraft J. A new time tree reveals Earth history's imprint on the evolution of modern birds. Sci Adv. 2015 [cited 2015 Dec 11];1: e1501005. Available from: http://www.ncbi.nlm.nih.gov/pubmed/26824065.

53. Gilbert C, Pace, II JK, Feschotte C. Horizontal SPINning of transposons. Commun. & Integr. Biol. [Internet]. Taylor & Francis; 2009 [cited 2009 Mar 1];2: 117–9. Available from: http://www.tandfonline.com/doi/abs/10.4161/cib.7720.

54. Gilbert C, Hernandez SS, Flores-Benabib J, Smith EN, Feschotte C. Rampant horizontal transfer of SPIN transposons in squamate reptiles. Mol Biol Evol. 2012 [cited 2012 Feb 18];29:503–15. Available from: http://www.ncbi.nlm.nih.gov/pubmed/21771716.

13

Diversity of *P*-element piRNA production among M' and Q strains and its association with P-M hybrid dysgenesis in *Drosophila melanogaster*

Keiko Tsuji Wakisaka[1*], Kenji Ichiyanagi[2], Seiko Ohno[3] and Masanobu Itoh[1,4*]

Abstract

Background: Transposition of *P* elements in the genome causes P–M hybrid dysgenesis in *Drosophila melanogaster*. For the P strain, the P–M phenotypes are associated with the ability to express a class of small RNAs, called piwi-interacting small RNAs (piRNAs), that suppress the *P* elements in female gonads. However, little is known about the extent to which piRNAs are involved in the P–M hybrid dysgenesis in M' and Q strains, which show different abilities to regulate the *P* elements from P strains.

Results: To elucidate the molecular basis of the suppression of paternally inherited *P* elements, we analyzed the mRNA and piRNA levels of *P* elements in the F1 progeny between males of a P strain and nine-line females of M' or Q strains (M' or Q progenies). M' progenies showed the hybrid dysgenesis phenotype, while Q progenies did not. Consistently, the levels of *P*-element mRNA in both the ovaries and F1 embryos were higher in M' progenies than in Q progenies, indicating that the M' progenies have a weaker ability to suppress *P*-element expression. The level of *P*-element mRNA was inversely correlated to the level of piRNAs in F1 embryos. Importantly, the M' progenies were characterized by a lower abundance of *P*-element piRNAs in both young ovaries and F1 embryonic bodies. The Q progenies showed various levels of piRNAs in both young ovaries and F1 embryonic bodies despite all of the Q progenies suppressing *P*-element transposition in their gonad.

Conclusions: Our results are consistent with an idea that the level of *P*-element piRNAs is a determinant for dividing strain types between M' and Q and that the suppression mechanisms of transposable elements, including piRNAs, are varied between natural populations.

Keywords: Ping-pong-paired piRNA, Natural populations, Hybrid sterility, Gonadal dysgenesis, *P*-element mRNA, Progenies

Background

Transposable elements (TEs) occupy a substantial fraction of eukaryotic genomes, and their mobilization causes insertional mutations. Therefore, although such mobilization could provide genetic variations and drive genome evolution [1, 2], TEs could also inflict deleterious effects on the host. Piwi-interacting small RNAs (piRNAs), which are generally 23–35 nucleotides (nt) in length, suppress the expression of TEs [3]. The piRNAs can be generated via primary pathways and ping-pong biogenesis [4]. In the primary pathway, long precursor RNAs are produced from genomic loci, chopped into 23- to 35-nt RNAs (called primary piRNAs), and loaded onto the Piwi-family of protein(s). In the ping-pong biogenesis, which is known as the ping-pong amplification cycle, the piRNA-bound Piwi-family of proteins cleaves an RNA that is complementary to the bound piRNA. The cleavage occurs at the site 10-nt away from the

* Correspondence: d2811509@edu.kit.ac.jp; k-wakisaka@mbn.nifty.com; mitoh@kit.ac.jp
[1]Department of Applied Biology, Kyoto Institute of Technology, Hashigamicyo, Matsugasaki, Sakyo-ku, Kyoto 606-8585, Japan
Full list of author information is available at the end of the article

5′ end of the guide piRNA, and the 3′ end of the cleaved RNA is trimmed to give a 23- to 35-nt RNA (ping-pong piRNA), which are loaded onto a Piwi-family protein to guide the next round of this complementarity-based RNA cleavage. Therefore, the two RNA species (ping-pong pairs) show a characteristic 10-nt complementarity in the respective 5′ regions, referred to as a "ping-pong signature." If a primary piRNA has a sequence antisense to a TE, it can guide the cleavage of the mRNA of the TE. Moreover, both primary and ping-pong piRNAs can guide the introduction of repressive chromatin modifications at genomic sites complementary to them. Both the primary and ping-pong biogenesis are active in germline cells in *Drosophila* [2, 5, 6] and in other organisms [4, 7]. However, in the *Drosophila* soma, only the primary pathway is utilized to generate piRNAs [8–13].

The *P* element is a DNA transposon, and their copies in the *Drosophila melanogaster* genome include structurally complete and incomplete variants. The autonomous complete elements, which are 2907 base pairs in length, encode an 87 kDa transposase that is expressed in the germline cells [14–16]. In *D. melanogaster*, crossing between females lacking *P* elements (M strain) and males carrying them (*P* strain) leads to the transposition of *P* elements in the F1 progeny (referred to as M progeny here), which causes abnormalities in the germline cells, such as gonadal dysgenesis (GD) with sterility, mutations, chromosomal breaks, and male recombination [17–20]. This phenomenon is known as P–M hybrid dysgenesis. In contrast, when *P*-strain females are mated with P-strain males, *P*-element mobilization is prevented by maternally deposited piRNAs in the germline cells and early embryos, which are laid by P-strain mothers but not P-progeny mothers (referred to as F1 embryos of P progenies) [21]. A female's capacity to allow *P*-element transposition is defined as *P* susceptibility, which is low in the P strain but high in the M strain.

M′ and Q strains, which show different P–M phenotypes from P strains, are currently the most common in the natural populations in Eurasia, Africa, Australia, and the Far East [22–24]. Although M′ progeny allows transposition of *P* elements in the germline cells (high *P* susceptibility), the M′ strains possess many copies of *P* elements in the genome [25–27]. The Q strain carries *P* elements and have an ability to repress *P* mobilization in their progenies (low *P* susceptibility) [28–30]. In contrast to the *P* strain, males of the M′ and Q strains have no ability to induce transposition of *P* elements in their progeny (low *P* inducibility). In wild-type strains, previous studies show that *KP* elements, which are nonautonomous incomplete elements, are associated with repression [31, 32]. It has been proven that *KP* polypeptides repress *P* transposition in M′ strains [33–36]. By contrast, in both M′ and Q strains, only a weak correlation was observed between the types of genomic *P* elements and the phenotypes of the P–M

system [37–40]. In our previous study, we proved that one line of M′ strain, named OM5 (see methods), have many *KP* elements in transcriptionally active sites and only a few autonomous *P* elements in inactive sites of their genomes [41]. *KP*-mediated repression and piRNA-mediated repression are also confounded [42]. Previously, it has been proved that weak piRNA-mediated repression enhances *KP*-mediated repression [43, 44]. Therefore, a major factor affecting the different *P* susceptibilities in the M′ and Q progenies remain unrevealed. It is possible that there are two hypotheses in the P–M system of M′ and Q strains as described below: (1) While neither strain contains active *P* elements to induce hybrid dysgenesis, the Q strains produce a greater number of piRNAs that enact maternal repression. (2) While M′ strains do not contain active *P* elements to induce hybrid dysgenesis, the Q strains repress dysgenesis both maternally and paternally through *KP*-mediated repression.

To study whether the production of piRNAs is involved in the difference in *P* susceptibility between M′ and Q progenies, we examined the expression levels of *P*-element piRNAs in the ovaries and whole F1 embryos. This was done by generating progenies from crossing males of a P strain and females of nine wild-type strains of the M′ or Q phenotype. We tested 2- to 3-day-old ovaries of the hybrids. These are considered to be affected by piRNAs derived from the maternally inherited *P* elements because Khurana et al. [45] showed that ovaries of 2- to 4-day-old hybrids generated by a cross between M-strain females and Har males produce no piRNAs. Moreover, the 2- to 3-day-old ovaries of hybrids were suitable for the evaluation of repression of *P* activity since they possess zygotic *P* elements from Har in their genome. Whole F1 embryos of hybrids were used for the same reasons as ovaries. The results revealed diversity in the expression levels of *P*-element piRNAs, which were correlated with mRNA expression. Importantly, we found that the production of *P*-element piRNAs was a factor dividing *P* susceptibility between the M′ and Q strains and that these piRNA production show different characters between natural strains.

Methods
Fly stocks
Nine isofemale *Drosophila melanogaster* lines were used: OM5, FIZ12 (FIZ-12-11), KY25 (KY-13-25), KY98 (KY-13-98), KY3 (KY-02-003), KY101 (KY-02-101), HKH (Hikone-H 1957), MSO12 (MSO-12-41), and KY74 (KY-02-074). Flies were maintained on a standard cornmeal medium at 25 °C in the laboratory throughout this investigation. The exception was for the GD test, where Harwich (Har) males and Canton S (CS) females were used as standard P and M strains, respectively. We used Har females as a control. These females had the capacity to repress paternal

P-element transposition by maternally deposited P-element piRNAs [21].

Gonadal dysgenesis (GD) test

GD tests were used to determine the strain types in the P–M system [18, 46]. Two kinds of crosses, A* (tested females × Har males) and A (CS females × tested males), were performed at 28 °C. By analyzing more than 50 F1 females for each line, the GD score was calculated as the percentage of females having dysgenic ovaries. The P–M strain type was determined based on GD scores in the cross A* (indicating susceptibility of P transposition) and those in the cross A (indicating P inducibility). The criteria for M′ strains were <10% GD in cross A and >10% GD in cross A*. The criteria for Q strains were <10% GD in both crosses [47] (see Table 1). KY25, KY98, MSO12, and FIZ12 were tested first. We retested KY3, HKH, KY101, KY74, and OM5, because these lines had undergone many generations since the previous GD tests [48].

RNA preparation

To accurately analyze the correlation between the number of P-element piRNAs and the expression level of P-element mRNA, both small RNAs and total RNAs were prepared from same sample, as described below. Total RNA was extracted from 2- to 3-day-old ovaries or 0- to 24-h F1 embryos with the miRNeasy kit (Qiagen). Small RNAs were separated using the RNeasy MinElute Cleanup Kit (Qiagen). 0- to 24-h embryos were generated by 30–40 couples of cross A* kept in bottles on dishes. Eight ovaries of 2- to 3-day-old F1 females were dissected. These ovaries were generated by approximately 20 couples kept in bottles for 4–7 days at the GD-inducing temperature of 28 °C [18, 46 were arranged]. In OM5 × Har, we used equal numbers of complete and dysgenic ovaries.

Small RNA sequencing

The small RNA libraries were produced using 1 μg of small RNAs with the Truseq small RNA sample preparation kit (Illumina). After PCR amplification, products of approximately 150 bp were collected from a 6%

Table 1 Strain types in the P-M system

P susceptibility high: >10%GD low: <10%GD	P inducibility low: <10%GD high: >10%GD	strain type
high	low	M′
low	low	Q
low	high	P
high	low	M (P-elements (−))

Drosophila melanogaster is divided into the four strain types by GD ratios. P susceptibility shows the regulatory capacity against the P-elements and P inducibility exhibits the ability to transpose P-elements in progeny

polyacrylamide gel. Single-end 50-bp sequencing of these libraries was carried out on MiSeq (Illumina).

Analysis of the obtained piRNA sequence was performed as previously described [21, 8, 45] using the CLC Genomics Workbench (detailed protocol is described in https://www.qiagenbioinformatics.com/support/manuals/). After trimming of the adaptor sequence by Transcriptomics Analysis in g_x, we removed the reads corresponding to 2SrRNA, which were included in a considerable ratios (average of 92% of total reads). To see how much of the sequencing libraries corresponded to 2SrRNAs, we examined the number of total reads, 23- to 30-nt piRNAs and 186 TE-derived 23- to 30-nt piRNAs (Additional file 1: Table S1). Reads that were mapped to rRNAs, tRNAs, and snoRNAs were removed. The remaining reads were mapped to the D. melanogaster genome (Release R22) using Download Genome in g_x. RNA reads of 23–35 nts that did not match miRNA sequences in miRBase [49] were defined as piRNAs. These sequences were then mapped to P-element sequences [14] and 186 transposons (total TEs) (Repbase) by Map Reads to Reference in g_x. For normalization across the samples, the read numbers of piRNAs mapped to P elements were divided by the total number of miRNA reads and multiplied by 1 million. This gave the reads per million (RPM miRNA reads). Ping-pong signatures were analyzed by per scripts [3, 50, 51].

RT-PCR and quantitative RT-PCR

cDNA was synthesized by superscript III reverse transcriptase (Invitrogen) using total RNA and oligo-dT primer. Quantitative amplification of cDNA was performed in duplicate using SYBR Green quantitation (Toyobo) on a 7000 HT Fast Real-Time PCR System (Applied Biosystems; forward and reverse primers: 5′-GTGGGAGTACACAAACA GAGTCCTG-3′ and 5′-CGTATCTGCGTGTCCGTGA AGA-3′). The level of P-element mRNA was normalized to that of RP49 mRNA (forward and reverse primers: 5′-CGGATCGATATGCTAAGCTGT and 5′-GCGCTT GTTCGATCCGTA) [52].

Statistical analysis

The Pearson product-moment correlation test and hierarchical cluster analysis were performed using R. For the hierarchical cluster analyses in Figs. 1e and 3b, we used the hclust function in R (ver. 3.0.2) with the furthest neighbor method.

Results

GD test revealed two lines of M′ and seven lines of Q strains

To test their capacity to regulate the paternally inherited P elements in F1 ovaries, females of nine natural strains were crossed to Har males (P strain) having high P inducibility (cross A*). The GD scores (fraction of their

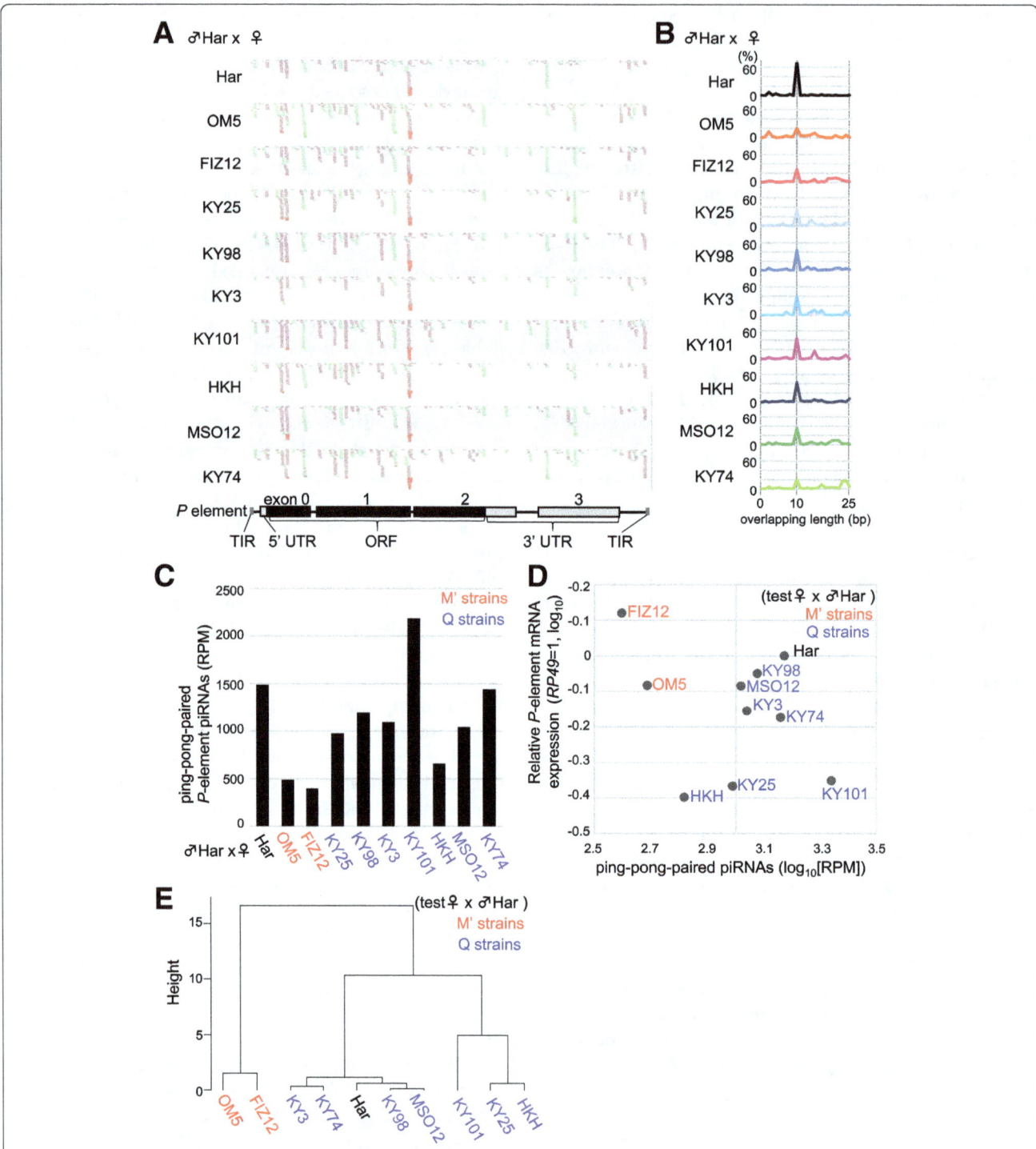

Fig. 1 Expression of piRNA and mRNA of *P* elements in adult ovaries of F1 progenies in cross A*. **a** Small RNA reads (23–35 nt in length) mapped to the sense (green) and antisense (red) strands of the *P* element are shown on the *P*-element structure (bottom). Har (top) was a P strain and used as a control. **b** Frequencies of length (0–25 bp, *x*-axis) of overlapping regions between sense and antisense small RNAs (23–35 nt) identified in ovaries of F1 progenies. An overlap of 10 bp is a signature of piRNA pairs produced via the ping-pong cycle. **c** The expression levels of ping-pong-paired piRNAs in F1 ovaries normalized by miRNA (reads per million [RPM] miRNAmiRNA reads). The strain names of mothers are shown in black (P), red (M'), and blue (Q). **d** The relationship between the log expression levels of mRNAs (*y*-axis) and ping-pong-paired piRNAs (*x*-axis) of *P* elements in F1 ovaries. The strain names of mothers are shown in black (P), red (M'), and blue (Q). The Pearson's correlation efficient is shown on the top. **e** A tree of hierarchical clustering of the nine natural strains and the Har strain based on the data shown in panel **c**. The strain names of mothers are shown in black (P), red (M'), and blue (Q). The M' strains are clustered together

daughters showing dysgenic ovaries, see Methods) in cross A* indicate the P susceptibility of the test strain (Table 1). F1 progeny of KY25, KY98, KY3, KY101, HKH, MSO12, and KY74 displayed GD scores of 0 to 10%, indicating that P-element transposition was highly repressed in their ovaries (Table 2). In contrast, OM5 and FIZ12 showed GD scores of more than 10%, indicating P-element transposition activity in their ovaries. We also analyzed ovaries of F1 progeny from cross A, where males of each strain were crossed to CS females (M stain) with P susceptibility. In all tests, F1 progeny displayed GD scores less than 1% (Table 2), indicating that P inducibility is very limited in the nine strains.

We classified these nine lines into two types according to the GD scores. Seven strains (KY25, KY98, KY3, KY101, HKH, MSO12, and KY74) showed low P susceptibility and low P inducibility, and thus they were Q strains. The other two strains (OM5 and FIZ12) were classified as M' strains due to their high P susceptibility and low P inducibility.

Various levels of ping-pong-paired piRNAs derived from P elements in ovaries of young dysgenic progenies

The GD test above showed that progenies from the M' strains (M' progenies) displayed higher P susceptibilities than those from the Q strain (Q progenies) and the P strain (P progenies). To examine the possibility that this variation is due to the difference in the expression level of P-element piRNAs in germline cells of the F1 progenies, we performed deep sequencing of small RNAs present in the ovaries of 2- to 3-day-old progenies of crosses between Har males and

M' or Q females. After removal of miRNAs and fragments of functional RNAs, small RNAs of 23- to 35 nt in length were mapped to the sequences of P elements to identify P-element piRNAs (Fig. 1a).

In all cases, we detected P-element-derived piRNAs in both sense and antisense directions. These piRNAs were mapped mainly to exons 0 and 1, showing that there is some sequence similarity between lines. The M' progenies (OM5 and FIZ15) produced the lowest numbers of piRNAs compared with the Q and P progenies, except for HKH. Such a low abundance was specific to the P element because the total TE-derived piRNAs in the M' progenies were comparable with those in others (Table 2). To study whether the detected piRNAs are generated via ping-pong biogenesis in germline cells, we analyzed the overlap between sense and antisense piRNAs (Fig. 1b). Indeed, a peak at 10 bp was evident in all cases, which suggested that a substantial fraction of the piRNAs were produced via ping-pong biogenesis. Interestingly, abundance of ping-pong-paired piRNAs were less in the M' progenies compared with the Q and P progenies, suggesting that the ability of M' progenies to amplify and maintain piRNAs in the germline cells is weaker than that of Q and P progenies (Fig. 1c). The Q progenies expressed various amounts of ping-pong piRNAs. These amounts were comparable with those in the P progenies and highlight that the higher ability to repress the P element is associated with a higher expression of ping-pong-paired piRNAs in the ovaries. In particular, KY101 progenies showed quite high amounts of ping-pong-paired piRNAs produced from P elements.

We next determined the levels of P-element mRNA in these ovaries by reverse transcription followed by

Table 2 GD ratios and total P-element piRNAs production in the progeny

Test strain	GD[a] (%) cross A* (♀test x ♂Har)	GD[a] (%) cross A (♀CS x ♂test)	Deduced strain type	P-element piRNAs (RPM)[bb] cross A* (♀test x ♂Har)		total-TE piRNAs (RPM)[bb] cross A* (♀test x ♂Har)	
				Ovaries	F1 embryos	Ovaries	F1 embryos
OM5	28.3	0	M'	5137	233	1,105,901	139,280
FIZ12	13.3	1	M'	6108	522	1,158,677	138,991
KY25	0	0	Q	7333	740	1,428,653	130,670
KY98	0	1	Q	9356	830	1,502,704	143,827
KY3	2.5	0	Q	8049	818	1,060,199	144,297
KY101	0	0	Q	18,009	1662	1,958,902	109,059
HKH	0	0	Q	4989	2421	1,048,057	417,233
MSO12	0	0	Q	8941	5200	1,351,448	521,266
KY74	0.8	0	Q	9077	6336	1,343,427	425,912
Har	0	100	P	7780	604	1,090,850	99,772

[a]Percentage of gysgenic ovaries from cross A* (test female x Har male) and cross A (CS female x test male). [bb]piRNA reads were divided by miRNA reads, expressed as reads per million miRNA reads (RPM) in the progeny from cross A*

quantitative PCR (qRT-PCR). The average expression levels of ovarian *P*-element mRNA was 0.1-fold lower than in embryonic *P*-element mRNA in 10 progenies. The mRNA levels varied between the progenies, with a tendency for the M′ progenies to show higher expression than the Q progenies (Fig. 1d). Furthermore, we repeated the qRT-PCR three to five times in four lines of M′ and Q strains and ensured that there was significantly higher expression of *P*-element mRNA in M′ (OM5) progenies compared with that in Q progenies (KY3, KY101 and KY74; p = 0.03, 0.003 and 0.05, respectively; Additional file 2: Figure S1). However, ovaries of KY3 (Q) progenies showed a high score of standard division (SD = 0.3). This suggests that individuals of KY3 progenies differ in their expression level of *P* elements. Importantly, the two M′ progenies were clustered in hierarchical clustering of *P*-element mRNA and *P*-element ping-pong piRNA expression levels (Fig. 1e). These results favor an idea that the level of ping-pong-paired piRNAs is one determining factor for the expression level of *P* elements in natural populations.

M′ progenies were characterized by a low ability to produce ping-pong-paired piRNAs and high levels of *P*-element expression in the ovaries. While Q progenies were distinguished from M′ progenies by the amount of ping-pong-paired piRNAs and the levels of *P*-element expression, they showed variable levels of expression of piRNAs and mRNA.

Various levels of ping-pong-paired piRNAs derived from *P* elements in F1 embryos of progenies

To study the possible involvement of piRNAs in the regulation of the paternally inherited *P* elements during embryogenesis of the F1 progeny, we next analyzed *P*-element piRNAs and mRNA in whole F1 embryos (<24 h after hatching) of progenies of cross A*. It has been proven that *P*-element piRNAs produced in F1 embryos of hybrids between M-strain females and Har are very limited [45]. In contrast, we detected *P*-element piRNAs in whole F1 embryos of M′, Q, and P progenies (Table 2). There was a considerable variation in the abundance. The M′ progenies again showed the lowest abundance of *P*-element piRNAs although they produced total TE-derived piRNAs at levels similar to those in the Q and P progenies (Table 2). Analysis of sense and antisense piRNAs revealed that ping-pong-paired piRNAs are generally lower in whole F1 embryonic bodies than in ovaries. In particular, the two M′ progenies, in addition to KY98, KY3, and HKH progenies, produced a fewer number of ping-pong-paired piRNAs (Fig. 2b and c). It is possible that some of the strange discrepancies with ovarian piRNAs from the same lines are caused by the limited power to accurately estimate

the ping-pong fraction. This could be due to the production level of total-TE-derived piRNAs in F1 embryonic bodies being less than those in the ovaries (Table 2). Therefore, the level of total *P*-element piRNAs was evaluated to compare differences between lines, as below.

We investigated whether the expression of *P*-element mRNA was associated with the production of piRNAs derived from *P* elements in whole F1 embryos of the natural strains. We quantified *P*-element mRNA in the F1 embryonic bodies. This revealed that *P*-element expression is somewhat higher (not significantly) in M′ progenies compared with Q progenies (Fig. 2d). We repeated qRT-PCR three times in five lines of M′, Q, and P strains and ensured that there was a significantly higher expression of *P*-element mRNA in M′ (OM5) progenies compared with those in the Q progenies (KY3, KY101, and KY74; $p < 0.05$) (Additional file 3: Figure S2A and B). Furthermore, 10 lines were classified into P, M′, and Q strains, and it was determined that the mRNA expression level was negatively correlated to the expression level of total *P*-element piRNAs (R = −0.88, $p < 0.01$; Fig. 2d). We made sure that this negative correlation between the total *P*-element piRNAs and the mRNA level was analyzed by three biological replicates for five progenies (R = −0.9, $p < 0.05$; Additional file 3: Figure S2). These results suggest that cells in the F1 embryonic bodies produce piRNAs mainly via the primary pathway and that these primary piRNAs play a role in *P*-element regulation during embryogenesis.

M′ strains were characterized by the lowest production of ping-pong-paired piRNAs in both young adult ovary and F1 embryonic bodies

The above results showed a tendency that ping-pong-paired *P*-element piRNAs in the ovary and the total *P*-element piRNAs in F1 embryos are less in the M′ progenies than in the Q and P progenies. To reveal whether there were clear differences in the amount of piRNAs derived from *P* elements between the M′ progenies and others, we did clustering analysis *P*-element ping-pong piRNAs production in the ovaries and total *P*-element piRNAs in F1 embryos of progenies. Actually, M′ progenies were characterized by the lowest production of *P*-element piRNAs in both the young adult ovary and in F1 embryonic bodies. For the Q and P progenies, KY101, Har, KY25, KY98, and KY3 showed higher production of *P*-element *P*-element piRNAs in young adult ovaries, while HKH, MSO12, and KY74 produced higher levels of *P*-element piRNAs in the F1 embryos (Fig. 3).

Discussion

Although the natural population of *D. melanogaster* generally carries *P* elements in their genome, the

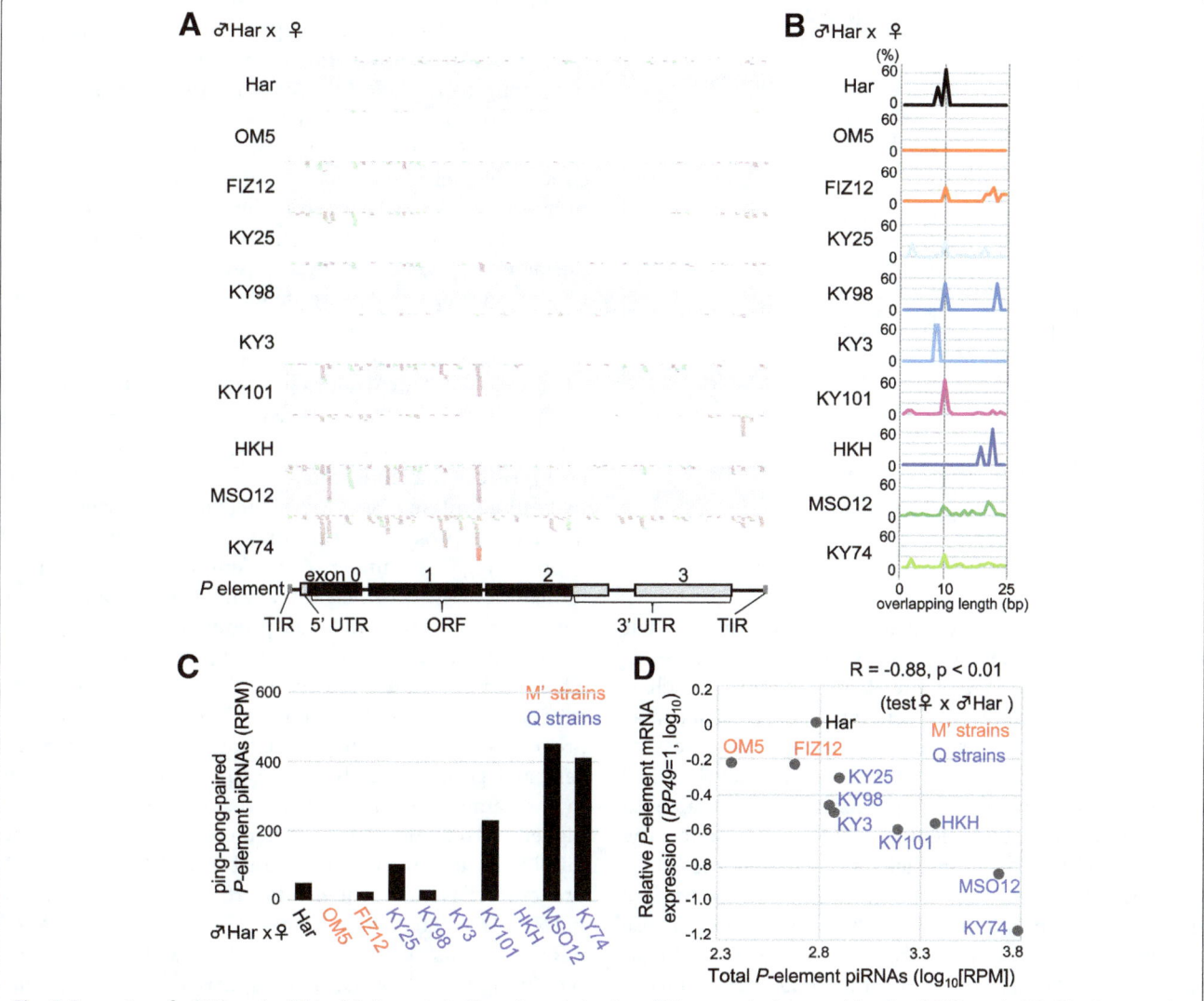

Fig. 2 Expression of piRNA and mRNA of P elements in F1 embryonic bodies of F1 progenies in cross A*. **a** Small RNA reads (23–35 nt) mapped to the sense (green) and antisense (red) strands of the P element are shown on the P-element structure (bottom). Har (top) was a P strain as a control. **b** Frequencies of length (0–25 bp, x-axis) of overlapping regions between sense and antisense small RNAs (23–35 nt) identified in F1 embryos. An overlap of 10 bp is a signature of piRNA pairs produced via the ping-pong cycle. **c** The expression levels of ping-pong-paired piRNAs in F1 ovaries (reads per million [RPM] miRNa reads). The strain names of mothers are shown in black (P), red (M'), and blue (Q). **d** The relationship between the log expression levels of mRNAs (y-axis) and piRNAs (x-axis) of P elements in F1 ovaries. The strain names of mothers are shown in black (P), red (M'), and blue (Q). The Pearson's correlation efficient is shown on the top

progeny displays a different resistance capacity against P elements as introduced upon hybridization with typical P strains. Here, we showed that the M' strains distinguished from the Q strains by low levels of P-element piRNA production in both the ovaries and the F1 embryos of dysgenic progenies, and that this is associated with a low ability to suppress P-element transcription. This character of M' strains is likely related to their high level of GD, which is linked to P-element transposition. In contrast, it was shown that the Q progenies produced various degrees of P-element piRNAs. This could confer the ability to resist P-element expression in embryonic bodies.

However, such varied production of P-element piRNAs among Q progenies did not induce different levels of GD.

Interestingly, M' progenies of the two lines, which showed moderate scores of GD in cross A*(10%–30%) indicating partial repression of P transposition, produced P-element piRNAs in young adult ovaries at some degree. In I–R hybrid dysgenesis, the levels of I-element piRNAs inversely correlated with dysgenic scores [53]. While it has been reported that other repressive factors for P-element transposition, such as proteins produced from full-length (type I, 66-kDa repressors) and internally deleted elements (type II, KP repressors), play a role

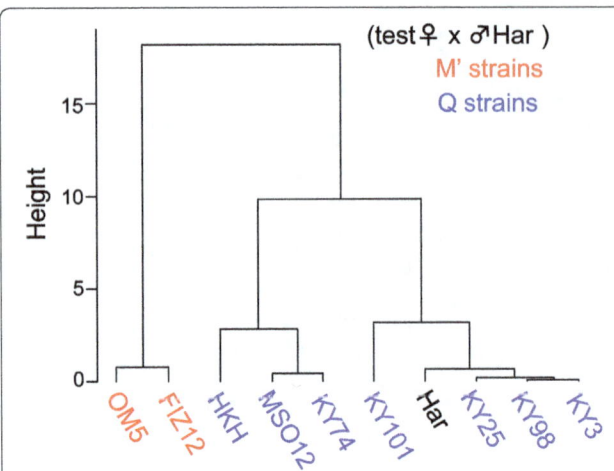

Fig. 3 Characterization of the natural strains based on piRNA levels in F1 progenies. Relationship between the expression levels (RPM) of *P*-element ping-pong-paired piRNAs in F1 ovaries and total *P*-element piRNAs in embryos. Hierarchical clustering of the nine strains and the Har

in germline cells to some degree, our results suggest that the level of *P*-element piRNAs in the M′ progenies is one major determinant of the *P* susceptibility, which is in addition to the P–M phenotype in M′ strains, as shown in the I–R system. Further studies are necessary to investigate M′ strains having various levels of *P* susceptibility. Why the M′ progenies are not able to produce abundant *P*-element piRNAs despite the presence of *P* elements in their maternal genomes? It is thought that piRNAs are inherited from the oocytes of the mothers and is imparted to the F1 progenies. These inherited piRNAs act to prime the ping-pong amplification cycle in the germline cells of the daughters. Thus, it is possible that the maternal lineage of the M′ strains does not produce abundant piRNAs. To produce both primary and ping-pong piRNAs, a genomic situation is required where *P* element(s) are located in the piRNA clusters [20]. Therefore, the copy number of *P* elements in the piRNA clusters is likely less in the genomes of the M′ strains, resulting in a reduced level of *P*-element piRNA production. Previously, it has been proven that autonomous complete *P* elements in M′ strains are transcriptionally inactive [41]. Therefore, the other possibility is that such *P* elements are repressed in M ′-strain parents and may not contribute to resistance against *P* elements introduced upon hybridization with typical P strains. Future studies, such as piRNA profiling of oocytes of mothers, will evaluate these possibilities.

For the Q strains, despite their resistance to paternal *P* elements, there was considerable variation in the mRNA and piRNA expression levels of *P* elements in both the ovary and the F1 embryonic bodies. Therefore, in Q strains, the molecular basis of production of *P*-element

piRNAs affecting the P–M phenotype is likely different from that in I–R hybrid dysgenesis. In particular, progeny of KY101 showed higher production of *P*-element ping-pong-paired piRNAs in the ovaries, suggesting that piRNAs act as a main suppressor during oogenesis. F1 embryos of MSO12 and KY74 progenies produced abundant *P*-element piRNAs, including ping-pong-paired piRNAs, and lower levels of *P*-element mRNA. This suggested that piRNAs act as one of the main suppressors during embryogenesis. Other Q progenies were classified into two groups that were characterized by KY101 and KY74, as discussed above. They allowed the expression of the *P*-element mRNA at levels similar to those in the M′ progenies. This would imply that other factors, such as protein repressors, are involved in the repression of *P*-element transposition in the Q progenies [54]. It is also possible that individuals could differ in their sensitivity to germline P activity (higher for M′ progenies and lower for Q progenies), resulting in different severities of hybrid dysgenesis under equivalent levels of transpositional activity. Furthermore, whole F1 embryos are composed of germ line cells producing ping-pong piRNAs and somatic cells producing antisense piRNAs. Thus, further studies are required to address the varied expression of both *P*-element piRNAs and mRNA in Q progenies, including the effect from embryonic somatic cells and germ line cells. Interestingly, Har progeny was in the same group as KY101 progeny, which showed a higher production of *P*-element ping-pong-paired piRNAs in the ovaries. It is possible that those Q and P progenies have *P* elements inserted into germ-specific piRNA clusters, which produce ping-pong-paired piRNAs. Thus, in the ovaries of Q and P progenies, ping-pong-paired piRNAs likely act to suppress *P* elements introduced upon hybridization with typical *P* strains. On the other hand, males of the P strain have a high ability to mobilize *P* elements in their progeny when they are mated with M-strain females; this is in contrast to what is found in the Q strain. Therefore, the P strain may possess many *P* elements in active expression sites of the genome. Another possibility is that the P strain produces lower levels of zygotic piRNAs derived from paternal *P* elements. More investigation into the insertion site of *P* elements and *P* inducibility is required. Furthermore, since *P*-element-derived piRNAs exhibited similar sequences in all lines, piRNA biogenesis may not differ between lines.

Conclusions

Our results suggest that piRNA abundance explains coarse phenotypic differences between M′ and Q cytotypes with respect to P-repression, but not more modest differences between Q strains. Whether this piRNA variation originates from genetic diversity, such as copy number and location of

P elements, or from long-term inheritance of small RNAs may be an interesting question. Moreover, our results evoke an interesting possibility that the suppression mechanisms of TEs including piRNAs are varied in natural populations.

Additional files

Additional file 1: Supplemental methods. Table S1. (DOCX 18 kb)

Additional file 2: Figure S1. Expression of mRNA of P elements in F1 ovaries of progenies of four lines. (PPTX 45 kb)

Additional file 3: Figure S2. Expression of piRNA and mRNA of P elements in F1 embryonic bodies of progenies of file lines. (PPTX 52 kb)

Abbreviations

cross A: CS females x tested males; cross A*: Tested females x Har males; CS: Canton S; GD: Gonadal dysgenesis; Har: Harwich; low P inducibility: The ability to induce transposition of P elements in the germline cells of progeny; M, P, M', or Q progeny: The F1 progeny between males of a P strain and females of each strain; P susceptibility: The capacity to allow transposition of P elements in the germline cells of progeny; piRNAs: Piwi-interacting small RNAs; qRT-PCR: Quantitative PCR; RPM: Reads per million miRNA reads; TEs: Transposable elements

Acknowledgements

We thank Minoru Horie and Akihiro Sekine for technical assistance. We also acknowledge Kuniaki Saito for helpful discussions.

Funding

This research did not receive any specific grant from funding agencies in the public, commercial, or not-for-profit sectors.

Authors' contributions

KTW designed the study; KTW, SO and MI generated experimental data; KTW, KI, and SO analyzed data; KTW, KI and MI wrote and edited the manuscript; All authors read and approved the final manuscript.

Competing interests

The authors declare that they have no competing interests.

Author details

[1]Department of Applied Biology, Kyoto Institute of Technology, Hashigamicyo, Matsugasaki, Sakyo-ku, Kyoto 606-8585, Japan. [2]Laboratory of Genome and Epigenome Dynamics, Department of Applied Molecular Biosciences, Graduate School of Bioagricultural Sciences, Nagoya University, Nagoya 464-8601, Japan. [3]Center for Epidemiologic Research in Asia, Shiga Univesity of Medical Science, Otsu, Shiga 520-2192, Japan. [4]Center for Advanced Insect Research Promotion (CAIRP), Kyoto Institute of Technology, Kyoto 606-8585, Japan.

References

1. Bennetzen JL. Transposable element contributions to plant gene and genome evolution. Plant Mol Biol. 2000;42:251–69.
2. Britten RJ. Transposable element insertions have strongly affected human evolution. Proc Natl Acad Sci U S A. 2010;107:19945–8.
3. Brennecke J, Aravin AA, Stark A, Dus M, Kellis M, et al. Discrete small RNA-generating loci as master regulators of transposon activity in Drosophila. Cell. 2007;128:1089–103.
4. Siomi MC, Sato K, Pezic D, Aravin AA. PIWI-interacting small RNAs: the vanguard of genome defence. Nat Rev Mol Cell Biol. 2011;4:246–58.
5. Gunawardane LS, Saito K, Nishida KM, Miyoshi K, Kawamura Y, et al. A slicer-mediated mechanism for repeat-associated siRNA 5' end formation in Drosophila. Science. 2007;315:1587–90.
6. Klattenhoff C, Theurkauf W. Biogenesis and germline functions of piRNAs. Development. 2008;135:3–9.
7. Kawaoka S, Izumi N, Katsuma S, Tomari Y. 3' end formation of PIWI-interacting RNAs in vitro. Mol Cell. 2011;43:1015–22.
8. Malone CD, Brennecke J, Dus M, Stark A, McCombie WR, et al. Specialized piRNA pathways act in germline and somatic tissues of the Drosophila ovary. Cell. 2009;137:522–35.
9. Saito K, Ishizu H, Komai M, Kotani H, Kawamura Y, et al. Roles for the Yb body components Armitage and Yb in primary piRNA biogenesis in Drosophila. Genes Dev. 2010;24:2493–8.
10. Olivieri D, Sykora MM, Sachidanandam R, Mechtler K, Brennecke J. An in vivo RNAi assay identifies major genetic and cellular requirements for primary piRNA biogenesis in Drosophila. EMBO J. 2010;29:3301–17.
11. Dennis C, Zanni V, Brasset E, Eymery A, Zhang L, et al. "Dot COM", a nuclear transit center for the primary piRNA pathway in Drosophila. PLoS One. 2013;8:e72752.
12. Ross RJ, Weiner MM, Lin H. PIWI proteins and PIWI-interacting RNAs in the soma. Nature. 2014;505:353–9.
13. Iwasaki YW, Murano K, Ishizu H, Shibuya A, Iyoda Y, Siomi MC, Siomi H, Saito K. Piwi Modulates Chromatin Accessibility by Regulating Multiple Factors Including Histone H1 to Repress Transposons. Mol Cell. 2016;3:408–19.
14. O'Hare K, Rubin GM. Structures of P transposable elements and their sites of insertion and excision in the Drosophila melanogaster genome. Cell. 1983;34:25–35.
15. Rio DC, Laski FA, Rubin GM. Identification and immunochemical analysis of biologically active Drosophila P element transposase. Cell. 1986;44:21–32.
16. Engels WR, Benz WK, Preston CR, Graham PL, Phillis RW, et al. Somatic effects of P element activity in Drosophila melanogaster: pupal lethality. Genetics. 1987;117:745–57.
17. Kidwell MG, Kidwell JF, Sved JA. Hybrid dysgenesis in Drosophila meranogaster: A syndrome of aberrant traits Including mutation, sterility and male recombination. Genetics. 1977;86:813–33.
18. Engels WR, Preston CR. Hybrid dysgenesis in Drosophila melanogaster: the biology of female and male sterility. Genetics. 1979;92:161–74.
19. Rubin GM, Kidwell MG, Bingham PM. The molecular basis of P-M hybrid dysgenesis: the nature of induced mutations. Cell. 1982;29:987–94.
20. Preston CR, Engels WR. P-element-induced male recombination and gene conversion in Drosophila. Genetics. 1996;144:1611–22.
21. Brennecke J, Malone CD, Aravin AA, Sachidanandam R, Stark A, et al. An epigenetic role for maternally inherited piRNAs in transposon silencing. Science. 2008;322:1387–92.
22. Bonnivard E, Higuet D. Stability of European natural populations of Drosophila melanogaster with regard to the P-M system: a buffer zone made up of Q populations. J Evol Biol. 1999;12:633–47.
23. Itoh M, Fukui T, Kitamura M, Uenoyama T, Watada M, Yamaguchi M. Phenotypic stability of the P-M system in wild populations of Drosophila melanogaster. Genes Genet. Syst. 2004;79:9–18.
24. Ignatenko OM, Zakharenko LP, Dorogova NV, Fedorova SA. P elements and the determinants of hybrid dysgenesis have different dynamics of propagation in Drosophila melanogaster populations. Genetica. 2015;143:751–9.
25. Anxolabéhère D, Kai H, Nouaud D, Périquet G, Ronsseray S. The geographical distribution of P-M hybrid dysgenesis in Drosophila melanogaster. Genet Sel Evol. 1984;16:15–26.
26. Kidwell MG. Hybrid dysgenesis in Drosophila melanogaster: nature and inheritance of P element regulation. Genetics. 1985;111:337–50.
27. Itoh M, Yu S, Watanabe TK, Yamamoto MT. Structural and genetic studies of the proliferation disrupter genes of Drosophila simulans and D. melanogaster. Genetica. 1999;106:23–229.
28. Kidwell MG. Hybrid dysgenesis in Drosophila melanogaster: the genetics of cytotype determination in a neutral strain. Genetics. 1981;98:275–90.
29. Bingham PM, Kidwell MG, Rubin GM. The molecular basis of P-M hybrid dysgenesis: the role of the P element, a P-strain-specific transposon family. Cell. 1982;29:995–1004.
30. O'Hare K, Driver A, McGrath S, Johnson-Schiltz DM. Distribution and structure of cloned P elements from the Drosophila melanogaster P strain pi 2. Genet Res. 1992;60:33–41.
31. Black DM, Jackson MS, Kidwell MG, Dover GA. KP elements repress P-induced hybrid dysgenesis in Drosophila melanogaster. EMBO J. 1987;6:4125–35.

32. Jackson MS, Black DM, Dover GA. Amplification of KP elements associated with the repression of hybrid dysgenesis in *Drosophila melanogaster*. Genetics. 1988;120:1003–13.

33. Rasmusson KE, Raymond JD, Simmons MJ. Repression of hybrid dysgenesis in *Drosophila melanogaster* by individual naturally occurring P elements. Genetics. 1993;133:605–22.

34. Lemaitre B, Ronsseray S, Coen D. Maternal repression of the P element promoter in the germline of *Drosophila melanogaster*: a model for the P cytotype. Genetics. 1993;135:149–60.

35. Andrews JD, Gloor GB. A role for the KP leucine zipper in regulating P element transposition in *Drosophila melanogaster*. Genetics. 1995;141:587–94.

36. Simmons MJ, Raymond JD, Grimes CD, Belinco C, Haake BC, et al. Repression of hybrid dysgenesis in *Drosophila melanogaster* by heat-shock-inducible sense and antisense *P*-element constructs. Genetics. 1996;144:1529–44.

37. Itoh M, Boussy IA. Full-size P and KP elements predominate in wild *Drosophila melanogaster*. Genes Genet Syst. 2002;77:259–67.

38. Itoh M, Takeuchi N, Yamaguchi M, Yamamoto MT, Boussy IA. Prevalence of full-size P and KP elements in North American populations of *Drosophila melanogaster*. Genetica. 2007;131:21–8.

39. Onder BS, Bozcuk AN. P – M phenotypes and their correlation with longitude in natural populations of *Drosophila melanogaster* from Turkey. Russ J Genet. 2012;48:1170–6.

40. Onder BS, Kasap OE. P element activity and molecular structure in *Drosophila melanogaster* populations from Firtina Valley, Turkey. J Insect Sci. 2014;14:16.

41. Fukui T, Inoue Y, Yamaguchi M, Itoh M. Genomic *P* elements content of a wild M' strain of *Drosophila melanogaster*: KP elements do not always function as type II repressor elements. Genes Genet Syst. 2008;83:67–54.

42. Kelleher ES. Reexamining the P-element invasion of Drosophila Melanogaster through the lens of piRNA silencing. Genetics. 2016;4:1513–31.

43. Simmons MJ, Thorp MW, Buschette JT, Becker JR. Transposon regulation in *Drosophila*: piRNA-producing P elements facilitate repression of hybrid dysgenesis by a P element that encodes a repressor polypeptide. Mol Gen Genomics. 2015;290:127–40.

44. Simmons MJ, Grimes CD, Czora CS. Cytotype,Regulation Facilitates Repression of Hybrid Dysgenesis by Naturally Occurring KP Elements in *Drosophila melanogaster*. G3 (Bethesda). 2016;6:1891–7.

45. Khurana JS, Wang J, Xu J, Koppetsch BS, Thomson TC, Nowosielska A, et al. Adaptation to P element transposon invasion in *Drosophila melanogaster*. Cell. 2011;147:1551–63.

46. Kidwell MG, Novy JB. Hybrid dysgenesis in *Drosophila meranogaster*: sterility resulting from gonadal dysgenesis in the P-M system. Genetics. 1979;92:1127–40.

47. Kidwell MG. Hybrid Dysgenesis in DROSOPHILA MELANOGASTER: Factors Affecting Chromosomal Contamination in the P-M System. Genetics. 1983;104:317–41.

48. Itoh M, Sasai N, Inoue Y, Watada M. P elements and P-M characteristics in natural populations of Drosophila melanogaster in the southernmost islands of Japan and in Taiwan. Heredity (Edinb). 2001;86:206–12.

49. Kozomara A, Griffiths-Jones S. miRBase: annotating high confidence microRNAs using deep sequencing data. Nucleic Acids Res. 2014;42:68–73.

50. Ichiyanagi T, Ichiyanagi K, Ogawa A, Kuramochi-Miyagawa S, Nakano T, et al. HSP90α plays an important role in piRNA biogenesis and retrotransposon repression in mouse. Nucleic Acids Res. 2014;42:11903–11.

51. Ichiyanagi K, Li Y, Watanabe T, Ichiyanagi T, Fukuda K, et al. Locus- and domain-dependent control of DNA methylation at mouse B1 retrotransposons during male germ cell development. Genome Res. 2011;21:2058–66.

52. Dourlen P, Bertin B, Chatelain G, Robin M, Napoletano F, et al. Drosophila fatty acid transport protein regulates rhodopsin-1 metabolism and is required for photoreceptor neuron survival. PLoS Genet. 2012;8:e1002833.

53. Ryazansky S, Radion E, Mironova A, Akulenko N, Abramov Y, et al. Natural variation of piRNA expression affects immunity to transposable elements. PLoS Genet. 2017;13:e1006731.

54. Castro JP, Carareto CM. Drosophila melanogaster *P* transposable elements: mechanisms of transposition and regulation. Genetica. 2004;121:107–118.

14

De-novo emergence of SINE retroposons during the early evolution of passerine birds

Alexander Suh[1,2]* (iD), Sandra Bachg[1], Stephen Donnellan[3,4], Leo Joseph[5], Jürgen Brosius[1,6], Jan Ole Kriegs[1,7] and Jürgen Schmitz[1]

Abstract

Background: Passeriformes ("perching birds" or passerines) make up more than half of all extant bird species. The genome of the zebra finch, a passerine model organism for vocal learning, was noted previously to contain thousands of short interspersed elements (SINEs), a group of retroposons that is abundant in mammalian genomes but considered largely inactive in avian genomes.

Results: Here we resolve the deep phylogenetic relationships of passerines using presence/absence patterns of SINEs. The resultant retroposon-based phylogeny provides a powerful and independent corroboration of previous sequence-based analyses. Notably, SINE activity began in the common ancestor of Eupasseres (passerines excluding the New Zealand wrens Acanthisittidae) and ceased before the rapid diversification of oscine passerines (suborder Passeri – songbirds). Furthermore, we find evidence for very recent SINE activity within suboscine passerines (suborder Tyranni), following the emergence of a SINE via acquisition of a different tRNA head as we suggest through template switching.

Conclusions: We propose that the early evolution of passerines was unusual among birds in that it was accompanied by *de-novo* emergence and activity of SINEs. Their genomic and transcriptomic impact warrants further study in the light of the massive diversification of passerines.

Keywords: Transposon, Retroposon, SINE, Birds, Passeriformes, Phylogenomics

Background

Short interspersed elements (SINEs) are the most abundant group of the reverse-transcribed retroposons in mammalian genomes [1]. They rely on *trans*-mobilization by the enzymatic machinery of long interspersed elements (LINEs) [2], a *parasitic* interaction so successful that the human genome contains >1,500,000 SINEs compared to <900,000 LINEs [3]. On the other hand, SINEs are scarce in avian genomes, and this has been noted as one of the most peculiar genomic features of birds [4–6]. While LINEs exhibit up to 700,000 copies in avian genomes,

there are only 6000–17,000 SINEs per avian genome [6], most of these being ancient and heavily degraded [7].

Presence/absence patterns of SINEs in orthologous genomic loci are rare genomic changes appreciated widely as virtually homoplasy-free phylogenetic markers [8, 9]. Given the aforementioned scarcity of SINEs, it is not surprising that the emergence and activity of SINEs has never been studied in birds. On the other hand, other types of retroposed elements (REs; LINEs from the chicken repeat 1 superfamily, CR1, and long terminal repeat elements, LTRs) have helped resolve the relationships of various groups of birds, such as Galliformes [10–12], Neoaves [13–15], Palaeognathae [16, 17], and others [18–21]. In the meantime, the sequencing of dozens of avian genomes has revealed SINEs with putative lineage specificity [5, 7, 22] and thus the potential

---

* Correspondence: alexander.suh@ebc.uu.se
[1]Institute of Experimental Pathology (ZMBE), University of Münster, D-48149 Münster, Germany
[2]Department of Evolutionary Biology (EBC), Uppsala University, SE-75236 Uppsala, Sweden
Full list of author information is available at the end of the article

for conducting phylogenetic presence/absence analyses in specific groups of birds.

Here we conduct, to our knowledge, the first study of the emergence and activity of SINEs in birds. We focus on the deep phylogenetic relationships of passerines, the largest radiation of birds with nearly 6000 extant species [23], using 44 presence/absence markers of SINEs and other REs. In contrast to the only previous study of retroposons in passerines with a single RE marker [24], our multilocus dataset permits the reassessment of sequence-based phylogenies (e.g., [23, 25, 26]) and, simultaneously, the reconstruction of the temporal activity of SINEs and other REs during early passerine evolution.

Results and discussion
Two CR1-mobilized SINEs in passerines
We initially chose RE marker candidates from selected retroposon families of the oscine passerine zebra finch *Taeniopygia guttata* (including TguSINE1, [5]; Additional file 1: Table S1) in October 2009, a time when genome assemblies were available only for chicken and zebra finch [4, 5]. Seventy four candidates for presence/absence loci were therefore identified via pairwise alignment of RE-flanking sequences from zebra finch to orthologous regions in chicken (Materials and Methods). This was followed by in-vitro presence/absence screening of RE marker candidates as detailed elsewhere [13, 27] using a representative taxon sampling of all major groups of passerines sensu Barker et al. [23] (Additional file 1: Table S2). We complemented this with a screening of GenBank [28] for additional SINEs, which identified a TguSINE1-like insertion in *myoglobin* intron 2 of the suboscine *Pitta anerythra* (accession number DQ785977) that is absent in the orthologous position of other *Pitta* species [29]. We termed this element "PittSINE" and identified PittSINE marker candidates in a DNA sample of *Pitta sordida* via inter-SINE PCR ([30]; Methods). This was followed by cloning of the 500-bp to 1000-bp fraction of PCR amplicons and sequencing of 24 clones, alignment to chicken and zebra finch genomes to reconstruct the left and right SINE-flanking regions, and then in-vitro presence/absence screening of nine PittSINE marker candidates.

Next, we characterized the structural organization of passerine SINEs (Fig. 1) using the available TguSINE1 consensus sequence [5] and after generating a majority-rule consensus of six PittSINE insertions in our sequenced presence/absence markers (Additional file 2). Both SINEs have highly similar, CR1-derived tails (Fig. 1) which exhibit the typical hairpin for putative binding by the CR1 reverse transcriptase and an 8-bp microsatellite at their very end for target-primed reverse transcription [31] (Additional file 3: Figure S1). However, the heads of these SINEs are derived from different tRNA genes, namely tRNA^{Ile} in TguSINE1 and tRNA^{Asp} in

PittSINE (Fig. 1). Sequence alignment suggests that the tRNA-derived SINE heads are more similar to the respective tRNA genes than they are to each other (Fig. 1c). However, the opposite is the case for the CR1-derived SINE tails, which exhibit four diagnostic nucleotides distinguishing them from the highly similar 3′ end of CR1-X1_Pass (Fig. 1c). To verify that these are specific to TguSINE1 and PittSINE, we screened the zebra finch genome assembly for the presence of the four diagnostic nucleotides in copies of CR1-X1_Pass. Among those copies most similar to CR1-X1_Pass, only one old copy (chr2:68,921,881–68,922,556) contained the four diagnostic nucleotides, suggesting that these were acquired randomly after the insertion event.

We further investigated this peculiar pattern using phylogenetic analyses of the CR1-derived SINE tails and avian CR1 subfamilies sensu ref. [32], which again suggests that TguSINE1 and PittSINE have a single SINE ancestor which derived its tail from CR1-X1_Pass (Fig. 2a). Assuming that SINEs are *trans*-mobilized by LINE reverse transcriptase enzymes due to high sequence similarity between SINE tails and LINE 3′ ends [2, 33] and thus depend on LINE activity, the most likely candidate for SINE mobilization is the CR1-X1_Pass subfamily. This is further supported by temporal overlap of TguSINE1 and CR1-X activity in RE landscapes of the zebra finch genome (Fig. 2b). Additionally, we detected direct evidence for temporal overlap of TguSINE1 and CR1-X1_Pass activity through our presence/absence analyses (Fig. 3a, Additional file 1: Table S2).

Retroposon-based phylogeny of passerines
Our extensive RE presence/absence analyses yielded 19 TguSINE1, 6 PittSINE, 13 CR1, and 6 LTR markers which we could trace across a representative taxon sampling of the major groups of passerines sensu Barker et al. [23] (cf. [34]). These RE markers are only those where we were able to obtain sequences for all taxa critical for a phylogenetic conclusion. Careful inspection of presence/absence alignments using strict criteria (see Materials and Methods) yielded a conflict-free set of RE markers (except for one marker potentially affected by incomplete lineage sorting; Fig. 3a), which we mapped on a maximum likelihood tree constructed from concatenated RE-flanking sequences from the same data set (Fig. 3a). For three of the deepest passerine branching events, we found a multitude of RE markers and thus statistically significant support in available RE marker tests [35, 36]. These relationships are the respective monophyly of passerines and oscines, as well as the monophyly of Eupasseres [37], a group comprising all passerines except the New Zealand wrens Acanthisittidae. The Eupasseres/Acanthisittidae split was first observed in sequence analyses of few nuclear genes [38,

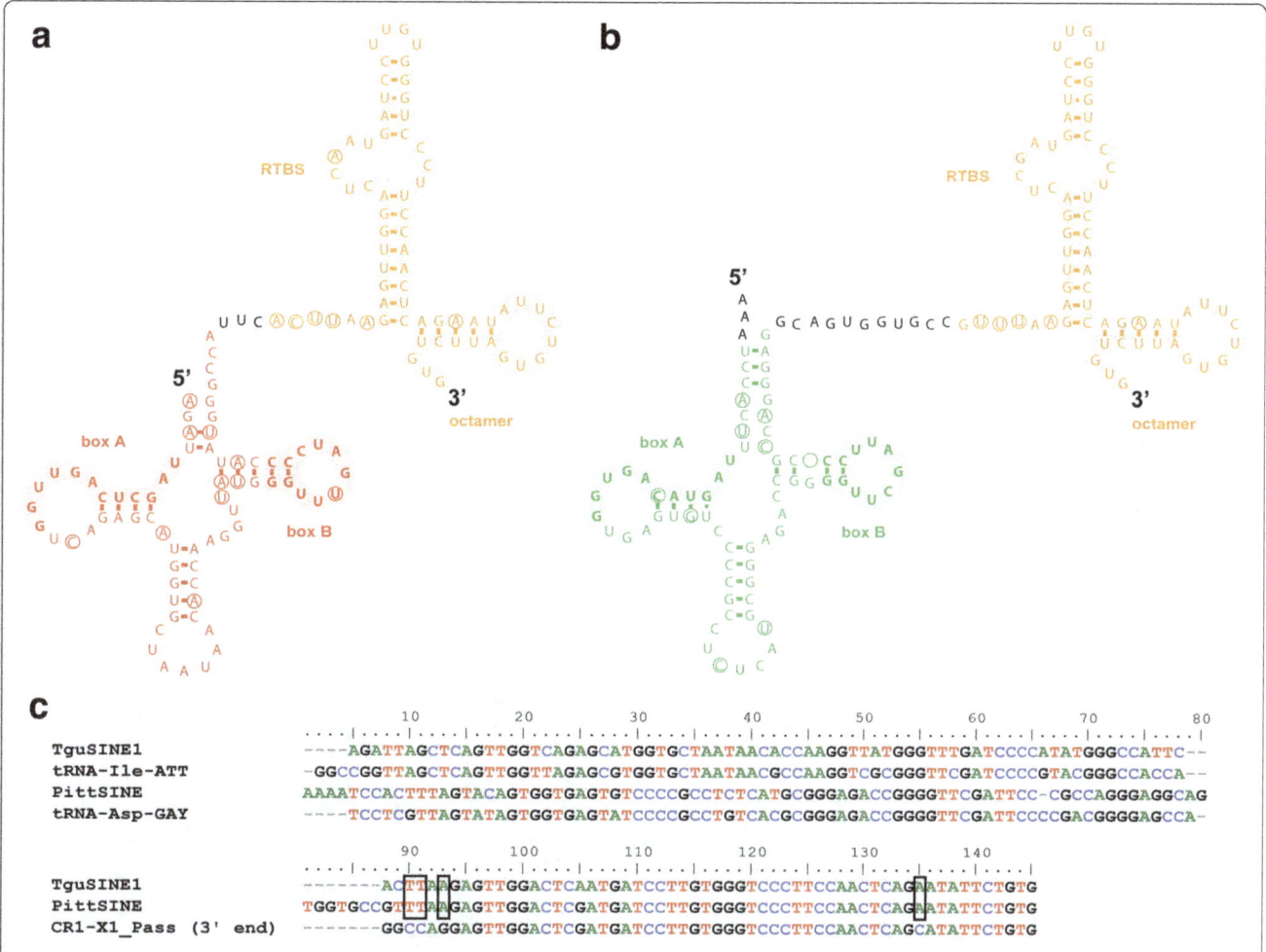

Fig. 1 Proposed RNA secondary structures of passerine SINEs with CR1-derived tails (orange) and tRNA-derived heads. The SINE heads are tRNA^Ile (red) in TguSINE1 (**a**) and tRNA^Asp (green) in PittSINE (**b**). Shaded regions denote promoter boxes A and B in tRNAs, as well as the reverse transcriptase binding site (RTBS) and 5'-AUUCURUG-3' microsatellite typical for CR1 elements of amniotes [31]. Circles indicate nucleotide differences between SINE consensus sequences and the respective tRNAs or CR1 they are derived from. The RTBS hairpin structure is also visible in mfold [57] predictions of SINE secondary structure (Additional file 3: Figure S1). **c** DNA sequence alignment of TguSINE1 and PittSINE with respective tRNA genes and the 3' end of CR1-X1_Pass. Black boxes denote diagnostic nucleotides present in the CR1-derived tails of TguSINE1 and PittSINE

39] and has since been confirmed in ever-growing nuclear sequence analyses (e.g., [23, 25, 26, 40]). Our analysis of rare genomic changes thus provides the first assessment of this group using an independent marker type and phylogenetic method. None of our RE markers inserted during the rapid radiation of oscine passerines, however, sequence analysis of the RE-flanking regions yielded a topology identical to the aforementioned previous studies. Of particular interest are the four deep-branching oscine lineages Menuridae (e.g., *Menura novaehollandiae*), Climacteridae (e.g., *Climacteris picumnus*), Maluridae/Meliphagidae (e.g., *Malurus cyaneus* and *Myzomela eques*), and Pomatostomidae (e.g., *Pomatostomus superciliosus*) because these four lineages together have been rarely included in passerine phylogenetic studies. We find a branching order (Fig. 3a) which recapitulates previous phylogenetic estimates

based on few nuclear genes [23] or ultraconserved elements [26]. This suggests that the rapid radiation of oscines can be congruently resolved even with non-genome-scale data. We note that this is in contrast to the neoavian radiation, which appears to be partially unresolvable even with genome-scale sequence analyses and thousands of retroposon markers (reviewed by [41]). Within passerines, we further note that the conflict between single-RE support for a Picathartidae/Corvidae clade [24] and sequence-based phylogenies [42] results from incorrect placing of this RE marker on the passerine Tree of Life due to methodological limitations (see legend of Fig. 4 for more information).

Emergence and activity of passerine SINEs

We then traced the emergence and activity of SINEs across the passerine Tree of Life. Given that RE marker

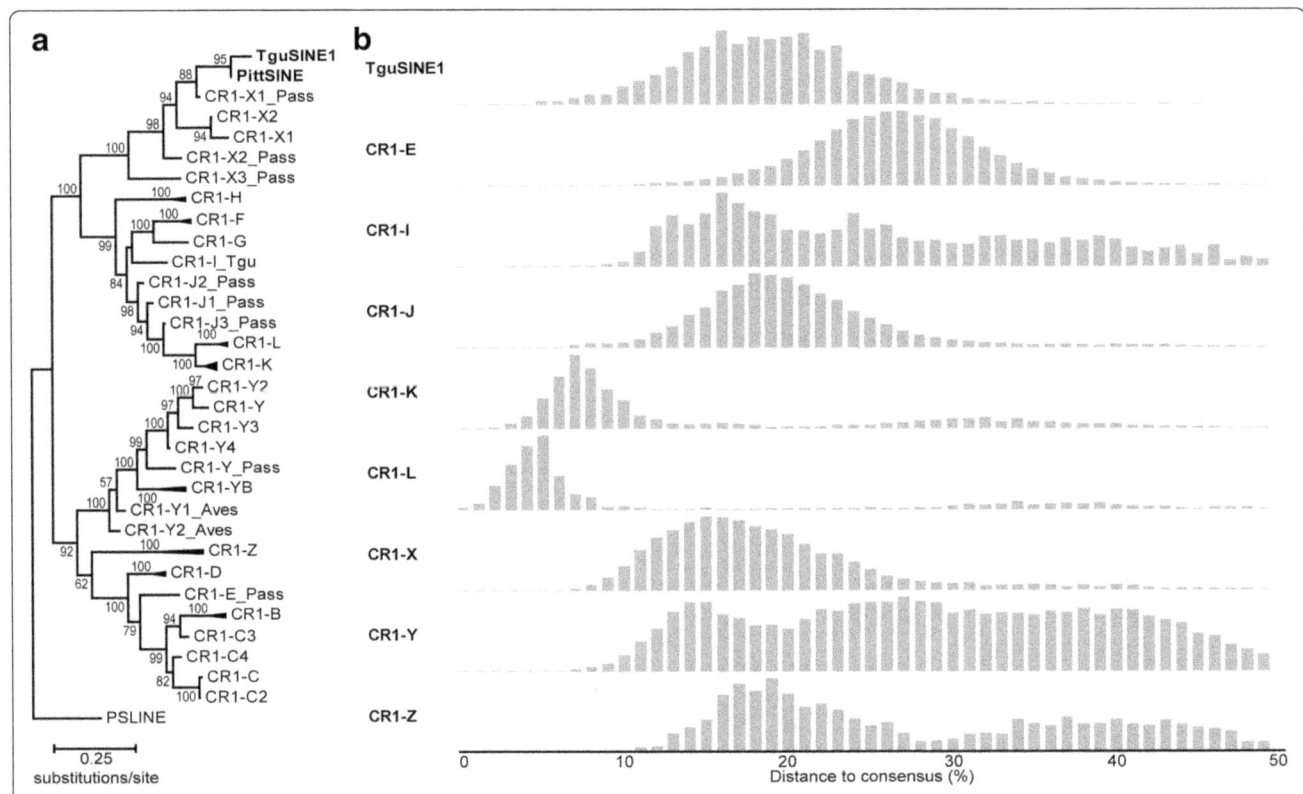

Fig. 2 Passerine SINEs share a common ancestor and are mobilized by CR1-X. **a** Maximum likelihood phylogeny of passerine SINE tails and avian CR1 subfamilies in Repbase [58] (GTRCAT model, 1000 bootstrap replicates) suggests that TguSINE1 and PittSINE arose from the same CR1-X subfamily (CR1-X1_Pass) and share a common SINE ancestor. Note that the topology of the CR1 phylogeny is identical to that of previous studies [20, 32]. **b** Comparison of the TguSINE1 landscape with landscapes of CR1 families (merged subfamilies from panel A) suggests temporal overlap of TguSINE1 and CR1-X activity in the zebra finch genome. RE landscapes were generated using the zebra finch assembly taeGut2 following methods detailed elsewhere [32]

candidates were initially chosen on chicken/zebra finch alignments, we expect no bias in the age distribution of RE markers on the lineage leading to zebra finch. TguSINE1 was mostly active in the ancestor of oscines and, to a lesser extent, in the ancestor of Eupasseres. Interestingly, we find no evidence for TguSINE1 activity in the common ancestor of passerines (cf. Additional file 3: Figure S2) or during/after the radiation of oscines and therefore hypothesize that Tgu-SINE1 emerged in Eupasseres and became extinct in the oscines' ancestor (Fig. 3a). The emergence of TguSINE1 is thus the first synapomorphic "genome morphology" character for Eupasseres and supplements support from skeletal morphology, which is limited to the presence of a 'six-canal pattern' in the hypotarsus [43].

In contrast to the situation in oscines, the activity of Tgu-SINE1 appears to have been longer in suboscines, postdating the divergence between Old World and New World suboscines (i.e., pitta and phoebe in Fig. 3a). This recent, potentially lineage-specific activity coincides with the putative restriction of PittSINEs to Old World suboscines (e.g., *Pitta* spp.), which is further supported by a much lower pairwise distance of PittSINE copies to the consensus (ranging from 0 to 11%, average 6.3%; Additional file 1: Table

S3) than in the case of TguSINE1 (Fig. 2b). As mentioned above, the CR1 phylogeny and four diagnostic nucleotides in the CR1-derived SINE tails (cf. Figs. 1c and 2a) indicate that TguSINE1 and PittSINE likely have a common SINE ancestor instead of being derived independently from a CR1-X_Pass LINE. This further suggests that the younger PittSINE emerged from the older TguSINE1 after acquisition of a new tRNA-derived head. Assuming that TguSINE1 and PittSINE were both active on the pitta lineage, we propose that the most plausible mechanism for PittSINE emergence was template switching from TguSINE1 to a nearby tRNA during reverse transcription (Fig. 3b). Slightly less parsimonious alternative explanations for PittSINE emergence might be gene conversion or genomic rearrangement between a TguSINE1 master gene and a tRNA^{Asp} gene, but these remain untestable in the absence of a pitta genome assembly. Template switching has been previously proposed in a wide range of chimeric retroposons (e.g., [44–47]) and appears to be a particularly common opportunity for SINEs to *parasitize* different LINEs via acquisition of new SINE tails [46, 48]. As previously observed for ancient amniote SINEs [49], our data show that template switching may also happen for SINE heads, whereby the

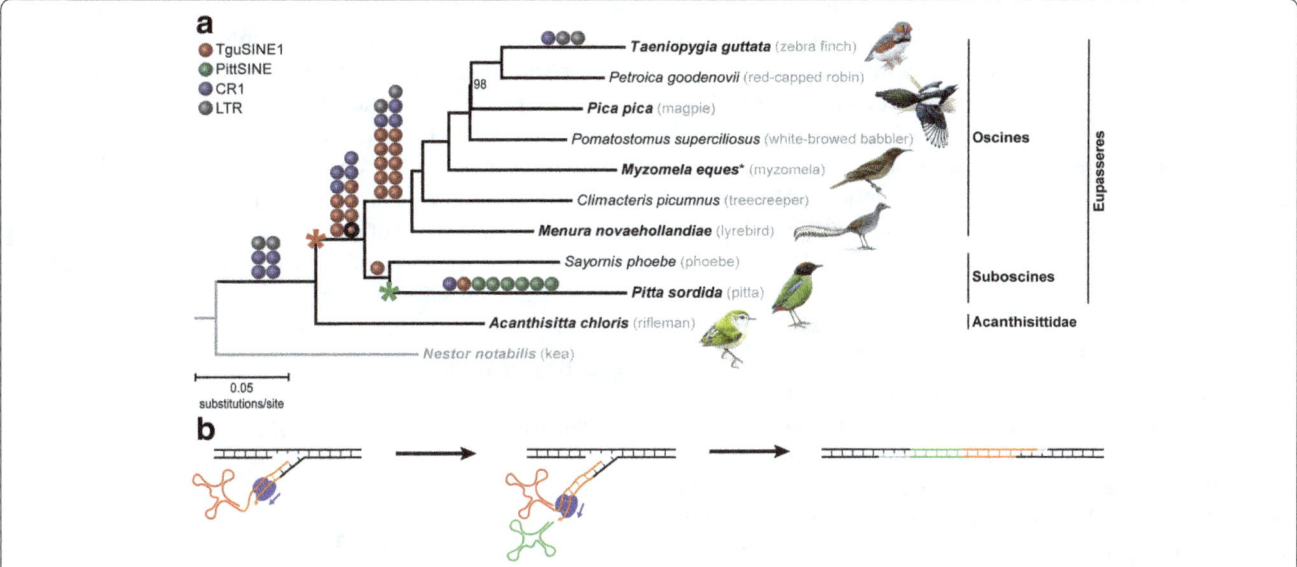

Fig. 3 Emergence and timing of CR1-mobilized SINE activity during early passerine evolution. **a** Phylogenomic analysis of early passerine relationships using retroposon presence/absence markers (colored balls) mapped on a maximum likelihood phylogeny of concatenated retroposon-flanking sequences (GTRCAT model, 1000 bootstrap replicates; Additional file 5). The single conflicting marker on the Eupasseres branch (Tgu10, cf. Additional file 1: Table S2) is indicated by a red ball with black circle and was likely affected by incomplete lineage sorting within Suboscines. Our sampling consists of the major deep passerine lineages sensu Barker et al. [23]. The later additions of two genome assemblies (*Corvus cornix* and *Manacus vitellinus*) were only included in the presence/absence table (Additional file 1: Table S2). Red and green asterisks indicate emergence of TguSINE1 and PittSINE, respectively. The black asterisk indicates that for some loci (Additional file 1: Table S2), *Malurus cyaneus* was sampled instead of *Myzomela eques* to represent the Maluridae/Meliphagidae clade [23]. Only bootstrap values <100% are shown and the names of pictured birds are emphasized in bold. **b** A scenario for the emergence of PittSINE. Template switching from TguSINE1 RNA (red, tRNA^Ile head; orange, CR1 tail) to tRNA^Asp (green) during target-primed reverse transcription by CR1 reverse transcriptase (blue). The resultant tRNA^Asp-CR1 chimaera was flanked by a target site duplication (grey) and transcriptional activation gave rise to the PittSINE family

acquisition of a new SINE head from a different tRNA and an appropriate upstream sequence close to the insertion site may provide intact and active promoter components for efficient transcription by RNA polymerase III.

Conclusion

To conclude, we reconstructed the deep phylogenetic relationships of passerines using presence/absence patterns of unusual SINE insertions and other REs. This permitted us to follow the emergence, activity, and extinction of TguSINE1 and PittSINE across the evolution of the most species-rich group of birds. While this SINE activity of ~2000 copies per oscine genome and ~2500 copies per suboscine genome (Additional file 3: Figure S2) was considerably lower than, for example, that in mammals, it nevertheless exemplifies that at least some birds have

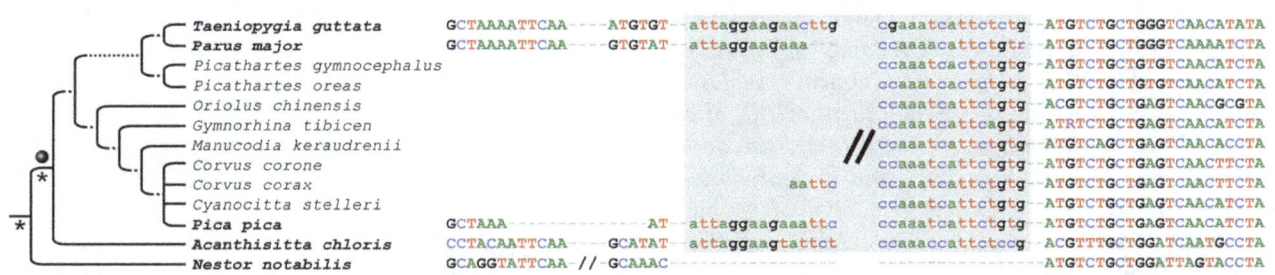

Fig. 4 A reassessment of the RE marker of Treplin & Tiedemann [24] through extended taxon sampling. Treplin & Tiedemann [24] inferred "*phylogenetic affinity of rockfowls (genus* Picathartes) *to crows and ravens (Corvidae)*" based on the Cor2 locus which they sequenced in rockfowls and corvids, and unsuccessful Cor2 PCR amplification in other passerines. We generated a nested oligonucleotide primer pair (5'- CAATACTTTGGAACACCTCAGAC-3' and 5'-GGCACCTGTCAATGGCTAC-3') and were able to amplify and sequence the Cor2 locus in additional species. Our extended phylogenetic sampling suggests that the RE insertion (lowercase nucleotides) occurred in the ancestor of all passerines (grey ball) due to RE presence in non-corvid passerines (*Taeniopygia guttata, Parus major, Acanthisitta chloris*) and RE absence in the parrot outgroup (*Nestor notabilis*). Taxa with bold names were sampled in the present study and the grey box denotes the 5' and 3' end of the CR1 insertion. Asterisks indicate branches in the avian Tree of Life which were previously recovered with significant support from retroposon markers [13]

a more diverse repetitive element landscape than previously anticipated. Furthermore, we note that the activity of TguSINE1 appears to coincide with the evolution of vocal learning during early passerine evolution [13]. Previous evidence suggests that ~4% of birdsong-associated transcripts in the zebra finch brain contain retroposons [5] and it thus remains to be seen whether SINE activity influenced the evolution of, for example, vocal learning in oscine passerines.

Methods

We identified candidates for presence/absence loci for TguSINE1 and other selected zebra finch retroposons via pairwise alignment of RE loci from zebra finch to orthologous regions in chicken. This was done by comparing and extracting the respective RE-flanking sequences in the UCSC Genome Browser [50], followed by automatic alignment using MAFFT version 6 [51]. In order to find the nine PittSINE marker candidates, we conducted inter-SINE PCR [30] using a single, PittSINE-specific oligonucleotide primer (5′-CTCGTTAGTATAGTGGT-GAGTGTC-3′) and standard PCR parameters of ref. [27] with 50 °C annealing temperature. Among the sampled passerines, inter-SINE PCR yielded strong amplification signal only in the pitta (data not shown). Additionally, we identified two TguSINE1 candidate loci in the pitta using a single TguSINE1-specific oligonucleotide primer (5′-CAGTTGGTTAGAGCGTGGTG-3′). All presence/absence screenings were done using oligonucleotide primers binding to conserved RE-flanking regions in chicken/zebra finch alignments (Additional file 1: Table S4), using the touchdown PCR and cloning protocols of ref. [13]. Two recently sequenced species (*Corvus cornix* and *Manacus vitellinus* [6, 52]) were added to reduce missing data in our presence/absence table (Additional file 1: Table S2).

For each presence/absence marker candidate, we first aligned all sequences automatically using MAFFT (E-INS-I option) and then manually inspected these for misalignments. We considered a marker candidate as phylogenetically informative and reliable *"if, in all species sharing this RE, it featured an identical orthologous genomic insertion point (target site), identical RE orientation, identical RE subtype, identical target site duplications (direct repeats, if present) and a clear absence in other species"* [13]. This led to a total of 44 high-quality RE presence/absence markers (Additional file 1: Table S2, Additional file 4).

All maximum likelihood sequence analyses were conducted using RAxML 8.1.11 [53] on the CIPRES Science Gateway [54]. For the CR1 phylogeny, we used the alignment from ref. [20], excluded grebe-specific CR1 elements, and added the CR1-derived tails of TguSINE1 and PittSINE (alignment length 710 bp). For the passerine phylogeny, we removed the RE sequences from our

presence/absence alignments and concatenated the remaining RE-flanking sequences into a multilocus alignment (Additional file 4; alignment length 22,410 bp).

Zebra finch TE landscapes were generated from RepeatMasker [55] '.align' files after CpG correction as detailed elsewhere [32]. For PittSINE copies and the PittSINE consensus, Kimura 2-parameter pairwise distances were estimated in MEGA6 ([56]; uniform rates among sites, pairwise deletion of gaps/missing data) after exclusion of CpG sites.

Additional files

> **Additional file 1: Tables S1–S4.** (PDF 312 kb)
>
> **Additional file 2:** Majority-rule consensus sequence for PittSINE as reconstructed from our PittSINE-bearing presence/absence patterns. (TXT 155 bytes)
>
> **Additional file 3: Figures S1–S2.** (PDF 447 kb)
>
> **Additional file 4:** Fasta-formatted alignments of all RE presence/absence markers. The presented loci are labeled corresponding to the markers listed in Additional file 1: Table S2. The names of transposed elements correspond to those in Repbase. (TXT 805 kb)
>
> **Additional file 5:** Fasta-formatted multilocus alignment of concatenated RE-flanking sequences used for generating the phylogenetic tree of Fig. 3a. (TXT 660 kb)

Abbreviations

CR1: Chicken repeat 1; LINE: Long interspersed element; Mb: Million basepairs; MY: Million years; MYA: Million years ago; RE: Retroposed element; RT: Reverse transcriptase; RTBS: Reverse transcriptase binding site; SINE: Short interspersed element

Acknowledgements

We thank Tim Pock, Meike Hüdig, and Felix Babatz for help with in-vitro experiments, and Gerald Mayr and Gennady Churakov for helpful discussions. We are grateful to Leanne Wheaton, Simone Schehka (Allwetterzoo Münster), Robert Palmer, Stephanie Hodges, Geoffrey E. Hill, Franziska A. Franke, Sharon Birks (Burke Museum), and Werner Beckmann (LWL-DNA- und Gewebearchiv) for providing blood and tissue samples, and to Jón Baldur Hlíðberg for generating the bird paintings. We thank three anonymous reviewers for their valuable comments. Some of the computations were performed on resources provided by the Swedish National Infrastructure for Computing (SNIC) through Uppsala Multidisciplinary Center for Advanced Computational Science (UPPMAX).

Funding

This research was funded by the Deutsche Forschungsgemeinschaft (KR3639 to J.O.K. and J.S.).

Authors' contributions

AS, JOK and JS conceived the project. AS designed the study. AS and JS performed in silico experiments. AS and SB performed in vitro experiments. SB, LJ, JB, JOK and JS contributed reagents, materials and analysis tools. AS analyzed the data and wrote the manuscript. JS, JOK and JB discussed and commented on the data and the manuscript. All authors read and approved the final manuscript.

Competing interests

The authors declare that they have no competing interests.

Author details

[1]Institute of Experimental Pathology (ZMBE), University of Münster, D-48149 Münster, Germany. [2]Department of Evolutionary Biology (EBC), Uppsala University, SE-75236 Uppsala, Sweden. [3]South Australian Museum, Adelaide, SA 5000, Australia. [4]School of Biological Sciences, The University of Adelaide, Adelaide 5005, Australia. [5]Australian National Wildlife Collection, CSIRO National Research Collections Australia, Canberra, ACT 2601, Australia. [6]Brandenburg Medical School (MHB), D-16816 Neuruppin, Germany. [7]LWL-Museum für Naturkunde, Westfälisches Landesmuseum mit Planetarium, D-48161 Münster, Germany.

References

1. Sotero-Caio C, Platt RN II, Suh A, Ray DA. Evolution and diversity of transposable elements in vertebrate genomes. Genome Biol Evol. 2017; 9(1):161–77.
2. Ohshima K, Hamada M, Terai Y, Okada N. The 3′ ends of tRNA-derived short interspersed repetitive elements are derived from the 3′ ends of long interspersed repetitive elements. Mol Cell Biol. 1996;16(7):3756–64.
3. Lander ES, Linton LM, Birren B, Nusbaum C, Zody MC, Baldwin J, Devon K, Dewar K, Doyle M, Fitzhugh W, et al. Initial sequencing and analysis of the human genome. Nature. 2001;409(6822):860–921.
4. Hillier LW, Miller W, Birney E, Warren W, Hardison RC, Ponting CP, Bork P, Burt DW, Groenen MA, Delany ME, et al. Sequence and comparative analysis of the chicken genome provide unique perspectives on vertebrate evolution. Nature. 2004;432(7018):695–716.
5. Warren WC, Clayton DF, Ellegren H, Arnold AP, Hillier LW, Künstner A, Searle S, White S, Vilella AJ, Fairley S, et al. The genome of a songbird. Nature. 2010;464(7289):757–62.
6. Zhang G, Li C, Li Q, Li B, Larkin DM, Lee C, Storz JF, Antunes A, Greenwold MJ, Meredith RW, et al. Comparative genomics reveals insights into avian genome evolution and adaptation. Science. 2014;346(6215):1311–20.
7. Kapusta A, Suh A. Evolution of bird genomes—a transposon's-eye view. Ann N Y Acad Sci. 2017;1389:164–85.
8. Ray DA, Xing J, Salem A-H, Batzer MA. SINEs of a nearly perfect character. Syst Biol. 2006;55(6):928–35.
9. Shedlock AM, Takahashi K, Okada N. SINEs of speciation: tracking lineages with retroposons. Trends Ecol Evol. 2004;19(10):545–53.
10. Kaiser VB, van Tuinen M, Ellegren H. Insertion events of CR1 retrotransposable elements elucidate the phylogenetic branching order in galliform birds. Mol Biol Evol. 2007;24(1):338–47.
11. Kriegs JO, Matzke A, Churakov G, Kuritzin A, Mayr G, Brosius J, Schmitz J. Waves of genomic hitchhikers shed light on the evolution of gamebirds (Aves: Galliformes). BMC Evol Biol. 2007;7:190.
12. Liu Z, He L, Yuan H, Yue B, Li J. CR1 retroposons provide a new insight into the phylogeny of Phasianidae species (Aves: Galliformes). Gene. 2012;502(2):125–32.
13. Suh A, Paus M, Kiefmann M, Churakov G, Franke FA, Brosius J, Kriegs JO, Schmitz J. Mesozoic retroposons reveal parrots as the closest living relatives of passerine birds. Nat Commun. 2011;2:443.
14. Suh A, Smeds L, Ellegren H. The dynamics of incomplete lineage sorting across the ancient adaptive radiation of neoavian birds. PLoS Biol. 2015; 13(8):e1002224.
15. Matzke A, Churakov G, Berkes P, Arms EM, Kelsey D, Brosius J, Kriegs JO, Schmitz J. Retroposon insertion patterns of neoavian birds: strong evidence for an extensive incomplete lineage sorting era. Mol Biol Evol. 2012;29(6):1497–501.
16. Baker AJ, Haddrath O, McPherson JD, Cloutier A. Genomic support for a moa-tinamou clade and adaptive morphological convergence in flightless ratites. Mol Biol Evol. 2014;31(7):1686–96.
17. Haddrath O, Baker AJ. Multiple nuclear genes and retroposons support vicariance and dispersal of the palaeognaths, and an early cretaceous origin of modern birds. Proc R Soc B. 2012;279(1747):4617–25.
18. Watanabe M, Nikaido M, Tsuda TT, Inoko H, Mindell DP, Murata K, Okada N: The rise and fall of the CR1 subfamily in the lineage leading to penguins. Gene 2006, 365(0):57-66.
19. Kuramoto T, Nishihara H, Watanabe M, Okada N. Determining the position of storks on the phylogenetic tree of waterbirds by retroposon insertion analysis. Genome Biol Evol. 2015;7(12):3180–9.
20. Suh A, Kriegs JO, Donnellan S, Brosius J, Schmitz J. A universal method for the study of CR1 retroposons in nonmodel bird genomes. Mol Biol Evol. 2012;29:2899–903.
21. St. John J, Cotter J-P, Quinn TW. A recent chicken repeat 1 retrotransposition confirms the Coscoroba–Cape barren goose clade. Mol Phylogenet Evol. 2005;37(1):83–90.
22. Suh A, Witt CC, Menger J, Sadanandan KR, Podsiadlowski L, Gerth M, Weigert A, McGuire JA, Mudge J, Edwards SV, et al. Ancient horizontal transfers of retrotransposons between birds and ancestors of human pathogenic nematodes. Nat Commun. 2016;7:11396.
23. Barker FK, Cibois A, Schikler P, Feinstein J, Cracraft J. Phylogeny and diversification of the largest avian radiation. Proc Natl Acad Sci U S A. 2004; 101(30):11040–5.
24. Treplin S, Tiedemann R. Specific chicken repeat 1 (CR1) retrotransposon insertion suggests phylogenetic affinity of rockfowls (genus Picathartes) to crows and ravens (Corvidae). Mol Phylogenet Evol. 2007;43(1):328–37.
25. Selvatti AP, Gonzaga LP, Russo CA. A Paleogene origin for crown passerines and the diversification of the oscines in the new world. Mol Phylogenet Evol. 2015;88:1–15.
26. Moyle RG, Oliveros CH, Andersen MJ, Hosner PA, Benz BW, Manthey JD, Travers SL, Brown RM, Faircloth BC. Tectonic collision and uplift of Wallacea triggered the global songbird radiation. Nat Commun. 2016;7:12709.
27. Suh A, Kriegs JO, Brosius J, Schmitz J. Retroposon insertions and the chronology of avian sex chromosome evolution. Mol Biol Evol. 2011;28:2993–7.
28. Benson DA, Cavanaugh M, Clark K, Karsch-Mizrachi I, Lipman DJ, Ostell J, Sayers EW. GenBank. Nucleic Acids Res. 2013;41(D1):D36–42.
29. Irestedt M, Ohlson JI, Zuccon D, Källersjö M, Ericson PGP. Nuclear DNA from old collections of avian study skins reveals the evolutionary history of the old world suboscines (Aves, Passeriformes). Zool Scripta. 2006;35(6):567–80.
30. Kaukinen J, Varvio S-L. Artiodactyl retroposons: association with microsatellites and use in SINEmorph detection by PCR. Nucleic Acids Res. 1992;20(12):2955–8.
31. Suh A. The specific requirements for CR1 retrotransposition explain the scarcity of retrogenes in birds. J Mol Evol. 2015;81:18–20.
32. Suh A, Churakov G, Ramakodi MP, Platt RN II, Jurka J, Kojima KK, Caballero J, Smit A, Vliet K, Hoffmann FG, et al. Multiple lineages of ancient CR1 retroposons shaped the early genome evolution of amniotes. Genome Biol Evol. 2015;7(1):205–17.
33. Kajikawa M, Okada N. LINEs mobilize SINEs in the eel through a shared 3′ sequence. Cell. 2002;111(3):433–44.
34. Cracraft J. Avian higher-level relationships and classification: Passeriforms. In: The Howard and Moore Complete Checklist of the Birds of the World 4th Edition. Edited by Dickinson EC, Christidis L, vol. 2. Eastbourne, U.K.: Aves Press; 2014: xvii-xlv.
35. Waddell PJ, Kishino H, Ota R. A phylogenetic foundation for comparative mammalian genomics. Genome Inform. 2001;12:141–54.
36. Kuritzin A, Kischka T, Schmitz J, Churakov G. Incomplete lineage sorting and hybridization statistics for large-scale retroposon insertion data. PLoS Comput Biol. 2016;12(3):e1004812.
37. Mayr G, Manegold A. The oldest European fossil songbird from the early Oligocene of Germany. Naturwissenschaften. 2004;91(4):173–7.
38. Ericson PG, Christidis L, Cooper A, Irestedt M, Jackson J, Johansson US, Norman JA. A Gondwanan origin of passerine birds supported by DNA sequences of the endemic New Zealand wrens. Proc R Soc B. 2002; 269(1488):235–41.
39. Barker FK, Barrowclough GF, Groth JG. A phylogenetic hypothesis for passerine birds: taxonomic and biogeographic implications of an analysis of nuclear DNA sequence data. Proc R Soc Lond Ser B Biol Sci. 2002;269(1488):295–308.
40. Ericson P, Klopfstein S, Irestedt M, Nguyen J, Nylander J. Dating the diversification of the major lineages of Passeriformes (Aves). BMC Evol Biol. 2014;14(1):8.
41. Suh A. The phylogenomic forest of bird trees contains a hard polytomy at the root of Neoaves. Zool Scripta. 2016;45(S1):50–62.
42. Han K-L, Braun EL, Kimball RT, Reddy S, Bowie RCK, Braun MJ, Chojnowski JL, Hackett SJ, Harshman J, Huddleston CJ, et al. Are transposable element insertions homoplasy free?: an examination using the avian tree of life. Syst Biol. 2011;60(3):375–86.
43. Manegold A, Mayr G, Mourer-Chauviré C. Miocene songbirds and the composition of the European passeriform avifauna. Auk. 2004;121(4):1155–60.
44. Gilbert N, Labuda D. Evolutionary inventions and continuity of CORE-SINEs in mammals. J Mol Biol. 2000;298(3):365–77.
45. Buzdin A, Ustyugova S, Gogvadze E, Vinogradova T, Lebedev Y, Sverdlov E. A new family of chimeric retrotranscripts formed by a full copy of U6 small

nuclear RNA fused to the 3' terminus of L1. Genomics. 2002;80(4):402–6.

46. Nishihara H, Plazzi F, Passamonti M, Okada N. MetaSINEs: broad distribution of a novel SINE superfamily in animals. Genome Biol Evol. 2016;8(3):528–39.

47. Brosius J. Genomes were forged by massive bombardments with retroelements and retrosequences. Genetica. 1999;107(1):209–38.

48. Ohshima K, Okada N. SINEs and LINEs: symbionts of eukaryotic genomes with a common tail. Cytogenet Genome Res. 2005;110(1–4):475–90.

49. Nishihara H, Smit AFA, Okada N. Functional noncoding sequences derived from SINEs in the mammalian genome. Genome Res. 2006;16(7):864–74.

50. Fujita P, Rhead B, Zweig A, Hinrichs A, Karolchik D, Cline MS, Goldman M, Barber G, Clawson H, Coelho ADM, Dreszer TR, Giardine BM, Harte RA, Hillman-Jackson J, Hsu F, Kirkup V, Kuhn RM, Learned K, Li CH, Meyer LR, Pohl A, Raney BJ, Rosenbloom KR, Smith KE, Haussler D, Kent WJ. The UCSC genome browser database: update 2011. Nucleic Acids Res. 2011;39:D876–82.

51. Katoh K, Toh H. Recent developments in the MAFFT multiple sequence alignment program. Brief Bioinform. 2008;9:286–98.

52. Poelstra JW, Vijay N, Bossu CM, Lantz H, Ryll B, Müller I, Baglione V, Unneberg P, Wikelski M, Grabherr MG, et al. The genomic landscape underlying phenotypic integrity in the face of gene flow in crows. Science. 2014;344(6190):1410–4.

53. Stamatakis A, Hoover P, Rougemont J. A rapid bootstrap algorithm for the RAxML web servers. Syst Biol. 2008;75:758–71.

54. Miller MA, Pfeiffer W, Schwartz T: Creating the CIPRES Science Gateway for inference of large phylogenetic trees. Proceedings of the Gateway Computing Environments Workshop (GCE). 2010:1–8.

55. Smit A, Hubley R, Green P. RepeatMasker Open-3.3.0. http://www.repeatmasker.org. 1996–2010.

56. Tamura K, Stecher G, Peterson D, Filipski A, Kumar S. MEGA6: molecular evolutionary genetics analysis version 6.0. Mol Biol Evol. 2013;30(12):2725–9.

57. Zuker M. Mfold web server for nucleic acid folding and hybridization prediction. Nucleic Acids Res. 2003;31(13):3406–15.

58. Jurka J, Kapitonov VV, Pavlicek A, Klonowski P, Kohany O, Walichiewicz J. Repbase update, a database of eukaryotic repetitive elements. Cytogenet Genome Res. 2005;110(1–4):462–7.

15

Locus-specific hypomethylation of the mouse IAP retrotransposon is associated with transcription factor-binding sites

Ken-ichi Shimosuga[1,2], Kei Fukuda[1,3], Hiroyuki Sasaki[1] and Kenji Ichiyanagi[1,4*] (iD)

Abstract

Background: Intracisternal A particle (IAP) is one of the most transpositionally active retrotransposons in the mouse genome, but its expression varies between cell types. This variation is believed to arise from differences in the epigenetic state (e.g., DNA methylation) of the 5′ long terminal repeat (LTR), where transcription starts. However, owing to the high copy number and high sequence similarity between copies, it is difficult to analyze the epigenetic states of individual IAP LTRs in a comprehensive manner.

Results: We have developed a method called Target Enrichment after Post-Bisulfite Adaptor Tagging (TEPBAT) to analyze the DNA methylation states of a large number of individual retrotransposon copies at once. Using this method, we determined the DNA methylation levels of >8500 copies of genomic IAP LTRs (almost all copies that we aimed to target by the PCR primers) in the sperm and tail. This revealed that the vast majority of the LTRs were heavily methylated in both sperm and tail; however, hypomethylated copies were more frequently found in the sperm than in the tail. Interestingly, most of these hypomethylated LTRs were solo-type, belonged to specific IAP subfamilies, and carried binding sites for transcription factors (TFs) that are active in male germ cells.

Conclusions: The current study revealed subfamily- and locus-specific hypomethylation of IAP LTRs, and suggests that binding of TFs is involved in the protection from DNA methylation, whereas the IAP internal sequence enhances methylation. Furthermore, the study demonstrated that TEPBAT offers a cost-effective method for a variety of DNA methylome studies that focus on retrotransposon sequences.

Keywords: DNA methylome, Intracisternal A particle, Endogenouse retrovirus, Transcription factor, Spermatogenesis, Mouse

Background

Approximately 40% of the mammalian genome comprises several million copies of retrotransposons [1], which include long terminal repeat (LTR) retrotransposons, long interspersed elements, and short interspersed elements. These retrotransposons are amplified by retrotransposition, a process in which their transcribed RNA is utilized to make a DNA copy by reverse transcription. It has been reported that retrotransposition causes heritable diseases, such as hemophilia A and B, muscular dystrophy, and X-linked agammaglobulinemia in humans [2]. In mice, 10–12% of spontaneous mutant alleles arise as a result of the retrotransposition of LTR retrotransposons, such as intracisternal A particle (IAP), early transposon (ETn), and MusD [3]. To diminish the transpositional activity, retrotransposon expression can be epigenetically regulated at the transcriptional level by DNA methylation at CpG sites and by repressive histone modifications. A knockout mutation of the mouse *Dnmt1* gene, which encodes a maintenance-type DNA methyltransferase, has been shown to cause derepression of IAP in whole embryos because of a passive loss of DNA methylation [4]. Likewise, several LTR retrotransposons are derepressed in mouse embryonic stem cells having deletion of the *Setdb1* gene, which encodes a protein methyltransferase acting on the lysine-9 residue of histone H3

* Correspondence: ichiyana@agr.nagoya-u.ac.jp
[1]Division of Epigenomics and Development, Medical Institute of Bioregulation, and Epigenome Network Research Center, Kyushu University, 3-1-1 Maidashi, Higashi-ku, Fukuoka 812-8582, Japan
[4]Laboratory of Genome and Epigenome Dynamics, Department of Applied Molecular Biosciences, Graduate School of Bioagricultural Sciences, Nagoya University, Nagoya 464-8601, Japan
Full list of author information is available at the end of the article

(H3K9). The derepression is attributable to the loss of H3K9 trimethylation at the LTRs [5, 6].

The mouse genome contains approximately 4000 copies of full-length or nearly full-length IAP, which consist of two LTRs at both ends and an internal sequence carrying *gag*, *pro*, and *pol* genes. In addition, the genome contains approximately 5000 copies of solo LTRs of IAP, where an LTR alone is present. The sequence of IAP LTR is 300- to 450-bp long, and contains 15–25 CpG sites. In somatic cells, the vast majority of these CpG sites are heavily methylated, and consequently, IAP expression level is very low [4]. However, it has been reported that a small fraction of LTRs escape methylation with significant variation between individuals; sometimes they behave as metastable epialleles (i.e., LTR copies in the A^{vy}, A^{iap}, and $Axin^{fu}$ loci) [7, 8]. Conversely, higher IAP expression is detected in preimplantation embryos (from the 8-cell to blastocyst stages) [9] and in prospermatogonia and spermatogonia [10], which are precursors of spermatozoa. Although DNA methylation levels in IAP copies are indeed low in blastocysts, those in spermatogonia are high [11–13]. Therefore, it is possible that in spermatogonia, some specific IAP copies are hypomethylated to serve as a source of IAP expression, whereas the vast majority remains highly methylated. It would be therefore interesting to analyze the methylation levels in individual IAP copies in a genome-wide manner, rather than to analyze them in bulk (via PCR using primers for the IAP consensus sequence). The development of the whole-genome bisulfite shotgun sequencing method offers an opportunity to analyze DNA methylation levels in individual IAP copies (i.e., the IAP methylome). However, the cost to obtain sufficient sequence depth is high. To reduce the cost while maintaining the comprehensive depth that is required, a method called high-throughput targeted repeat element bisulfite sequencing (HT-TREBS) [14] was recently developed. The study confirmed the high level of DNA methylation at vast majority of IAP LTRs, while a small subset are hypomethylated. However, the mechanism underlying their hypomethylation remains unknown.

In the present study, we determined the IAP methylomes in germ and somatic cells using a method where IAPs and their flanking sequences were selectively amplified using a random primer and an IAP-specific primer after bisulfite treatment of genomic DNA. Deep sequencing analysis revealed that specific IAP subfamilies were hypomethylated, whereas a vast majority were highly methylated. Most of the hypomethylated copies were solo LTRs and carried binding motifs for specific transcription factors (TFs). We discuss a possible role of TFs in protecting these copies from methylation and a role for the internal sequence in recruiting methylation enzymes.

Results

Determination of DNA methylation levels of IAP LTR copies by the TEPBAT method

To selectively obtain bisulfite sequencing data for the IAP LTR sequences, we developed a method that was modified from the PBAT library preparation method designed for whole-genome bisulfite sequencing [15]. In our method (Fig. 1a), designated as TEPBAT (Target Enrichment after Post-Bisulfite Adaptor Tagging), the first DNA strand was synthesized using bisulfite-treated DNA as a template and the *tag-plus-random* primer, which consisted of a random tetramer and a specific tag sequence. The random tetramer end of the primer enabled genome-wide synthesis of the first DNA strand. In the second step, IAP-containing regions were selectively amplified by PCR with the IAP-specific primer and tag-sequence primer (which was almost same as the *tag-plus-random* primer, but did not contain the random tetramer sequence).

Genomic copies of the IAPLTR1_Mm_LTR (referred to as IAPLTR1), IAPLTR1a_MM (IAPLTR1a), IAPLTR2_Mm (IAPLTR2), IAPLTR2a2_Mm (IAPLTR2a2), and IAPLTR2b subfamilies comprise more than half of the total IAP LTRs in the genome (see Fig. 2a) and are less diverged because of recent retrotransposition. Therefore, in the present study, we targeted these copies for methylation analysis. To selectively amplify these copies, the IAP-BS1 and IAP-BS2 primers were designed in regions that are highly conserved among the subfamilies (Fig. 1b). IAP-BS1 was designed in the reverse orientation to amplify the 5′ region and the upstream flanking sequence, whereas IAP-BS2 was designed in the forward orientation to amplify the 3′ region and the downstream flanking sequence. We note that the IAP-BS2 primer could hybridize to IAPEY_LTR (IAPEY) and IAPEY2_LTR (IAPEY2) weakly. The PCR products were ligated to the sequencing adaptor, and paired-end deep sequencing was performed on HiSeq2500 so that one of the paired reads facilitated mapping uniquely to the genome, whereas the other contained the IAP sequence.

To investigate and compare the IAP methylation profiles of germ and somatic cells, we prepared genomic DNAs from the sperm and tail of the same male mouse (the C57BL6/J strain). Using these DNAs, we obtained 86 and 81 million sequencing read pairs, respectively. After removing the primer sequences (including the region of the random tetramer) and low-quality nucleotides, the read pairs were mapped to the mouse genome sequence to call the methylation state of each CpG site (see Methods). On average, about 10 CpG sites in an LTR were covered by the reads with an average sequencing depth of about 320 (yielding about $320 \times 10 = 3200$ methylation calls). To calculate the methylation level at individual IAP LTRs (expressed as a fractional value

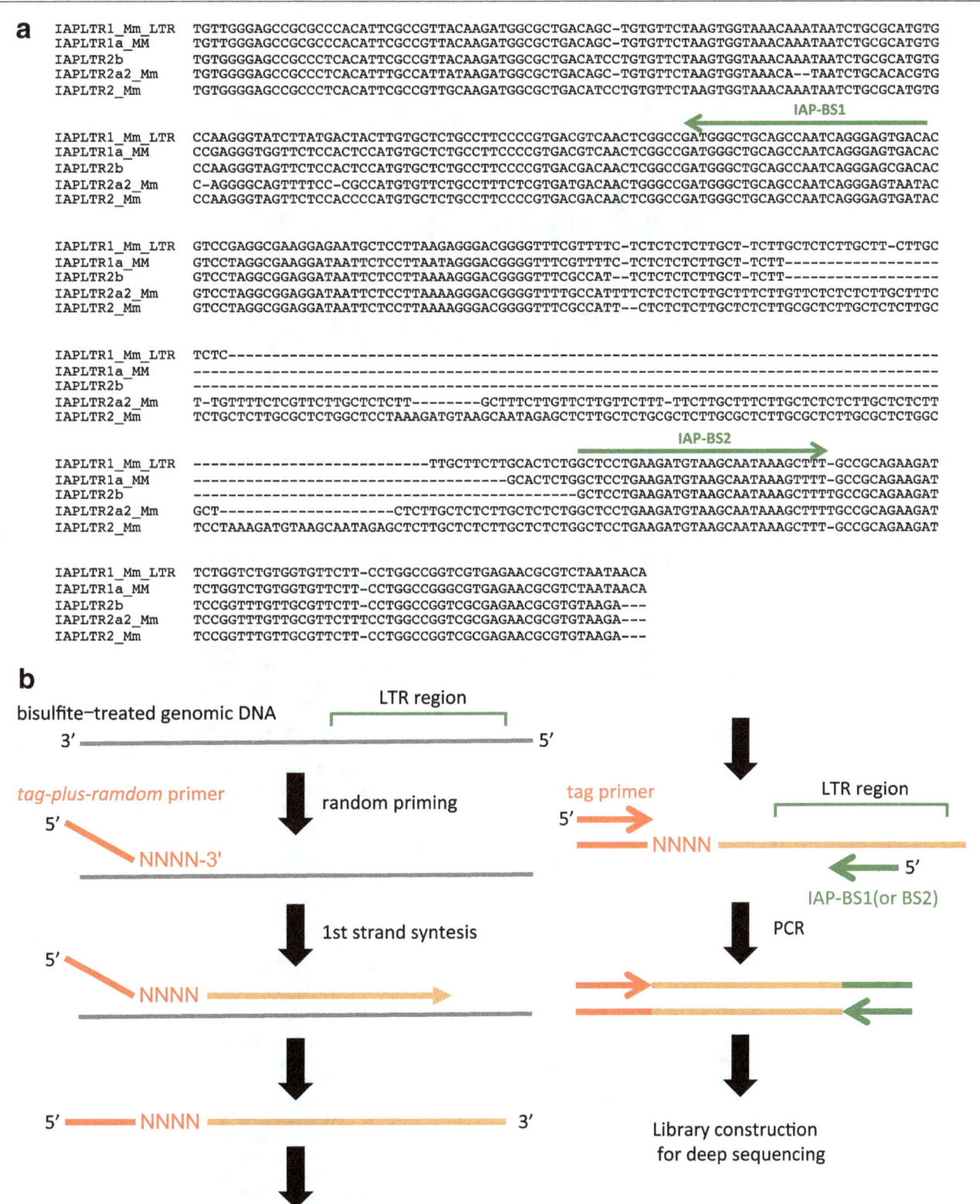

Fig. 1 Experimental design of TEPBAT. **a** Nucleotide sequence alignment of the consensus sequences of IAPLTR1, IAPLTR1a, IAPLTR2, IAPLTR2a2, and IAPLTR2b. The primer regions and orientations (IAP-BS1 and IAP-BS2) used in TEPBAT are indicated as green arrows. **b** Experimental design to enrich IAP regions in the bisulfite-treated genomic DNA. After bisulfite treatment, the first DNA strand was synthesized using the *tag-plus-random* primer, and the IAP regions were amplified by PCR using the specific primer (green arrow) and tag primer (red arrow)

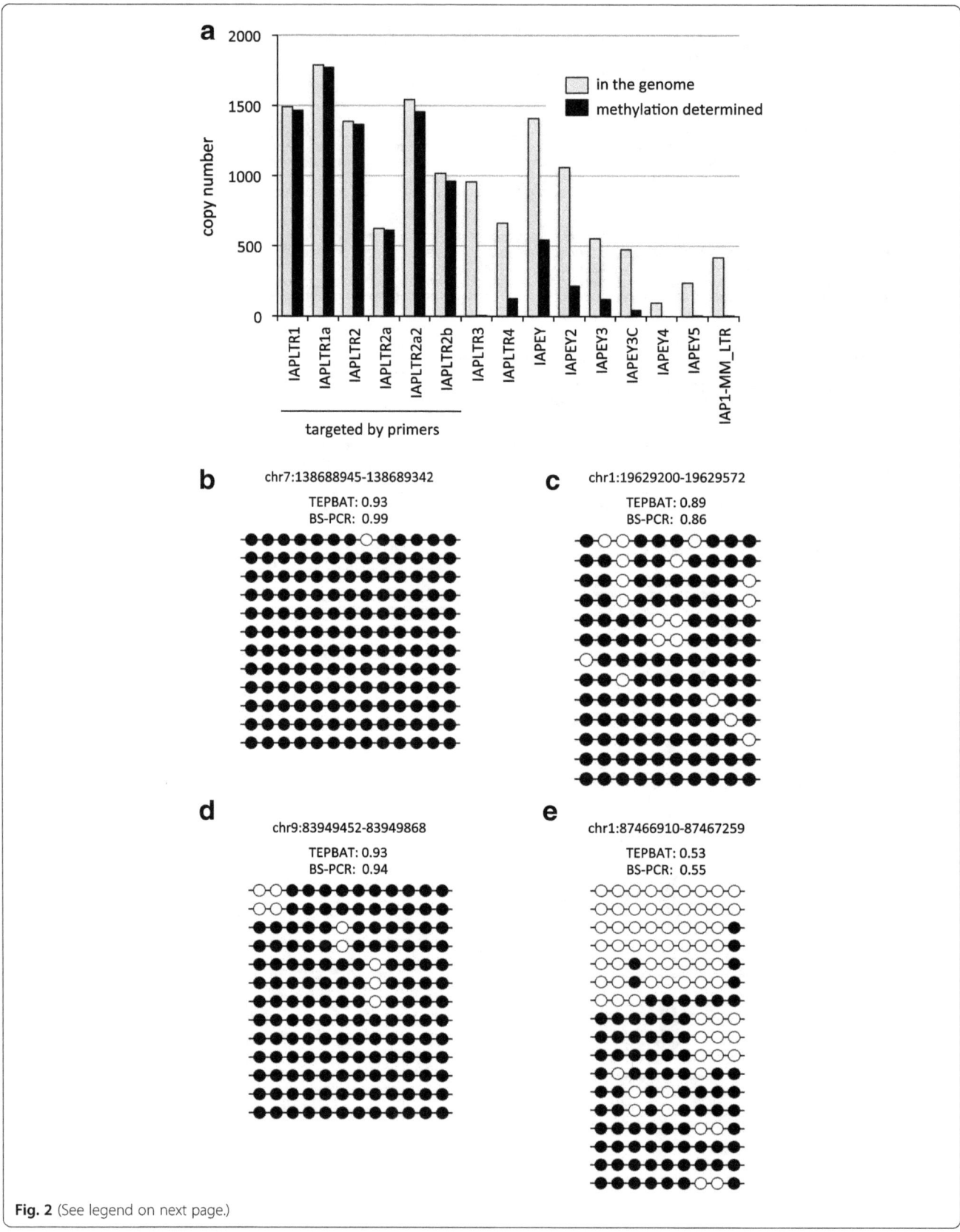

Fig. 2 (See legend on next page.)

(See figure on previous page.)
Fig. 2 DNA methylation levels of individual IAP LTR copies. **a** The number of LTR copies of the subfamilies. The numbers of the total genomic copies are shown in gray and those for which methylation levels were analyzed (total CpG methylation calls of ≧20) are shown in black. **b-e** Bisulfite-PCR results for selected loci in the tail DNA. Closed and open circles indicate methylated and unmethylated CpG sites, respectively. Each row represents a clone of bisulfite-PCR products. The genomic location (mm10) of the loci and their methylation levels (expressed as a fractional value) determined by TEPBAT and bisulfite-PCR (BS-PCR) are shown

[0.00–1.00] and calculated by dividing the methylated cytosine calls by a sum of methylated and unmethylated cytosine calls), the methylation levels at CpG sites within an LTR were averaged. LTRs with <20 CpG methylation calls were excluded from the analysis, retaining the methylation data of 8698 and 8517 LTRs for the sperm and tail, respectively. For comparative analysis, we focused on 8153 LTRs whose methylation levels could be analyzed in both samples. These LTRs included >97% of genomic copies of the targeted IAP subfamilies (IAPLTR1/1a/2/2a/2a2/2b), and 40 and 20% of genomic IAPEY and IAPEY2 LTRs, respectively (Fig. 2a). To validate the acquired methylation data, we determined the methylation levels of selected LTR loci by bisulfite-PCR using the tail DNA. This data was consistent with the TEPBAT data (Fig. 2b-e).

High-throughput analysis revealed that 6677 (80% of the total LTRs analyzed) and 7726 (93%) LTRs exhibited methylation levels of >0.8 in the sperm and tail, respectively (Fig. 3a, b). This is consistent with the notion that IAP LTRs are generally methylated. However, a small fraction of LTRs showed medium (0.2–0.8) and low (≤0.2) methylation levels. Hypomethylation (methylation level ≤0.2) was tissue-specific, and only three LTR copies were hypomethylated in both the sperm and tail (Fig. 3c). In the sperm, 43 and 1612 LTR copies showed low and medium methylation levels, respectively (Fig. 3a), whereas 14 and 612 copies showed low and medium methylation levels in the tail, respectively (Fig. 3b). Therefore, although generally hypermethylated, the IAP LTR methylation level is relatively low in germ cells as compared to somatic cells.

The IAP LTR methylation levels could vary between individuals. Therefore, different individuals were analyzed for several LTR loci in the tail and sperm by bisulfite-PCR (Fig. 3d). The methylation levels at locus3 (IAPLTR2) in tail showed significant variation between individuals (0.47 in individual #2, whereas 0.03 in individual #1 and <0.01 in individuals #3 and #4). However, the methylation levels at other loci were largely conserved among the four individuals analyzed.

Solo LTRs of specific subfamilies display hypomethylation
We investigated whether each subfamily showed specific methylation profiles. The median methylation level was >0.8 for all subfamilies (Fig. 4a, b); however, we found that IAPEY and IAPLTR2a2 in the sperm and IAPLTR2

in the tail contained some hypomethylated loci. Indeed, out of the 43 hypomethylated LTRs in the sperm (methylation level of ≤0.2), 23 (53%), and 13 (30%) belonged to IAPEY and IAPLTR2a2, respectively (Fig. 4c). In the tail, 7 (50%) out of the 14 hypomethylated LTRs belonged to IAPLTR2 (Fig. 4d). Therefore, even though most subfamilies showed high methylation levels, a fraction of copies of the specific subfamilies were hypomethylated. IAPEY is a relatively young subfamily accumulated in the Y chromosome [16] with 60% of ~1400 genomic copies being in the Y chromosome (Additional file 1: Figure S1). Because the fraction of analyzed IAPEY copies was small (Fig. 2a), it should be noted that the number of hypomethylated IAPEY copies may be underestimated. The Y-chromosome copies are more divergent than those in autosomes (Additional file 1: Figure S1), and most (75%) of the copies of which methylation levels were determined reside in autosomes. Consequently, almost all of the copies showing hypomethylation in sperm are also autosomal.

Even in these subfamilies, most loci were heavily methylated. To elucidate the mechanism of hypomethylation, we analyzed the sequence features of the hypomethylated loci. First, because LTR sequences exist as a terminal repeat or as a solo LTR, we determined whether these features affect methylation levels. We manually annotated all IAP LTRs in the genome as 5′ LTR, 3′ LTR, or solo LTR (Additional file 2: Table S1). The 8153 LTRs, for which methylation levels were determined in both tissues, included comparable numbers of 5′ LTRs, 3′ LTRs, and solo LTRs (2820, 2760, and 2573 copies, respectively). However, solo LTRs were significantly enriched with hypomethylated loci; for example, 36 (84%) and 10 (71%) of hypomethylated loci in the sperm and tail, respectively, were solo LTRs (Fig. 5a, b). On the other hand, 5′ and 3′ LTRs were less frequently hypomethylated. For these hypomethylated copies, it was generally observed in both sperm and tail that only one LTR was hypomethylated whereas the other in the same element was hypermethylated (Fig. 5c, d), suggesting that methylation levels of the 5′ and 3′ LTRs in an element are regulated independently of each other.

It should be noted that even for solo LTRs of the specific subfamily, not all loci were hypomethylated. Therefore, for these solo LTRs, we investigated whether some genomic and epigenomic features were associated with hypomethylation. However, we did not find a strong correlation while analyzing their locations relative to genes (i.e., promoter, exon, intron, and intergenic), the distance to the nearest transcriptional start site, GC content of their neighboring regions (in 1-, 10-, or 100-kb bins), the number of CpG sites in their neighboring regions (in 1-, 10-, or 100-kb bins), the nucleotide divergence compared to the respective consensus sequences, methylation levels of their flanking regions in sperm [17], or their methylation levels in primordial germ cells at embryonic day 13.5 [18] (data not shown).

Hypomethylated loci have specific sequences

The sequences of LTR copies even of the same subfamily display slight variations. We analyzed the LTR sequences to see whether specific sequences are associated with hypomethylation. We aligned the sequences of the 11 IAPLTR2a2 sequences determined to be hypomethylated in the sperm, with randomly selected 34 hypermethylated (methylation levels of about 0.92) IAPLTR2a2 sequences. Clustering of the sequences by the neighbor-joining method using Mega5 [19] revealed that these 45 sequences were divided into two major clades (Fig. 6a), one of which contained most of the hypomethylated LTRs. These were designated as the hypomethylated clade. Likewise, the 20 hypomethylated (in the sperm) and 22 randomly selected hypermethylated IAPEY sequences were divided into two clades, where the hypomethylated sequences were clustered together (Fig. 6b). The five hypomethylated (in the tail) and 20 randomly selected hypermethylated IAPLTR2 sequences were divided into two clades; the hypomethylated sequences were again clustered together (Fig. 6c).

In each case, the clear clustering (bootstrap values >86) suggested the presence of single nucleotide variations (SNVs) or sequence blocks that discriminate the two clades. Indeed, we identified several nucleotide positions that were conserved in the hypomethylated clade but not in the other clade. Given the possibility that sequence differences may result in differences in binding of TFs, we searched sequence motifs for TF binding by FIMO [20]. This revealed that eight TF-binding motifs (Prop1, Spi1, Ubp1, Hnf4g, Mitf, Maz, Mafk, and Nf2l2) were significantly enriched in the hypomethylated clade of IAPLTR2a2. In particular, the motifs for Maz and Ubp1 were absent in all (Maz) or most (Ubp1) of hypermethylated sequences, whereas all hypomethylated sequences carried multiple motif sequences. These hypomethylated clade-specific Ubp1 and Maz-binding motifs were all located in the R region of the LTR (Fig. 7), which is known to play a role in transcriptional regulation and shows extensive sequence variation [21]. It is noteworthy that out of the eight TFs mentioned above, Maz and Ubp1 are expressed in spermatogenic cells in the published transcriptome data [22]. Therefore, it is

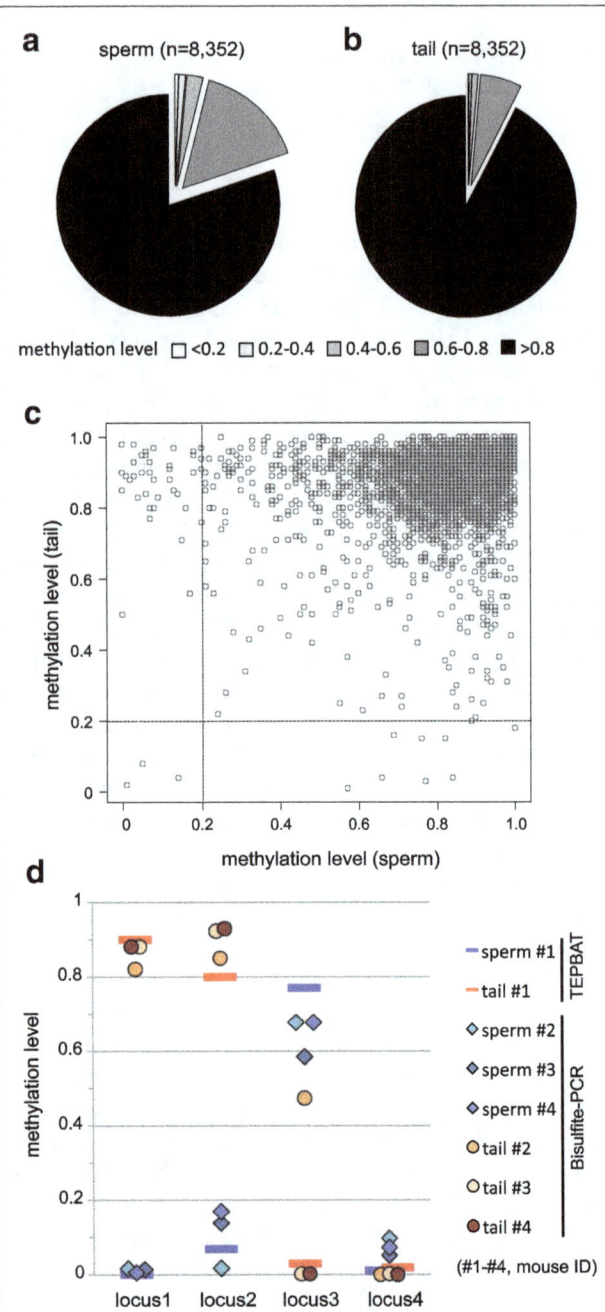

Fig. 3 IAP LTR methylation landscape. **a, b** Pie charts of LTR copies categorized by their methylation levels in the sperm (**a**) and in the tail (**b**). Methylation levels are indicated at the bottom. **c** Scatter plot of DNA methylation levels of IAP LTR copies (n = 8153) in the sperm and tail (x- and y-axis, respectively). **d** Bisulfite-PCR results for selected loci in the sperm and tail in different individuals (mouse ID #2, #3, and #4). These individuals are different from the one (mouse ID #1) used for TEPBAT analysis. Brue (sperm) and red (red) bars indicate the methylation levels determined by TEPBAT. Blue (sperm) and red (tail) circles indicate the methylation levels of three individuals determined by sequencing of 10 to 16 PCR clones. The genomic location (mm10) of the loci are as follows: locus1, chr8:42,217,148-42,217,495 (IAPEY); locus2, chr18:87,502,570-87,503,025 (IAPLTR2a2); locus3, chr13:4,942,652-4,943,122 (IAPLTR2); locus4, chr3:96,489,247-96,489,584 (IAPLTR1)

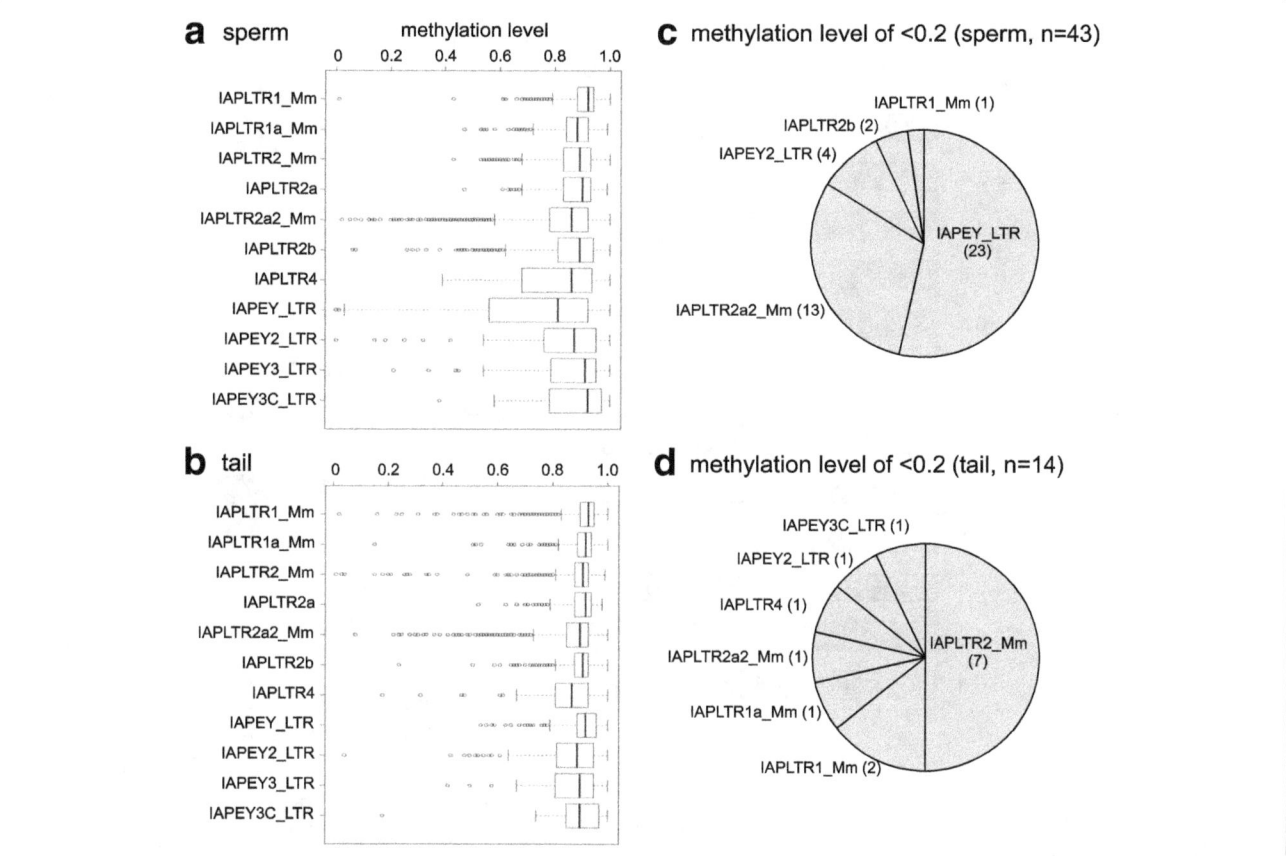

Fig. 4 Specific subfamilies are enriched in hypomethylated copies. **a, b** Box plots of DNA methylation levels of IAP subfamilies in the sperm (**a**) and in the tail (**b**). **c, d** Subfamily distributions of hypomethylated copies (methylation level < 0.2) in the sperm (**c**) and in the tail (**d**) are shown in pie charts. The numbers in parentheses indicate actual copy numbers

conceivable that the binding of these TFs during spermatogenesis reduces the methylation level of their binding sites in these cells and spermatozoa. Interestingly, the hypomethylated sequences tend to have more CpG sites than hypermethylated sites, especially around the TF binding motifs (Fig. 7).

For the IAPEY subfamily, we also identified clade-specific SNVs, but could not identify clade-specific TF-binding motifs. In contrast, for the IAPLTR2 subfamily, we identified Plag1 and Sp1 binding motifs that were specific to the hypomethylated clade. The expression of these TFs in the adult tail is unknown.

Discussion

In the present study, we developed a method called TEPBAT to effectively analyze DNA methylation levels of genomic copies of a specific retrotransposon in bulk. In this method, a single lane of a HiSeq run was sufficient to obtain methylation data of thousands of genomic IAP LTR copies. It seems conceivable that other interspersed repeats can also be analyzed by TEPBAT if appropriately designed primers are used.

Although most of the genomic IAP LTR copies were heavily methylated in the sperm and tail, we found that LTRs of a specific subfamily exhibited a tendency to be hypomethylated, especially when specific sequences were present. This implies that the epigenetic state of IAP is dictated by the genetic sequence. Indeed, we found that several TF-binding motifs are specifically associated with hypomethylated sequences, which suggests that TF binding interferes with DNA methyltransferases, and subsequent methylation modifications (Fig. 8). A similar association between variation of TF-binding motifs and methylation levels has been observed for unique sequences as well [23–26]. However, not all copies in the hypomethylated clade were hypomethylated, suggesting that additional factor(s) may also be involved in the regulation of IAP methylation.

In contrast to solo LTRs, hypomethylation of 5′ and 3′ LTRs was very infrequent. A possible explanation would be that the internal sequence of IAP (*gag, pro, pol,* and 5′ and 3′ untranslated regions) carry a sequence(s) bound to a protein(s) important for de novo and/or maintenance of DNA methylation. This type of methylation regulation may include Krüppel-associated

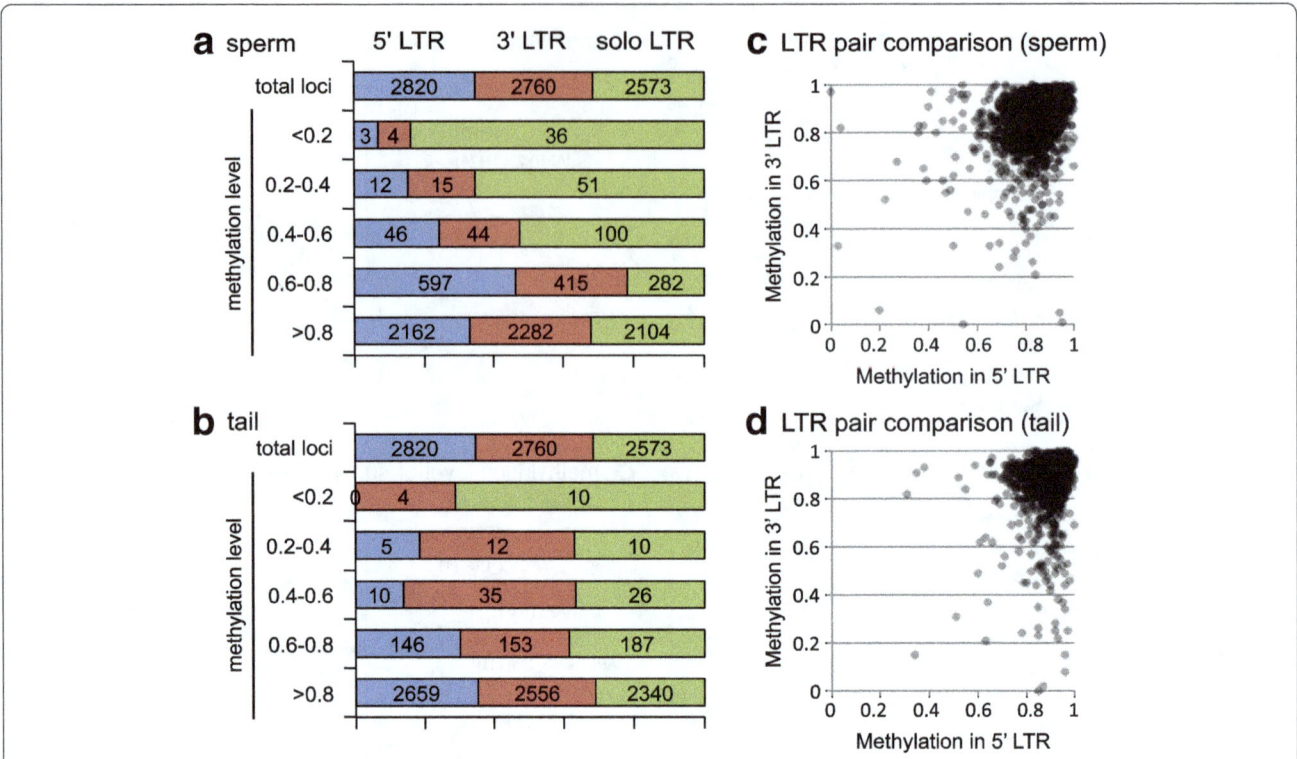

Fig. 5 LTR features and DNA methylation. The LTR copies (n = 8153) are categorized into 5' LTR (blue), 3' LTR (red) and solo LTR (green) and into five groups according to their methylation levels in the sperm (**a**) and in the tail (**b**). The numbers indicate actual copy numbers. (**c, d**) Comparison of methylation levels in sperm (**c**) and tail (**d**) in the two LTRs within the same elements. For IAP copies having both 5' and 3' LTRs, the methylation levels in the 5' LTRs (x axis) are plotted against those in the 3' LTRs (y axis)

box containing zinc finger proteins (KRAB-ZFPs) that regulate the epigenetic state of their binding sequences. KRAB-ZFPs can bind DNA in a sequence-specific manner via the ZF domain, while they can also bind the KAP1 protein via the KRAB domain [27, 28]. The KAP1 protein in turn binds SetDB1, histone deacetylases, and DNA methyltransferases; therefore, the KRAB-ZFP/KAP1 complex induces a repressive chromatin state in the KRAB-ZFP binding regions. For example, among several hundred mouse KRAB-ZFPs, ZFP809 binds an endogenous copy of Mouse Leukemia Virus to induce DNA and histone H3K9 methylation in the internal sequence and LTRs [29]. Likewise, the human KRAB-ZFPs, ZNF91 and ZNF93, bind to SVA and L1 retrotransposons, respectively, leading to an induction of histone H3K9 methylation [30]. It is possible that a certain KRAB-ZFP binds to the IAP internal sequence to induce DNA methylation. ZFP819 may be a candidate because the protein has been shown to bind the IAP internal sequence (5' UTR and the *pol* region of IAPEZ-int) and LTR (IAPLTR1a), and to regulate IAP expression [31]. In addition, not mutually exclusive, it is also possible that the internal sequence inhibits binding of the TET1, TET2, and/or TET3 enzymes, which catalyze oxidation of methylcytosine leading to loss of methylation [32].

It has been proposed that some genomic copies of retrotransposons can serve as a source of epigenetic and phenotypic diversity within a population [33]. IAP is of particular interest in this regard because several IAP copies have been shown to behave as metastable epialleles [7, 8]. These copies in the A^{vy}, A^{iap}, and $Axin^{fu}$ alleles belong to the IAPLTR1 subfamily and not present in our mouse strain (C57BL6/J) [7, 8]. Thus, the mechanism for their occasional hypomethylation remains unknown. To understand the retrotransposon-mediated heritable epigenetic changes more profoundly, genome-wide analyses of epigenetic states of these sequences must be performed for different tissues of many individuals with the discrimination of individual copies. The results here demonstrate that TEPBAT offers a cost-effective method for such studies.

Conclusions

Using TEPBAT, we revealed that solo LTRs of specific subfamilies (IAPLTR2a2 and IAPEY in sperm and IAPLTR2 in tail) tend to be hypomethylated. Binding of TFs seems to account for hypomethylation of these copies, emphasizing the importance of TFs in regulating the epigenome. It should be noted, however, that the hypomethylation of IAPLTR2a2 and IAPEY copies in the

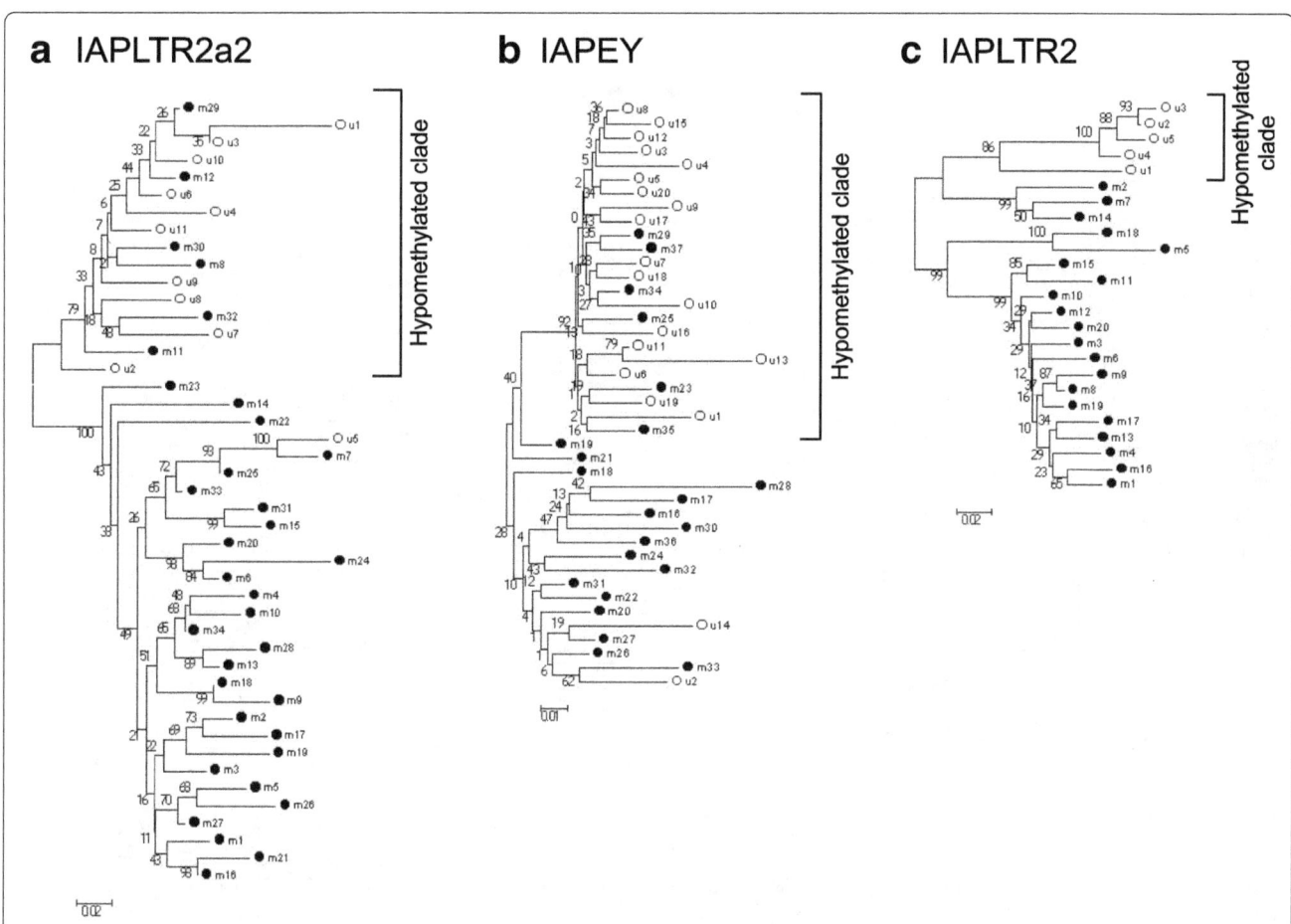

Fig. 6 Clustering of hypermethylated LTR sequences in the phylogenetic trees. Neighbor-joining trees of members of (**a**) IAPLTR2a2, (**b**) IAPEY, and (**c**) IAPLTR2. Nucleotide sequences of the hypomethylated copies and randomly selected hypermethylated copies of the same subfamily were aligned using Clustal X [37] to generate a neighbor-joining tree on Mega5 [19]. Open and closed circles indicate hypomethylated and hypermethyated copies, respectively. In each panel, u1, u2, u3, etc. are locus names of hypomethylated copies, and m1, m2, m3, etc. are locus names of hypermethylated copies. The clade that includes the most hypomethylated copies (designated as hypomethylated clade) is indicated. The numbers on the nodes indicate bootstrap values (1000 replicates)

sperm does not explain the higher expression of IAP in spermatogenic cells because most of the hypomethylated copies were solo LTRs lacking the internal sequence. Although further studies will be required to solve the issue, our method can be applied to a variety of DNA methylome studies that focus on retrotransposon sequences.

Methods
Sequence library construction and paired-end sequencing
High-molecular-weight genomic DNA was prepared from tail and sperm samples of the same adult mouse (C57BL6/J) by a standard procedure. The DNA was treated with 10 M bisulfite (sodium and ammonium salt), as described previously [34]. The resultant DNA (25 ng) was incubated with 80 nM *tag-plus-random* primer (5'-GCAGTGAACT-GACTACAGGNNNN-3') in Klenow buffer (10 mM TrisHCl, 10 mM $MgCl_2$, 1 mM DTT, and 125 μM each of dNTPs) at 94 °C for 5 min, then quickly cooled down to

4 °C. After 7.5 U of Klenow Fragment exo⁻ (New England Biolab) was added, the reaction mixture was incubated at 4 °C for 15 min. The temperature was then increased to 37 °C at the rate of 1 °C per minute and maintained at 37 °C for 90 min. After heat inactivation at 70 °C for 10 min, DNA was purified using AMPure XP (Beckman Coulter). IAP-containing genomic regions were amplified by PCR (25 cycles of 94 °C for 15 s, 55 °C for 30 s, and 68 °C for 30 s) using KOD-plus neo (Toyobo), tag primer (5'-CAGTGAAC- TGACTACAGG-3'), and either of IAP-BS1 (5'-GGGGAAGGTAGAGTATAWG-3') or IAP-BS2 (5'-GGTTTTTGAAGATGTAAGTAATAAAGTTTT-3') primers. The 350–450-bp long PCR products were purified by electrophoresis using a 2% agarose gel, and Illumina sequencing adaptors were added to their ends using the TruSeq DNA Sample Prep kit (Illumina). Paired-end 100-bp sequencing was carried out on a HiSeq2500 in the highoutput mode (one lane per tissue sample with equal

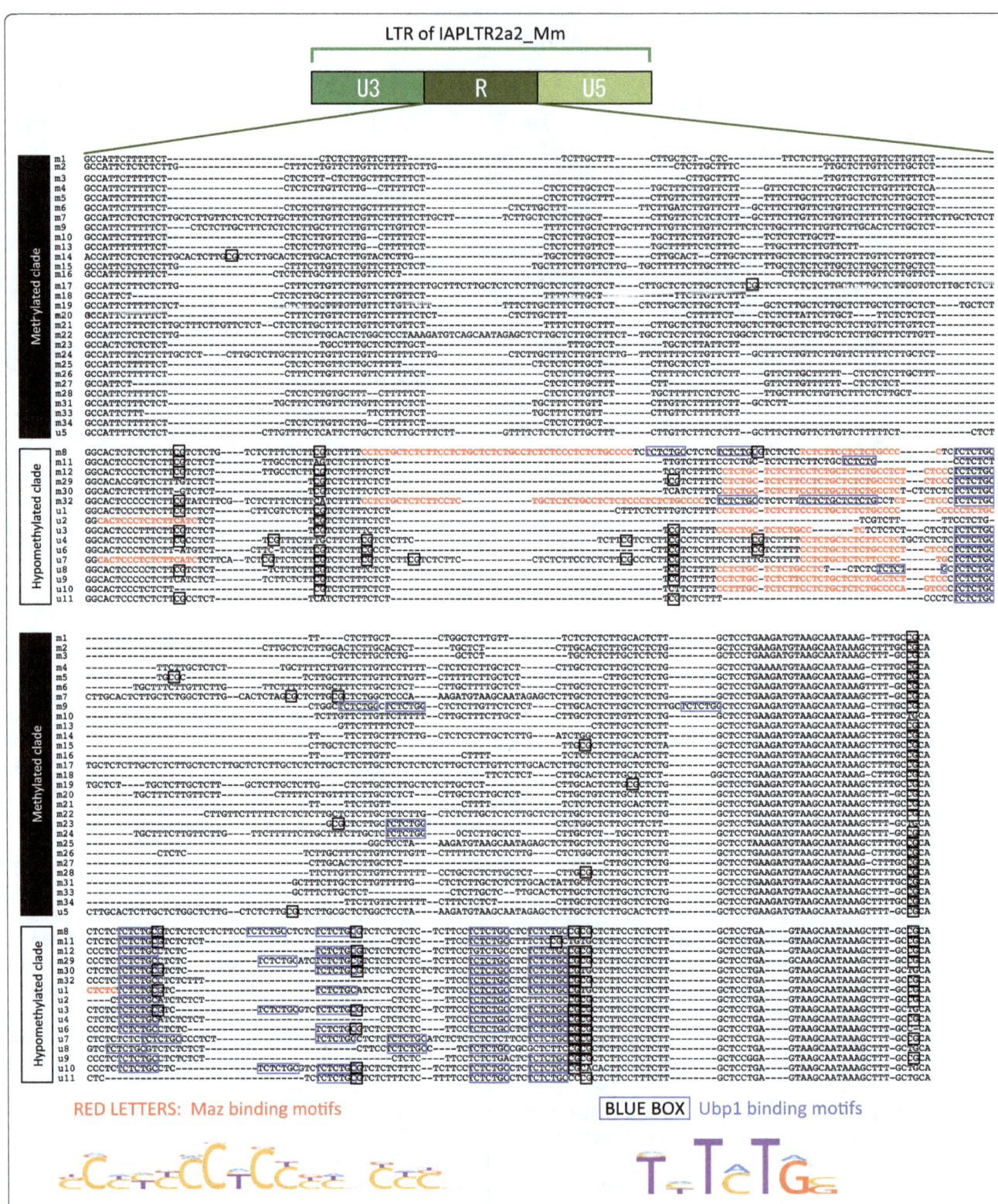

Fig. 7 Hypomethylated clade of IAPLTR2a2 is characterized by the presence of Maz- and Ubp1-binding motifs. The nucleotide sequences of the R region of the LTR (see top for a schematic view of the LTR sub-regions) of IAPLTR2a2 copies belonging to the methylated and hypomethylated clades were aligned by Clustal X [37]. Maz-binding motifs are highlighted in red. Ubp1-binding sites are indicated by a blue box. CpG sites are indicated by a black box. Members of the hypomethylated clade carry multiple Maz and Ubp1 sites, whereas almost all members of the hypermethylated clade carry none. Sequence Logo representation of the binding motifs is shown at the bottom

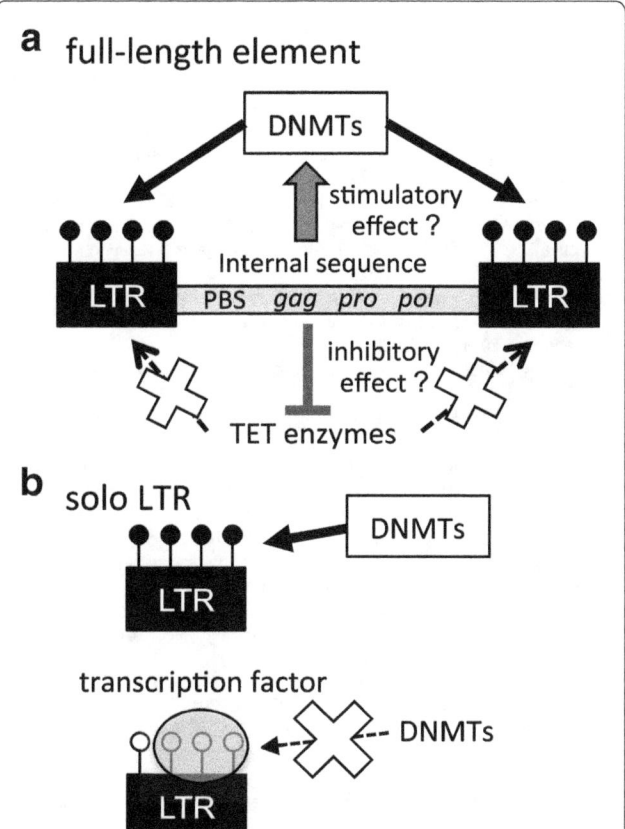

a full-length element

b solo LTR

Fig. 8 Models for molecular mechanisms that regulate IAP methylation. **a** In copies containing the internal sequence, some parts of the internal sequence may stimulate DNA methylation in their flanking LTR sequences. Closed circles represent methylated CpG sites in LTRs. DNMTs, DNA methyltransferases. TET enzymes, Ten-Eleven translocation enzymes (catalyzing oxidation of methylcytosine). PBS, primer binding site. **b** Most solo LTR copies are methylated by DNMTs (top). However, copies carrying TFBSs bind to these TFs if the TFs are expressed in the cell. This binding inhibits the action of DNMTs (bottom). The bound transcription factor is represented as a gray oval. Open circles represent unmethylated CpG sites in the LTRs

specific primers (forward and reverse primers were designed for the LTR and flanking region, respectively). PCR fragments were cloned into the pGEM-Teasy vector (Promega) and sequenced using 3730 DNA Analyzer (Applied Biosystems).

Methylation analysis of IAP LTR sequences
The genomic positions of IAP sequences were obtained from the repeatmasker table from the UCSC genome browser [36]. For each LTR copy, their features (5′ LTR, 3′ LTR, or solo LTR) were manually determined (Table S1). To calculate LTR methylation levels, the methylation levels of CpG sites within the LTRs were averaged. Only those LTRs having ≥20 CpG methylation calls (sum of sequencing depth at CpG sites) were used for further analysis. Only LTRs present in the reference sequence were analyzed.

Abbreviations
H3K9: The lysine-9 residue of histone H3; IAP: Intracisternal A particle; KRAB-ZFP: Krüppel-associated box containing zinc finger protein; LTR: Long terminal repeat; PCR: Polymerase chain reaction; SNV: Single nucleotide variation; TEPBAT: Target Enrichment after Post-Bisulfite Adaptor Tagging; TF: Transcription factor

Acknowledgements
We thank Dr. Hidehiro Toh and Ms. Miho Miyake for assistance in DNA sequencing.

Funding
Research grants from the Ministry of Education, Culture, Sports, Science, and Technology of Japan to KI (grant numbers 25,503,003, 16H04817, and 16 K14784) and to HS (JP25112010).
Research grant from the Takeda Science Foundation to KI.

Authors' contributions
KS performed experiments and data analysis. KF performed experiments. HS supervised the study. KI designed the study, performed sequence analysis, and wrote the manuscript. All authors read and approved the final manuscript.

Competing interests
The authors declare that they have no competing interests.

Author details
[1]Division of Epigenomics and Development, Medical Institute of Bioregulation, and Epigenome Network Research Center, Kyushu University, 3-1-1 Maidashi, Higashi-ku, Fukuoka 812-8582, Japan. [2]Trygroup Incorporated, 1-8-10 Kudankita, Chiyoda-ku, Tokyo 102-0073, Japan. [3]Cellular Memory Laboratory, RIKEN, Wako, Saitama 351-0198, Japan. [4]Laboratory of Genome and Epigenome Dynamics, Department of Applied Molecular Biosciences, Graduate School of Bioagricultural Sciences, Nagoya University, Nagoya 464-8601, Japan.

amounts of the IAP-BS1 and IAP-BS2 libraries being mixed). About 80% of read pairs contained the tag and IAP-BS1/BS2 sequences. After removing the primer sequences (including the random tetramer region) and low-quality bases by a perl script, the read pairs were mapped to the genome (mm10) by Bismark [35] with default parameters to call the methylation state at each CpG position. With this set of parameters, read pairs with multiple hits were discarded. The mapping efficiency was 22–25%. The low mapping efficiency in comparison to PBAT (typically 50–70%) is likely due to the fact that many LTR copies are flanked by repeat elements such as IAP internal sequences.

Bisulfite-PCR analysis
Using the bisulfite-treated tail DNA, touch-down PCR was performed as described previously [34] with locus-

References
1. Mouse Genome Sequencing, C, Waterston RH, Lindblad-Toh K, Birney E, Rogers J, Abril JF, Agarwal P, Agarwala R, Ainscough R, Alexandersson M, et al. Initial sequencing and comparative analysis of the mouse genome. Nature. 2002;420:520–62.

2. Belancio VP, Hedges DJ, Deininger P. Mammalian non-LTR retrotransposons: for better or worse, in sickness and in health. Genome Res. 2008;18:343–58.

3. Maksakova IA, Romanish MT, Gagnier L, Dunn CA, van de Lagemaat LN, Mager DL. Retroviral elements and their hosts: insertional mutagenesis in the mouse germ line. PLoS Genet. 2006;2:e2.

4. Walsh CP, Chaillet JR, Bestor TH. Transcription of IAP endogenous retroviruses is constrained by cytosine methylation. Nat Genet. 1998;20:116–7.

5. Karimi MM, Goyal P, Maksakova IA, Bilenky M, Leung D, Tang JX, Shinkai Y, Mager DL, Jones S, Hirst M, et al. DNA methylation and SETDB1/H3K9me3 regulate predominantly distinct sets of genes, retroelements, and chimeric transcripts in mESCs. Cell Stem Cell. 2011;8:676–87.

6. Matsui T, Leung D, Miyashita H, Maksakova IA, Miyachi H, Kimura H, Tachibana M, Lorincz MC, Shinkai Y. Proviral silencing in embryonic stem cells requires the histone methyltransferase ESET. Nature. 2010;464:927–31.

7. Morgan HD, Sutherland HG, Martin DI, Whitelaw E. Epigenetic inheritance at the agouti locus in the mouse. Nat Genet. 1999;23:314–8.

8. Rakyan VK, Chong S, Champ ME, Cuthbert PC, Morgan HD, Luu KV, Whitelaw E. Transgenerational inheritance of epigenetic states at the murine Axin(fu) allele occurs after maternal and paternal transmission. Proc Natl Acad Sci U S A. 2003;100:2538–43.

9. Svoboda P, Stein P, Anger M, Bernstein E, Hannon GJ, Schultz RM. RNAi and expression of retrotransposons MuERV-L and IAP in preimplantation mouse embryos. Dev Biol. 2004;269:276–85.

10. Dupressoir A, Heidmann T. Germ line-specific expression of intracisternal A-particle retrotransposons in transgenic mice. Mol Cell Biol. 1996;16:4495–503.

11. Lane N, Dean W, Erhardt S, Hajkova P, Surani A, Walter J, Reik W. Resistance of IAPs to methylation reprogramming may provide a mechanism for epigenetic inheritance in the mouse. Genesis. 2003;35:88–93.

12. Reik W, Dean W, Walter J. Epigenetic reprogramming in mammalian development. Science. 2001;293:1089–93.

13. Sasaki H, Matsui Y. Epigenetic events in mammalian germ-cell development: reprogramming and beyond. Nat Rev Genet. 2008;9:129–40.

14. Ekram MB, Kim J. High-throughput targeted repeat element bisulfite sequencing (HT-TREBS): genome-wide DNA methylation analysis of IAP LTR retrotransposon. PLoS One. 2014;9:e101683.

15. Miura F, Enomoto Y, Dairiki R, Ito T. Amplification-free whole-genome bisulfite sequencing by post-bisulfite adaptor tagging. Nucleic Acids Res. 2012;40:e136.

16. Fennelly J, Harper K, Laval S, Wright E, Plumb M. Co-amplification to tail-to-tail copies of MuRVY and IAPE retroviral genomes on the Mus Musculus Y chromosome. Mamm Genome. 1996;7:31–6.

17. Hammoud SS, Low DH, Yi C, Carrell DT, Guccione E, Cairns BR. Chromatin and transcription transitions of mammalian adult germline stem cells and spermatogenesis. Cell Stem Cell. 2014;15:239–53.

18. Kobayashi H, Sakurai T, Miura F, Imai M, Mochiduki K, Yanagisawa E, Sakashita A, Wakai T, Suzuki Y, Ito T, et al. High-resolution DNA methylome analysis of primordial germ cells identifies gender-specific reprogramming in mice. Genome Res. 2013;23:616–27.

19. Tamura K, Peterson D, Peterson N, Stecher G, Nei M, Kumar S. MEGA5: molecular evolutionary genetics analysis using maximum likelihood, evolutionary distance, and maximum parsimony methods. Mol Biol Evol. 2011;28:2731–9.

20. Grant CE, Bailey TL, Noble WS. FIMO: scanning for occurrences of a given motif. Bioinformatics. 2011;27:1017–8.

21. Christy RJ, Brown AR, Gourlie BB, Huang RC. Nucleotide sequences of murine intracisternal A-particle gene LTRs have extensive variability within the R region. Nucleic Acids Res. 1985;13:289–302.

22. Inoue K, Ichiyanagi K, Fukuda K, Glinka M, Sasaki H. Switching of dominant retrotransposon silencing strategies from posttranscriptional to transcriptional mechanisms during male germ-cell development in mice. PLoS Genet. 2017;13:e1006926.

23. Banovich NE, Lan X, McVicker G, van de Geijn B, Degner JF, Blischak JD, Roux J, Pritchard JK, Gilad Y, Methylation QTL. Are associated with coordinated changes in transcription factor binding, histone modifications, and gene expression levels. PLoS Genet. 2014;10:e1004663.

24. Fukuda K, Ichiyanagi K, Yamada Y, Go Y, Udono T, Wada S, Maeda T, Soejima H, Saitou N, Ito T, et al. Regional DNA methylation differences between humans and chimpanzees are associated with genetic changes, transcriptional divergence and disease genes. J Hum Genet. 2013;58:446–54.

25. Fukuda K, Inoguchi Y, Ichiyanagi K, Ichiyanagi T, Go Y, Nagano M, Yanagawa Y, Takaesu N, Ohkawa Y, Imai H, et al. Evolution of the sperm methylome of primates is associated with retrotransposon insertions and genome instability. Hum Mol Genet. 2017;26:3508–19.

26. Gutierrez-Arcelus M, Lappalainen T, Montgomery SB, Buil A, Ongen H, Yurovsky A, Bryois J, Giger T, Romano L, Planchon A, et al. Passive and active DNA methylation and the interplay with genetic variation in gene regulation. elife. 2013;2:e00523.

27. Lupo A, Cesaro E, Montano G, Zurlo D, Izzo P, Costanzo P. KRAB-zinc finger proteins: a repressor family displaying multiple biological functions. Curr Genomics. 2013;14:268–78.

28. Urrutia R. KRAB-containing zinc-finger repressor proteins. Genome Biol. 2003;4:231.

29. Wolf D, Goff SP. Embryonic stem cells use ZFP809 to silence retroviral DNAs. Nature. 2009;458:1201–4.

30. Jacobs FM, Greenberg D, Nguyen N, Haeussler M, Ewing AD, Katzman S, Paten B, Salama SR, Haussler D. An evolutionary arms race between KRAB zinc-finger genes ZNF91/93 and SVA/L1 retrotransposons. Nature. 2014;516:242–5.

31. Tan X, Xu X, Elkenani M, Smorag L, Zechner U, Nolte J, Engel W, Pantakani DV. Zfp819, a novel KRAB-zinc finger protein, interacts with KAP1 and functions in genomic integrity maintenance of mouse embryonic stem cells. Stem Cell Res. 2013;11:1045–59.

32. Wu X, Zhang Y. TET-mediated active DNA demethylation: mechanism, function and beyond. Nat Rev Genet. 2017;18:517–34.

33. Whitelaw E, Martin DI. Retrotransposons as epigenetic mediators of phenotypic variation in mammals. Nat Genet. 2001;27:361–5.

34. Ichiyanagi K, Li Y, Watanabe T, Ichiyanagi T, Fukuda K, Kitayama J, Yamamoto Y, Kuramochi-Miyagawa S, Nakano T, Yabuta Y, et al. Locus- and domain-dependent control of DNA methylation at mouse B1 retrotransposons during male germ cell development. Genome Res. 2011; 21:2058–66.

35. Krueger F, Andrews SR. Bismark: a flexible aligner and methylation caller for Bisulfite-Seq applications. Bioinformatics. 2011;27:1571–2.

36. Tyner C, Barber GP, Casper J, Clawson H, Diekhans M, Eisenhart C, Fischer CM, Gibson D, Gonzalez JN, Guruvadoo L, et al. The UCSC genome browser database: 2017 update. Nucleic Acids Res. 2017;45:D626–34.

37. Thompson, J. D., Gibson, T. J. and Higgins, D. G. Multiple sequence alignment using ClustalW and ClustalX. Curr Protoc Bioinformatics. 2002; Chapter 2:Unit 2 3.

16

The *Drosophila mojavensis Bari3* transposon: distribution and functional characterization

Antonio Palazzo[†], Roberta Moschetti[†], Ruggiero Caizzi and René Massimiliano Marsano[*]

Abstract

Background: *Bari*-like transposons belong to the *Tc1-mariner* superfamily, and they have been identified in several genomes of the *Drosophila* genus. This transposon's family has been used as paradigm to investigate the complex dynamics underlying the persistence and structural evolution of transposable elements (TEs) within a genome. Three structural *Bari* variants have been identified so far and can be distinguished based on the organization of their terminal inverted repeats. *Bari3* is the last discovered member of this family identified in *Drosophila mojavensis*, a recently emerged species of the Repleta group of the genus *Drosophila*.

Results: We studied the insertion pattern of *Bari3* in different *D. mojavensis* populations and found evidence of recent transposition activity. Analysis of the transposase domains unveiled the presence of a functional nuclear localization signal, as well as a functional binding domain. Using luciferase-based assays, we investigated the promoter activity of *Bari3* as well as the interaction of its transposase with its left terminus. The results suggest that *Bari3* is transposition-competent. Finally we demonstrated transposase transcript processing when the transposase gene is overexpressed *in vivo* and *in vitro*.

Conclusions: *Bari3* displays very similar structural and functional features with its close relative, *Bari1*. Our results strongly suggest that *Bari3* is an independent element that has generated genomic diversity in *D. mojavensis*. It can autonomously transcribe its transposase gene, which in turn can localize in the nucleus and bind the terminal inverted repeats of the transposon. Nevertheless, the identification of an unpredicted spliced form of the *Bari3* transposase transcript allows us to hypothesize a control mechanism of its mobility based on mRNA processing. These results will aid the studies on the *Bari* family of transposons, which is intriguing for its widespread diffusion in Drosophilids coupled with a structural diversity generated during the evolution of *Bari*-like elements in their host genomes.

Keywords: Transposon, Transposase, *Tc1*-like elements, *Bari3*, *Drosophila mojavensis*, Transposase transcript splicing, Nuclear localization signal, Luciferase promoter assay

Background

A consistent fraction of eukaryotic genomes is composed of transposable elements (TEs). Although they were originally considered as 'selfish' or 'junk' elements [1,2] and as potentially representing endogenous mutagens, they are now believed to represent one of the major forces driving the evolution of genes and genomes [3-5].

DNA-based TEs belong to the Class II of transposons and use a DNA-mediated mode of transposition and self-encoded transposases to catalyze the transposition reaction, unlike Class I elements that move via reverse transcription of RNA intermediates. Seventeen cut-and-paste DNA transposons superfamilies have been discovered so far [6], with the best studied undoubtedly being the *Tc1-mariner* superfamily.

The *IS630-Tc1-mariner* (or ItmDx(D/E superfamily)) [7] constitutes the largest group of cut-and-paste Class II transposons. These elements are up to 2 Kbp in length and usually contain a single transposase-encoding gene, typically flanked by two short terminal inverted repeats (TIRs). The transposase of these elements is sufficient to catalyze the transposition reaction *in vitro* [8] by recognition of the TIRs, explaining in part the wide phylogenetic occurrence of *Tc1/mariner*-like elements [9].

---

* Correspondence: renemassimiliano.marsano@uniba.it

[†]Equal contributors

Dipartimento di Biologia, Università degli Studi di Bari "Aldo Moro", Via Orabona 4, 70125 Bari, Italy

The complex dynamics underlying the invasion and the persistence of TEs in a genome could be better understood by studying different elements belonging to the same family and hosted in genomes of different species [10]. Furthermore this kind of approach could give clues in improving the transposition efficiency of TEs in order to establish new transposon-based integration tools [11]. As an example, the *mos1* element discovered in *Drosophila mauritiana* has been used as starting point to isolate the *Himar1* element in the horn fly *Haematobia irritans,* which transposition efficiency has been further improved *in vitro* [8,12].

In this view the *Bari* family potentially represents an interesting case study in the *Drosophila* genus.

Three related *Bari* sub-families (*Bari1*, *Bari2* and *Bari3*), differing in their structural organization and their potential transposition ability, are known to exist in different *Drosophila* species [13,14]. While elements related to *Bari1* and *Bari3* can be either potentially autonomous or not, elements related to *Bari2* are all non-autonomous [13,14]. *Bari*-like elements belong to the IR-DR group of the *Tc1* lineage, comprising elements with terminal ends of about 250 bp in length. This group also includes other *Drosophila*-related TEs such as *S* [15], *Minos* [16], and *Paris* [17], as well as non-insect members like the *Sleeping Beauty (SB)* [11] and the *Frog Prince (FP)* [18] transposons, reconstructed from fish and amphibian genomes, respectively. These elements encode transposases containing a predicted functional bipartite nuclear localization signal (NLS), two helix-turn-helix (HTH) motifs in the N-terminal region and an acidic DD34E triad in the C-terminal region [19-21].

Most of the information on the *Bari*-like elements is related to the *Bari1* element probably due to its presence into the *D. melanogaster* genome in a putatively active form, as demonstrated by direct [22] and indirect [23] evidence.

Recently the NLS and the DNA binding site of the transposase encoded by the *Bari1* element have been functionally characterized [24]. The TIRs of *Tc1-mariner* elements possess two or three direct repeats (DRs), that are the putative binding sites for the transposase and are necessary for the transposition of autonomous elements [21,25,26]. *Bari1* has three DRs in its terminal sequences that are all bounded, although with different efficiency, by the *Bari1* transposase [24].

Bari3 is the last discovered member of the *Bari* family. It has been identified in the genome of the emerging species *D. mojavensis*, but homologous sequences can be also identified in the sequenced genomes of the phylogenetically distant species *D. pseudoobscura, D. persimilis and D. willistoni* [14]. Its structural characteristics, that is, long TIRs with three DRs bracketing a transposase coding region, allowed the determination of the evolutionary

dynamics acting on the transposon termini [14]. Furthermore, at least ten identical copies of this element can be detected in the sequenced genome *D. mojavensis*, suggesting its very recent transposition activity. Previous studies concerning the phylogenetic distribution of the *Bari*-like elements have disclosed inconsistencies with the species phylogeny that have demonstrated [27] or postulated [14] ancient horizontal gene transfer events.

These observations along with our previous functional study of the *Bari1* transposon [24], prompted us to investigate and compare this new member of the *Bari* family in order to gain insight into the biology of this transposon family.

Here, we show that *Bari3* is a widely distributed transposon in the *D. mojavensis* populations with a variable copy number within the genome of different subspecies. Similarly to *Bari1*, the *Bari3* transposase is able to bind the TIRs of the transposon and localizes in the nucleus of *Drosophila* and human cells. We have also investigated the internal promoter of *Bari3* and the transposon-transposase interaction. Furthermore, transient transposase gene overexpression allowed the isolation of an unexpected spliced transcript in cultured cells and in embryos. These data are discussed in the light of previous studies concerning a putative transposition control of the *Bari* family.

Results
The distribution of *Bari3* in the genome of *Drosophila mojavensis*

We previously reported, using *in silico* approaches, the recent invasion of the transposon *Bari3* in the genome of the emerging *Drosophila* species, *D. mojavensis*, [14]. *D. mojavensis* is endemic to the Sonoran Desert of North America, with different subpopulations specialized in feeding on different necrotic cactus tissues and showing both genetic differentiation and reproductive isolation [28-30].

In order to estimate the activity of the *Bari3*, we analyzed its distribution in the population of *D. mojavensis* collected in different geographical regions of California and Mexico (Figure 1 and Table 1).

A full length *Bari3* element was cloned from the genome of the sequenced *D. mojavensis* strain (pT/moja11) using a PCR-based strategy (see Methods section) [32]. Sequence and structure of this element are described in Additional file 1.

The DNA extracted *en masse* from ten *D. mojavensis* populations was digested with the endonuclease EcoRI and analyzed by Southern blot hybridization. We used an internal 592-bp fragment (Figure 2A) as a probe, subcloned from the full-length *Bari3* element. To avoid nonspecific detection of divergent sequences related to transposon relics, we applied high-stringency conditions for our hybridization experiments. The pattern obtained is shown in Figure 2 (panel B) and clearly indicates

Figure 1 Geographical origin of the *Drosophila mojavensis* strains analyzed in this study. The prefix 15081 has been omitted for space restriction (see Table 1). *D. mojavensis* subspecies are indicated according to the color code showed.

variability in both the copy number and genomic distribution of the *Bari3* elements among the populations analyzed. We estimate that the *baja* and *wrigleyi* subspecies contain from 5 to 11 copies of the transposon, while the *mojavensis* and *sonorensis* subspecies contain 1 to 3 copies of *Bari3*. As expected, only very faint bands can be detected in the distant species *D. melanogaster* and *D. pseudoobscura*, confirming that the *Bari3* element of *D. mojavensis* is quite divergent from the

Bari3 element of *D. pseudoobscura* and from the *Bari1* and *Bari2* elements of *D. melanogaster* [14].

We used the sequence of *Bari3* as our query, to perform a BLAST analysis against the WGS database of *D. mojavensis*. These experiments revealed ten full-length copies of *Bari3* and at least ten defective ones, slightly divergent in sequence and bearing mostly terminal truncations. This result, summarized in Additional file 2, is in line with the hybridization pattern observed for the

Table 1 *Drosophila mojavensis* strains used in this study

DSSC code	Subspecies[a] (race[b])	Collection place[a]	Collection date[a]	Bari3 Southern/FISH signals detected[c]
15081-1352.00	mojavensis (A)	Chocolate Mountains, Riverside County, California	N/D	1
15081-1352.06	mojavensis (A)	Chocolate Mountains, Riverside County, California	N/D	½
15081-1351.01	sonorensis (BI)	Tiburon Island, Gulf of California Mexico	(1964)	1 (faint)
15081-1351.17	sonorensis (BI)	Punta Onah Sonora, Mexico	(1988)	2
15081-1352.02	wrigleyi (C)	USC marine station, Catalina Island, California	(1991)	9/9
15081-1352.14	wrigleyi (C)	Santa Catalina Island, California	(2002)	11/16
15081-1352.22	wrigleyi (C)	Catalina Island, California	(2002)	10
15081-1352.29	wrigleyi (C)	Little Harbor, Catalina Island, California	(2004)	9
15081-1352.30	wrigleyi (C)	Catalina Island, California	(2002)	5
15081-1352.03	baja (BII)	San Esteban Island Gulf of California Mexico	(1965)	ND/5
15081-1352.20	baja (BII)	Cape Region, Santiago, Baja California South Mexico	(1996)	5/7

[a]Data from Drosophila Species Stock Center (DSSC).
[b]Race definition is according to Pfeiler *et al.* [31].
[c]This study.
N.D., not determined.

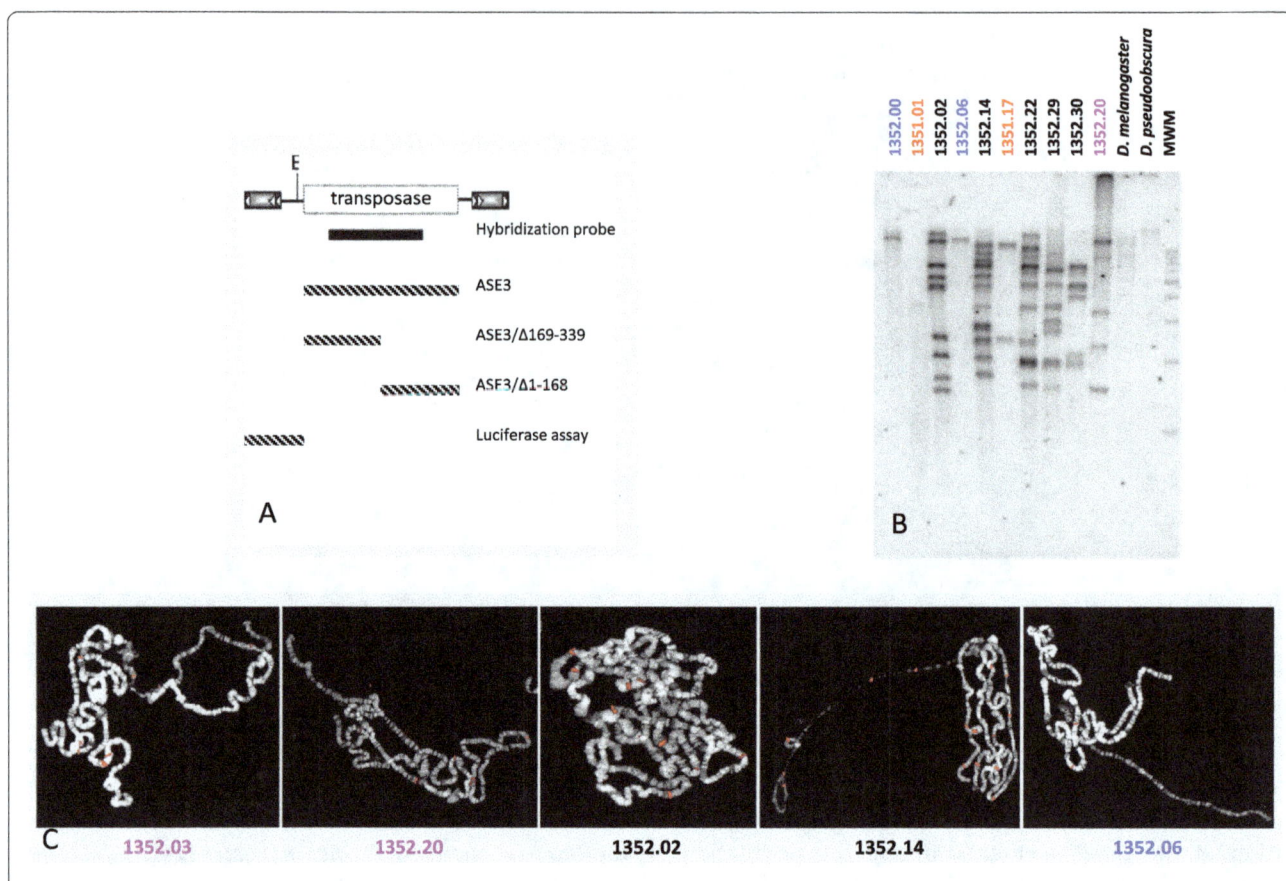

Figure 2 *Bari3* **distribution in the genome of** *Drosophila mojavensis*. **A)** Schematic representation of the *Bari3* transposon. The EcoRI site used for the genomic analyses and the position of the probe (black bar) are showed. Dashed bars represent the transposon fragments tested in this work. **B)** Southern blot hybridization of DNA samples extracted from ten *D. mojavensis* populations MWM, 1Kb DNA molecular weight marker (Promega). **C)** Fluorescence In Situ Hybridization (FISH) on polytene chromosomes prepared from five *D. mojavensis* strains. Merged images (DAPI and Cy3) are shown. Hybridization signals are pseudo-colored in red. The subspecies color code legend reported in the bottom of the figure refers to the hybridization experiments.

1352.22 strain and suggests that at least part of the differences observed are due to degenerated *Bari3* copies.

We further characterized the *Bari3* insertion sites within the genome of *D. mojavensis* populations by analyzing the *in situ* hybridization pattern over polytene chromosomes of five different strains. As shown in Figure 2 (panel C), a variable number of hybridization sites were revealed that are substantially in accord with the number of polymorphic bands seen in Southern hybridization experiments. Taken together, these results strongly indicate a recent transposition activity of *Bari3*.

Analysis of the *Bari3* transposase domains

To gain further insight into the *Bari3* transposon, we started a preliminary characterization of its transposase. Typically, the NLS is present at the N-terminus of the transposase in *Tc1-mariner* elements, although other elements may present the NLS at the C-terminus. The presence of a functional nuclear import domain was firstly

assayed because it represents a necessary condition for the mobility of a transposon.

Immuno-detection was performed in cells transiently overexpressing a V5-His tagged *Bari3* transposase. Subcellular localization was assayed in two model cellular systems, the *Drosophila* S2R + and the human HepG2 cells. With the aim to localize the NLS domain within the transposase protein we tested either the full-length (ASE3) or truncated versions (ASE3/Δ169-339 and ASE3/Δ1-168) of the transposase fused to the V5-His tag in the above mentioned cell types. A schematic representation of the transposase gene fragments tested in these experiments is shown in Figure 2A. The cellular localization of the expressed proteins was then visualized by using a monoclonal anti-V5 antibody.

The results are showed in Figure 3. Full-length *Bari3* transposase localizes to the nucleus in both cell types (Figure 3-C and 3-L), indicating that a nuclear import signal is contained within the protein and is functional in

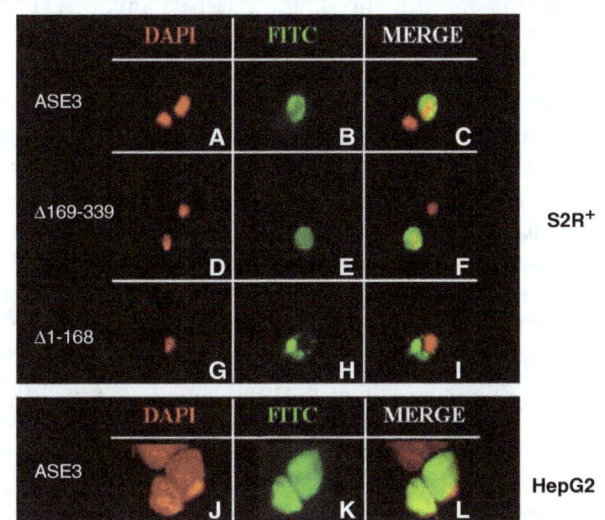

Figure 3 Subcellular localization of the *Bari3* transposase.
Upper Panel. Localization of the full-length (ASE3), the N-terminal (Δ169-339) and the carboxyl terminal (Δ1-168) portion of the *Bari3* transposase in S2R + cells. Lower Panel. Localization of the full-length *Bari3* transposase in HepG2 cells. With 4',6-diamidino-2-phenylindole (DAPI) signal **(A, D, G, J)**; Fluorescein isothiocyanate (FITC) signal **(B, E, H, K)**; merged signals **(C, F, I, L)**.

both insect and mammalian cells. Furthermore, we mapped the NLS signal within the N-terminal half portion of the protein since a deleted C-terminal construct (Δ169-339) retains its nuclear localization (Figure 3, panel F), while the deleted N-terminal part of the transposase (Δ1-168) does not (Figure 3, panel I).

The presence of additional canonical motifs in the *Bari3* transposase also has been investigated using a combination of *in silico* methods. The primary sequence of the *Bari3* transposase was compared to other functional *Tc1-mariner* like transposase sequences including *SB*, *FP*, *minos*, *Hsmar* and *mos1* and the recently characterized *Bari1* in a multialignment.

The identification of the HTH structure of the analyzed transposase was performed by *in silico* prediction with PredictProtein [33] and the predicted alpha helices were annotated on a multiple alignment generated with Multalin [34] (Figure 4).

A bipartite DNA binding domain thought to be responsible for recognition of the transposon termini can be easily detected at the N-terminus of the protein. This domain is divergent in sequences among the compared transposases, but the predicted alpha helices of both HTH motifs occupy a similar position with respect to each other, suggesting the functional conservation of these divergent sequences. As demonstrated for other *Tc1*-like elements, the N-terminal domain of the transposase may also contain motifs mediating dimerization (or tetramerization) of the transposase [21].

A GRPR-like motif (GRKP) motif characteristic of the homeo-domain proteins [35] is also present at position 59 of *Bari3* and between the two HTH motifs. This domain precedes an additional HTH region (that is, the homeo-like domain) in all the transposases aligned. The multiple alignment also highlights the presence of a putative bipartite NLS rich in basic amino acids, whose functionality has been experimentally demonstrated for

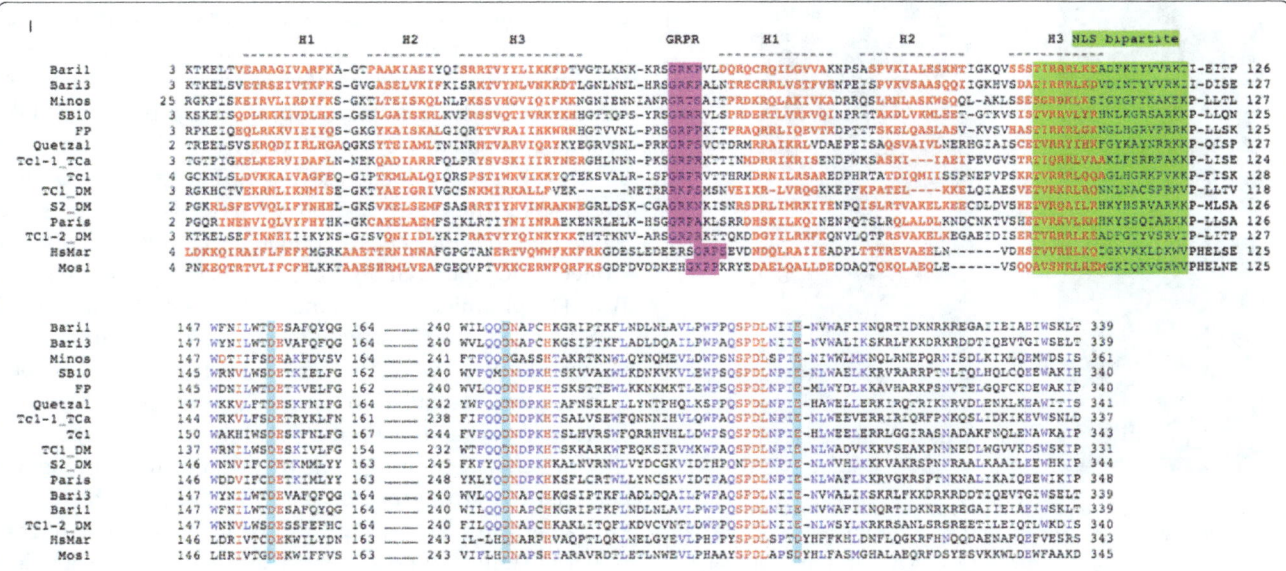

Figure 4 Multiple alignment of *Tc1-mariner* transposases. Residues of the DNA binding domain (consisting of the H1-H3 alpha helices and indicated above the alignment) are red boldfaced, the GRPR domain is highlighted in purple, nuclear localization signal (NLS) is highlighted in green and the acidic triad of the catalytic domains (DDE) is highlighted in turquoise.

Bari3 transposase (see Results above and Figure 3). Finally, the catalytic domain, characterized by the typical DDE motif, is also recognizable in the primary sequence of *Bari3*.

Overexpression of *Bari3* transposase produces spliced transcripts

We recently reported that *Bari1* transcripts can be subjected to post-transcriptional processing under specific experimental conditions [24]. The *Bari1* processed transcripts could be theoretically involved in the regulation of the transposition as they potentially encode for truncated transposase molecules, which can poison the active transposon-transposase complex [24].

We have investigated the possibility that *Bari3* could also generate similar processed transcripts.

RT-PCR experiments performed after transient overexpression of *Bari3* transposase (pAC/ASE3 plasmid) in S2R + cells led to the identification of a transcript of unexpected size in addition to the expected full-length transcript (Figure 5 left panel). Sequence comparison of the cloned short cDNA with the full-length transcript sequence reveals a deletion of 699 bp bracketed by canonical GT-AG consensus of the splicing sites (see Additional file 3, panel A). Interestingly, the short cDNA

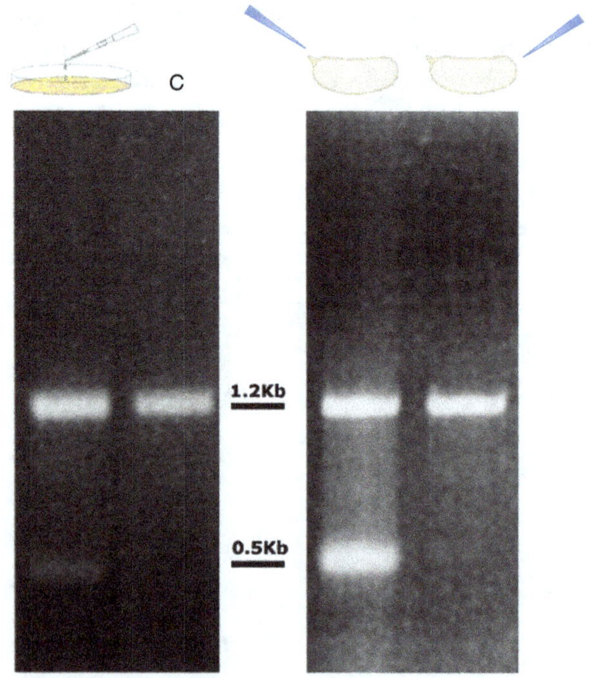

Figure 5 Reverse-transcription polymerase chain reaction (RT-PCR) results. Left panel, RT-PCR results from transfected cells. C = control indicating the expected full-length transcript. Right panel, RT-PCR results from embryos injected in the anterior (left-most lane) or in the posterior (right-most lane) pole. Position of bands relative to the 1Kb DNA Ladder (New England Biolabs) is indicated.

still displays an ORF encoding the last 98 amino acids of the wild type *Bari3* transposase (see Additional file 3, panel B). Therefore, a canonical splicing event is likely to generate an uncommon short transcript of *Bari3* upon overexpression in S2R + cells.

With the aim to confirm this result *in vivo*, we have transiently overexpressed *Bari3* in *D. melanogaster* wild type embryos. We performed two parallel sets of experiments in which embryos were microinjected with the pAC/ASE3 plasmid either in the posterior pole or in the anterior pole. We reasoned that this strategy could give us the chance to analyze the transposase expression in two very different cellular environments of the embryo. Somatic cells reside in the anterior part of the embryo whereas the posterior part is enriched in precursors of germinal cells, that is, the pole cells.

Two transcripts differing in size were detected upon transient overexpression of *Bari3* in the anterior pole of *D. melanogaster* wild-type embryos (Figure 5 right panel). The pattern obtained looks identical to the pattern observed in cultured cell experiments. Interestingly, only embryos injected in their anterior pole produced the additional short transcript, while in embryos injected in the posterior pole only a single band, corresponding in size to the expected full-length *Bari3* transcript, is detectable. Sequence comparison of the two short cDNA cloned respectively from transfected S2R + cells and from embryos reveals that they are 100% identical and harbor the same spliced fragment.

Bari3 transposon harbors an endogenous promoter and interacts with the transposase

Tc1-mariner transposable elements usually contain a single gene encoding transposase. To ensure their mobility they need to autonomously drive transcription, and therefore must contain a promoter element in their left (5') terminus. We have tested the promoter activity of a 356-bp *Bari3* fragment (-1 to -356 relative to the translational start site) using a luciferase assay. The tested fragment overlapping the entire 256-bp left TIR of *Bari3* (plus the 99 bp long spacer, [see Additional file 1]) was directionally cloned into the pGL3B vector, obtaining the pGL3B-Ba3LTIR plasmid. The plasmid was transiently transfected in S2R + cells and the luciferase activity measured. The values obtained were then compared to the values obtained after transfection of the 'empty' pGL3B vector (that is, carrying a promoter-less luciferase gene) and to the luciferase activity in cell transfected with a plasmid carrying the strong promoter of the transposable element *copia* [36]. The results shown in Figure 6A suggest that the sequence tested has a detectable promoter activity, roughly 15% with respect to the *copia* promoter. The promoter activity of *Bari3* is also detectable in HeLa cells (data not shown).

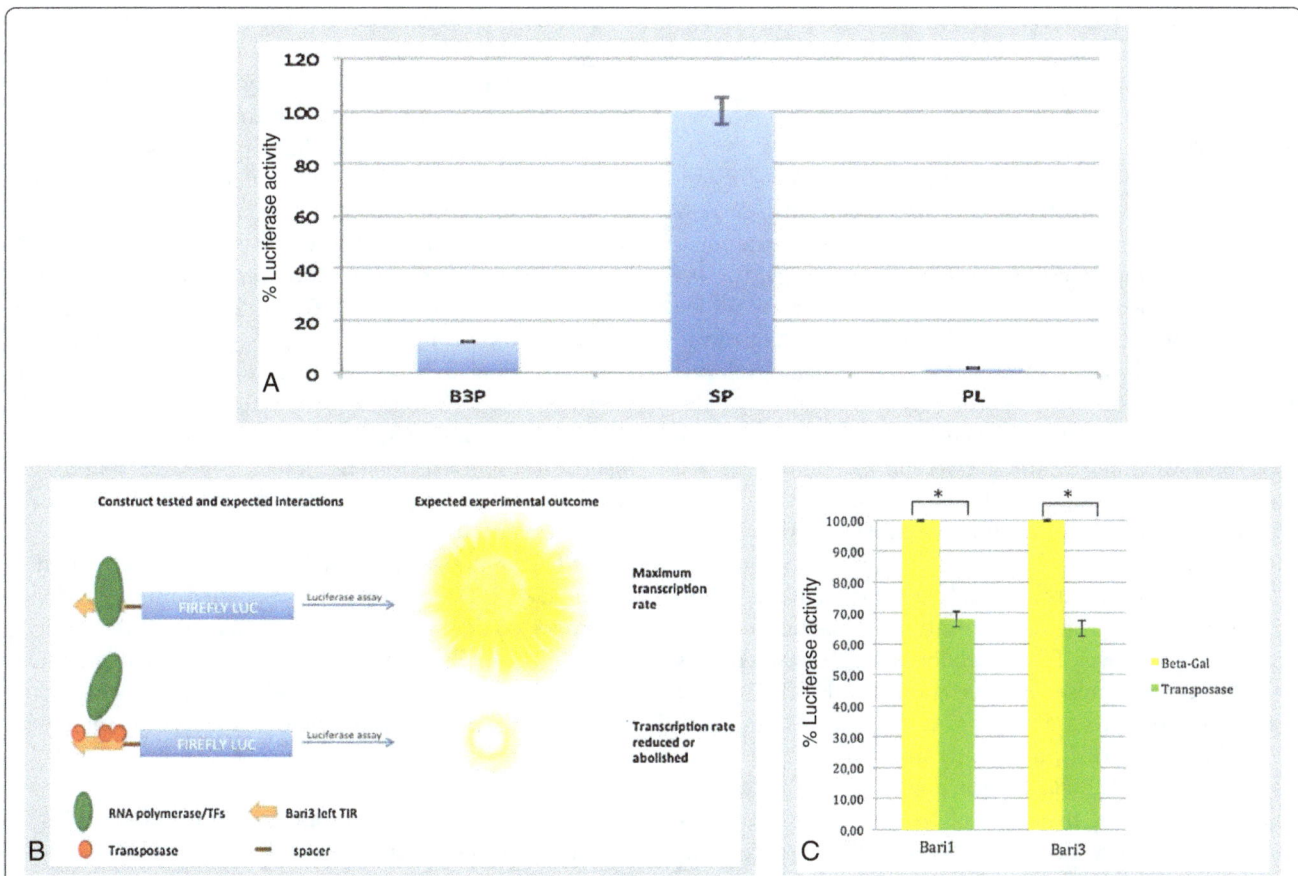

Figure 6 Luciferase promoter assay and the transposon-transposase interaction. A) Luciferase promoter assay. B3P, *Bari3* promoter; SP, strong promoter (*copia* promoter); PL, promoter-less. **B**: Rationale of the luciferase activity suppression assay (see main manuscript text for additional details). **C**: Luciferase activity suppression assay results. Asterisks denote $P < 0.05$.

We have developed a simple assay based on the luciferase transcriptional suppression to detect the transposase-transposon interaction. The rationale of this procedure is depicted in Figure 6B. Briefly, since the TIR sequence of *Bari3* harbors the transposase binding sites [14] that can overlap the promoter region, we hypothesized that the promoter activity could be negatively affected, totally or at least in part, if *Bari3* transposase is expressed in the same cell, thus disturbing the interaction between transcription factors and their binding sites. The advantages of this method with respect to well-established procedures for *in vitro* (EMSA, CHIP) or *in vivo* (One Hybrid) studies, already used in the characterization of the TIR sequence of *Bari1* [24], are the low costs and fast experiments. We performed this test in HeLa cells due to their greater tractability in terms of transfection efficiency and growth respect to S2R + cells, and because we observed *Bari3* promoter activity also in this experimental system (not shown).

In order to validate this procedure we used the previously validated interaction of *Bari1* left TIR and the *Bari1* transposase as a positive control [24].

HeLa cells were transfected with the pcDNA/ASE3 plasmid expressing the *Bari3* transposase. Then, they were further transfected with the pGL3B-Ba3LTIR plasmid 8 hours after the first transfection. The luciferase activity was measured after 24 hours and compared to the luciferase activity measured in cells transfected with the pGL3B-Ba3LTIR plasmid alone. Assuming that cells transfected with pGL3B-Ba3LTIR, in the absence of transposase protein, represent the 100% level of luciferase expression, any significant decrease in luciferase activity can be ascribed to the presence of transposase binding to the DRs on the *Bari3* left TIR. As a negative control, we measured the luciferase activity in cells transfected with a β-Galactosidase-expressing plasmid the protein (in place of the pcDNA/ASE3 plasmid), and then further transfected with the pGL3B-Ba3LTIR plasmid, as described above. The results show a significant lowering of the luciferase activity in cells overexpressing *Bari1* transposase or *Bari3* transposase if compared to the luciferase activity measured in the presence of β-Galactosidase expressed in the same conditions (Figure 6C). Taken together the results obtained indicate that the reduced promoter activity observed can be ascribed to the transposase interaction with the *Bari3* terminal inverted repeat, probably at the DR sites.

Discussion

The post-genomic era allows identification of novel transposable elements, which can be ascribed to known or new families of the major TEs clades. Besides understanding the potential impact of TEs in genome plasticity, the increasing knowledge on TE biology has found applications both in biotechnology and medicine [37,38]. A growing number of TE-based integration tools have been developed in the past 30 years either starting from reconstructed elements [11] or from intact elements isolated from the more diverse organisms [39,40]. New genomic sequences are promising sources of novel transposons, and their functional characterization would give hints for their use in genetics and biotechnology.

Emerging species are probably a mine of information concerning TEs. The reorganization, repositioning and acquisition of novel TEs by genomes are considered as one of the main pulses in speciation [3,41]. *Bari3* might represent one such case, as it has been isolated in the genome of *D. mojavensis*, a recently diverged species of the Repleta group [42]. *Bari3* has novel structural features compared to other members of the *Bari* families, *Bari1* and *Bari2*. *Bari1* has imperfect short TIRs bracketing the transposase gene [43], while *Bari2* has identical long TIRs but mutated transposase [13]. Contrary to older elements, such as *Bari1*, that lost TIRs identity but retained transposition activity [22], or such as *Bari2* which accumulated deleterious mutations that impaired its transposition activity, the *Bari3* element present in *D. mojavensis* appears to be a 'young' *Bari*-like element possessing a transposase coding region and perfect long TIRs.

While the diffusion of *Bari*-like elements through a wide range of *Drosophila* strains is intriguing, the functional and structural features underpinning the success of these elements to colonize different species remain unknown. Here we focus on four informative aspects of this process, that is, 1) the genomic distribution of *Bari3* across different *D. mojavensis* populations; 2) the presence of an internal promoter able to drive the transcription of the transposase gene; 3) the cellular localization of the transposase and its physical interaction with the transposon; 4) the existence of a post-transcriptional regulation mechanism based on alternative splicing in the control of the transposition of the *Bari3* element.

The genomic distribution across different *Drosophila mojavensis* populations suggests that *Bari3* is an active element

Based on molecular, morphological and ethological data, which support the differentiation across the geographical distribution of the species, *D. mojavensis* consists of four recognized races. Albeit the limited sample size, our results reflect the genetic variability of different populations of *D. mojavensis* observed in previous population studies [44]. We found that the copy number of *Bari3* is related to the *D. mojavensis* subspecies, and to the distinct geographic region they occupy. For instance the subspecies *mojavensis* (breeding in barrel cactus in Mojave Desert and the Grand Canyon) and *sonorensis* (breeding in organ pipe cactus in Sonora and Southern Arizona) contain few *Bari3* copies. On the contrary the *baja* (breeding in agria in Baja) and *wrigleyi* (breeding in prickly pear in Santa Catalina) subspecies (Table 1 and Figure 2) are characterized by a higher number of insertions. These evidences, taken together, could suggest that environmental factors might have a role in the determination of strain-specific copy number [45].

In silico analyses performed in the sequenced strain of *D. mojavensis* (15081-1352.22) identified multiple identical *Bari3* copies, [14] [see Additional file 2], as well as several terminally truncated *Bari3* copies, that may have originated by repair of DNA-breaks induced during transposition [46]; both types of elements are compatible with recent activity of *Bari3*.

Bari3 harbors an internal promoter and encodes a putatively active transposase

Transposons need to express their own transposase in order to move within the genome. We have demonstrated that the sequence upstream the translational start site of *Bari3* is able to drive the transcription of downstream sequences, thus behaving as a promoter (Figure 6A). As a member of the *Tc1-mariner* superfamily, *Bari3* has a weak promoter, ensuring low transposase levels. In fact, high transposase activity would probably be deleterious for the host genome, or would trigger inhibitory mechanisms to block transposition (for example, overexpression inhibition). The presence of a promoter in the analyzed sequence suggests that the transcription of the transposase gene is a possible event *in vivo*, further supporting the hypothesis that *Bari3* is an active element.

The presence of a functional Bari3 transposase was tested both by *in silico* and molecular approaches. Nuclear localization of the transposase is essential for the mobilization of chromosomal copies of the transposon. Here, we used a deletion approach and found that a NLS motif is present within the first 168 amino acids of *Bari3* transposase (Figure 3). We mapped this domain in position 103 to 121 of the transposase's primary sequence, based on comparative analysis of 14 transposases encoded by transposons of the *Tc1-mariner* superfamily (Figure 4). Furthermore, by combining multiple alignment and protein motif detection analysis, we present clear evidence that the transposase present in *Bari3* possesses all typical domains of the *Tc1-mariner* transposases (Figure 3), including a correctly spaced DDE amino acidic triad involved in the catalysis. Similar analysis suggested that *Bari3* transposase contains a N-terminal DNA binding domain, and this finding was further investigated by a new experimental strategy presented in this paper (see below).

The transposon-transposase interaction is also a necessary condition for the transposition reaction to occur. Taking advantage of the dual properties of the left terminal sequence of TIR-containing transposons (that is, to act as a promoter and as binding site for the transposase), we described a new approach based on a modified promoter luciferase assay and demonstrated the transposase-left TIR interaction. This assay is based on the assumption that if the transposase/left TIR interaction occurs, then a reduction of the reporter activity (that is, luciferase) should be observed (Figure 6 B). Indeed, the presence of *Bari3* transposase resulted in a significant reduction of the reporter activity, suggesting the presence of transposase binding sites within the left TIR. The left TIRs of *Bari3* and *Bari1* present 62% of sequence similarity (RC and RMM unpublished observation), and share also share three highly conserved stretches of DNA in the transposon termini [14]. In *Bari1*, these stretches represent the transposase binding sites [24], and their high similarity strongly suggests that these sequences are also genuine binding sites for the *Bari3* transposase.

The possible role of transposase-processed transcripts in *Bari3* regulation

Nothing is currently known about the regulation of *Bari3* in *D. mojavensis*, but it is likely that it must be subjected to regulatory mechanisms that contain its transposition.

A number of transposition repressive mechanisms, regulating *Tc1-mariner* elements, have been discovered to date, starting from self-regulation (overexpression inhibition [47-49], post-translational modifications of the transposase [50,51], self-encoded repressors [52,53]) to the cell-developed control systems (siRNA [54] and piRNA [55] pathways, chromatin-level transcriptional repression [56]), or simply stochastic accumulation of detrimental mutations in the transposase-coding gene [57]. Some of these control mechanisms have been demonstrated for *Bari*-like elements [58-60]. In light of our results, similar controlling mechanisms can be hypothesized for *Bari3*.

Similarly to other transposons, including its closest relative *Bari1*, epigenetic regulation of *Bari3* mediated by piRNA could be expected due to the presence of small RNA in the genome of *D. mojavensis* (generated in unidentified genomic loci). Furthermore, the integrity of the left and right TIRs suggests that both could drive transcription, which might result in the formation of dsRNA molecules able to trigger the siRNA/piRNA response.

In addition, our observation that the transposase gene transcripts may undergo processing could be also taken in consideration in future studies concerning additional regulation mechanism controlling *Bari3*.

We have recently reported that the *Bari1* element is subjected to transcript processing when the transposase is overexpressed in cultured cells or *in vivo* in an unrepressed genetic background due to mutations in key genes controlling the piRNA pathway [24]. Here we have investigated the possibility that *Bari3* transcripts could have similar post-transcriptional processing in similar experimental conditions. *In vitro* analyses performed by transient overexpression of the plasmid pAC/ASE3 in S2R + cells revealed the presence of a cDNA of unexpected size, which is the result of a canonical splicing process and potentially encodes for a transposase lacking the binding domain (Figure 5 and see Additional file 3).

Interestingly, a processed transcript sharing the same structural features has been identified also after overexpression in *Drosophila* embryos. It is worth noting that embryos differentially process the *Bari3* transposase transcript in the anterior pole or in the posterior pole, suggesting that the transposase RNA processing is probably soma-specific or it relies on the presence of splicing factors not uniformly distributed along the longitudinal axis of the embryo.

The finding that embryos process transposase transcripts in the anterior pole is slightly surprising as processing would be more likely to occur in the posterior pole of the embryo where the germ line is going to be developed. The somatic post-transcriptional control is somehow reminiscent of the somatic splicing of *P-element* in *D. melanogaster* [61]. We cannot hypothesize obvious functions for the protein encoded by the processed *Bari3* transcript, which is formally a N-terminal truncated version of the wild type *Bari3* transposase, thus lacking the DNA binding function and part of the catalytic domain [see Additional file 3].

The presence of splicing sites in transposase encoding genes has been reported for other well-studied transposons like the *Ac* element [62], whereas *Tc1-mariner* elements do not usually contain introns in their transposase coding genes. However cryptic splicing sites can be activated following transposon insertions within the host genes' coding regions [63,64], a process that allows genes to acquire novel exons and to evolve new splicing and expression patterns [65].

Our results demonstrated the presence of cryptic splice sites in *Bari3*, probably activated upon overexpression in cultured cells and in *D. melanogaster* embryos. It is possible that activation of these sites could constitute an additional, or an alternative, method of protection against transposition. The hypothetical protein product encoded by the detected spliced transcript should have lost the

DNA binding activity, the protein-protein interaction domain, the NLS and part of the catalytic domain (see Figure 5 and Additional file 3), and consequently, it should not negatively influence transposition efficiency. By contrast, the partial depletion of the full-length transposase-encoding mRNAs, resulting from its splicing, could have an impact on *Bari3* transposition, due to the lower transposase mRNA amount that can be translated.

Interestingly, in a recent paper the splicing process has been linked to the siRNA pathway in the regulation of transposons in the encapsulated yeast *Cryptococcus neoformans*. The presence of suboptimal splice sites in transposons' transcripts could lead to stalling of the spliceosome, which produces partial or incomplete mRNA precursors and consequent triggering of the siRNA/piRNA response [66]. It can be speculated that similar mechanisms could be involved in the control of *Bari3* transposition.

Conclusions

The characterization of the *Bari3* transposon presented in this paper increases the current knowledge on the *Tc1*-like elements. Our results justify further studies on the *Bari* family of transposons. These elements are intriguing both for their widespread diffusion in Drosophilids and for their structural diversity. To fully understand the biology of these TEs, it will be necessary to undertake studies connecting structural (for example, short versus long TIRs) to functional features of different *Bari* subfamilies (namely *Bari1* and *Bari3*). In this context, a remarkable result is the transcript processing, which appears as a recurrent feature in active elements of the *Bari* family [24]. A probable scenario could be the existence of a pathway leading to the depletion of the mRNA transposase source in response to a defined threshold, blocking transposition upon failure of other control mechanisms.

Methods

Drosophila stocks and cell culture maintenance

Drosophila mojavensis stocks were obtained from the Drosophila Species Stock Center (University of California, San Diego) and reared on banana/*Opuntia* medium. Fly stocks from different species were maintained on standard cornmeal-agar medium at 24°C.

S2R$^+$ cells (Drosophila Genomics Resource Center, Bloomington, USA) were cultured in Schneider's insect medium supplemented with 10% FBS, 1% penicillin/streptomycin, at 26°C. HeLa and HepG2 cells were grown in Dulbecco's Minimum Essential Medium supplemented with 10% FBS, 200 mM glutamine, 1% penicillin/streptomycin, and maintained at 37°C with 5% CO_2.

Plasmid construction

Standard cloning procedures were used to obtain the plasmids used in this study [67]. A list of the oligonucleotides used in PCR steps is provided as additional file [see Additional file 4]. The full length *Bari3* element (pT/moja11) was PCR-isolated from *D. mojavensis* DNA using the FL2_for/FL2_rev primers targeting the element in the *D. mojavensis* scaffold_6540 and was cloned into the pGEM-T easy vector (Promega, Madison, WI, USA).

Bari3_UP/Bari3_Low, Bari3_UP/Bari3_N-Ter Low, and Bari3_C-Ter Up/Bari3_Low were used to amplify and were subsequently cloned into the KpnI and NotI sites of pAC5.1/V5-His vector (Invitrogen, Carlsbad, CA, USA), DNA sequences encoding respectively the full length *Bari3* transposase gene (pAC/ASE3), the first 168 (pAC/Δ169-339) or the last 171 amino acids of the transposase (pAC/Δ1-168). The fusion constructs were subcloned in pcDNA3.1 (Invitrogen, Carlsbad, CA, USA) using EcoRI and BamHI restriction sites, obtaining the plasmids pcDNA/ASE3, pcDNA/Δ169-339, and pcDNA/Δ1-168. The plasmid, pcDNA/ASE1 has been described in [24].

The Ba3TIR was amplified from the pT/moja11 plasmid with the TERBa3_UP/TERBa3_LOW primers and cloned into the XhoI and NcoI sites of the pGL3B vector (Promega, Madison, WI, USA) to obtain the pGL3B-Ba3LTIR plasmid.

The Ba1TIR was amplified from the p28/47D [43] plasmid with the TERBa1_UP/TERBa1_LOW primers and cloned into the XhoI and NcoI sites of the pGL3B vector (Promega, Madison, WI, USA) to obtain the pGL3B-Ba1LTIR plasmid.

The *copia* promoter was amplified from the pCoBLAST vector (Promega, Madison, WI, USA) with the copia_for/copia_rev primers and cloned into the XhoI and NcoI sites of the pGL3B vector to obtain the pGL3B-copia plasmid.

pcDNA3.1/myc-His(−)/lacZ (Life Technologies, Grand Island, NY, USA) was used to express β-Galactosidase.

All plasmids were sequence-verified.

DNA extraction, Southern blotting and fluorescence *in situ* hybridization

Genomic DNA was prepared according to [68]. DNA samples were digested with the EcoRI restriction enzyme (New England Biolabs Inc, Ipswich, MA, USA), which cuts once in the reference sequence of *Bari3* (see Figure 2A), electrophoresed, blotted onto Hybond N filters and hybridized under high stringency hybridization conditions [67]. Probes used in Southern blot hybridization were labeled with [α-32P] dATP by random priming.

Polytene chromosomes were prepared from salivary glands of third instar *D. mojavensis* larvae essentially as described in [69]. Probes used in fluorescence *in situ* hybridization were labeled by nick-translation with the Cy3-dCTP fluorescent precursor (GE Healthcare Life Sciences, Pittsburgh, PA, USA), and chromosomes were counterstained with 4,6-diamidino-2-phenylindole-dihydrochloride

(DAPI). Finally, digital images were obtained using an Olympus epifluorescence microscope equipped with a cooled CCD camera. Gray scale images, obtained separately for Cy3 and DAPI fluorescence using specific filters, were pseudo-colored and merged to produce the final image using Adobe Photoshop.

A 592-bp probe was amplified from the pT/moja11 clone using primers Moj11_534Up/Moj11_1126Low, and used for all hybridization experiments.

Embryo microinjection and post-injection care

Microinjection of pre-blastoderm embryos was performed essentially as described in [70] with little modifications. Females of the *Oregon-R* strain were allowed to lay eggs for one hour on grape juice agar plates. Eggs were washed with a 70% ethanol (v/v) solution, and aligned manually on a coverslip, mounted on a microscope slide, briefly desiccated, covered with halocarbon oil and injected at either their posterior or anterior pole with a capillary needle attached to an Eppendorf Femtojet microinjector. Needles for microinjection were obtained from borosilicate glass capillaries, pulled with a Narishige PC-10 puller. Concentration of injected DNA was usually 0.5 to 0.8 mg/ml. After injection, the cover slip containing the embryos were carefully removed from the slides and transferred to grape juice plates. After incubation at 18°C for 24 hours, embryos were further subjected to RNA extraction.

Plasmid transfection and immuno-detection of recombinant proteins

One day prior to transfection cells were seeded and let grow into 6-well plates containing sterile glass coverslips. Respectively 1×10^6 and 5×10^5 S2R$^+$ and HepG2 cells were transfected with 1 µg of purified plasmids DNA using TransIt LT1 (Mirus Bio, Madison, WI, USA).

For immunofluorescence staining, the cells attached to slides were washed with phosphate-buffered saline and fixed with 4% formaldehyde for 10 minutes at room temperature followed by three washes in PBS. Blocking was performed with a solution containing 10% fetal bovine serum and 0.5% of Triton X-100 for 30 minutes followed by two washes in PBS for 2 minutes each.

Cells were incubated with a dilution 1:500 of V5 antibody (Invitrogen, Carlsbad, CA, USA) conjugated with fluorescein isothiocyanate (FITC) fluorochrome for 2 hours. After three washes in PBS, the cells were stained with DAPI (4',6-diamidino-2-phenylindole) and mounted with antifade 1,4-diazabicyclo[2.2.2]octane (DABCO).

Slides were imaged under an Olympus (Tokyo, Japan) epifluorescence microscope equipped with a cooled CCD camera. At least 100 positive cells per slide were observed. Grey-scale images, obtained by separately recording FITC and DAPI fluorescence, were pseudo-colored and merged to obtain the final image using Adobe Photoshop program.

Promoter luciferase assay

S2R + cells were transfected with 1 µg of the appropriate plasmid (either pGL3B-Ba1LTIR, pGL3B-Ba3LTIR pGL3B-copia or the empty pGL3B). *Renilla* luciferase construct (pRL-SV40; Promega, Madison, WI, USA) was used for normalization. Luciferase expression was measured by the detection of luminescence using the dual luciferase reporter assay system (Promega, Madison, WI, USA) according to the manufacturer instructions. Measurements were recorded on GLOMAX 20/20 luminometer (Promega, Madison, WI, USA). The average expression level from three replicate transfections was normalized to the *Renilla* luciferase co-transfection control. This value was further normalized to the average expression level from three normalized replicates of the pGL3B-copia plasmid to yield a relative luciferase activity estimate.

For the luciferase activity suppression assay HeLa cells were previously transfected with plasmid expressing either transposase (pcDNA/ASE3, pcDNA/ASE1) or β-Galactosidase (pcDNA3.1/myc-His(–)/lacZ) (Invitrogen, Carlsbad, CA, USA).

Error bars represent the standard deviation. Student's t test was used to evaluate statistical significance.

Transcriptional analysis

RNA was extracted with TRIzol® Reagent (Invitrogen, Carlsbad, CA, USA). Cultured cells were directly processed after two washes in PBS 1X. Quantitation and estimation of RNA purity were performed using a NanoDrop spectrophotometer.

A total of 1 µg RNA was converted to cDNA using the QIAQuick reverse transcription kit (Qiagen, Hilden, Germany) and following the manufacturer's instruction. cDNA samples from transfected S2R + cells and from injected embryos were amplified with the AC5_forward/BGH_Rev primers. Nested PCR was performed using the Bari3_Up1/V5_rev primers.

In silico methods

Pairwise alignments were performed using either the NCBI online tools or the LALIGN tool (http://embnet.vital-it.ch/software/LALIGN_form.html).

Multiple alignments were performed using the Multalin tool (http://multalin.toulouse.inra.fr/) [34]. Protein secondary structures predictions were performed using the PhD secondary structure prediction method (https://www.predictprotein.org/) [71]. Sequences used for construction of the multiple alignment in Figure 4 were retrieved from the Repbase database (www.girinst.org) [72].

Additional files

Additional file 1: Sequence and main features of *Bari3*.

Additional file 2: *Bari3* in the reference genome of *Drosophila mojavensis*.

Additional file 3: Structure of the spliced *Bari3* transcript and its encoded protein.

Additional file 4: List of the primers used in this work.

Abbreviations

CCD: charge-coupled device; CHIP: chromatin immunoprecipitation; DAPI: 4',6-diamidino-2-phenylindole; DR: direct repeat; EMSA: electrophoretic mobility shift assay; FITC: fluorescein isothiocyanate; HTH: helix-turn-helix; IR-DR: inverted repeat-direct repeat; Kbp: kilobase pairs; NLS: nuclear localization signal; ORF: open reading frame; PBS: phosphate-buffered saline; PCR: polymerase chain reaction; TE: transposable element; TIR: terminal inverted repeat.

Competing interests

AP, RC, RMM have applied for a patent related to part of the content of this manuscript. The remaining authors declare that they have no competing interests.

Authors' contributions

AP, RM, and RMM, performed the experiments. RC and RMM conceived the study, participated in its design and coordination, and drafted the manuscript. All authors read and approved the final manuscript.

Acknowledgements

We thank Dr. Konstantinos Lefkimmiatis for critical reading of the manuscript and useful suggestions. This work was supported by 'Progetti di ricerca di Ateneo' from Universita' degli Studi di Bari 'Aldo Moro' to RC and RMM. Universita' degli Studi di Bari 'Aldo Moro' is also gratefully acknowledged for its contribution to support the Open Access costs of this article.

References

1. Doolittle WF, Sapienza C: **Selfish genes, the phenotype paradigm and genome evolution.** *Nature* 1980, **284**:601–603.
2. Orgel LE, Crick FH: **Selfish DNA: the ultimate parasite.** *Nature* 1980, **284**:604–607.
3. Bohne A, Brunet F, Galiana-Arnoux D, Schultheis C, Volff JN: **Transposable elements as drivers of genomic and biological diversity in vertebrates.** *Chromosome Res* 2008, **16**:203–215.
4. Rebollo R, Romanish MT, Mager DL: **Transposable elements: an abundant and natural source of regulatory sequences for host genes.** *Annu Rev Genet* 2012, **46**:21–42.
5. Tollis M, Boissinot S: **The evolutionary dynamics of transposable elements in eukaryote genomes.** *Genome Dyn* 2012, **7**:68–91.
6. Yuan YW, Wessler SR: **The catalytic domain of all eukaryotic cut-and-paste transposase superfamilies.** *Proc Natl Acad Sci U S A* 2011, **108**:7884–7889.
7. Shao H, Tu Z: **Expanding the diversity of the IS630-Tc1-mariner superfamily: discovery of a unique DD37E transposon and reclassification of the DD37D and DD39D transposons.** *Genetics* 2001, **159**:1103–1115.
8. Lampe DJ, Churchill ME, Robertson HM: **A purified mariner transposase is sufficient to mediate transposition in vitro.** *EMBO J* 1996, **15**:5470–5479.
9. Plasterk RH: **The Tc1/mariner transposon family.** *Curr Top Microbiol Immunol* 1996, **204**:125–143.
10. Cizeron G, Biemont C: **Polymorphism in structure of the retrotransposable element 412 in Drosophila simulans and D. melanogaster populations.** *Gene* 1999, **232**:183–190.
11. Ivics Z, Hackett PB, Plasterk RH, Izsvak Z: **Molecular reconstruction of Sleeping Beauty, a Tc1-like transposon from fish, and its transposition in human cells.** *Cell* 1997, **91**:501–510.
12. Robertson HM, Lampe DJ: **Recent horizontal transfer of a mariner transposable element among and between Diptera and Neuroptera.** *Mol Biol Evol* 1995, **12**:850–862.

13. Moschetti R, Caggese C, Barsanti P, Caizzi R: **Intra- and interspecies variation among Bari-1 elements of the melanogaster species group.** *Genetics* 1998, **150**:239–250.
14. Moschetti R, Chlamydas S, Marsano RM, Caizzi R: **Conserved motifs and dynamic aspects of the terminal inverted repeat organization within Bari-like transposons.** *Mol Genet Genomics* 2008, **279**:451–461.
15. Merriman PJ, Grimes CD, Ambroziak J, Hackett DA, Skinner P, Simmons MJ: **S elements: a family of Tc1-like transposons in the genome of Drosophila melanogaster.** *Genetics* 1995, **141**:1425–1438.
16. Franz G, Savakis C: **Minos, a new transposable element from Drosophila hydei, is a member of the Tc1-like family of transposons.** *Nucleic Acids Res* 1991, **19**:6646.
17. Petrov DA, Schutzman JL, Hartl DL, Lozovskaya ER: **Diverse transposable elements are mobilized in hybrid dysgenesis in Drosophila virilis.** *Proc Natl Acad Sci U S A* 1995, **92**:8050–8054.
18. Miskey C, Izsvak Z, Plasterk RH I, Ivics Z: **The Frog Prince: a reconstructed transposon from Rana pipiens with high transpositional activity in vertebrate cells.** *Nucleic Acids Res* 2003, **31**:6873–6881.
19. Brillet B, Bigot Y, Auge-Gouillou C: **Assembly of the Tc1 and mariner transposition initiation complexes depends on the origins of their transposase DNA binding domains.** *Genetica* 2007, **130**:105–120.
20. Ivics Z, Izsvak Z, Minter A, Hackett PB: **Identification of functional domains and evolution of Tc1-like transposable elements.** *Proc Natl Acad Sci U S A* 1996, **93**:5008–5013.
21. Izsvak Z, Khare D, Behlke J, Heinemann U, Plasterk RH, Ivics Z: **Involvement of a bifunctional, paired-like DNA-binding domain and a transpositional enhancer in Sleeping Beauty transposition.** *J Biol Chem* 2002, **277**:34581–34588.
22. Marsano RM, Caizzi R, Moschetti R, Junakovic N: **Evidence for a functional interaction between the Bari1 transposable element and the cytochrome P450 cyp12a4 gene in Drosophila melanogaster.** *Gene* 2005, **357**:122–128.
23. Caggese C, Pimpinelli S, Barsanti P, Caizzi R: **The distribution of the transposable element Bari-1 in the Drosophila melanogaster and Drosophila simulans genomes.** *Genetica* 1995, **96**:269–283.
24. Palazzo A, Marconi S, Specchia V, Bozzetti MP, Ivics Z, Caizzi R, Marsano RM: **Functional characterization of the bari1 transposition system.** *PLoS One* 2013, **8**:e79385.
25. Cui Z, Geurts AM, Liu G, Kaufman CD, Hackett PB: **Structure-function analysis of the inverted terminal repeats of the sleeping beauty transposon.** *J Mol Biol* 2002, **318**:1221–1235.
26. Fischer SE, van Luenen HG, Plasterk RH: **Cis requirements for transposition of Tc1-like transposons in C. elegans.** *Mol Gen Genet* 1999, **262**:268–274.
27. Dias ES, Carareto CM: **Ancestral polymorphism and recent invasion of transposable elements in Drosophila species.** *BMC Evol Biol* 2012, **12**:119.
28. Markow TA, Castrezana S, Pfeiler E: **Flies across the water: genetic differentiation and reproductive isolation in allopatric desert Drosophila.** *Evolution* 2002, **56**:546–552.
29. Hocutt GDG: *Reinforcement of Premating Barriers to Reproduction Between Drosophila arizonae and Drosophila mojavensis*, PhD thesis: Arizona State University; 2000.
30. Zouros E, d'Entremont CJ: **Sexual isolation among populations of drosophila mojavensis: response to pressure from a related species.** *Evolution* 1980, **34**:421–430.
31. Pfeiler E, Reed LK, Markow TA: **Inhibition of alcohol dehydrogenase after 2-propanol exposure in different geographic races of Drosophila mojavensis: lack of evidence for selection at the Adh-2 locus.** *J Exp Zool B Mol Dev Evol* 2005, **304**:159–168.
32. Clark AG, Eisen MB, Smith DR, Bergman CM, Oliver B, Markow TA, Kaufman TC, Kellis M, Gelbart W, Iyer VN, Pollard DA, Sackton TB, Larracuente AM, Singh ND, Abad JP, Abt DN, Adryan B, Aguade M, Akashi H, Anderson WW, Aquadro CF, Ardell DH, Arguello R, Artieri CG, Barbash DA, Barker D, Barsanti P, Batterham P, Batzoglou S, Begun D, *et al*: **Evolution of genes and genomes on the Drosophila phylogeny.** *Nature* 2007, **450**:203–218.
33. Rost B, Yachdav G, Liu J: **The PredictProtein server.** *Nucleic Acids Res* 2004, **32**:W321–W326.
34. Corpet F: **Multiple sequence alignment with hierarchical clustering.** *Nucleic Acids Res* 1988, **16**:10881–10890.
35. Gehring WJ, Qian YQ, Billeter M, Furukubo-Tokunaga K, Schier AF, Resendez-Perez D, Affolter M, Otting G, Wuthrich K: **Homeodomain-DNA recognition.** *Cell* 1994, **78**:211–223.

36. Sinclair JH, Sang JH, Burke JF, Ish-Horowicz D: Extrachromosomal replication of copia-based vectors in cultured Drosophila cells. *Nature* 1983, **306**:198–200.

37. VandenDriessche T, Ivics Z, Izsvak Z, Chuah MK: Emerging potential of transposons for gene therapy and generation of induced pluripotent stem cells. *Blood* 2009, **114**:1461–1468.

38. Palazzoli F, Testu FX, Merly F, Bigot Y: Transposon tools: worldwide landscape of intellectual property and technological developments. *Genetica* 2010, **138**:285–299.

39. Di Matteo M, Matrai J, Belay E, Firdissa T, Vandendriessche T, Chuah MK: PiggyBac toolbox. *Methods Mol Biol* 2012, **859**:241–254.

40. Spradling AC, Stern DM, Kiss I, Roote J, Laverty T, Rubin GM: Gene disruptions using P transposable elements: an integral component of the Drosophila genome project. *Proc Natl Acad Sci U S A* 1995, **92**:10824–10830.

41. Hua-Van A, Le Rouzic A, Boutin TS, Filee J, Capy P: The struggle for life of the genome's selfish architects. *Biol Direct* 2011, **6**:19.

42. Reed LK, Markow TA: Early events in speciation: polymorphism for hybrid male sterility in Drosophila. *Proc Natl Acad Sci U S A* 2004, **101**:9009–9012.

43. Caizzi R, Caggese C, Pimpinelli S: Bari-1, a new transposon-like family in Drosophila melanogaster with a unique heterochromatic organization. *Genetics* 1993, **133**:335–345.

44. Ross CL, Markow TA: Microsatellite variation among diverging populations of Drosophila mojavensis. *J Evol Biol* 2006, **19**:1691–1700.

45. Capy P, Gasperi G, Biemont C, Bazin C: Stress and transposable elements: co-evolution or useful parasites? *Heredity (Edinb)* 2000, **85**(Pt 2):101–106.

46. Witsell A, Kane DP, Rubin S, McVey M: Removal of the bloom syndrome DNA helicase extends the utility of imprecise transposon excision for making null mutations in Drosophila. *Genetics* 2009, **183**:1187–1193.

47. Lohe AR, Hartl DL: Autoregulation of mariner transposase activity by overproduction and dominant-negative complementation. *Mol Biol Evol* 1996, **13**:549–555.

48. Lampe DJ, Grant TE, Robertson HM: Factors affecting transposition of the Himar1 mariner transposon in vitro. *Genetics* 1998, **149**:179–187.

49. Izsvak Z, Ivics Z: Sleeping beauty transposition: biology and applications for molecular therapy. *Mol Ther* 2004, **9**:147–156.

50. Germon S, Bouchet N, Casteret S, Carpentier G, Adet J, Bigot Y, Auge-Gouillou C: Mariner Mos1 transposase optimization by rational mutagenesis. *Genetica* 2009, **137**:265–276.

51. Bouchet N, Jaillet J, Gabant G, Brillet B, Briseno-Roa L, Cadene M, Auge-Gouillou C: cAMP protein kinase phosphorylates the Mos1 transposase and regulates its activity: evidences from mass spectrometry and biochemical analyses. *Nucleic Acids Res* 2014, **42**:1117–1128.

52. Robertson HM, Engels WR: Modified P elements that mimic the P cytotype in Drosophila melanogaster. *Genetics* 1989, **123**:815–824.

53. Misra S, Rio DC: Cytotype control of Drosophila P element transposition: the 66 kd protein is a repressor of transposase activity. *Cell* 1990, **62**:269–284.

54. Ghildiyal M, Seitz H, Horwich MD, Li C, Du T, Lee S, Xu J, Kittler EL, Zapp ML, Weng Z, Zamore PD: Endogenous siRNAs derived from transposons and mRNAs in Drosophila somatic cells. *Science* 2008, **320**:1077–1081.

55. McCue AD, Slotkin RK: Transposable element small RNAs as regulators of gene expression. *Trends Genet* 2012, **28**:616–623.

56. Cernilogar FM, Onorati MC, Kothe GO, Burroughs AM, Parsi KM, Breiling A, Lo Sardo F, Saxena A, Miyoshi K, Siomi H, Siomi MC, Carninci P, Gilmour DS, Corona DF, Orlando V: Chromatin-associated RNA interference components contribute to transcriptional regulation in Drosophila. *Nature* 2011, **480**:391–395.

57. Lohe AR, Moriyama EN, Lidholm DA, Hartl DL: Horizontal transmission, vertical inactivation, and stochastic loss of mariner-like transposable elements. *Mol Biol Evol* 1995, **12**:62–72.

58. Specchia V, Piacentini L, Tritto P, Fanti L, D'Alessandro R, Palumbo G, Pimpinelli S, Bozzetti MP: Hsp90 prevents phenotypic variation by suppressing the mutagenic activity of transposons. *Nature* 2010, **463**:662–665.

59. Wang SH, Elgin SC: Drosophila Piwi functions downstream of piRNA production mediating a chromatin-based transposon silencing mechanism in female germ line. *Proc Natl Acad Sci U S A* 2011, **108**:21164–21169.

60. Zamparini AL, Davis MY, Malone CD, Vieira E, Zavadil J, Sachidanandam R, Hannon GJ, Lehmann R: Vreteno, a gonad-specific protein, is essential for germline development and primary piRNA biogenesis in Drosophila. *Development* 2011, **138**:4039–4050.

61. Laski FA, Rio DC, Rubin GM: Tissue specificity of Drosophila P element transposition is regulated at the level of mRNA splicing. *Cell* 1986, **44**:7–19.

62. Lisson R, Hellert J, Ringleb M, Machens F, Kraus J, Hehl R: Alternative splicing of the maize Ac transposase transcript in transgenic sugar beet (Beta vulgaris L.). *Plant Mol Biol* 2010, **74**:19–32.

63. Rushforth AM, Anderson P: Splicing removes the Caenorhabditis elegans transposon Tc1 from most mutant pre-mRNAs. *Mol Cell Biol* 1996, **16**:422–429.

64. Menssen A, Hohmann S, Martin W, Schnable PS, Peterson PA, Saedler H, Gierl A: The En/Spm transposable element of Zea mays contains splice sites at the termini generating a novel intron from a dSpm element in the A2 gene. *EMBO J* 1990, **9**:3051–3057.

65. Stower H: Alternative splicing: Regulating Alu element 'exonization'. *Nat Rev Genet* 2013, **14**:152–153.

66. Dumesic PA, Madhani HD: The spliceosome as a transposon sensor. *RNA Biol* 2013, **10**:1653–1660.

67. Sambrook J, Russell DW: *Molecular Cloning: A Laboratory Manual.* Woodbury, NY: Cold Spring Harbor Laboratory Press; 2001.

68. Roberts DB: *Drosophila: A Practical Approach.* 2nd edition. Oxford: IRL Press at Oxford University Press; 1998.

69. Labrador M, Naveira H, Fontdevila A: Genetic mapping of the Adh locus in the repleta group of Drosophila by in situ hybridization. *J Hered* 1990, **81**:83–86.

70. Rubin GM, Spradling AC: Genetic transformation of Drosophila with transposable element vectors. *Science* 1982, **218**:348–353.

71. Rost B, Sander C: Prediction of protein secondary structure at better than 70% accuracy. *J Mol Biol* 1993, **232**:584–599.

72. Jurka J, Kapitonov VV, Pavlicek A, Klonowski P, Kohany O, Walichiewicz J: Repbase Update, a database of eukaryotic repetitive elements. *Cytogenet Genome Res* 2005, **110**:462–467.

17

Whole genome sequencing in *Drosophila virilis* identifies *Polyphemus*, a recently activated Tc1-like transposon with a possible role in hybrid dysgenesis

Justin P Blumenstiel

Abstract

Background: Hybrid dysgenic syndromes in *Drosophila* have been critical for characterizing host mechanisms of transposable element (TE) regulation. This is because a common feature of hybrid dysgenesis is germline TE mobilization that occurs when paternally inherited TEs are not matched with a maternal pool of silencing RNAs that maintain transgenerational TE control. In the face of this imbalance TEs become activated in the germline and can cause F1 sterility. The syndrome of hybrid dysgenesis in *Drosophila virilis* was the first to show that the mobilization of one dominant TE, the *Penelope* retrotransposon, may lead to the mobilization of other unrelated elements. However, it is not known how many different elements contribute and no exhaustive search has been performed to identify additional ones. To identify additional TEs that may contribute to hybrid dysgenesis in *Drosophila virilis*, I analyzed repeat content in genome sequences of inducer and non-inducer lines.

Results: Here I describe *Polyphemus*, a novel Tc1-like DNA transposon, which is abundant in the inducer strain of *D. virilis* but highly degraded in the non-inducer strain. *Polyphemus* expression is also increased in the germline of progeny of the dysgenic cross relative to reciprocal progeny. Interestingly, like the *Penelope* element, it has experienced recent re-activation within the *D. virilis* lineage.

Conclusions: Here I present the results of a comprehensive search to identify additional factors that may cause hybrid dysgenesis in *D. virilis*. *Polyphemus*, a novel Tc1-like DNA transposon, has recently become re-activated in *Drosophila virilis* and likely contributes to the hybrid dysgenesis syndrome. It has been previously shown that the *Penelope* element has also been re-activated in the inducer strain. This suggests that TE co-reactivation within species may synergistically contribute to syndromes of hybrid dysgenesis.

Keywords: Hybrid dysgenesis, *Drosophila virilis*, Transposable element, *Penelope*, piRNA, Genome instability, Epigenetics

Background

Hybrid dysgenesis, a syndrome of sterility and increased mutation in crosses between different strains of the same species, was first shown in *Drosophila melanogaster* to be driven by *P* elements that are inherited paternally, but not maternally [1-4]. Activation of this DNA transposon subsequently leads to germline DNA damage and sterility [5-8]. In *D. melanogaster*, transposable element

(TE) mediated syndromes of hybrid dysgenesis are also driven by the *I* element retrotransposon [9] and the *hobo* DNA transposon [10]. An important syndrome of hybrid dysgenesis has also been characterized in *D. virilis* [11-14]. This syndrome is significant as it is accompanied by mobilization of different, unrelated transposable element families [13,15]. Critically, even elements such as the *Ulysses* retrotransposon that are evenly distributed between strains become mobilized in this cross. Recent studies using genome sequencing approaches indicate that *P* elements also induce the mobilization of other

Correspondence: jblumens@ku.edu
Department of Ecology and Evolutionary Biology, University of Kansas, 1200 Sunnyside Avenue, Lawrence KS 66049, USA

elements in *D. melanogaster* [5]. Thus, co-mobilization of TEs may be a common feature of dysgenic syndromes but the mechanism by which TE co-mobilization occurs is poorly understood. One mechanism that has been proposed is that DNA damage caused by the activation of one TE family disrupts piRNA silencing mechanisms in the germline *via* the DNA damage response [5]. In this model, disrupted piRNA silencing in turn leads to activation of normally repressed and unrelated TEs. Alternatively, DNA damage arising from transposition may drive activation of TEs through other mechanisms. Additionally, it has been proposed that, like viruses, TEs may encode suppressors of RNA silencing [16]. In this case, the expression of a suppressor of RNA silencing encoded by a single activated TE could lead to global TE de-repression. Finally, it is important to consider the possibility that multiple TE families may be more abundant within inducer strains [17]. Thus, the activation of multiple TE families in a dysgenic syndrome may also be explained by independent mechanisms acting across each family.

To distinguish among these hypotheses, it is critical to define the landscape of TE copy number imbalance between inducer and non-inducer strains in hybrid dysgenic syndromes. Many previous studies indicate that the *Penelope* element is likely to be the main driver of hybrid dysgenesis in *Drosophila virilis*. It is the only known element with multiple, active copies in the inducer strain and for which active copies are entirely absent from the non-inducer strain [11,17,18]. In addition, its expression is greatly increased in the gonads of dysgenic progeny and injection of embryos with *Penelope* constructs can lead to increased incidence of TE mediated mutation [11,14,19]. However, additional studies indicate that while *Penelope* may be the dominant cause of sterility, other factors may also contribute. For one, some strains of *D. virilis* that behave as neutral strains - maternally protecting against dysgenesis but not inducing it paternally - lack piRNAs from the *Penelope* element in their ovaries [20]. If *Penelope* is the sole cause of dysgenesis, it is difficult to explain how these strains protect against the induction of dysgenesis since mothers would be unable to provide *Penelope* piRNA to the next generation. Second, two additional TEs - *Helena* and *Paris* - also show high abundance of active, euchromatic copies in the inducer strain and lower abundance in the non-inducer strain. Evidence suggests these two elements also contribute to the sterility phenotype [17]. Whether these three elements act synergistically to cause sterility is not known. It is also not known whether they jointly contribute to the mobilization of other elements such as *Ulysses*.

The discovery of these candidate inducer elements - *Penelope*, *Helena*, and *Paris* - was facilitated by the recovery of TE insertions that gave rise to visible mutations during dysgenic co-mobilization. Thus, it has not been clear whether additional elements may contribute to the dysgenic syndrome in *D. virilis*. Here, I present the first systematic effort to identify additional TEs that may cause hybrid dysgenesis in *D. virilis*.

Results and discussion

To identify additional TEs that may contribute to the hybrid dysgenesis syndrome of *D. virilis*, I performed whole genome, 100 bp paired-end Illumina sequencing of DNA collected from inducer (strain 160) and non-inducer (strain 9) flies. Based on a genome size of 364 Mb estimated from flow cytometry [21,22], sequencing yielded approximately 24 X and approximately 21 X coverage for strain 160 and strain 9, respectively. After trimming for quality (https://github.com/najoshi/sickle), reads (one single end from each pair) were then mapped using BWA-MEM [23] to a library of *D. virilis* repeat sequences computationally predicted by the PILER program [24]. Figure 1 indicates the ratio for the number of reads (160:9) mapping to each PILER centroid, normalized by total number of reads mapped to the reference genome. From this, I identified centroid.25.39 to be enriched about 27-fold in strain 160 relative to strain 9 (P <0.001, chi-squared test), in a ratio similar to that observed with the centroid corresponding to the *Penelope* element (about 32-fold; Figure 1). This suggested that centroid.25.39 may correspond to an element that, like *Penelope*, is in excess in the inducer strain. Centroid.25.39 was therefore further characterized.

To determine the consensus repeat sequence corresponding to centroid.25.39, I performed blastn with this computationally predicted repeat against the *Drosophila virilis* reference genome (the reference strain also induces hybrid dysgenesis). After performing several rounds of iterated blast, I identified and extracted the consensus sequence of a highly repeated Tc1-like transposon with 235-bp inverted flanking repeats that I have designated *Polyphemus* (Figure 2A and Additional file 1). Within this sequence there is an open reading frame that corresponds to a 344 amino-acid sequence with 65% identity (beginning to end) to the *S* element previously identified in *D. melanogaster* [25] and 59% identity (beginning to end) to the *Paris* element identified in *D. virilis* [15]. Both of these elements belong to the Tc1/*mariner* superfamily of cut-and-paste DNA transposons. Conservation of the catalytic DDE domain is noted in the comparison to the Tc1 element (Figure 2B). Phylogenetic analysis indicates that *Polyphemus* is located within the Tc1 clade of the larger Tc1/*mariner* group of DNA transposons (Figure 2C). Furthermore, it is most closely related to the *S* element and *Paris*. Interestingly, there are two alternate translation start codons that extend the putative reading frame up to 57 codons and into the first inverted repeat. However, the extended 57 amino

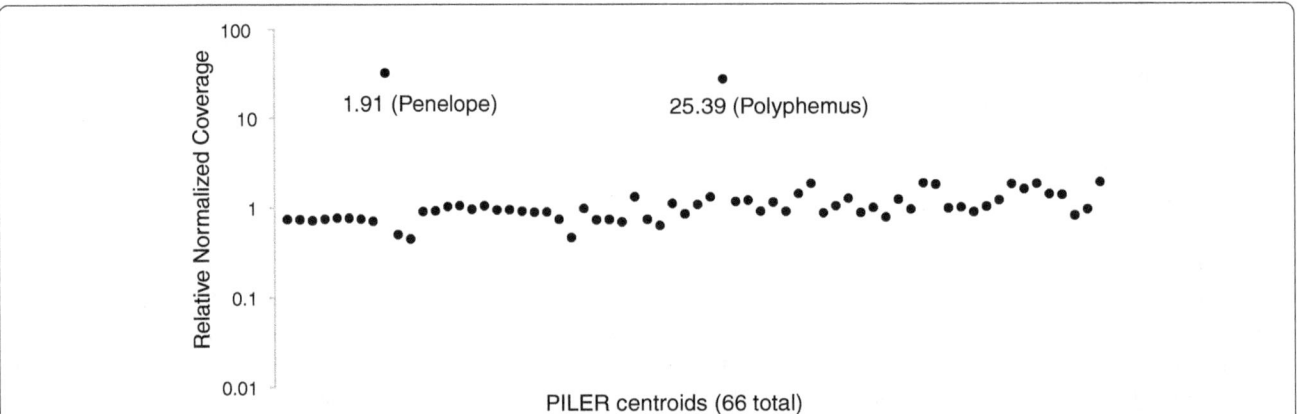

Figure 1 Relative mapping abundance for all 66 PILER centroids from genome sequence reads (100 bp reads) of strains 160 and 9.
Mapped reads were normalized to all reads mapped to the reference genome using BWA-MEM. Two PILER centroids show high abundance in strain 160: 1.91, which corresponds to the *Penelope* element, and 25.39, which corresponds to *Polyphemus*.

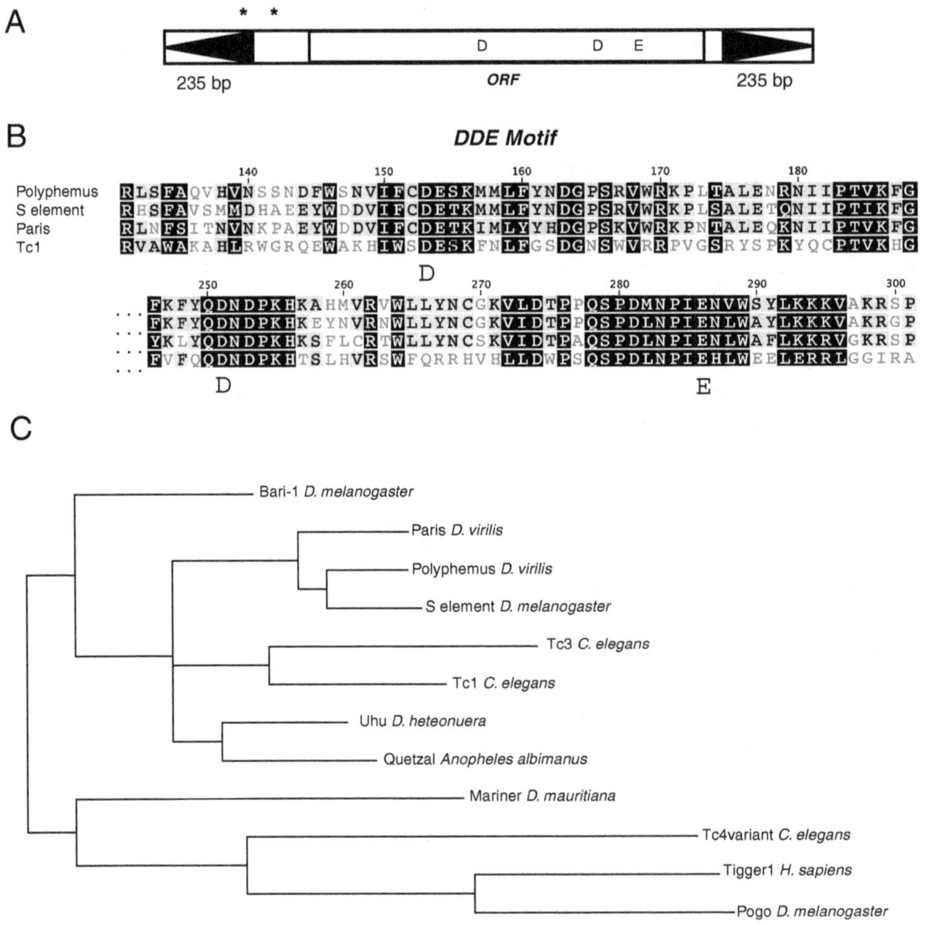

A

235 bp   ORF   235 bp

D   D   E

B

DDE Motif

	140	150	160	170	180
Polyphemus	RLSFAQVHVNSSNDFWSNVIFCDESKMMLFYNDGPSRVWRKPLTALENRNIIPTVKFG
S element	RHSFAVSMMDHAEEYWDDVIFCDETKMMLFYNDGPSRVWRKPLSALETQNIIPTIKFG
Paris	RLNFSITNVNKPAEYWDDVIFCDETKIMLYYHDGPSKVWRKPNTALEQKNIIPTVKFG
Tc1	RVAWAKAHLRWGRQEWAKHIWSDESKFNLFGSDGNSWVRRPVGSRYSPKYQCPTVKHG

D

	250	260	270	280	290	300
...	FKFYQDNDPKHKAHMVRVWLLYNCGKVLDTPPQSPDMNPIENVWSYLKKKVAKRSP
...	FKFYQDNDPKHKEYNVRNWLLYNCGKVIDTPPQSPDLNPIENLWAYLKKKVAKRGP
...	YKLYQDNDPKHKSFLCRTWLLYNCSKVIDTPAQSPDLNPIENLWAFLKKRVGKRSP
...	FVFQQDNDPKHTSLHVRSWFQRRHVHLLDWPSQSPDLNPIEHLWEELERRLGGIRA

D   E

C

Bari-1 *D. melanogaster*
Paris *D. virilis*
Polyphemus *D. virilis*
S element *D. melanogaster*
Tc3 *C. elegans*
Tc1 *C. elegans*
Uhu *D. heteonuera*
Quetzal *Anopheles albimanus*
Mariner *D. mauritiana*
Tc4variant *C. elegans*
Tigger1 *H. sapiens*
Pogo *D. melanogaster*

Figure 2 *Polyphemus* is a cut and paste transposon belonging to the Tc1 family. (A) Overall structure of *Polyphemus* with 235-bp inverted repeats indicated by black arrows. Asterisks indicate putative alternate translation start sites. Position of the DDE motif that catalyzes the transposition reaction is indicated. **(B)** Alignment of amino acids that contain the DDE motif from closely related members of the Tc1 family: the founding Tc1 as well as the *S* element (from *D. melanogaster*) and *Paris* (from *D. virilis*). **(C)** Phylogeny of the Tc1/mariner family. *Polyphemus, Paris,* and *S* element form a clade within the Tc1 group.

acid sequence shares no sequence similarity to any known protein and therefore the transcription start site is likely downstream of these alternate translation start sites.

Based on coverage across the entire length of this element, representation of *Polyphemus* is greater in strain 160 than strain 9 (Figure 3). A similar analysis for the *Penelope* element confirms an even greater difference between strain 160 and strain 9. I next sought to determine sequence heterogeneity within the mapped reads since high sequence similarity among copies is often indicative of recent activity. To determine this, I extracted element specific mappings and used piledriver (https://github.com/arq5x/piledriver) to analyze sequence heterogeneity by counting the frequency, at each nucleotide position, of the most common variant (Figure 3). In strain 160, *Polyphemus* shows very little heterogeneity among mapped reads, suggesting recent activity of a single lineage. In contrast, there is great heterogeneity among mapped reads for *Polyphemus* in strain 9. Similar results are also observed for *Penelope*, for which strain 9 is known to only have degraded, non-functional copies [26,27].

Using available genome assemblies, I then investigated the presence of this element across all available arthropod genomes to determine if it may have been recently derived from another known species, analogous to the way the *P* element in *D. melanogaster* was derived from *D. willistoni*. Using blastn with default match and mismatch scores (Match: 1, Mismatch: -3, Gap Open: 5, Gap Extension: 2) no hits were identified with an E-value cutoff of less than E-10 in any other species. Thus, it is unlikely to have entered *D. virilis* via recent horizontal transfer from any of these species with sequenced genomes. Using blastn solely on the available *D. virilis* reference genome I found that, in addition to the many nearly identical copies, many fragments were identified with E-values ranging from E-40 to E-180. Thus, while no hits with similar levels of significance were found outside *D. virilis*, a wide range of divergent fragments were identified within *D. virilis*. This suggests that lineages of *Polyphemus* have been residing within the *D. virilis* lineage for a significant period of time. Therefore, I investigated the evolutionary history of *Polyphemus* within the *Drosophila virilis* genome by generating a phylogenetic tree of all *Polyphemus* fragments (coding sequence only) in the assembled *Drosophila virilis* genome using GARLI [28] with a GTR model and no rate heterogeneity with empirical base frequencies. From the phylogenetic analysis (Figure 4), it is apparent that there is an active clade that has recently proliferated on a background of highly divergent fragments. Considering

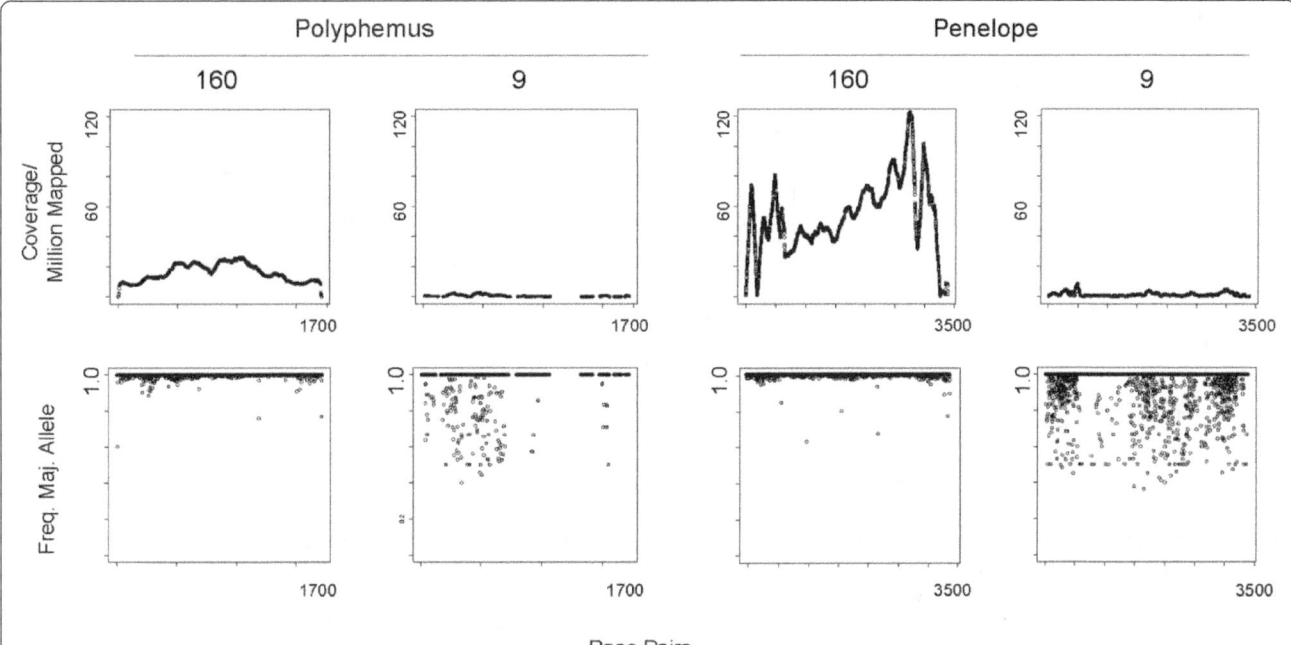

Figure 3 Mapping coverage and sequence heterogeneity for *Polyphemus* and *Penelope* from strain 160 and strain 9 genomic reads. For all plots, mapping coverage and sequence heterogeneity is shown along the length of the element. For *Polyphemus*, this is 1,704 bp. For *Penelope*, this is 3,394 bp. Read mapping coverage is measured on a per nucleotide basis, normalized by 1 million mapped reads. Based on coverage/million reads mapped, both *Polyphemus* and *Penelope* are enriched in strain 160. *Penelope* shows greater excess than *Polyphemus*. Using piledriver, I also determined sequence heterogeneity among mapped reads for *Polyphemus* and *Penelope* by scoring the frequency of the most common variant at each nucleotide position. In strain 160, *Polyphemus* and *Penelope* mapped reads are highly similar. In strain 9, mapped reads show great heterogeneity.

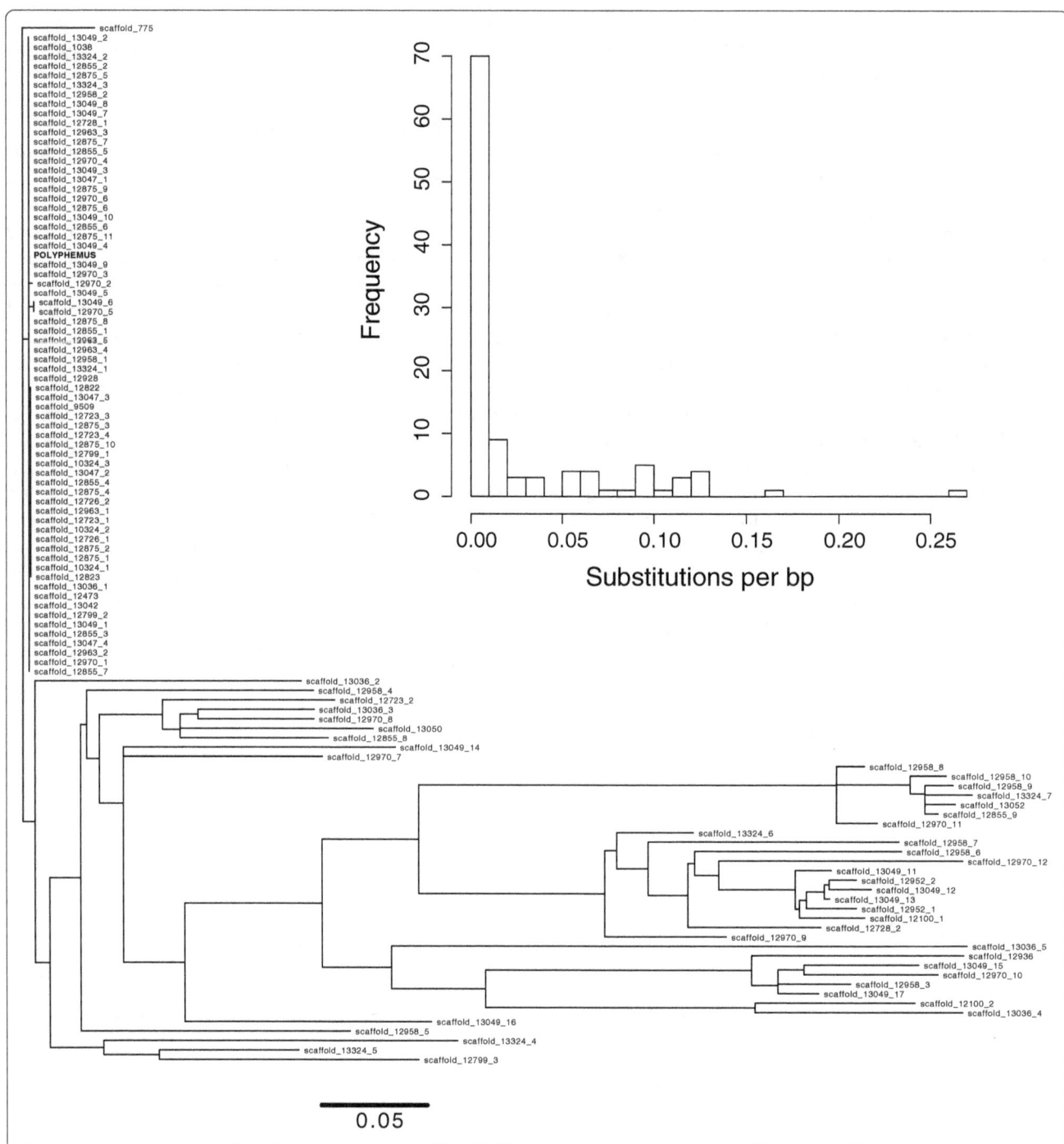

Figure 4 Phylogenetic analysis of *Polyphemus* fragments identified by blastn from the reference *D. virilis* genome. Phylogenetic tree was generated among all aligned fragments using GARLI. Distribution of branch lengths, showing a broad distribution of older fragments, is shown within the inset, indicating an older time of activity centered around 0.10 subs/bp. Note: since fragments were used, not all represent full length elements.

the distribution of terminal branch lengths, many are around 0.10 substitutions per base pair long. Assuming a per nucleotide substitution rate of 1.45×10^{-9}/bp/gen [29] and 10 generations per year, many of these fragments are about 7 million years old. Thus, it appears that at least one *Polyphemus* lineage has resided in the *D. virilis* genome for a long time and has become recently activated within the

species, including lines that induce hybrid dysgenesis. A similar pattern has previously been demonstrated for the *Penelope* element [26]. One possibility is that *Polyphemus* re-invaded *D. virilis* via horizontal transfer from another member of the *D. virilis* group.

To investigate whether *Polyphemus*, like *Penelope*, shows increased expression when inherited paternally but not maternally, I analyzed RNA-seq data from 0 to 2-hour-old embryos laid by reciprocal F1 females of the dysgenic and non-dysgenic crosses. The sterility phenotype of dysgenesis is not fully penetrant and these embryos from this direction of the cross are therefore derived from F1 females that escape sterility. F1 females of two different ages were used to examine the dynamics of expression over lifetime and embryos rather than ovaries were used to avoid problems associated with measuring gene expression in dysgenic ovaries that may be skewed in representation of somatic and germline material. Being 0 to 2 hours old, these embryos provide a measure of strictly germline expression in the F1 female. As has been previously demonstrated, *Penelope* expression is significantly higher in the germline of females from the dysgenic direction of the cross (Figure 5). Interestingly, this difference depends on the age of the F1 female (Table 1). *Penelope* germline expression is decreased in older F1 females from the dysgenic direction of the cross. *Polyphemus* expression is also higher in the germline of F1 females from the dysgenic cross, though the level of expression and magnitude of difference is smaller compared to *Penelope* (Figure 5). Interestingly, this effect does not depend on the age of the F1 female (Table 2). Thus, like *Penelope*, *Polyphemus* shows increased expression when inherited paternally.

Conclusions

Here I describe *Polyphemus*, a new Tc1-like transposable element in *D. virilis* that may contribute to the hybrid dysgenic syndrome. Whereas highly similar copies are

Table 1 *Penelope* RNA-seq analysis: ANOVA results

	Df	Sum Sq	Mean Sq	F	Pr (>F)
Treat	1	26600	26600	1594.6615	2.32E-12
Age	1	1523.7	1523.7	91.3473	2.40E-06
Barcode	2	137	68.5	4.1071	0.04988
Treat X age	1	1722.3	1722.3	103.2484	1.37E-06
Residuals	10	166.8	16.7		

Results were from two treatments (dysgenic and non-dysgenic), which were each collected from two different ages (12 to 16 days old and 19 to 21 days old). For each of these four samples, two different RNA-seq libraries were constructed with different barcodes and libraries were run in duplicate.

abundant in the inducer strain, only degraded copies are found in the non-inducer strain. Nonetheless, the lack of active copies in the non-inducer strain does not suggest an entirely new invasion of *Polyphemus* into *D. virilis*. Instead, phylogenetic analysis indicates that different lineages of *Polyphemus* have persisted in *D. virilis* for many years. Against this history, it appears that a *Polyphemus* variant has now become re-activated. This re-activation may have occurred via a horizontal transfer event from another member of the *D. virilis* group that has maintained an active *Polyphemus* lineage since divergence from *D. virilis*. Alternatively, an active lineage of *Polyphemus* may have continuously persisted in *D. virilis*. In this case, individuals within the species would have been segregating with respect to rare active copies of *Polyphemus*. In this scenario, strains or populations that maintain rare active copies may have functioned as reservoirs for later re-activation of *Polyphemus*. Interestingly, the *Penelope* element shows a similar pattern. Active copies

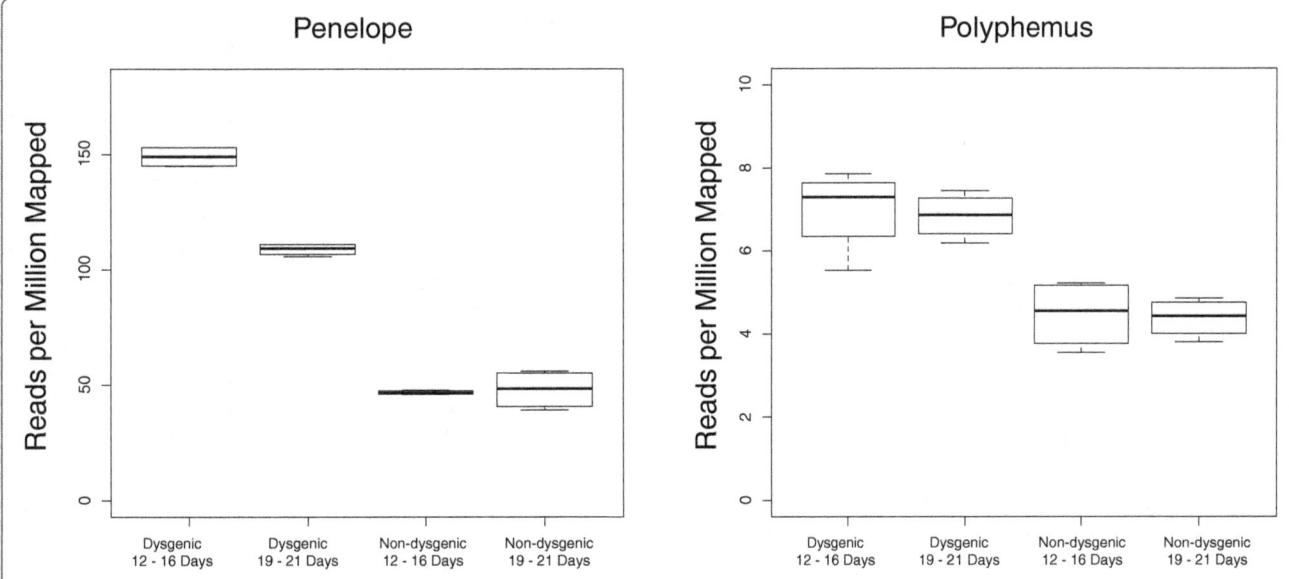

Figure 5 Expression of *Penelope* and *Polyphemus* measured in RNA-seq reads per million mapped. RNA was collected from 0 to 2-hour-old embryos of F1 females (at 12 to 16 days and 19 to 21 days) from both directions of the cross. Error indicates error derived from technical replicates (Barcode and Lane effects).

Table 2 _Polyphemus_ RNA-seq analysis: ANOVA results

	Df	Sum Sq	Mean Sq	F	Pr (>F)
Treat	1	24.9251	24.9251	43.4233	6.16E-05
Age	1	0.0564	0.0564	0.0983	0.7604
Barcode	2	0.9773	0.4886	0.8513	0.4556
Treat X age	1	0.0039	0.0039	0.0068	0.9359
Residuals	10	5.74	0.574		

Results were from two treatments (dysgenic and non-dysgenic), which were each collected from two different ages (12 to 16 days old and 19 to 21 days old). For each of these four samples, two different RNA-seq libraries were constructed with different barcodes and libraries were run in duplicate.

of _Penelope_ are highly abundant in the inducer strain, but this is not because _Penelope_ is entirely new to the _D. virilis_ lineage. The non-inducer strain 9 also possesses old, inactive copies and different variant lineages of _Penelope_ are seen in different species of the _D. virilis_ group. Thus, it appears that both of these elements have become re-activated from within the _D. virilis_ group and now may jointly cause hybrid dysgenesis.

In _D. melanogaster_, three different element families are known to cause hybrid dysgenesis. In the most well understood P-M system, the _P_ element invaded via horizontal from the distant _D. willistoni_ species [30]. However, for the _I-R_ system and the _hobo_ system, both _I_ elements [31] and _hobo_ elements [32] have remnant copies residing in the _D. melanogaster_ genome. This suggests that dysgenic syndromes may frequently result from re-activation of TE lineages that, in contrast to the _P_ elements, have been long-term genomic residents. Interestingly, the presence of multiple different elements, such as _Paris_ and _Helena_, that are also in excess in the inducer strain of _D. virilis_ seems to indicate that TE control has been diminished in strain 160. Perhaps this has occurred by the same mechanism that leads to TE co-mobilization. If so, _Penelope_ might be the cause of this TE excess in strain 160, but might not be the sole proximate cause of sterility in hybrid dysgenesis. For this reason, the mechanisms that are responsible for TE co-mobilization in dysgenic syndromes may also be relevant to understanding global TE dynamics within species after one or more TEs becomes re-activated.

Expression of _Polyphemus_ is higher in the germline of dysgenic progeny, though not to the same magnitude as _Penelope_. In light of this, it is important to note that the sterility syndrome is evident early in development [33], not at the time that gonadal expression is typically measured. For this reason, TE expression in adult females that have escaped sterility - a necessary condition for measuring germline gene expression - may not be a perfect proxy for understanding TE expression early in development. The change in _Penelope_ expression during the aging process indicates that TE activity is likely to be dynamic in the life of a dysgenic F1. For this reason, it

will be critical to determine the patterns of germline activity for all four TEs early in development. Combined with genetic approaches, this may elucidate the causal factors of sterility and TE co-mobilization in the hybrid dysgenic syndrome of _Drosophila virilis_.

Methods

Genome sequencing and analysis of Polyphemus

DNA was collected from wandering third instar larvae from strain 160 (the inducer strain) and strain 9 (the non-inducer strain). DNA was then sonicated and fragments between 400 to 500 bp were selected for Illumina library preparation. Each library was 100 bp, paired-end sequenced on an individual lane of a GAII using, yielding 43.7 million (strain 160) and 37.6 million (strain 9) read pairs. Since pairs are not independent samples, only single ends of each pair were selected for this analysis. Single reads were quality trimmed using the Sickle application with default settings. Subsequent to quality trimming, reads were mapped using BWA-MEM to a pre-computed PILER library of repeat sequences from _D. virilis_ (ftp://ftp.flybase.net/genomes/aaa/transposable_-elements/PILER-DF). Total read counts for each centroid were normalized to the total number of reads mapping back to the reference and the ratio of 160:9 normalized reads was determined. From this, the 1720 bp PILER centroid.25.39 was identified as highly abundant in strain 160 but not strain 9. This centroid was used in a blastn search of the _D. virilis_ reference genome and many nearly exact copies were identified. Three of these elements were extracted with 1000 bp of flanking sequence from each side. Reciprocal pairwise blast between these three fragments identified a core sequence 1,708 bp long of near identify among these fragments. Further annotation of this sequence was performed using Geneious. Phylogenetic analysis of Tc1/_mariner_ members was performed using MrBayes on a MUSCLE amino acid alignment until the average standard deviation of split frequencies was less than 0.01. Analysis of heterogeneity among fragments was performed using piledriver to examine sequence heterogeneity of reads mapping to the respective TE. From the piledriver output, the frequency of the most common base at each position was calculated. Phylogenetic analysis was performed by collecting all blastn fragments from the _D. virilis_ reference genome (excluding inverted repeats) with E values better than E-5. Fragments smaller than 100 bp were removed from the blast output and the blast output anchored to the active sequence was used as an alignment. Tree searching was performed using GARLI on the CIPRES server (http://www.phylo.org/) with a GTR model, no rate heterogeneity and empirical base frequencies. Terminal branch length distributions were extracted from the resulting tree file.

Expression analysis

RNA was collected from 0 to 2-hour-old embryos that were laid by F1 females of the dysgenic (strain 9 mothers and strain 160 fathers) and non-dysgenic (strain 160 mothers and strain 9 fathers) crosses. Embryos of F1 females were chosen to avoid the confounding effects that are presented by ovaries that may differ in somatic *vs.* germline tissue representation in dysgenic crosses. Dysgenic and non-dysgenic crosses were set up *en masse* and hundreds of F1 females were collected soon after eclosion. Hundreds of reciprocal F1 males from the same crosses were also collected, combined in equal proportions, then reallocated equally to the collected dysgenic and non-dysgenic F1 females in mating cages. This was done to ensure a sufficient egg lay from the F1 females escapers of the dysgenic cross. Such females, if only provided their dysgenic brothers, lay few eggs. Providing equivalent but mixed populations of reciprocal males to the female pools also ensures proper genetic control over paternal effects *en masse* since reciprocal males of the dysgenic and non-dysgenic crosses are genetically different. From these cages, eggs were collected over 0 to 2-hour egg lay durations and each collection was immediately flash frozen in liquid nitrogen. Collections were pooled into two different age classes based on the age of the mother: 12 to 16 days old (about 1 week after sexual maturity) and 19 to 21 days old.

Pooled RNA was collected and Illumina libraries were generated for single-end, 50 bp RNA-seq. To control for index effects each RNA sample was used to generate two index libraries for a total of eight libraries, each of which were run in replicate on two different lanes. Trimming and filtering was performed using the Galaxy server (https://usegalaxy.org/). Up to 16 bp were quality trimmed from the 3' end from each read and remaining reads with more than 2 bp with quality less than 20 were removed. Trimmed and filtered reads were mapped to TE sequences using CLC with mismatch, insertion and deletion scores equal to 2.3 and 3, respectively. Expression levels were measured by the number of reads that mapped each TE normalized by the total number of reads mapping to the reference genome. ANOVA was performed in R using the aov command and a model that included the effects of treatment (dysgenic *vs.* non-dysgenic), age, index, and an interaction between age and treatment.

Abbreviations

GTR: Generalised time reversible; TE: Transposable element.

Competing interests

The author declares that he has no competing interest.

Acknowledgements
I thank Chris Harrison for generating the RNA-seq libraries. DNA library prep and sequencing was performed by the David H. Murdock Research Institute.

Funding was provided by NSF MCB-1022165, an internal grant through NIH-P20GM103638, and the University of Kansas.

References

1. Engels WR: The P-Family of transposable elemens in *Drosophila*. *Annu Rev Genet* 1983, 17:313–344.
2. Bingham PM, Kidwell MG, Rubin GM: The molecular basis of P-M dysgenesis - the role of the P-element, a P-strain-specific transposon family. *Cell* 1982, 29:995–1004.
3. Engels WR: Hybrid dysgenesis in *Drosophila melanogaster* - rules of inheritance of female sterility. *Genet Res* 1979, 33:219–236.
4. Kidwell MG, Novy JB: Hybrid dysgenesis in *Drosophila melanogaster* - sterility resulting from gonadal-dysgenesis in the P-M system. *Genetics* 1979, 92:1127–1140.
5. Khurana JS, Wang J, Xu J, Koppetsch BS, Thomson TC, Nowosielska A, Li C, Zamore PD, Weng Z, Theurkauf WE: Adaptation to P element transposon invasion in *Drosophila melanogaster*. *Cell* 2011, 147:1551–1563.
6. Margulies L: A high level of hybrid dysgenesis in *Drosophila*: high thermosensitivity, dependence on DNA repair, and incomplete cytotype regulation. *Mol Gen Genet* 1990, 220:448–455.
7. Engels WR: In Mobile DNA. Edited by Berg DE, Howe MH. Washington, DC: American Society for Microbiology; 1989.
8. Engels WR, Preston CR: Identifying P factors in *Drosophila* by means of chromosome breakage hotspots. *Cell* 1981, 26:421–428.
9. Bucheton A, Paro R, Sang HM, Pelisson A, Finnegan DJ: The molecular basis of the I-R hybrid dysgenesis syndrome in *Drosophila melanogaster* - identification, cloning and properties of the I-factor. *Cell* 1984, 38:153–163.
10. Yannopoulos G, Stamatis N, Monastirioti M, Hatzopoulos P, Louis C: *hobo* is responsible for the induction of hybrid dysgenesis by strains of *Drosophila melanogater* bearing the male recombination factor 23.5MRF. *Cell* 1987, 49:487–495.
11. Evgenev MB, Zelentsova H, Shostak N, Kozitsina M, Barskyi V, Lankenau DH, Corces VG: *Penelope*, a new family of transposable elements and its possible role in hybrid dysgenesis in *Drosophila virilis*. *Proc Natl Acad Sci U S A* 1997, 94:196–201.
12. Lozovskaya ER, Scheinker VS, Evgenev MB: A hybrid dysgenesis syndrome in *Drosophila virilis*. *Genetics* 1990, 126:619–623.
13. Scheinker VS, Lozovskaya ER, Bishop JG, Corces VG, Evgenev MB: A long terminal repeat-containing retrotransposon in mobilized during hybrid dysgenesis in *Drosophila virilis*. *Proc Natl Acad Sci U S A* 1990, 87:9615–9619.
14. Rozhkov NV, Aravin AA, Zelentsova ES, Schostak NG, Sachidanandam R, McCombie WR, Hannon GJ, Evgen'ev MB: Small RNA-based silencing strategies for transposons in the process of invading *Drosophila* species. *RNA* 2010, 16:1634–1645.
15. Petrov DA, Schutzman JL, Hartl DL, Lozovskaya ER: Diverse transposable elements are mobilized in hybrid dysgenesis in *Drosophila virilis*. *Proc Natl Acad Sci U S A* 1995, 92:8050–8054.
16. Blumenstiel JP, Hartl DL: Evidence for maternally transmitted small interfering RNA in the repression of transposition in *Drosophila virilis*. *Proc Natl Acad Sci U S A* 2005, 102:15965–15970.
17. Vieira J, Vieira CP, Hartl DL, Lozovskaya ER: Factors contributing to the hybrid dysgenesis syndrome in *Drosophila virilis*. *Genet Res* 1998, 71:109–117.
18. Evgen'ev MB: What happens when Penelope comes? An unusual retroelement invades a host species genome exploring different strategies. *Mobile Genet Elem* 2013, 3:e24542.
19. Pyatkov KI, Shostak NG, Zelentsova ES, Lyozin GT, Melekhin MI, Finnegan DJ, Kidwell MG, Evgen'ev MB: Penelope retroelements from *Drosophila virilis* are active after transformation of *drosophila melanogaster*. *Proc Natl Acad Sci U S A* 2002, 99:16150–16155.
20. Rozhkov NV, Schostak NG, Zelentsova ES, Yushenova IA, Zatsepina OG, Evgen'ev MB: Evolution and dynamics of small RNA response to a retroelement invasion in *Drosophila*. *Mol Biol Evol* 2013, 30:397–408.
21. Drosphila 12 Genomes Consortium, Clark AG, Eisen MB, Smith DR, Bergman CM, Oliver B, Markow TA, Kaufman TC, Kellis M, Gelbart W, Iyer VN, Pollard DA, Sackton TB, Larracuente AM, Singh ND, Abad JP, Abt DN, Adryan B, Aguade M, Akashi H, Anderson WW, Aguadro CF, Ardell DH, Arguello R, Artieri CG, Barbash DA, Barker D, Barsanti P, Batterham P, Batzoglu S, et al:

Evolution of genes and genomes on the *Drosophila* phylogeny. *Nature* 2007, **450**:203–218.

22. Bosco G, Campbell P, Leiva-Neto JT, Markow TA: **Analysis of *Drosophila* species genome size and satellite DNA content reveals significant differences among strains as well as between species.** *Genetics* 2007, **177**:1277–1290.

23. Li H, Durbin R: **Fast and accurate long-read alignment with burrows-wheeler transform.** *Bioinformatics* 2010, **26**:589–595.

24. Smith CD, Edgar RC, Yandell MD, Smith DR, Celniker SE, Myers EW, Karpen GH: **Improved repeat identification and masking in dipterans.** *Gene* 2007, **389**:1–9.

25. Merriman PJ, Grimes CD, Ambroziak J, Hackett DA, Skinner P, Simmons MJ: **S elements: a family of Tc1-like transposons in the genome of *Drosophila melanogaster*.** *Genetics* 1995, **141**:1425–1438.

26. Lyozin GT, Makarova KS, Velikodvorskaja VV, Zelentsova HS, Khechumian RR, Kidwell MG, Koonln EV, Evgen'ev MB: **The structure and evolution of *penelope* in the *virilis* species group of *Drosophila*: an ancient lineage of retroelements.** *J Mol Evol* 2001, **52**:445–456.

27. Morales-Hojas R, Vieira CP, Vieira J: **The evolutionary history of the transposable element *penelope* in the *Drosophila virilis* group of species.** *J Mol Evol* 2006, **63**:262–273.

28. Zwickl D: *Genetic algorithm approaches for the phylogenetic analysis of large biological sequence datasets under the maximum likelihood criterion.* Austin, TX: The University of Texas; 2006.

29. Li HP, Stephan W: **Inferring the demographic history and rate of adaptive substitution in *Drosophila*.** *PLoS Genet* 2006, **2**:1580–1589.

30. Daniels SB, Peterson KR, Strausbaugh LD, Kidwell MG, Chovnick A: **Evidence for horizontal transmission of the P transposable element between *Drosophila* species.** *Genetics* 1990, **124**:339–355.

31. Bucheton A, Simonelig M, Vaury C, Crozatier M: **Sequences similar to the I transposable element involved in I-R hybrid dysgenesis in *D. melanogaster* occur in other *Drosophila* species.** *Nature* 1986, **322**:650–652.

32. Boussy IA, Itoh M: **Wanderings of *hobo*: a transposon in *Drosophila melanogaster* and its close relatives.** *Genetica* 2004, **120**:125–136.

33. Sokolova MI, Zelentsova ES, Shostak NG, Rozhkov NV, Evgen'ev MB: **Ontogenetic consequences of dysgenic crosses in *Drosophila virilis*.** *Int J Dev Biol* 2013, **57**:731–739.

18

Horizontal transfer of transposons between and within crustaceans and insects

Mathilde Dupeyron[1], Sébastien Leclercq[1], Nicolas Cerveau[1,2], Didier Bouchon[1] and Clément Gilbert[1*]

Abstract

Background: Horizontal transfer of transposable elements (HTT) is increasingly appreciated as an important source of genome and species evolution in eukaryotes. However, our understanding of HTT dynamics is still poor in eukaryotes because the diversity of species for which whole genome sequences are available is biased and does not reflect the global eukaryote diversity.

Results: In this study we characterized two *Mariner* transposable elements (TEs) in the genome of several terrestrial crustacean isopods, a group of animals particularly underrepresented in genome databases. The two elements have a patchy distribution in the arthropod tree and they are highly similar (>93% over the entire length of the element) to insect TEs (Diptera and Hymenoptera), some of which were previously described in *Ceratitis rosa* (*Crmar2*) and *Drosophila biarmipes* (*Mariner-5_Dbi*). In addition, phylogenetic analyses and comparisons of TE versus orthologous gene distances at various phylogenetic levels revealed that the taxonomic distribution of the two elements is incompatible with vertical inheritance.

Conclusions: We conclude that the two *Mariner* TEs each underwent at least three HTT events. Both elements were transferred once between isopod crustaceans and insects and at least once between isopod crustacean species. *Crmar2* was also transferred between tephritid and drosophilid flies and *Mariner-5* underwent HT between hymenopterans and dipterans. We demonstrate that these various HTTs took place recently (most likely within the last 3 million years), and propose iridoviruses and/or *Wolbachia* endosymbionts as potential vectors of these transfers.

Keywords: Horizontal transfer, Transposable elements, Isopod crustaceans, Hexapods

Background

Horizontal transfer (HT) of genetic material is the transmission of DNA between non-mating organisms [1]. Most known eukaryote-to-eukaryote HT events are transfers of transposable elements (TEs) [2]. Given the profound impact TEs have on the genome architecture of their hosts, HT of TEs (HTT) is increasingly recognized as an important force in eukaryote genome evolution [3]. On the TE side, spreading between genomes via HT may be viewed as a strategy to escape vertical extinction due to purifying selection, mutational decay and/or host defense mechanisms. Among the over 330 cases of eukaryote-to-eukaryote HTT events characterized so far, the vast majority involve DNA transposons

(n = 188 cases) and LTR retrotransposons (n = 118 cases) [4], indicating that the long-term survival of these TEs may rely more on HT than that of non-LTR retrotransposons. Yet, while whole genomes are sequenced at an exponential pace, the global diversity of eukaryote genomes is still poorly represented, precluding any strong generalization on HTT dynamics. Even in animals, whole genome sequencing efforts are biased towards species closely related to model organisms or species of economic interest, and whole genome sequences are lacking for many large taxonomic groups. Our current understanding of the global HTT dynamics and impacts is therefore incomplete, both at the host and TE level.

With only one genome fully sequenced [5] out of over 50,000 species described [6], crustaceans are particularly

---
* Correspondence: clement.gilbert@univ-poitiers.fr
[1]Université de Poitiers, UMR CNRS 7267 Ecologie et Biologie des Interactions, Equipe Ecologie Evolution Symbiose, 86022 Poitiers, Cedex, France
Full list of author information is available at the end of the article

underrepresented in genome databases. The order Isopoda (Vericrustacea clade according to [7]) is unique among crustaceans in that the colonization of landmasses by one of its lineages (belonging to the suborder Oniscidea) during the Mesozoic yielded a large diversity of terrestrial species (>3,600 [8]) now distributed all over the world in every biotope (except for the poles) [9]. In this study we report new cases of HTT involving terrestrial isopod crustaceans and hexapods. We used a combination of cross-species PCR screening of TEs, phylogenetic and other evolutionary analyses to characterize in detail these HTT and to shed light on the evolutionary dynamics of the first two TEs described in isopod crustaceans.

Results and discussion

Characterization of two *Mariner* elements in the isopod crustacean *Armadillidium vulgare*

In order to detect TEs that underwent horizontal transfer between isopod crustaceans and other taxa, we used all consensus sequences deposited in Repbase [10] as of May 2013 as queries to perform BLASTn searches on draft genomic contigs and on a transcriptome of the pill bug *Armadillidium vulgare* that have been generated in our lab as part of other ongoing projects. Importantly, the contigs generated by these projects are too short to carry out a comprehensive *de novo* mining of *A. vulgare* TEs. The BLASTn searches yielded two TEs belonging to the Tc1/Mariner superfamily of Class II DNA transposons that show more than 90% identity over more than 500 bp to *A. vulgare* sequences. The first one (*Crmar2*) was originally characterized in the tephritid fly *Ceratitis rosa* based on a PCR/sequencing screening [11], and the second one (*Mariner-5_Dbi*) was described by Kojima and Jurka [12] in *Drosophila biarmipes* based on whole genome sequence data mining. We reconstructed an *A. vulgare* consensus sequence of both elements (named *Crmar2_Avul* and *Mariner-5_Avul*) using 100 to 1300 bp-long fragments resulting from our various BLAST outputs, such that the entire sequence of the consensus was covered by at least five different copies. *Crmar2_Avul* is 1304 bp in length, has 39-bp terminal inverted repeats (TIRs) and encodes a 361 amino acid (aa) transposase while *Mariner-5_Avul* is 1013 bp in length, has 28-bp TIRs and encodes a 200 aa transposase. Both elements are flanked by TA target site duplications, which is characteristic of the Tc1/Mariner superfamily [13]. The evolution of this superfamily has yielded a large number of elements which have colonized the genome of many eukaryote taxa [14,15] and have been classified in various subfamilies (for example, [16]). *Crmar2* belongs to the *rosa* subfamily [11] and our phylogenetic analysis of the transposase revealed that *Mariner-5_Dbi* belongs to the irritans subfamily [see Additional file 1: Figure S1].

Taxonomic distribution of the two *Mariner* elements in eukaryotes

Next we sought to assess the taxonomic distribution of *Crmar2_Avul* and *Mariner-5_Avul* in eukaryotes by performing BLASTn searches on all eukaryotic genomes that were available in Genbank as of May 2013. In addition to the species in which the two elements had previously been described (*C. rosa*, *Anastrepha ludens* and *Anastrepha suspensa* for *Crmar2*; *D. biarmipes* for *Mariner-5_Dbi*) we found TEs highly similar (>90% identity over >500 bp) to *Crmar2_Avul* in *Drosophila ananassae* and *Drosophila bipectinata*, and to *Mariner-5_Avul* in the ant *Harpegnathos saltator*. Interestingly, the taxonomic distribution of the two elements is patchy, not only at the level of the arthropod phylogeny, but also within the lower level taxa in which we found them (Figure 1), a pattern likely indicative of horizontal transfer [17]. For example, *Crmar2_Avul* is only present in two closely related *Drosophila* species out of the 13 that we searched, and *Mariner-5_Avul* was identified only in one of the three hymenopteran genomes available.

Horizontal transfer of the two *Mariner* elements between hexapods and isopods and within hexapods

To formally assess whether the distribution of the two elements in arthropods is the result of HT, we compared TE genetic distances calculated between hexapod species and *A. vulgare* to distances calculated for 46 orthologous genes available for both *A. vulgare* and *Drosophila melanogaster* [7]. As illustrated in Figure 2, distances between orthologous genes (average = 35%, min = 21%, and max = 49%) are much higher than distances between TEs (average = 6%, min = 4.9%, and max = 7.4%). Under vertical transmission of the TEs, TE distances between taxa are expected to be higher than distances between orthologous genes because TEs are known to evolve neutrally after insertion in a given genome [22], that is, faster than host genes that evolve under purifying selection due to functional constraints. The high similarity between *A. vulgare* and hexapoda TEs coupled with the deep divergence time between these two arthropod taxa (>400 million years) and to the patchy distribution of the two elements in arthropods allows us to confidently conclude that the presence of both *Crmar2* and *Mariner-5* in isopod crustaceans and hexapods results from HT. Interestingly, we found that gene distances between tephritid (*Ceratitis capitata*) and drosophilid (*D. melanogaster*) flies on one hand (average = 27%, min = 16%, and max = 45%) and between *Drosophila* and the ant *H. saltator* on the other hand (average = 34%, min = 18%, and max = 56%) are also much higher than TE distances (4% for both elements; Figure 2). This pattern suggests that in addition to transferring horizontally between hexapods and isopods, *Crmar2* also underwent HT

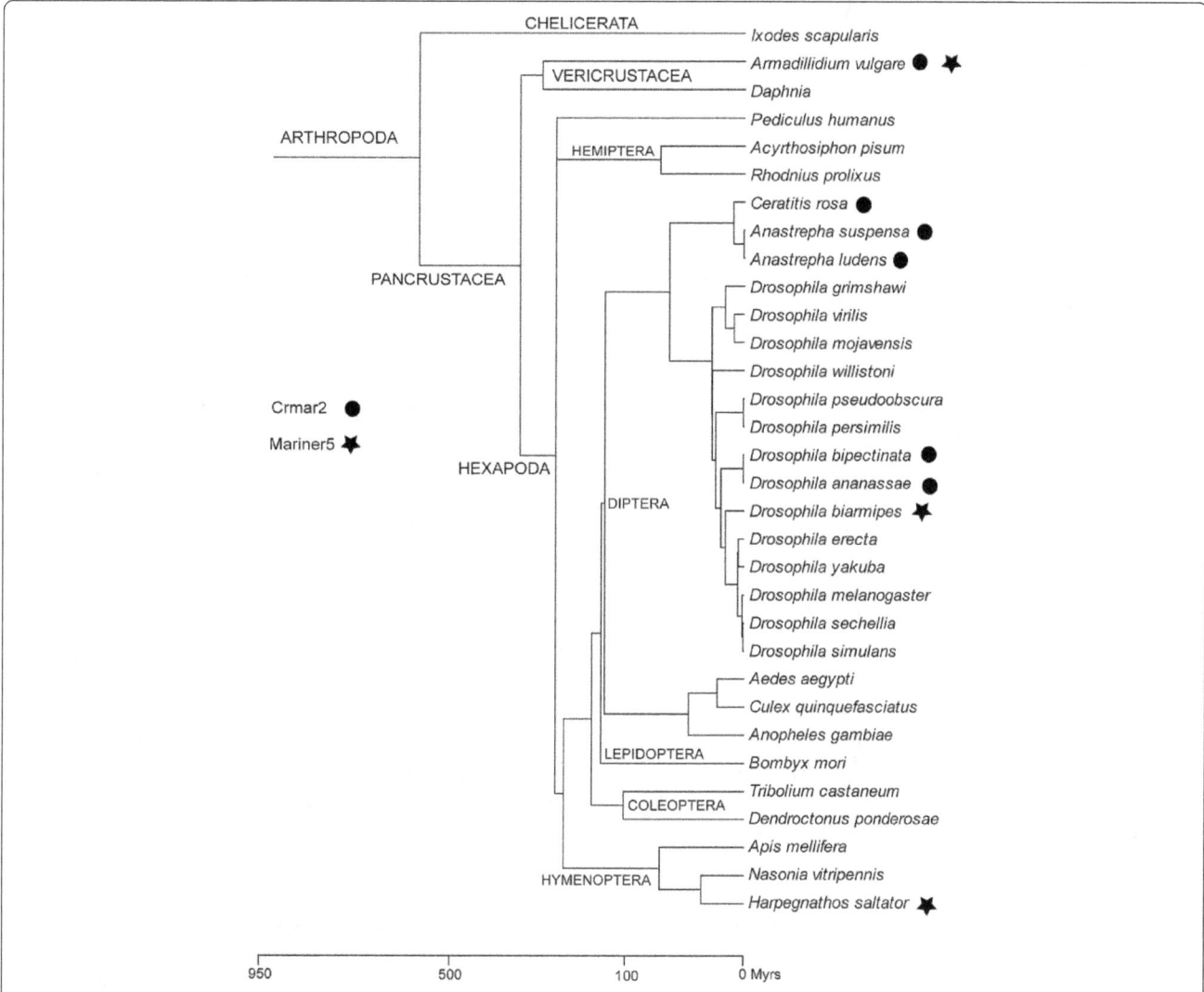

Figure 1 Timetree of Arthropoda. The tree includes all species in which *Crmar2* and *Mariner-5* were found as well as most closely related species for which whole genome sequences are available in Genbank. Phylogenetic relationships are taken from Tamura *et al.* [18], Regier *et al.* [19], Sharkey [20] and Regier *et al.* [7]. Divergence times are taken from Hedges *et al.* [21]. Divergence times between *D. bipectinata* and *D. ananassae*, between *D. biarmipes* and other *Drosophila* species, between *Dendroctonus ponderosae* and *Tribolium castaneum*, between *A. vulgare* and *Daphnia pulex*, and between *H. saltator* and *Nasonia vitripennis* are unknown and have therefore been set at arbitrary values for illustrative purposes.

between the two dipteran lineages and *Mariner-5* also transferred horizontally between dipterans and hymenopterans.

Recent horizontal transfer of the two *Mariner* elements within isopods

To shed light on the evolutionary dynamics of *Crmar2_-Avul* and *Mariner-5_Avul* in terrestrial isopod crustaceans we carried out two sets of PCR screenings in 14 species. The first screening involved primer pairs designed in the internal region of the elements in order to check for the presence of each TE in the various species (type 1 primers in Additional file 2: Figure S2). The second screening aimed at finding specific copies of the two elements that would be shared at orthologous loci

between the various isopod species. For the latter screen we used primer pairs for which one primer was designed in the 5' or 3' end of *Crmar2_Avul* and *Mariner-5_Avul* and the other primer was designed in the flanking region of several copies of each element (type 2 primers in Additional file 2: Figure S2; n = 4 for *Crmar2_Avul* and 3 for *Mariner-5_Avul*). The first screening (internal primers) uncovered *Crmar2_Avul* and *Mariner-5_Avul*, respectively, in six and nine of the 14 species (Figure 3). The second screening (search for orthologous copies) did not reveal any shared copies between *A. vulgare* and any of the other 13 isopod species. This absence of amplification could be due to a lack of conservation of the regions flanking the various copies in the different

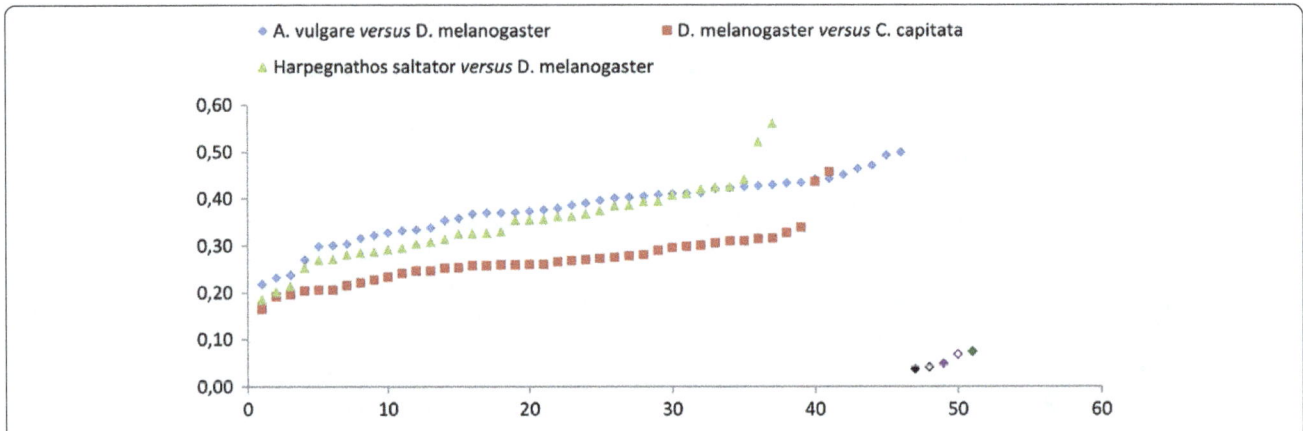

Figure 2 Graph illustrating the pairwise corrected distances between arthropod orthologous genes and between *Crmar2* and *Mariner-5* consensus sequences. Orthologous gene distances between *A. vulgare* and *D. melanogaster*, between *D. melanogaster* and *C. capitata*, and between *H. saltator* and *D. melanogaster* are illustrated with blue lozenges, red squares, and green triangles, respectively. Other symbols correspond to transposable element (TE) distances: *Mariner-5* between *H. saltator* and *D. biarmipes* (filled black lozenge), *Crmar2* between *C. rosa* and *D. bipectinata* (empty black lozenge), *Mariner-5* between *A. vulgare* and *D. biarmipes* (filled purple lozenge), *Crmar2* between *A. vulgare* and *C. rosa* (empty purple lozenge), and *Crmar2* between *A. vulgare* and *D. bipectinata* (filled green lozenge). A detailed list of genes and distances is provided in Additional file 8: Table S3. Before the distance values were plotted, they were sorted by ascending order.

species. But perhaps more interestingly, we also noticed that two of the four *Crmar2* copies failed to amplify in one of the two *A. vulgare* individuals in which we searched them, suggesting that they are polymorphic in terms of presence/absence in *A. vulgare* populations. Overall, these data indicate that both elements likely underwent recent HT in isopods and may still be active.

To further test the possibility that both elements invaded isopod genomes recently via HT and are still actively transposing, we cloned and sequenced two to five different copies of *Crmar2_Avul* and *Mariner-5_Avul* in all isopod species in which we found them (except for *Eluma purpurascens* in which all ten clones that we sequenced contained an identical copy of *Crmar2*).

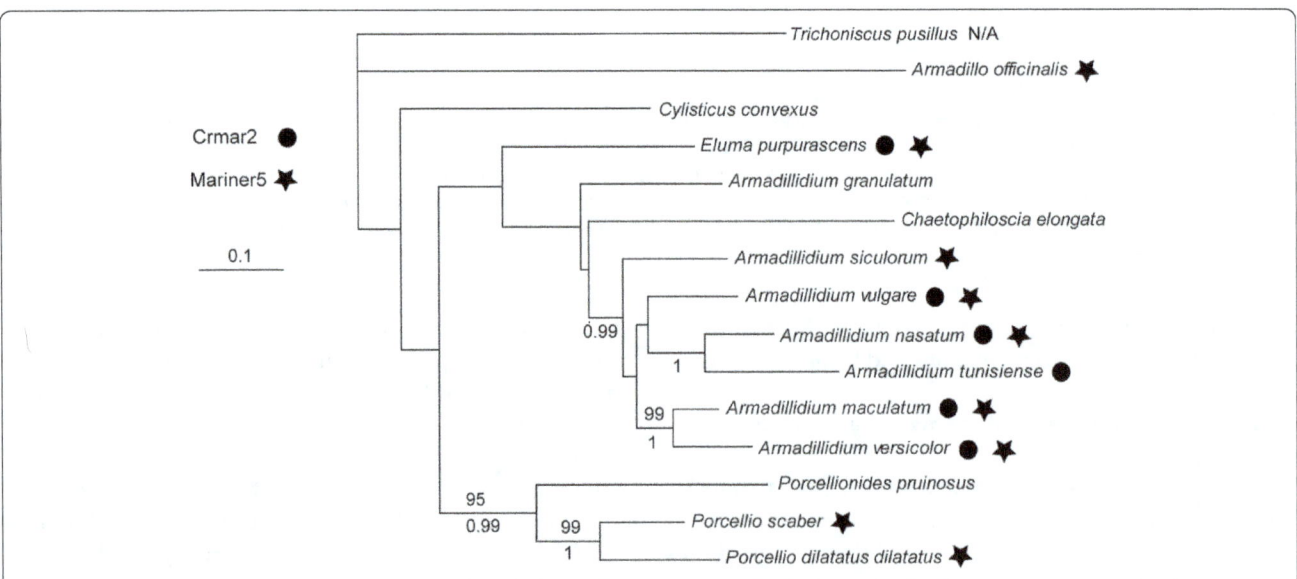

Figure 3 Phylogenetic tree of the various terrestrial isopod crustacean species included in this study. The topology of the tree corresponds to the consensus of the bootstrap analysis performed under the maximum likelihood criteria. Bootstrap values above 70% and Bayesian posterior probabilities above 0.9 are shown above and below branches respectively. Species in which *Crmar2* and/or *Mariner-5* were uncovered by PCR are marked with a filled circle and/or a star. *Trichoniscus pusillus* was used as an outgroup to root the present phylogeny based on the topology obtained in Michel-Salzat and Bouchon [23] and was not screened for the presence of the two TEs (N/A: not applicable). The other species were PCR screened but none of the two elements were amplified.

Pairwise genetic distances between these copies within each genome are all very low (average = 1.2%, min = 0.9%, and max = 2% for *Crmar2_Avul* and average = 4.5%, min = 2%, and max = 6% for *Mariner-5_Avul*). In addition, the between-species distances for both TEs are also much lower than distances calculated for the androgenic gland hormone gene (average = 34%; Figure 4). Following the same reasoning as for the comparisons between TE and orthologous gene distances discussed above, we believe these results strongly suggest that both *Crmar2_Avul* and *Mariner-5_Avul* underwent one or more recent HTs within isopods. Interestingly, using RT-PCR, we verified that *Crmar2_Avul* and *Mariner-5_Avul* are transcribed both in somatic and germ cells in *A. vulgare* [see Additional file 3: Figure S3]. Furthermore, seven of the eight *Crmar2_Avul* transcripts that we uncovered in the *A. vulgare* transcriptome contain a full-length and intact (devoid of non-sense mutations) ORF (sequences provided in Additional file 4: Dataset 1), suggesting that at least one source of functional transposase is transcribed for this element in *A. vulgare*.

To provide an estimate of the absolute age of the activity burst of *Mariner-5* and *Crmar2*, we divided the average copy/consensus distance calculated for each element in *D. biarmipes* (11%) and *D. bipectinata* (6.9%) by the experimentally derived neutral substitution rate of *D. melanogaster* (0.0346 substitutions per base per million years (myr); [24-26]). This yielded a burst age of 3.2 myrs for *Mariner-5* in *D. biarmipes* and 2 myrs for *Crmar2* in *D. bipectinata*. The age of *Mariner-5_Avul* and *Crmar2_Avul* cannot be precisely estimated because nuclear substitution rates are not available for isopods. Together with the absence of shared orthologous copies

of both elements between the various isopods species and the seemingly polymorphic state of *Crmar2* in *A. vulgare*, the fact that an intact *Crmar2_Avul* transposase is transcribed in this species is consistent with a recent invasion of isopod genomes by *Mariner-5_Avul* and *Crmar2_Avul* and suggest both elements are active sources of genomic variation in this major crustacean group. In addition, given that isopod *Crmar2* and *Mariner-5* are highly similar to *Drosophila Crmar2* and *Mariner-5* (95% and 93% identity, respectively; Figure 2), we believe these HTTs most likely took place within the past few million years at most.

Number of horizontal transfer of transposon events

To assess the number of HTTs that occurred between hexapods and crustacean isopods and within each of the two taxa, we reconstructed a phylogeny of both elements based on an alignment including all copies of *Crmar2* and *Mariner-5* that we sequenced from the various isopod species and those that we found in the other sequenced arthropod genomes. In the resulting *Crmar2* tree both hexapod and isopod *Crmar2* elements are monophyletic (Figure 5). Therefore, a single HTT event between isopods and hexapods needs to be inferred to explain the taxonomic distribution of this element in arthropods. Within hexapods, *Crmar2* TEs from the tephritid *C. rosa* are more closely related to *Drosophila* elements than they are to the other tephritid elements found in *A. ludens* and *A. suspensa* by Gomulski *et al.* [11]. This topology upholds our TE versus orthologous gene distance analysis (see above), indicating that *Crmar2* also underwent HT within dipterans. In the *Mariner-5* tree (Figure 6), isopod elements fall within

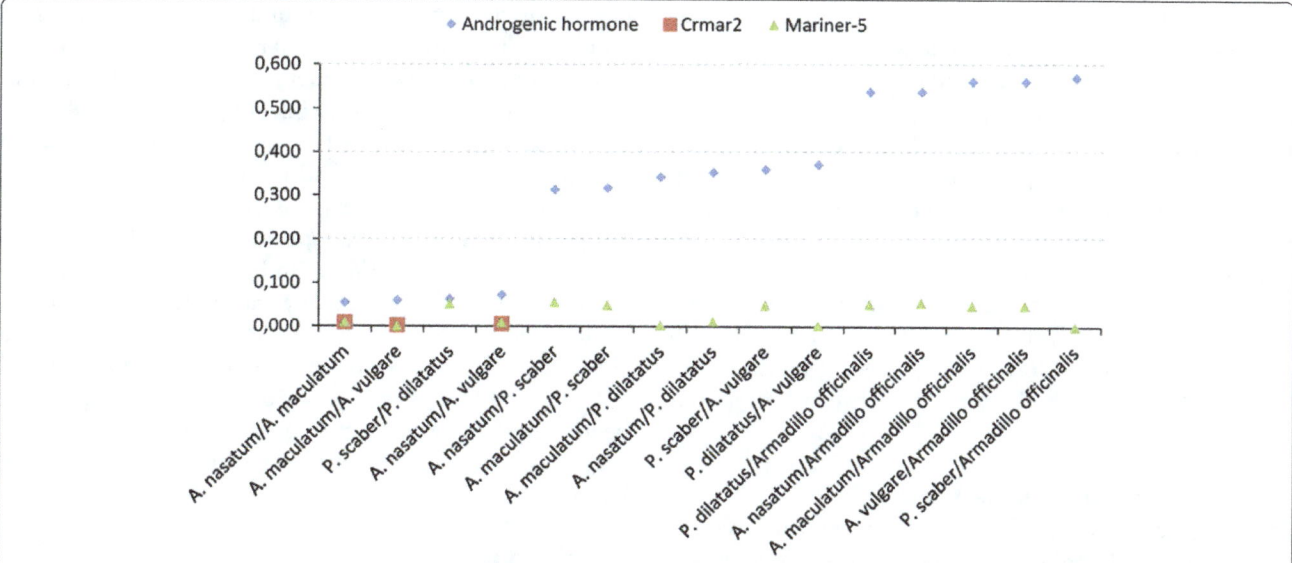

Figure 4 Graph illustrating the pairwise corrected distances between the androgenic gland hormone gene and between *Crmar2* and *Mariner-5* consensus sequences within terrestrial isopods.

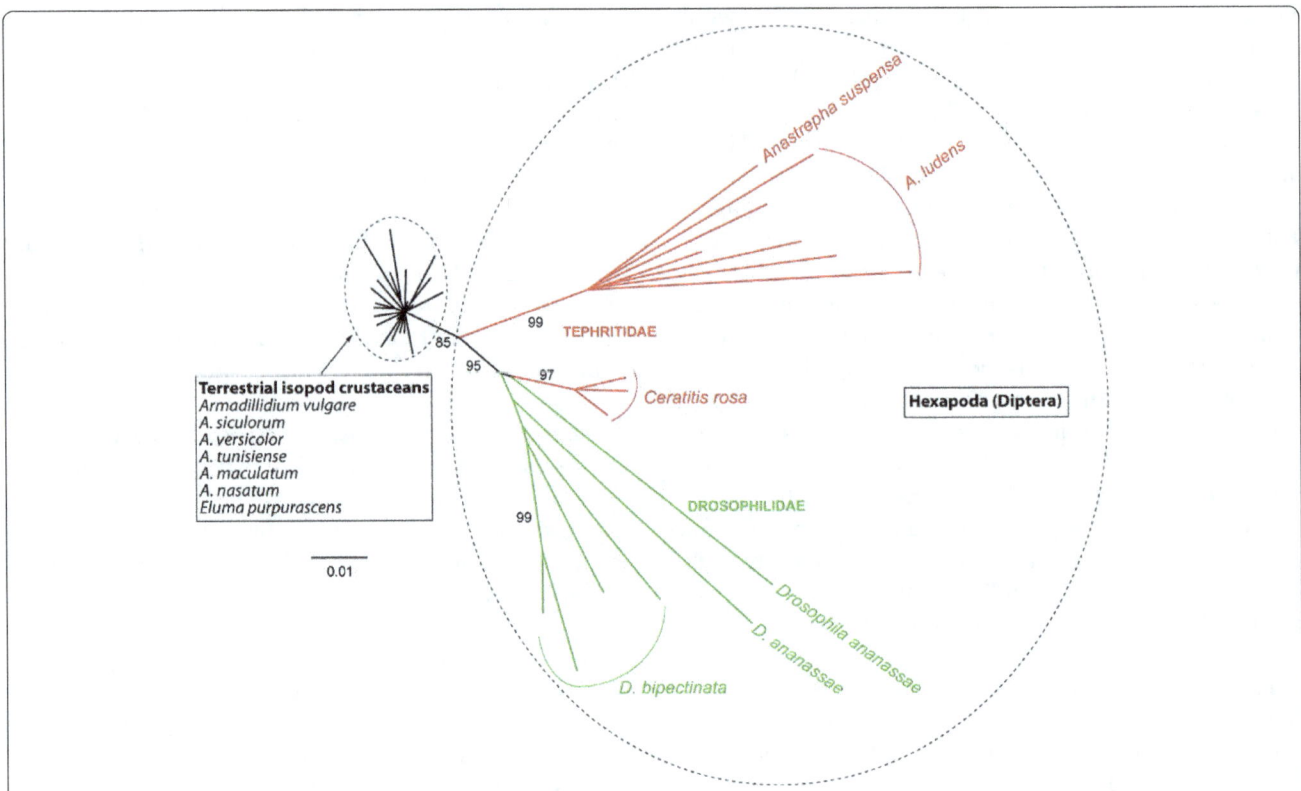

Figure 5 Phylogenetic tree of *Crmar2* copies. Maximum likelihood bootstrap values above 70% are shown on branches. Given the absence of phylogenetic support for the branching of *Crmar2* copies within isopods, the name of the species is shown only once in the black rectangle to facilitate the reading of the figure.

two relatively distantly related clusters. However, given that the tree is unrooted, we cannot conclude on whether *Mariner-5* was transferred once or more than once between hexapods and isopods. Interestingly, the topology of isopod *Mariner-5* copies (Figure 6) is clearly incongruent with that of the isopod tree (Figure 3). For example, *Porcellio dilatatus dilatatus* and *Porcellio scaber* form a strongly supported clade in the species phylogeny, but *Crmar2* copies from the former fall within the cluster that groups all *Armadillidium* and *Eluma purpurascens* copies and those of *P. scaber* group with *Crmar2* copies from *Armadillo officinalis*. This phylogenetic incongruence between host and TE phylogenies, together with the general lack of phylogenetic resolution within isopod *Crmar2* and *Mariner-5* clusters (Figures 5 and 6), and the fact that copies from each isopod species do not form monophyletic groups, further supports the HT of both elements between the various isopod species.

Potential vectors of horizontal transfer

Though the question of the mechanisms and vectors underlying HTT between multicellular eukaryotes remains largely open, growing evidence suggests that host-parasite relationships likely facilitate such transfers

[27-31]. In particular, viruses are often cited as ideal HTT vectors due to their capacity to inject DNA/RNA into host cells [32-35]. Though the viral fauna infecting the various species involved in *Crmar2* and *Mariner-5* HTT is poorly known, it is noteworthy that members of the Iridoviridae have been found in several species of dipterans, hymenopterans and terrestrial isopods [36,37]. In addition, a recent study identified two TEs inserted in the genome of an iridovirus infecting dipteran [38], emphasizing the potential of this type of viruses to shuttle transposons between their hosts. Another possible route for HTT to occur in arthropods is via transfers of endosymbiotic bacteria. Several species of isopods as well as *D. ananassae*, *D. bipectinata* and *A. suspensa* are known to bear intracellular, maternally transmitted alphaproteobacteria called *Wolbachia* [39-43]. The fact that isopod *Wolbachia* strains are known to have undergone several HT [44] and that several genes of eukaryotic origin have been found integrated in *Wolbachia* genomes [45-47] suggest that endosymbionts could also facilitate HT of DNA between hosts.

Conclusions

In this study, we have characterized the evolutionary dynamics of two Tc1/Mariner elements in isopod crustaceans

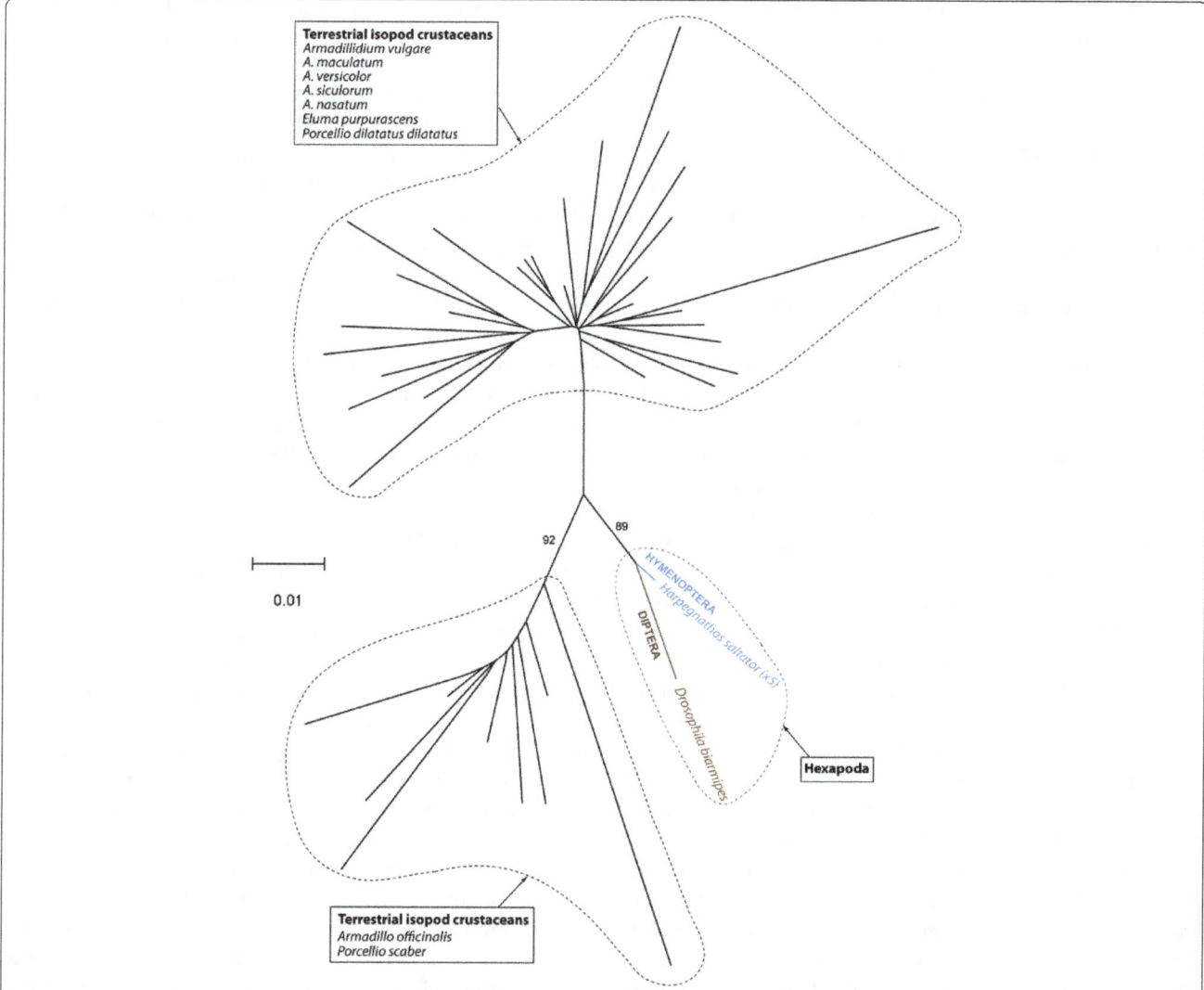

Figure 6 Phylogenetic tree of *Mariner-5* copies. Maximum likelihood bootstrap values above 70% are shown on branches. Given the absence of phylogenetic support for the branching of *Mariner-5* copies within the two clusters of isopod sequences, the name of the species is shown only once in the two black rectangles to facilitate the reading of the figure.

and shown that their current taxonomic distribution in arthropods results from at least one HT between hexapods and isopods as well as one or more HTs within isopods. Furthermore, we have demonstrated that *Crmar2* transferred horizontally between drosophilid and tephritid flies and that *Mariner-5* underwent HT between Diptera and Hymenoptera. Conservatively, and assuming that *Crmar2* and *Mariner-5* transferred horizontally only once within the isopods, we have uncovered a total of six new HTT events in this study. Together with 70 previously known cases (for example, [48-50]; reviewed in [4]) our results bring to 76 the number of HT events described in metazoans for the Tc1/Mariner superfamily, further emphasizing the indifference of these elements to host factors to transpose [51]. Of note, HT of two other *Mariner* elements have previously been characterized in marine crustaceans [52]

(one between two decapods and one between a decapod and an amphipod), but our study is the first to report HTT involving terrestrial crustaceans. Finally, we have shown that these newly described HTT events most likely took place within the last 3 myrs, and we propose iridoviruses and/or *Wolbachia* endosymbionts as the potential vectors of transfer, a hypothesis that will be interesting to test in future studies.

Methods

Mining of available eukaryote genomes

In order to identify transposable elements similar to *Crmar2* and *Mariner-5_Dbi* that could have been horizontally transferred between isopods and other taxa we used the nucleotide sequence of the two elements to carry out BLASTn searches against the nr (non-redundant nucleotide), EST (expressed sequence tag) and

WGS (whole genome sequence) databases available on the NCBI website. We considered only those elements that showed more than 90% nucleotide identity over more than 500 bp of our query sequences.

DNA extraction, PCR, cloning, and sequencing

Genomic DNA was extracted from 14 species of terrestrial isopods (Figure 3) using the Qiagen™ DNeasy blood and tissue extraction kit (Hilden, Germany). PCRs were carried out using four types of primer pairs: 1) one pair designed on the internal region of *Mariner-5_Avul* and *Crmar2_Avul*, 2) three and four pairs designed to screen specific copies of the two elements at orthologous position in the 14 isopod species, 3) one pair designed to amplify the mitochondrial cytochrome oxidase I (*Co1*) for the six species for which this gene is not available in Genbank, and 4) one pair designed to amplify the mitochondrial 16S gene for all species included in this study except for *T. pusillus* (taken from Genbank). The list of PCR primers used in this study, together with their respective melting temperatures, is given in Additional file 5: Table S1. PCR reactions were conducted using the following temperature cycling: initial denaturation at 94°C for 5 min, followed by 30 cycles of denaturation at 94°C for 30 s, annealing at 52 to 58°C (depending on the primer pair) for 30 s, and elongation at 72°C for 45 sec, ending with a 10 min elongation step at 72°C. PCR products obtained with the *Co1* primers were purified and directly sequenced using ABI BigDye sequencing mix (1.4 µl template PCR product, 0.4 µl BigDye, 2 µl manufacturer supplied buffer, 0.3 µl primer, and 6 µl H2O). Sequencing reactions were ethanol precipitated and run on an ABI 3730 sequencer. PCR products obtained with the primers internal to *Crmar2_Avul* and *Mariner-5_Avul* were cloned into pGEM-T easy vector (Promega, USA, Madison, WI) and several clones were Sanger-sequenced as described above until we obtained five different copies of each element in the various species in which we found them.

Sequencing of androgenic gland hormone (*Agh*) cDNA

Total RNA was isolated from androgenic glands of fifteen males (6 glands per individual) using the RNeasy Mini kit (Qiagen, Hilden, Germany). The cDNA was synthesized using the M-MLV-RT kit (Promega). PCR amplification was performed using several degenerated primer pairs designed on the consensus sequences of *Agh* cDNAs of *A. vulgare*, *P. scaber* and *P.dilatatus* [see Additional file 6: Table S2] [53]. PCR and direct sequencing were performed as described above.

RT-PCR

The expression of *Crmar2_Avul* and *Mariner-5_Avul* was assessed in both dissected ovaries and somatic tissues (head + nervous chain) of *A. vulgare* females using the SuperScript™ III First-Strand Synthesis System for RT-PCR» (Invitrogen, Eugene, OR, USA).

Transposon distances versus gene distances

All *Crmar2* and *Mariner-5* consensus sequences reconstructed in this study are provided in Additional file 7: Dataset 2. In order to test whether *Crmar2* and *Mariner-5* TEs were inherited vertically or horizontally we compared the distances calculated between TE consensus sequences and between several orthologous genes for several pairs of taxa. To calculate gene distances between Isopoda and Hexapoda, we used the 57 *A. vulgare* genes sequenced by Regier *et al.* [7] as queries to perform BLASTn searches against the *D. melanogaster* genome. We chose the *D. melanogaster* genome rather than the genome of *Drosophila* species involved in the HTT characterized in this study because it is the most completely sequenced and best assembled *Drosophila* genome available. We found 46 *A. vulgare* orthologs in *D. melanogaster*, which we aligned at the nucleotide level between the two species. For the *Ceratitis/Drosophila* gene distances we used the 46 genes resulting from the above search as queries to perform BLASTn searches against the whole genome sequence of *C. capitata*. This search yielded 41 genes that we aligned at the nucleotide level with those of *D. melanogaster*. The same approach was used to find genes orthologous between *D. melanogaster* and *H. saltator* which resulted in the alignment of 37 genes. Genetic distances between *Crmar2* and *Mariner-5* consensus sequences as well as between each pair of orthologous genes were calculated using the Jukes Cantor model in MEGA 5 [54]. The name of all genes and the distances between them are provided in Additional file 8: Table S3. Jukes Cantor distances were also calculated between the various copies of both TEs found within each genome in which we found them.

Phylogenetic analyses

The phylogeny of the 14 terrestrial isopod species understudy was reconstructed using *16S*, *Co1*, and *Agh* sequences. All sequences produced in this study were deposited in Genbank under accession numbers KF957774-KF957833. Each gene was aligned manually using BioEdit 7.0.5.3 [55], and ambiguous regions were removed. All alignments are provided in Additional file 9: Dataset 3. A bootstrapped neighbor joining phylogeny was first reconstructed using MEGA 5 for each alignment with the maximum likelihood distance option. Given that no incongruence supported by bootstrap values >50% was observed between the three resulting trees (not shown), we then concatenated the three alignments and reconstructed a bootstrapped maximum likelihood phylogeny of the three combined markers using PhyML 3 [56]. A bayesian analysis of this alignment was also performed using MrBayes [57] in order to obtain posterior probabilities for each node of the tree. The model of nucleotide evolution best fitting the combined alignment (GTR + I + G) and used for the phylogenetic analyses was chosen based on the Akaike information criterion (AIC) in jModeltest 2 [58].

The phylogeny of *Crmar2* and *Mariner-5* was reconstructed based on alignments of all different copies of each element from each species in which we found them. Ambiguous regions and regions absent in more than 25% of the sequences were removed. Alignments are provided in Additional file 9: Dataset 3. The model of nucleotide evolution best fitting each alignment (TPM1uf + G for both elements) was chosen based on the Akaike information criterion (AIC) in jModeltest 2 and each alignment was analyzed using PhyML 3. In order to assess the phylogenetic position of *Mariner-5_Dbi* in the mariner tree, we have aligned the amino acid sequence of a transposase representative of most described mariner subfamily and performed a neighbor joining analysis using MEGA 5 (JTT model, 1000 bootstrap replicates). The accession numbers of the sequences we have used are provided in Additional file 1: Figure S1.

Additional files

Additional file 1: Figure S1. Phylogenetic relationships between various mariner transposases showing that *Mariner-5_Dbi* belongs to the *irritans* subfamily.

Additional file 2: Figure S2. Illustration of the position of the two types of primer sets we used to screen for *Mariner-5_Avul* and *Crmar2_Avul* elements in the various isopod species. The sequence of the primers is provided in Additional file 5: Table S1. TIR: Terminal inverted repeat.

Additional file 3: Figure S3. Pictures of agarose gels showing the results of the reverse transcription PCR experiments on *Crmar2* (**A**) and *Mariner-5* (**B**) in *Armadillidium vulgare* ovaries and somatic tissues (head + nervous chain). A band of the expected size was obtained for all reactions showing that both elements are transcribed in *A. vulgare* soma and germ line. RT, reverse transcriptase; L, size ladder.

Additional file 4: Dataset 1. Sequences of the *Crmar2* transcripts encoding a full length, intact, and thus potentially functional transposase aligned together with the *Crmar2_Avul* consensus sequence (fasta format).

Additional file 5: Table S1. List of primers used to amplify and sequence *Crmar2* and *Mariner-5* elements.

Additional file 6: Table S2. List of primers used to amplify and sequence the androgenic gland hormone.

Additional file 7: Dataset 2. Consensus sequences of *Crmar2* and *Mariner-5* elements reconstructed in this study (fasta format).

Additional file 8: Table S3. List of genes sequenced by Regier *et al.* [7] used to calculate genetic distances between the various species included in this study.

Additional file 9: Dataset 3. Sequence alignments used to reconstruct the phylogeny of crustacean isopod species and *Crmar2* and *Mariner-5* copies (fasta format).

Abbreviations

HTT: horizontal transfer of transposons; LTR: long terminal repeat; myr: million year; RT-PCR: reverse-transcription polymerase chain reaction; TE: transposable element.

Competing interests

The authors declare that they have no competing interests.

Authors' contributions

MD and CG designed the study, carried out the experiments and analyses and drafted the manuscript. SL generated the draft assembly of the *Armadillidium vulgare* genome and contributed to the writing of the manuscript. NC sequenced the androgenic gland hormone gene in the various isopod species and contributed to the writing of the manuscript. DB provided the draft transcriptome assembly of *A. vulgare* and contributed to the writing of the manuscript. All authors read and approved the final manuscript.

Acknowledgements

We thank all the technical staff of the UMR CNRS 7267. We acknowledge Catherine Debenest for assistance with the dissection of terrestrial isopods and Richard Cordaux and Pierre Grève for insightful comments on an earlier version of the manuscript. This work was supported by the CNRS, the Agence Nationale de la Recherche (ANR ImmunSymbArt 10-BLAN-1701), and a European Research Council Starting Grant (FP7/2007-2013, grant 260729 EndoSexDet) to Richard Cordaux.

Author details

[1]Université de Poitiers, UMR CNRS 7267 Ecologie et Biologie des Interactions, Equipe Ecologie Evolution Symbiose, 86022 Poitiers, Cedex, France. [2]Present address: Courant Research Center Geobiology, Geomicrobiology and Symbiosis Group, University of Göttingen, Goldschmidtstraße 3, 37077 Göttingen, Germany.

References

1. Keeling PJ, Palmer JD: **Horizontal gene transfer in eukaryotic evolution.** *Nat Rev Genet* 2008, **9**:605–618.
2. Schaack S, Gilbert C, Feschotte C: **Promiscuous DNA: horizontal transfer of transposable elements and why it matters for eukaryotic evolution.** *Trends Ecol Evol* 2010, **25**:537–546.
3. Ivancevic AM, Walsh AM, Kortschak RD, Adelson DL: **Jumping the fine LINE between species: Horizontal transfer of transposable elements in animals catalyses genome evolution.** *Bioessays* 2013, **35**:1071–1082.
4. Wallau GL, Ortiz MF, Loreto EL: **Horizontal transposon transfer in eukarya: detection, bias, and perspectives.** *Genome Biol Evol* 2012, **4**:689–699.
5. Colbourne JK, Pfrender ME, Gilbert D, Thomas WK, Tucker A, Oakley TH, Tokishita S, Aerts A, Arnold GJ, Basu MK, Bauer DJ, Cáceres CE, Carmel L, Casola C, Choi JH, Detter JC, Dong Q, Dusheyko S, Eads BD, Fröhlich T, Geiler-Samerotte KA, Gerlach D, Hatcher P, Jogdeo S, Krijgsveld J, Kriventseva EV, Kültz D, Laforsch C, Lindquist E, Lopez J, *et al*: **The ecoresponsive genome of Daphnia pulex.** *Science* 2011, **331**:555–561.
6. Martin JW, Davis GE: *An updated classification of the recent Crustacea. Science Series 39.* Los Angeles, CA: Natural History Museum of Los Angeles County; 2001.
7. Regier JC, Shultz JW, Zwick A, Hussey A, Ball B, Wetzer R, Martin JW, Cunningham CW: **Arthropod relationships revealed by phylogenomic analysis of nuclear protein-coding sequences.** *Nature* 2010, **463**:1079–1083.
8. Schmalfuss H: *World catalog of terrestrial isopods (Isopoda: Oniscidea),* Stuttgarter Beiträge zur Naturkunde Serie A, Volume Serie A, Nr. 654. Stuttgart; 2003:341.
9. Broly P, Deville P, Maillet S: **The origin of terrestrial isopods (Crustacea: Isopoda: Oniscidea).** *Evol Ecol* 2013, **27**:461–476.
10. Jurka J: **Repbase update: a database and an electronic journal of repetitive elements.** *Trends Genet* 2000, **16**:418–420.
11. Gomulski LM, Torti C, Bonizzoni M, Moralli D, Raimondi E, Capy P, Gasperi G, Malacrida AR: **A new basal subfamily of Mariner elements in Ceratitis rosa and other tephritid flies.** *J Mol Evol* 2001, **53**:597–606.
12. Kojima KK, Jurka J: **DNA transposons from the Drosophila biarmipes genome.** *Repbase Reports* 2012, **12**:740.
13. Wicker T, Sabot F, Hua-Van A, Bennetzen JL, Capy P, Chalhoub B, Flavell A, Leroy P, Morgante M, Panaud O, Paux E, SanMiguel P, Schulman AH: **A unified classification system for eukaryotic transposable elements.** *Nat Rev Genet* 2007, **8**:973–982.

14. Hartl DL, Lohe AR, Lozovskaya ER: **Modern thoughts on an ancyent marinere: function, evolution, regulation.** *Annu Rev Genet* 1997, **31**:337–358.

15. Plasterk RH, Izsvák Z, Ivics Z: **Resident aliens: the Tc1/mariner superfamily of transposable elements.** *Trends Genet* 1999, **15**:326–332.

16. Rouault JD, Casse N, Chénais B, Hua-Van A, Filée J, Capy P: **Automatic classification within families of transposable elements: application to the mariner Family.** *Gene* 2009, **448**:227–232.

17. Silva JC, Loreto EL, Clark JB: **Factors that affect the horizontal transfer of transposable elements.** *Curr Issues Mol Biol* 2004, **6**:57–71.

18. Tamura K, Subramanian S, Kumar S: **Temporal patterns of fruit fly (Drosophila) evolution revealed by mutation clocks.** *Mol Biol Evol* 2004, **21**:36–44.

19. Regier JC, Shultz JW, Kambic RE: **Pancrustacean phylogeny: hexapods are terrestrial crustaceans and maxillopods are not monophyletic.** *Proc Biol Sci* 2005, **272**:395–401.

20. Sharkey MJ: **Phylogeny and classification of hymenoptera.** *Zootaxa* 2007, **1668**:521–548.

21. Hedges SB, Dudley J, Kumar S: **TimeTree: a public knowledge-base of divergence times among organisms.** *Bioinformatics* 2006, **22**:2971–2972.

22. Lampe DJ, Witherspoon DJ, Soto-Adames FN, Robertson HM: **Recent horizontal transfer of mellifera subfamily *Mariner* transposons into insect lineages representing four different orders shows that selection acts only during horizontal transfer.** *Mol Biol Evol* 2003, **20**:554–562.

23. Michel-Salzat A, Bouchon D: **Phylogenetic analysis of mitochondrial LSU rRNA in oniscids.** *C R Acad Sci III* 2000, **323**:827–837.

24. Cutter AD: **Divergence times in Caenorhabditis and Drosophila inferred from direct estimates of the neutral mutation rate.** *Mol Biol Evol* 2008, **25**:778–786.

25. Keightley PD, Trivedi U, Thomson M, Oliver F, Kumar S, Blaxter ML: **Analysis of the genome sequences of three Drosophila melanogaster spontaneous mutation accumulation lines.** *Genome Res* 2009, **19**:1195–1201.

26. Obbard DJ, Maclennan J, Kim KW, Rambaut A, O'Grady PM, Jiggins FM: **Estimating divergence dates and substitution rates in the Drosophila phylogeny.** *Mol Biol Evol* 2012, **29**:3459–3473.

27. Houck MA, Clark JB, Peterson KR, Kidwell MG: **Possible horizontal transfer of *Drosophila* genes by the mite *Proctolaelaps regalis*.** *Science* 1991, **253**:1125–1128.

28. Yoshiyama M, Tu Z, Kainoh Y, Honda H, Shono T, Kimura K: **Possible horizontal transfer of a transposable element from host to parasitoid.** *Mol Biol Evol* 2001, **18**:1952–1958.

29. Gilbert C, Schaack S, Pace JK II, Brindley PJ, Feschotte C: **A role for host-parasite interactions in the horizontal transfer of transposons across phyla.** *Nature* 2010, **464**:1347–1350.

30. Kuraku S, Qiu H, Meyer A: **Horizontal transfers of Tc1 elements between teleost fishes and their vertebrate parasites, lampreys.** *Genome Biol Evol* 2012, **4**:929–936.

31. Wijayawardena BK, Minchella DJ, DeWoody JA: **Hosts, parasites, and horizontal gene transfer.** *Trends Parasitol* 2013, **29**:329–338.

32. Routh A, Domitrovic T, Johnson JE: **Host RNAs, including transposons, are encapsidated by a eukaryotic single-stranded RNA virus.** *Proc Natl Acad Sci U S A* 2012, **109**:1907–1912.

33. Jehle JA, Fritsch E, Nickel A, Huber J, Backhaus H: **TCl4.7: a novel lepidopteran transposon found in Cydia pomonella granulosis virus.** *Virology* 1995, **207**:369–379.

34. Jehle JA, Nickel A, Vlak JM, Backhaus H: **Horizontal escape of the novel Tc1-like lepidopteran transposon TCp3.2 into Cydia pomonella granulovirus.** *J Mol Evol* 1998, **46**:215–224.

35. Piskurek O, Okada N: **Poxviruses as possible vectors for horizontal transfer of retroposons from reptiles to mammals.** *Proc Natl Acad Sci U S A* 2007, **104**:12046–12051.

36. Williams T: **Natural invertebrate hosts of iridoviruses (Iridoviridae).** *Neotrop Entomol* 2008, **37**:615–632.

37. Lupetti P, Montesanto G, Ciolfi S, Marri L, Gentile M, Paccagnini E, Lombardo BM: **Iridovirus infection in terrestrial isopods from Sicily (Italy).** *Tissue Cell* 2013, **45**:321–327.

38. Piégu B, Guizard S, Spears T, Cruaud C, Couloux A, Bideshi DK, Federici BA, Bigot Y: **Complete genome sequence of invertebrate iridescent virus 22 isolated from a blackfly larva.** *J Gen Virol* 2013, **94**:2112–2116.

39. Fenn K, Conlon C, Jones M, Quail MA, Holroyd NE, Parkhill J, Blaxter M: **Phylogenetic relationships of the *Wolbachia* of nematodes and arthropods.** *PLoS Pathog* 2006, **2**:e94.

40. Bouchon D, Cordaux R, Grève P: **Feminizing *Wolbachia* and the evolution of sex determination in isopods.** In *Insect Symbiosis*. 3rd edition. Edited by Bourtzis K, Miller T. Boca Raton, FL: Taylor and Francis Group LLC; 2008:273–294.

41. Coscrato VE, Braz AS, Perondini ALP, Selivon D, Marino CL: *Wolbachia* in *Anastrepha* fruit flies (Diptera: Tephritidae). *Curr Microbiol* 2009, **59**:295–301.

42. Cordaux R, Bouchon D, Grève P: **The impact of endosymbionts on the evolution of host sex-determination mechanisms.** *Trends Genet* 2011, **27**:332–341.

43. Ravikumar H, Prakash BM, Sampathkumar S, Puttaraju HP: **Molecular subgrouping of *Wolbachia* and bacteriophage WO infection among some Indian *Drosophila* species.** *J Genet* 2011, **90**:507–510.

44. Cordaux R, Michel-Salzat A, Bouchon D: *Wolbachia* infection in crustaceans: novel hosts and potential routes for horizontal transmission. *J Evol Biol* 2001, **14**:237–243.

45. Klasson L, Kambris Z, Cook PE, Walker T, Sinkins SP: **Horizontal gene transfer between *Wolbachia* and the mosquito *Aedes aegypti*.** *BMC Genomics* 2009, **10**:33.

46. Woolfit M, Iturbe-Ormaetxe I, McGraw EA, O'Neill SL: **An ancient horizontal gene transfer between mosquito and the endosymbiotic bacterium *Wolbachia pipientis*.** *Mol Biol Evol* 2009, **26**:367–374.

47. Duplouy A, Iturbe-Ormaetxe I, Beatson SA, Szubert JM, Brownlie JC, McMeniman CJ, McGraw EA, Hurst GD, Charlat S, O'Neill SL, Woolfit M: **Draft genome sequence of the male-killing *Wolbachia* strain wBol1 reveals recent horizontal gene transfers from diverse sources.** *BMC Genomics* 2013, **14**:20.

48. Robertson HM: **The mariner transposable element is widespread in insects.** *Nature* 1993, **362**:241–245.

49. Robertson HM: **Evolution of DNA transposons in eukaryotes.** In *Mobile DNA II*. Edited by Craig NL, Robert Craigie R, Gellert M, Lambowitz A. Washington, DC: ASM Press; 2002:1093–1110.

50. Lorite P, Maside X, Sanllorente O, Torres MI, Periquet G, Palomeque T: **The ant genomes have been invaded by several types of *Mariner* transposable elements.** *Naturwissenschaften* 2012, **99**:1007–1020.

51. Kidwell MG: **Evolutionary biology, voyage of an ancient mariner.** *Nature* 1993, **362**:202.

52. Casse N, Bui QT, Nicolas V, Renault S, Bigot Y, Laulier M: **Species sympatry and horizontal transfers of *Mariner* transposons in marine crustacean genomes.** *Mol Phylogenet Evol* 2006, **40**:609–619.

53. Greve P, Braquart-Varnier C, Strub JM, Felix C, Van Dorsselaer A, Martin G: **The glycosylated androgenic hormone of the terrestrial isopod Porcellio scaber (Crustacea).** *Gen Comp Endocrinol* 2004, **136**:389–397.

54. Tamura K, Peterson D, Peterson N, Stecher G, Nei M, Kumar S: **MEGA5: molecular evolutionary genetics analysis using maximum likelihood, evolutionary distance, and maximum parsimony methods.** *Mol Biol Evol* 2011, **28**:2731–2739.

55. Hall T: *BioEdit version 5.0.6*. http://www.mbio.ncsu.edu/BioEdit/bioedit.html.

56. Guindon S, Gascuel O: **A simple, fast, and accurate algorithm to estimate large phylogenies by maximum likelihood.** *Syst Biol* 2003, **52**:696–704.

57. Ronquist F, Teslenko M, van der Mark P, Ayres DL, Darling A, Höhna S, Larget B, Liu L, Suchard MA, Huelsenbeck JP: **MrBayes 3.2: efficient Bayesian phylogenetic inference and model choice across a large model space.** *Syst Biol* 2012, **61**:539–542.

58. Darriba D, Taboada GL, Doallo R, Posada D: **jModelTest 2: more models, new heuristics and parallel computing.** *Nat Methods* 2012, **9**:772.

PERMISSIONS

The contributors of this book come from diverse backgrounds, making this book a truly international effort. This book will bring forth new frontiers with its revolutionizing research information and detailed analysis of the nascent developments around the world.

We would like to thank all the contributing authors for lending their expertise to make the book truly unique. They have played a crucial role in the development of this book. Without their invaluable contributions this book wouldn't have been possible. They have made vital efforts to compile up to date information on the varied aspects of this subject to make this book a valuable addition to the collection of many professionals and students.

This book was conceptualized with the vision of imparting up-to-date information and advanced data in this field. To ensure the same, a matchless editorial board was set up. Every individual on the board went through rigorous rounds of assessment to prove their worth. After which they invested a large part of their time researching and compiling the most relevant data for our readers.

The editorial board has been involved in producing this book since its inception. They have spent rigorous hours researching and exploring the diverse topics which have resulted in the successful publishing of this book. They have passed on their knowledge of decades through this book. To expedite this challenging task, the publisher supported the team at every step. A small team of assistant editors was also appointed to further simplify the editing procedure and attain best results for the readers.

Apart from the editorial board, the designing team has also invested a significant amount of their time in understanding the subject and creating the most relevant covers. They scrutinized every image to scout for the most suitable representation of the subject and create an appropriate cover for the book.

The publishing team has been an ardent support to the editorial, designing and production team. Their endless efforts to recruit the best for this project, has resulted in the accomplishment of this book. They are a veteran in the field of academics and their pool of knowledge is as vast as their experience in printing. Their expertise and guidance has proved useful at every step. Their uncompromising quality standards have made this book an exceptional effort. Their encouragement from time to time has been an inspiration for everyone.

The publisher and the editorial board hope that this book will prove to be a valuable piece of knowledge for researchers, students, practitioners and scholars across the globe.

LIST OF CONTRIBUTORS

France Dufresne
Département de Biologie, Chimie et Géographie, Université du Québec à Rimouski, Rimouski, Québec G5L 3A1, Canada
Centre d'Études Nordiques, Université Laval, Québec G1V 0A6, Canada

Roland Vergilino
Département de Biologie, Chimie et Géographie, Université du Québec à Rimouski, Rimouski, Québec G5L 3A1, Canada
Department of Integrative Biology, University of Guelph, Guelph, Ontario N1G 2W1, Canada
Centre d'Études Nordiques, Université Laval, Québec G1V 0A6, Canada

Shannon HC Eagle and Teresa J Crease
Department of Integrative Biology, University of Guelph, Guelph, Ontario N1G 2W1, Canada

Zhou Tang, Min-Jin Han and Ze Zhang
School of Life Sciences, Chongqing University, Chongqing 400044, China
College of Forestry and Life Science, Chongqing University of Sciences and Arts, Yongchuan, Chongqing 40216, China

Hua-Hao Zhang and Xiao-Gu Zhang
College of Pharmacy and Life Science, Jiujiang University, Jiujiang 332000, China

Ke Huang
College of Forestry and Life Science, Chongqing University of Sciences and Arts, Yongchuan, Chongqing 40216, China

David A Ray, Roy N Platt II and Sarah F. Mangum
Department of Biological Sciences, Texas Tech University, Lubbock, TX 79409, USA

Heidi JT Pagan
Harbor Branch Oceanographic Institute, Florida Atlantic University, Fort Pierce, FL, USA

Ashley R Kroll and Sarah Schaack
Department of Biology, Reed College, Portland, OR 97202, USA

Richard D Stevens
Department of Natural Resources Management and the Museum, Texas Tech University, Lubbock, TX 79409, USA

Bo Gao, Dan Shen, Songlei Xue, Cai Chen, Hengmi Cui and Chengyi Song
Institute of Epigenetics & Epigenomics, College of Animal Science & Technology, Yangzhou University, Yangzhou, Jiangsu 225009, China

Stefania Mantziou, Soteroula Thrasyvoulou, Theodore Tzavaras and Theodore Tzavaras
Laboratory of General Biology, Faculty of Medicine, School of Health Sciences, University of Ioannina, Ioannina 45110, Greece

Georgios Markopoulos and Evangelos Kolettas
Laboratory of General Biology, Faculty of Medicine, School of Health Sciences, University of Ioannina, Ioannina 45110, Greece
Biomedical Research Division, Institute of Molecular Biology and Biotechnology, Foundation of Research and Technology (IMBB-FORTH), University Campus, Ioannina 45110, Greece

Dimitrios Noutsopoulos
Laboratory of Molecular Biology and Genetics, Department of Biological Applications and Technology, School of Health Sciences, University of Ioannina, Ioannina 45110, Greece

Demetrios Gerogiannis
Department of Computer Science, School of Sciences, University of Ioannina, Ioannina 45110, Greece

Georgios Vartholomatos
Hematology Laboratory, Unit of Molecular Biology, University Hospital of Ioannina, Ioannina 45110, Greece

Shannon H. C. Eagle and Teresa J. Crease
Department of Integrative Biology, University of Guelph, Guelph, ON N1G 2 W1, Canada

Kun Liu
Graduate Program in Botany and Plant Sciences, University of California, Riverside, CA 92521, USA

Susan R. Wessler
Department of Botany and Plant Sciences, University of California, Riverside, CA 92521, USA

M. Murshida Mahbub and Shawn M. Christensen
Department of Biology, University of Texas at Arlington, 501 S. Nedderman Drive, Room 337, Arlington, TX 76010, USA

Saiful M. Chowdhury
Department of Chemistry and Biochemistry, University of Texas at Arlington, 700 Planetarium Place, Room 130, Arlington, TX 76010, USA

Jinmin Lee, Seyoung Mun and Kyudong Han
Department of Nanobiomedical Science & BK21 PLUS NBM Global Research Center for Regenerative Medicine, Dankook University, Cheonan 330-714, Republic of Korea

Dong Hee Kim
Department of Anesthesiology and Pain Management, College of Medicine, Dankook University, Cheonan 330-714, Republic of Korea

Chun-Sung Cho
Department of Neurosurgery, College of Medicine, Dankook University, Cheonan 330-714, Republic of Korea

Dong-Yep Oh
Gyeongsangbuk-Do Livestock Research Institution, Yeongju 750-871, Republic of Korea

Nicolas Bargues and Emmanuelle Lerat
CNRS, UMR 5558, Laboratoire Biométrie et Biologie Evolutive, Université de Lyon, Université Claude Bernard Lyon 1, F-69622 Villeurbanne, France

Natasha Avila Bertocchi, Fabiano Pimentel Torres, Analía del Valle Garnero and Ricardo José Gunski
Programa de Pós-graduação em Ciências Biológicas, Universidade Federal do Pampa (Unipampa), São Gabriel, Rio Grande do sul 97300-000, Brazil
Laboratório de Diversidade Genética Animal, Universidade Federal do Pampa (Unipampa), São Gabriel, Rio Grande do sul 97300-000, Brazil

Gabriel Luz Wallau
Departamento de Entomologia, Instituto Aggeu Magalhães – FIOCRUZ-CPqAM, Recife, Pernambuco, Brazil

Keiko Tsuji Wakisaka
Department of Applied Biology, Kyoto Institute of Technology, Hashigamicyo, Matsugasaki, Sakyo-ku, Kyoto 606-8585, Japan

Masanobu Itoh
Department of Applied Biology, Kyoto Institute of Technology, Hashigamicyo, Matsugasaki, Sakyo-ku, Kyoto 606-8585, Japan
Center for Advanced Insect Research Promotion (CAIRP), Kyoto Institute of Technology, Kyoto 606-8585, Japan

Seiko Ohno
Center for Epidemiologic Research in Asia, Shiga Univesity of Medical Science, Otsu, Shiga 520-2192, Japan

Sandra Bachg and Jürgen Schmitz
Institute of Experimental Pathology (ZMBE), University of Münster, D-48149 Münster, Germany

Alexander Suh
Institute of Experimental Pathology (ZMBE), University of Münster, D-48149 Münster, Germany
Department of Evolutionary Biology (EBC), Uppsala University, SE-75236 Uppsala, Sweden

Jürgen Brosius
Institute of Experimental Pathology (ZMBE), University of Münster, D-48149 Münster, Germany
Brandenburg Medical School (MHB), D-16816 Neuruppin, Germany

Jan Ole Kriegs
Institute of Experimental Pathology (ZMBE), University of Münster, D-48149 Münster, Germany
LWL-Museum für Naturkunde, Westfälisches Landesmuseum mit Planetarium, D-48161 Münster, Germany

Stephen Donnellan
South Australian Museum, Adelaide, SA 5000, Australia
School of Biological Sciences, The University of Adelaide, Adelaide 5005, Australia

Leo Joseph
Australian National Wildlife Collection, CSIRO
National Research Collections Australia, Canberra,
ACT 2601, Australia

Hiroyuki Sasaki
Division of Epigenomics and Development,
Medical Institute of Bioregulation, and Epigenome
Network Research Center, Kyushu University, 3-1-
1 Maidashi, Higashi-ku, Fukuoka 812-8582, Japan

Ken-ichi Shimosuga
Division of Epigenomics and Development,
Medical Institute of Bioregulation, and Epigenome
Network Research Center, Kyushu University, 3-1-
1 Maidashi, Higashi-ku, Fukuoka 812-8582, Japan
Trygroup Incorporated,1-8-10 Kudankita, Chiyoda-
ku, Tokyo 102-0073, Japan

Kei Fukuda
Division of Epigenomics and Development,
Medical Institute of Bioregulation, and Epigenome
Network Research Center, Kyushu University, 3-1-
1 Maidashi, Higashi-ku, Fukuoka 812-8582, Japan
Cellular Memory Laboratory, RIKEN, Wako,
Saitama 351-0198, Japan

Kenji Ichiyanagi
Division of Epigenomics and Development,
Medical Institute of Bioregulation, and Epigenome
Network Research Center, Kyushu University, 3-1-
1 Maidashi, Higashi-ku, Fukuoka 812-8582, Japan

Laboratory of Genome and Epigenome Dynamics,
Department of Applied Molecular Biosciences,
Graduate School of Bioagricultural Sciences,
Nagoya University, Nagoya 464-8601, Japan

**Antonio Palazzo, Roberta Moschetti, Ruggiero
Caizzi and René Massimiliano Marsano**
Dipartimento di Biologia, Università degli Studi di
Bari "Aldo Moro", Via Orabona 4, 70125 Bari, Italy

Justin P Blumenstie
Department of Ecology and Evolutionary Biology,
University of Kansas, 1200 Sunnyside Avenue,
Lawrence KS 66049, USA

**Mathilde Dupeyron, Sébastien Leclercq, Didier
Bouchon and Clément Gilbert**
Université de Poitiers, UMR CNRS 7267 Ecologie et
Biologie des Interactions, Equipe Ecologie Evolution
Symbiose, 86022 Poitiers, Cedex, France Göttingen,
Germany

Nicolas Cerveau
Université de Poitiers, UMR CNRS 7267 Ecologie et
Biologie des Interactions, Equipe Ecologie Evolution
Symbiose, 86022 Poitiers, Cedex, France Göttingen,
Germany
Research Center Geobiology, Geomicrobiology
andm Symbiosis Group, University of Göttingen,
Goldschmidtstraße 3, 37077

Index